D1717672

Analytical Techniques in Forensic Science

Analytical Techniques in Forensic Science

Edited by

Rosalind Wolstenholme
Sheffield Hallam University
Sheffield, UK

Sue Jickells
Retired Analytical Chemist, formerly University of East Anglia
and King's College London

Shari Forbes
Université du Québec à Trois-Rivières
Québec, Canada

This edition first published 2021
© 2021 John Wiley & Sons Ltd

All rights reserved. No part of this publication may be reproduced, stored in a retrieval system, or transmitted, in any form or by any means, electronic, mechanical, photocopying, recording or otherwise, except as permitted by law. Advice on how to obtain permission to reuse material from this title is available at http://www.wiley.com/go/permissions.

The right of Rosalind Wolstenholme, Sue Jickells and Shari Forbes to be identified as the authors of the editorial material in this work has been asserted in accordance with law.

Registered Offices
John Wiley & Sons, Inc., 111 River Street, Hoboken, NJ 07030, USA
John Wiley & Sons Ltd, The Atrium, Southern Gate, Chichester, West Sussex, PO19 8SQ, UK

Editorial Office
The Atrium, Southern Gate, Chichester, West Sussex, PO19 8SQ, UK

For details of our global editorial offices, customer services, and more information about Wiley products visit us at www.wiley.com.

Wiley also publishes its books in a variety of electronic formats and by print-on-demand. Some content that appears in standard print versions of this book may not be available in other formats.

Limit of Liability/Disclaimer of Warranty
In view of ongoing research, equipment modifications, changes in governmental regulations, and the constant flow of information relating to the use of experimental reagents, equipment, and devices, the reader is urged to review and evaluate the information provided in the package insert or instructions for each chemical, piece of equipment, reagent, or device for, among other things, any changes in the instructions or indication of usage and for added warnings and precautions. While the publisher and authors have used their best efforts in preparing this work, they make no representations or warranties with respect to the accuracy or completeness of the contents of this work and specifically disclaim all warranties, including without limitation any implied warranties of merchantability or fitness for a particular purpose. No warranty may be created or extended by sales representatives, written sales materials or promotional statements for this work. The fact that an organization, website, or product is referred to in this work as a citation and/or potential source of further information does not mean that the publisher and authors endorse the information or services the organization, website, or product may provide or recommendations it may make. This work is sold with the understanding that the publisher is not engaged in rendering professional services. The advice and strategies contained herein may not be suitable for your situation. You should consult with a specialist where appropriate. Further, readers should be aware that websites listed in this work may have changed or disappeared between when this work was written and when it is read. Neither the publisher nor authors shall be liable for any loss of profit or any other commercial damages, including but not limited to special, incidental, consequential, or other damages.

Library of Congress Cataloging-in-Publication Data

Names: Wolstenholme, Rosalind, editor. | Jickells, Sue, editor. | Forbes, Shari, editor.
Title: Analytical techniques in forensic science / edited by Dr. Rosalind Wolstenholme, Dr. Sue Jickells, Prof. Shari Forbes.
Description: First edition. | Hoboken, NJ : Wiley, 2021. | Includes bibliographical references and index.
Identifiers: LCCN 2020022864 (print) | LCCN 2020022865 (ebook) | ISBN 9781119978282 (hardback) | ISBN 9781119033813 (adobe pdf) | ISBN 9781119033820 (epub)
Subjects: MESH: Forensic Sciences–methods | Chemistry Techniques, Analytical
Classification: LCC RA1051 (print) | LCC RA1051 (ebook) | NLM W 700 | DDC 614/.1–dc23
LC record available at https://lccn.loc.gov/2020022864
LC ebook record available at https://lccn.loc.gov/2020022865

Cover Design: Wiley
Cover Image: © Forensic science graph - CReative Commons - source MDPI, https://doi.org/10.3390/separations3030026

Set in 9.5/12.5pt STIXTwoText by SPi Global, Chennai, India
Printed and bound in Singapore by Markono Print Media Pte Ltd

10 9 8 7 6 5 4 3 2 1

Contents

List of Contributors *xvii*
Preface *xix*
Acknowledgements *xxi*

Part I   Preparing for Analysis *1*

1   **Introduction to Forensic Science**   3
        Sue Jickells, Rosalind Wolstenholme and Shari Forbes
1.1   Forensic Science   3
1.2   The Forensic Process   6
1.2.1   Forensic Principles and the Crime Scene   6
1.2.2   Preparatory Issues in Laboratory Analysis   11
1.2.3   Interpretation of Forensic Evidence   13
1.2.3.1 The Expert Witness and Interpretation   14
1.2.3.2 Evidential Value   15
1.2.3.3 Statistical Interpretation   18
1.2.3.4 Bayesian Statistics   20
1.3   Judicial Systems   22
1.3.1   Criminal vs. Civil Law   22
1.3.2   Adversarial vs. Inquisitorial System   24
1.3.3   Rules of Evidence   25
1.3.3.1 Admissibility of Evidence   25
1.3.4   Types of Evidence   26
1.3.5   Opinion and Expert Testimony   28
1.3.5.1 Admissibility of Scientific and Technical Evidence   28
1.4   The Role of Analytical Chemistry in Forensic Science   30
1.4.1   Techniques Used for Chemical Analysis   31
        References   32

2	**Analytical Methodology and Experimental Design** *35*
	Florian Wulfert and Rosalind Wolstenholme
2.1	Scientific Method *35*
2.2	What Do We Mean by Analysis? *36*
2.3	The Stages of Analysis *36*
2.3.1	Quantification *37*
2.3.1.1	External Standards *37*
2.3.1.2	Internal Standards *38*
2.3.1.3	Standard Addition *38*
2.4	Analysis Development *39*
2.4.1	Error Estimation *39*
2.4.2	Quality Assurance and Quality Control *40*
2.4.3	Method Development and Experimental Designs *41*
2.4.4	Selecting Critical Variables with Factorial Designs *42*
2.4.4.1	Categorical Variables *43*
2.4.4.2	Reduced Designs *44*
2.4.4.3	Final Practical Experimental Considerations *44*
2.4.4.4	Deciding on Significance *44*
2.4.4.5	Interpretation *45*
2.4.5	Modelling the Significant Variables Using Response Surface Designs *46*
2.4.5.1	Sparse Response Surface Designs *48*
2.4.5.2	Analysing Response Surface Models *48*
2.4.5.3	Validation *49*
2.4.5.4	Optimisation *49*
3	**Presumptive Testing** *51*
	Rosalind Wolstenholme and Shari Forbes
3.1	Introduction *51*
3.2	Drugs *52*
3.2.1	Drugs Seizure Sampling *52*
3.2.2	Major Drug Classes *52*
3.2.2.1	Marijuana *52*
3.2.2.2	Opioids, Cocaine, and Amphetamines *53*
3.2.2.3	Barbiturates and Benzodiazepines *53*
3.2.2.4	LSD *53*
3.2.2.5	New Psychoactive Substances *55*
3.2.3	Presumptive Tests for Drugs *56*
3.2.3.1	Colour Tests *56*
3.2.3.2	Thin Layer Chromatography *56*
3.2.3.3	Microcrystal Tests *56*
3.3	Firearms Discharge Residue *57*
3.3.1	Firearms Discharge Residue Sampling *57*
3.3.2	Firearms Discharge Residue Presumptive Tests *58*
3.4	Explosives *59*
3.4.1	Explosive Residue Sampling *60*

3.4.2	Explosive Residue Presumptive Tests	60
3.4.2.1	Colour Tests	60
3.4.2.2	Thin Layer Chromatography	61
3.4.2.3	Portable Instruments	61
3.5	Ethanol (Ethyl Alcohol)	61
3.5.1	Breath Alcohol Testing	61
3.5.1.1	Electronic Devices	62
3.5.1.2	Chemical Test Devices	63
3.5.2	Saliva-Based Testing	63
3.6	Ignitable Liquid Residues	64
3.7	Non-Chemical Presumptive Tests	65
3.7.1	Electronic Detectors	65
3.7.1.1	Electronic Detectors for Fire Investigations	65
3.7.1.2	Electronic Detectors for Explosives and Illicit Drugs	66
3.7.2	Canine Detection	67
	References	68
4	**Sample Preparation**	*71*
	Sue Jickells	
4.1	Sample Preparation	71
4.2	Extraction	75
4.2.1	Solvent Extraction	76
4.2.2	Liquid–Liquid Extraction	77
4.2.3	Solid Phase Extraction	82
4.2.3.1	Stationary Phases	85
4.2.3.2	Normal Phase	92
4.2.3.3	Reversed Phase	93
4.2.3.4	Ion Exchange	95
4.2.3.5	Molecularly Imprinted Polymers	95
4.2.3.6	Immunoaffinity SPE	97
4.2.4	Solid-Phase Microextraction	97
4.2.5	QuEChERS	101
4.2.6	Sample Handling Post Extraction	101
4.2.6.1	Solvent Evaporation	101
4.2.6.2	Derivatisation	102
4.3	Sample Preparation for Inorganic Analyses	102
4.3.1	Total Analysis	103
4.3.2	Chemical Speciation	105
4.4	DNA Profiling	105
4.5	Conclusion	106
	References	106

Part II Spectroscopic and Spectrometric Techniques 109

5 The Electromagnetic Spectrum 111
Rosalind Wolstenholme
Reference 114

6 Ultraviolet–Visible and Fluorescence Spectroscopy 115
Rosalind Wolstenholme
6.1   Forensic Introduction 115
6.2   Theory 115
6.2.1   Electronic Transitions 115
6.2.2   Photoluminescence and Fluorescence 118
6.2.3   Quantification 120
6.2.3.1   UV-Vis Quantification 120
6.2.3.2   Fluorescence Quantification 121
6.3   Instrumentation 122
6.3.1   UV-Vis Spectrometers 122
6.3.2   Fluorescence Spectrometers/Fluorometers 123
6.3.3   Coupling Techniques 126
6.3.4   Microspectrophotometers 126
6.3.5   Hyperspectral Imaging 126
6.3.6   Filtered Light Examination 127
6.4   Application to Analyte 128
6.4.1   Transmission Analysis in Solution 128
6.4.1.1   UV-Vis Solution Analysis 128
6.4.1.2   Fluorescent Solution Analysis 129
6.4.2   MSP Sample Preparation 129
6.4.3   Acquiring a Spectrum 130
6.4.3.1   Capture of Spectra in Solution 130
6.4.3.2   MSP and HSI Sample Analysis 131
6.4.4   Forensic Applications 131
6.4.4.1   Writing Ink Examination 132
6.4.4.2   Fibre Examination 133
6.5   Interpretation and Law 134
6.5.1   Interpreting UV-Vis Spectra 135
6.5.2   Interpreting Fluorescence Spectra 137
6.5.3   UV-Vis and Fluorescence Spectroscopy in Court 138
6.6   Case Studies 138
6.6.1   Case Study 1 138
6.6.2   Case Study 2 139
6.7   Forensic Developments 140
References 140

7	**Infrared Spectroscopy** *145*
	Barbara Stuart
7.1	Introduction *145*
7.2	Theory of the Technique *145*
7.2.1	Basis of the Technique *145*
7.2.2	Instrumentation *146*
7.2.3	Transmission Spectroscopy *148*
7.2.4	Reflectance Spectroscopy *148*
7.2.5	Infrared Microspectroscopy *150*
7.2.6	Handheld and Portable Instruments *151*
7.3	Application to Analyte *151*
7.3.1	Sampling *151*
7.3.2	Spectrum Analysis *152*
7.4	Interpretation and Law *155*
7.5	Case Studies – Discrimination of Acrylic Fibres *157*
7.6	Forensic Developments *158*
	References *159*

8	**Raman Spectroscopy** *161*
	Rosalind Wolstenholme
8.1	Forensic Introduction *161*
8.2	Theory *161*
8.2.1	Raman Scattering *161*
8.2.2	Modes of Vibration *163*
8.2.3	Raman Shift *165*
8.2.4	Raman Instrumentation *166*
8.2.4.1	Lasers, Fluorescence, and Resolution *166*
8.2.4.2	Dispersive versus FT *167*
8.2.4.3	Dispersive Raman Spectrometers *168*
8.2.4.4	FT-Raman Spectrometers *169*
8.2.4.5	Polarisers *169*
8.2.4.6	Microscopes and Imaging *169*
8.2.4.7	Portable Instruments and Probes *170*
8.2.4.8	Quantitation *170*
8.2.5	Advanced Techniques *171*
8.2.5.1	Resonance Raman Spectroscopy *171*
8.2.5.2	SERS/SERRS *171*
8.2.5.3	SORS *172*
8.2.6	Advantages and Disadvantages of Raman Spectroscopy *173*
8.3	Application to Analyte *174*
8.3.1	Acquiring a Spectrum *174*
8.3.2	Forensic Applications *175*

8.3.2.1	Pen Ink *175*	
8.3.2.2	Paint *175*	
8.3.2.3	Drugs of Abuse *176*	
8.4	Interpretation and Law *177*	
8.4.1	Interpreting Raman Spectra *177*	
8.4.2	Raman Spectroscopy in Court *179*	
8.5	Case Studies *180*	
8.5.1	Case Study 1 *180*	
8.5.2	Case Study 2 *180*	
8.6	Forensic Developments *181*	
	References *181*	
9	**Scanning Electron Microscopy** *185*	
	Grzegorz Zadora and Aleksandra Michalska	
9.1	Introduction *185*	
9.2	Theory of the Technique *186*	
9.2.1	Scanning Electron Microscope *186*	
9.2.2	X-Ray Detection *191*	
9.2.3	Operating Conditions *192*	
9.2.4	Specimen Preparation *193*	
9.2.4.1	Vacuum Evaporation *194*	
9.3	Application to Analyte(s) *195*	
9.3.1	Gunshot Residue *196*	
9.3.2	Glass *200*	
9.3.3	Other Samples *203*	
9.4	Interpretation and Law *203*	
9.4.1	Evidence Evaluation on Source Level *203*	
9.4.2	Evidence Evaluation on Activity Level *206*	
9.5	Case Study *207*	
9.5.1	GSR – Case Study *207*	
9.5.2	Glass – Comparison and Classification Problem *209*	
9.5.3	Glass – Was the Car Bulb Switched on During the Accident? *212*	
	References *214*	
10	**Mass Spectrometry** *219*	
	Mark C. Parkin and Alan Brailsford	
10.1	Introduction *219*	
10.1.1	Forensic Application of Mass Spectrometry *221*	
10.2	Theory of the Technique *223*	
10.2.1	Principles of Mass Spectrometry *223*	
10.2.2	Sample Introduction *224*	
10.2.3	Modes of Sample Ionisation *225*	
10.2.3.1	Electron Ionisation *225*	

10.2.3.2	Chemical Ionisation	*227*
10.2.3.3	Electrospray Ionisation	*230*
10.2.3.4	Atmospheric Pressure Chemical Ionisation	*231*
10.2.3.5	Desorption and Ambient Methods	*232*
10.2.3.6	Matrix-Assisted Laser Desorption/Ionisation	*232*
10.2.3.7	Secondary Ion Mass Spectrometry	*234*
10.2.3.8	Desorption Electrospray Ionisation	*234*
10.2.3.9	Direct Analysis in Real Time	*234*
10.2.4	Ion Separation – Mass Analysers	*235*
10.2.4.1	Mass Range, Resolution and Accuracy	*235*
10.2.4.2	Magnetic Sector	*236*
10.2.4.3	Quadrupoles – Quadrupole Mass Filter	*236*
10.2.4.4	Quadrupole Ion Trap	*237*
10.2.4.5	Time of Flight	*238*
10.2.4.6	Fourier Transform Instruments – Ion Cyclotron Resonance	*239*
10.2.4.7	Fourier Transform Instruments – Orbitrap	*240*
10.2.4.8	Tandem Mass Spectrometry – Ion Fragmentation by Collision Induced Dissociation	*241*
10.2.4.9	Tandem Mass Analysers – Ion Traps	*242*
10.2.4.10	Tandem Mass Analysers – Triple Quadrupoles	*242*
10.2.4.11	Tandem Mass Analysers – Hybrid Instruments	*242*
10.2.5	Ion Detection	*243*
10.2.5.1	Electron Multipliers	*243*
10.2.5.2	Faraday Cup	*244*
10.2.6	Anatomy of a Mass Spectrum	*244*
10.2.6.1	The Molecular or Quasi-Molecular Ion	*245*
10.2.6.2	The Fragment Region	*247*
10.2.6.3	Full Scan Mass Spectra	*247*
10.2.6.4	Product Ion Spectra	*248*
10.2.6.5	Extracted Ion Chromatograms	*248*
10.2.6.6	Selected Ion Chromatograms and Multiple Reaction Monitoring	*249*
10.2.6.7	Precursor Ion Detection and Neutral Loss Scanning	*252*
10.3	Application to Analytes	*252*
10.4	Interpretation and Law	*254*
10.4.1	Chain of Custody	*254*
10.4.2	New Forensic Regulations	*255*
10.4.3	ID Criteria – Screen and Confirmation	*255*
10.4.4	Chromatographic Criteria	*256*
10.4.5	Mass Spectrometric Identification Criteria	*256*
10.5	Case Studies	*257*
10.5.1	Serial Killing by Poisoning	*257*
10.5.2	Surreptitious Insulin Administration	*257*
10.6	Forensic Developments	*258*

10.6.1	Beyond Blood and Urine	*258*
10.6.2	High Mass Accuracy Mass Spectrometry	*259*
10.6.3	Mobile Mass Spectrometers	*260*
	References	*261*

11 Isotope Ratio Mass Spectrometry *267*
Sarah Benson and Kylie Jones

11.1	Forensic Introduction	*267*
11.2	Basis of the Technique	*268*
11.2.1	Isotopes	*268*
11.2.2	Isotopic Abundance and Delta Notation	*268*
11.2.3	Standards and Reference Materials	*269*
11.2.4	Isotopic Variability – Fractionation and Mixing	*270*
11.2.5	Isotopic Variability of Natural Materials	*272*
11.2.6	Instrumentation: Stable Isotope Ratio Mass Spectrometers	*272*
11.3	Introduction to the Isotope Ratio Mass Spectrometer	*276*
11.3.1	IRMS – Detection and Measurement	*276*
11.3.2	Sample Preparation	*277*
11.3.3	Bulk Stable Isotope Analysis	*277*
11.3.4	Bulk Measurements by Quantitative High Temperature Combustion	*278*
11.3.5	Bulk Measurements by Quantitative High Temperature Conversion	*279*
11.3.6	Compound Specific Isotope Analysis	*279*
11.4	Interpretation	*280*
11.5	Case Studies	*281*
11.6	Applications in Forensic Science	*283*
11.6.1	Distinguishing between Naturally Occurring and Synthetic Materials in Doping, e.g. Endogenous and Exogenous (Synthetic) Testosterone	*284*
11.6.2	Determining Authenticity and Predicting Geographical Origin of Food, Pharmaceuticals and Other Materials, e.g. Counterfeiting	*284*
11.6.3	Tracing the Geographic Origin and Movement of Wildlife, Persons and Materials	*284*
11.6.4	Identifying the Source of Environmental Contaminants	*285*
11.6.5	Determining the Geographical Origin of Plant Materials, e.g. Natural Illicit Drugs – Cannabis, Cocaine, and Heroin	*285*
11.6.6	Characterising Microorganisms	*286*
11.6.7	Determining Synthetic Pathways Used to Manufacture Illicit Drugs, e.g. Ecstasy and MDMA, Methamphetamine, and Amphetamine	*286*
11.6.8	Distinguishing between Two or More Samples of a Material to Infer Source or a Common Origin	*287*
11.6.9	Distinguishing Between Two or More Samples of Ignitable Liquids and Chemicals	*287*
11.6.10	Determining Source Through Association of Starting Materials and End Products, e.g. Explosives	*288*
11.7	Future of IRMS and Stable Isotopic Comparisons	*288*
	References	*288*

Part III Chromatographic Techniques 295

12   Chromatographic Separation and Theory 297
Sue Jickells and Shari Forbes
12.1   Introduction 297
12.2   Chromatography 298
12.2.1   Planar Chromatography 299
12.2.2   Column Chromatography 300
12.3   The Separation Process 300
12.3.1   Distribution Constant 303
12.3.2   Hold-Up Time (or Volume) 304
12.3.3   Retention Time (or Volume) 305
12.3.3.1   Retention Time and Sample Concentration 306
12.3.4   Retention Factor 306
12.3.5   Separation Factor 307
12.4   Separation Theory 307
12.4.1   Plate Theory 307
12.4.2   Theory versus Practice: Band Broadening 308
12.4.3   Rate Theory 311
12.4.3.1   Eddy Diffusion (A) 312
12.4.3.2   Longitudinal Diffusion (B) 313
12.4.3.3   Mass Transfer (C) 314
12.4.3.4   Non-Column Parameters Contributing to Band Broadening 316
12.5   Practical Applications of Chromatographic Theory 316
12.5.1   Optimising Chromatographic Separations 317
12.5.1.1   Resolution 317
12.5.1.2   GC 319
12.5.1.3   Mobile Phase 320
12.6   Conclusion 323
References 323

13   Gas Chromatography 327
Shari Forbes
13.1   Introduction 327
13.2   Gas Chromatography Components 327
13.2.1   Mobile Phase System 328
13.2.2   Sample Injection System 329
13.2.2.1   Liquid Samples 330
13.2.2.2   Gases and Volatile Compounds 334
13.2.2.3   Gas Samples 334
13.2.2.4   Volatile Compounds: Headspace Analysis 335
13.2.2.5   Static Headspace Analysis 335
13.2.2.6   Dynamic Headspace Analysis 336
13.2.2.7   Pyrolysis GC 338
13.2.3   Columns and Chromatographic Separation 338

13.2.3.1 Column Selection *340*
13.2.3.2 Column Temperature and Programming *341*
13.2.4 Detectors and Detection Systems *343*
13.2.4.1 Flame Ionisation Detectors *344*
13.2.4.2 Electron Capture Detectors *345*
13.2.4.3 Nitrogen–Phosphorous Detectors *345*
13.2.4.4 Mass Spectrometric Detection Systems *346*
13.3 Application to Analyte *348*
13.3.1 Sample Derivatisation *348*
13.3.2 Qualitative Analysis *350*
13.3.3 Quantitative Analysis *351*
13.3.3.1 Methods of Quantitative Analysis *353*
13.4 Interpretation and Law *354*
13.5 Case Studies *356*
13.5.1 Case Study 1 *356*
13.5.2 Case Study 2 *357*
13.6 Forensic Developments *358*
13.6.1 Multidimensional GC *358*
13.6.2 Portable GC *361*
References *362*

14 High Performance Liquid Chromatography and Ultra-High Performance Liquid Chromatography Including Liquid Chromatography–Mass Spectrometry *365*
Sophie Turfus and Luke N. Rodda
14.1 Introduction *365*
14.2 Components of an HPLC instrument and their Optimisation *368*
14.2.1 Pump and Mixer *368*
14.2.2 Autosampler and Inlet *370*
14.2.3 Injector *370*
14.2.4 Column *370*
14.2.4.1 Stationary Phase *371*
14.2.4.2 Column Dimensions *373*
14.2.4.3 Particle Size *373*
14.2.4.4 Pre-Column/Guard Column *373*
14.2.5 Fittings *374*
14.2.6 Mobile Phase *375*
14.2.6.1 Mobile Phase A *376*
14.2.6.2 Mobile Phase B *376*
14.2.7 Effect of Temperature/Flow Rate *379*
14.2.8 Detector *380*
14.2.8.1 Mass Spectrometer *380*
14.2.8.2 UV Detector *382*

14.2.8.3	PDA Detector *383*	
14.3	Related Techniques *384*	
14.3.1	Ion Chromatography *384*	
14.3.2	Affinity Chromatography *384*	
14.3.3	Chiral Chromatography *385*	
14.4	Chromatography Theory *385*	
14.5	Detection *386*	
14.6	Coupling of Liquid Chromatography to Mass Spectrometry *388*	
14.7	Types of Analytes *390*	
14.7.1	Basic Analytes *390*	
14.7.2	Acidic Analytes *390*	
14.7.3	Proteins *391*	
14.7.4	DNA *391*	
14.7.5	Chiral Compounds *392*	
14.7.6	Bulk Drugs and High-Concentration Analytes *392*	
14.7.7	Low-Concentration Analytes *392*	
14.8	Accreditation and Method Validation *393*	
14.8.1	Use of Internal Standards *393*	
14.8.2	Effect of Sample Matrix *394*	
14.8.3	Ion Ratios *394*	
14.9	Interpretation of Results in the Forensic and Legal Context *394*	
14.10	Case Studies *396*	
14.10.1	Case Study 1: Post-Mortem Death Investigation – Poly-Drug Overdose *396*	
14.10.2	Case Study 2: Post-Mortem Death Investigation – No Derivatisation Needed for LC-MS *397*	
14.10.3	Case Study 3: Driving Under the Influence of Drugs – Increased Sensitivity with LC-MS *398*	
14.11	Forensic Developments *399*	
14.11.1	Column Switching and Two-Dimensional HPLC *399*	
14.11.2	Capillary Liquid Chromatography *401*	
14.11.3	Column-on-a-Chip Technologies *401*	
14.12	Conclusion *402*	
	References *402*	
15	**Capillary and Microchip Electrophoresis** *407*	
	Lucas Blanes, Ellen Flávia Moreira Gabriel, Renata Mayumi Saito, Wendell Karlos Tomazelli Coltro, Nerida Cole, Philip Doble, Claude Roux and Robson Oliveira dos Santos	
15.1	Capillary Electrophoresis: Introduction *407*	
15.2	Microchip-Capillary Electrophoresis *410*	
15.2.1	Sample Injection Modes in ME *410*	
15.3	Detection Systems *411*	
15.4	CE and ME in Forensic Analysis *412*	

15.5	Case Study: Lab-on-a-Chip Screening of Methamphetamine and Pseudoephedrine in Clandestine Laboratory Samples *412*	
15.5.1	Screening of Methamphetamine and Pseudoephedrine from Clandestine Laboratories *416*	
15.5.2	Interferents *416*	
15.5.3	Simulated Surface Swabs *418*	
15.6	Conclusions *418*	
	Acknowledgements *419*	
	References *419*	

Index *425*

List of Contributors

Sarah Benson
AFP (Australian Federal Police)
Canberra
Australia

Lucas Blanes
Oswaldo Cruz Foundation
Rio De Janeiro
Brazil

Alan Brailsford
King's College London
London
UK

Nerida Cole
Swinburne University of Technology
Melbourne
Australia

Wendell Karlos Tomazelli Coltro
Universidade Federal de Goiás
Goiás
Brazil

Philip Doble
University of Technology Sydney
Sydney
Australia

Shari Forbes
Université du Québec à Trois-Rivières
Québec
Canada

Ellen Flávia Moreira Gabriel
Universidade Federal de Goiás
Goiás
Brazil

Sue Jickells
Retired Analytical Chemist, formerly University of East Anglia

and

King's College London
UK

Tim Jickells
University of East Anglia
Norwich
UK

Kylie Jones
AFP (Australian Federal Police)
Canberra
Australia

Aleksandra Michalska
Institute of Forensic Research
Kraków
Poland

Robson Oliveira dos Santos
University of Santa Cruz do Sul
Santa Cruz do Sul
Brazil

Mark Parkin
Eurofins Forensic Services
London
UK

Luke N. Rodda
Office of the Chief Medical Examiner
San Francisco
USA

and

University of California
San Francisco
USA

Claude Roux
University of Technology Sydney
Sydney
Australia

Renata Mayumi Saito
Brazilian Navy Technology Center in São Paulo
São Paulo
Brazil

Barbara Stuart
University of Technology Sydney
Sydney
Australia

Sophie Turfus
University of Huddersfield
Huddersfield
UK

Rosalind Wolstenholme
Sheffield Hallam University
Sheffield
UK

Florian Wulfert
Sheffield Hallam University
Sheffield
UK

Grzegorz Zadora
Institute of Forensic Research
Kraków
Poland

and

University of Silesia in Katowice
Katowice
Poland

Preface

Forensic science is a fascinating and important subject. It provides vital information in many criminal and civil cases and allows courts to address the questions: what has happened and who is responsible? There are three main stages in the forensic science process: crime scene examination, where evidence is collected and preserved; laboratory examination of recovered evidence to determine if there are any associations between pieces of evidence; and court reporting, where the scientist presents their findings to the court. Scene examination and court reporting are well represented in forensic textbooks. However, as university lecturers interested in forensic chemistry, we saw a need for a textbook that combines analytical theory and forensic application. That is to say, for a book that looks at the interface between analytical chemistry and (primarily) trace evidence, where trace evidence means materials found in small quantities such as paint chips, fibres, and firearm discharge residue. We hope to provide the depth of technique theory and application required for students on science based forensic science undergraduate and postgraduate courses, bridging the gap between introductory and advanced texts. We also hope the book will serve as a resource for forensic scientists who may wish to broaden their knowledge of analytical forensic science or for analytical scientists who wish to apply their knowledge in a forensic context.

Sheffield, November 2019

Rosalind Wolstenholme
Sue Jickells
Shari Forbes

Acknowledgements

This textbook has been a long time in the making. One of us has had two children, one has retired, and one has moved continent, twice! I am very grateful to my co-authors and all contributors for agreeing to be involved and for their incredible patience throughout the process and while we dealt with unexpected delays, including skydiving accidents!

Very many thanks are due to Sue who, as my MSc tutor, was the first to put it into my head that I could read for a PhD. She was my first thought when I considered this project and soon realised that I couldn't do it on my own. Sue has always been enthusiastic about the book and what we could do with it. We have managed more than one ice cream on the North Norfolk coast while editing the book and I hope there will be more. I am also grateful that when I asked Sue if she would like to be involved in the book, she had the good sense to bring in Shari. As well as her immensely valuable editorial contribution, Shari has been our pacemaker, keeping us going by consistently meeting the deadlines we set ourselves in our regular conference calls. I am eternally thankful that you both stuck with the book and me.

Thanks are due to friends and colleagues who have commented on various chapters and to my elder sister who is my word processing guru and provided invaluable help with formatting in an emergency. Thanks also to the rest of my family for support at various times along the way.

Finally, my love and gratitude go to my husband, Matthew Roberts, who has been involved in many one-sided book related conversations and got the children breakfast during early morning conference calls to the other side of the world, and my love always to my children, Francesca and Bronwen. They have not known a time when I wasn't fretting in one way or another about 'the book'. I can finally give them the weekend away in a camper van that I have been promising they could have when the book was completed!

Rosalind Wolstenholme

Thanks go to all the analytical chemists that I've worked with through my career. You have all taught me/forced me to learn more about analysis. Thanks also to Roz and Shari for the journey of this book. It was tough at times but you both made it much more enjoyable. What will I do with all the extra time?

Sue Jickells

Thank you to Roz and Sue for inviting me to join this journey, even before they knew me that well. It was a rewarding (and at times challenging!) experience and I am glad we got through it together. I appreciate having the opportunity to work with you as colleagues and to know you better as friends. Thanks to all my other colleagues who I called on to help with the book, whether it was writing a chapter, editing, or reviewing content. Your efforts were greatly appreciated.

Shari Forbes

Part I

Preparing for Analysis

1

Introduction to Forensic Science

Sue Jickells, Rosalind Wolstenholme and Shari Forbes

1.1 Forensic Science

Forensic science is typically defined as the application of science to the law; both criminal and civil law. Most people tend to associate forensic science with the investigation of crimes such as burglary; arson; possession of illegal drugs; drug trafficking; drink and drug driving offences; and attacks against the person including murder and sexual assault. However, forensic science is applied to the investigation of a far wider range of potential prosecutions including war crimes; fraud; medical incidents; doping offences in sport; environmental pollution incidents; road traffic accidents; maritime and aviation incidents; industrial incidents; and issues relating to food authenticity.

Potential scenarios which may result in prosecution through criminal or civil justice systems and which require some sort of forensic examination are almost limitless. (Note that, unless otherwise stated, the terms civil justice, civil law etc. will be used to describe the processes of being sued rather than meaning constitution based legal systems.)

Similarly, different countries have different judicial systems and different systems for investigating cases that may result in prosecution. Hence, it is difficult to discuss all possible types of scenarios, authorities and personnel involved in investigation and the processes to be used. Thus, the discussions which follow are based on some of the major types of crime prosecuted under the major types of criminal judicial system and how such crimes would be investigated, emphasising the analytical chemistry techniques associated with such investigations.

It is the job of those working in the field of forensic science to consider whether there is evidence that can provide information about a particular incident or situation. The Oxford Dictionary defines 'evidence' as 'Information drawn from personal testimony, a document, or a material object, used to establish facts in a legal investigation or admissible as testimony in a law court' and 'The available body of facts or information indicating whether a belief or proposition is true or valid' (https://en.oxforddictionaries.com/definition/evidence). When we use the term 'evidence' in this book, it implies these definitions. We have deliberately included both definitions because not all evidence that a scientist finds through their investigations will result in judicial proceedings. The evidence may have intelligence or *inves-*

tigative value proving, for example, that a suspect could not have committed a crime and hence no prosecution is brought against them. It might also be that the evidence obtained is sufficient in civil cases to persuade parties to make a settlement out of court or, in criminal cases, for a suspect to enter a 'guilty' plea such that there is no requirement for evidence obtained by a forensic scientist to be presented in court.

Evidence may be obtained through analysis, which is defined as 'Detailed examination of the elements or structure of something' (https://en.oxforddictionaries.com/definition/analysis). Such analysis could be as simple as visual examination, for example, identifying flakes of paint on a car involved in a fatal 'hit and run' incident. In some cases, the shape of the paint flake may provide sufficient evidence of provenance; fitting exactly the piece of paint missing from a car suspected of being involved in the incident (termed a physical fit). What if there is no physical fit but the surface colour of the paint flake appears to be similar to the colour of the suspect car? More in depth evidence is required, which entails a more sophisticated analysis to provide chemical information about the paint. This, together with information about the colours of the layers present and their thicknesses, will provide evidence as to whether or not the paint flake may originate from the suspect car.

This book is concerned with the analytical techniques used to provide information about the nature of the sample being investigated, with emphasis on techniques that provide information about the chemical nature of samples, or techniques that involve measuring a chemical property of the sample to provide evidence. These analytical techniques are used exceedingly widely in other fields but we concentrate on forensic science in this book because, as analytical chemists involved in the teaching of forensic science, we believe that a forensic scientist using such techniques should have a good understanding of them. This includes how the techniques work, the nature of the information obtained, as well as what this means in terms of interpreting the information provided through analysis and coming to a conclusion about evidential value. Sophisticated and complex instrumentation is often used as part of these analytical techniques and discussion of such instrumentation forms one of the main elements of this book. Table 1.1 gives some examples of common types of evidence that might be received in a forensic laboratory and analytical techniques appropriate for use in their examination.

Table 1.1 Selected evidence types and common sources for them. The focus here is on trace evidence.

Evidence types	Common sources
Fibres	Garments, furnishings
Hairs	Human hair (head, facial, etc.), animal hair
Paint	Cars, door and window frames, walls
Glass	Windows, car windscreens, drinking glasses, bottles
Documents	Notes, cheques, wills
Firearm discharge residue/explosives	Firearms, spent ammunition, improvised devices

Note the distinction between interpreting information obtained through analysis and coming to a conclusion about evidential value as a result. A good example of this is analysis of ethanol in blood, associated with driving under the influence of alcohol. Interpretation of data resulting from analysis provides a quantitative result giving the concentration of ethanol in blood but this information alone is not sufficient. A decision has then to be made as to whether this value is above or below the legal restriction when driving, taking into account the precision of the analytical method, resulting in a conclusion about the evidential value.

You might have noticed that we use the term 'forensic scientist' in the previous paragraph. What defines a forensic scientist? In this book we are using it to refer to a scientist who works in an organisation whose primary business is forensic science and where the expectation is that analyses will result in evidence that may be presented in court. We recognise that there will be many more scientists where the results from their analyses may result in evidence presented in court but where the primary business of the organisation for which they work is not forensic science. An example of this would be an analytical chemist analysing blood or urine samples in a hospital to identify and quantify drugs for diagnostic and treatment purposes. From time to time, analysis may identify a sample that indicates that something illegal has taken place and this may result in legal proceedings. Such an analyst is far more likely to refer to themselves as an analytical chemist, or in this specific case as a toxicologist, rather than a forensic scientist (or, in this case, a forensic toxicologist). Thus, although the emphasis in this book is on the use of analytical techniques used in forensic science, we hope that analytical chemists working in other areas will find this text useful because, they too, should have a good understanding of the analytical techniques that they use.

Many of the chemical analytical techniques used in forensic science cannot be used 'at scene', i.e. where the sample to be investigated originated. This is generally because the techniques are too complex to bring to scenes; the nature of the scene may not be an appropriate environment in which to carry out the analysis; or the instrumentation used is not portable. Hence it is common practice to bring the sample to a specialist laboratory for analysis. A scientist working in a forensic laboratory needs to have knowledge of what has happened to the sample before it came into their possession because this can have important implications for analysis and in interpretation of analytical results. Such information may include the environment from which the sample came; how long it was in this environment; who collected the sample and how; how was it packaged and transported from the point of collection to the forensic laboratory; and how and where it has been stored since it was received. It is not our intention to cover in detail the steps that take place before a sample is analysed in a forensic science laboratory – other authors have already covered this aspect thoroughly (Gardner 2011; Horswell 2016) – but we will give a brief overview here to give context to the content of individual chapters in this book and context to the place of analytical evidence in an investigation. We will also give a brief overview of the basic principles that allow forensic investigations to be carried out, the forensic issues and questions pre and post laboratory analysis, and the judicial systems within which forensic science operates.

1.2 The Forensic Process

A simplified flow chart showing the stages in a forensic investigation is given in Figure 1.1. The specific personnel and protocols will vary according to jurisdiction but the basic process will remain the same.

The usual starting point for an investigation involving forensic science is where there is the suspicion that a possible offence has been committed or there are suspicious circumstances requiring investigation. Depending on the circumstances, the police may or may not be involved in the investigation. If it is obvious that a crime has been committed, or there are suspicious circumstances resulting in harm or loss of life, the police will be involved early in the investigation and will take a coordinating role.

The scene must be managed so that the police and specialist investigators can do their job but also so that they do not contaminate or destroy potential evidence (see Section 1.2.1). It is usual in most jurisdictions to employ specialist investigators. These may have different titles in different jurisdictions and depending on the nature of the potential incident they are investigating. Most police forces in the UK employ crime scene investigators whose role it is to examine crime scenes. They will evaluate the evidence at the scene, taking into account possible information from victims, suspects and eyewitnesses, and carry out analyses at the scene which will aid the police in their investigations. Where more specialised analysis is required, crime scene investigators will collect samples and package them for transport to specialist forensic science providers, usually working in what are referred to as 'forensic science laboratories'. In the UK, these laboratories are generally operated independently of the police authorities but in some countries and jurisdictions, the police may have direct responsibility for crime scene investigation and forensic laboratories. Whatever system is in place, there should be close cooperation between the crime scene investigators and forensic laboratories because in many jurisdictions, the scientist working in a forensic laboratory is unlikely to have visited the scene and hence needs information about the nature of the crime suspected to have taken place, the origin of samples and the conditions to which the sample may have been exposed before collection. Was the sample exposed to strong sunlight; was it subject to rain or covered with water; could animals have had access to it, etc.? Scene investigators must record this information and make it available to the analyst.

Depending on circumstances, the police will call in other specialists to help with investigation at the scene. Such specialists may include photographers, fingerprint recovery experts, forensic pathologists, road traffic incident investigators, arson investigators, firearms and ballistics experts, and blood pattern analysts. The role of all these specialists is to investigate the scene and to provide expert opinion to the police (and, ultimately, to the courts) on what has taken place and evidence to support judicial proceedings.

1.2.1 Forensic Principles and the Crime Scene

There are four main types of physical evidence that may be recovered from a crime scene: chemical (e.g. drugs and alcohol); biological (e.g. DNA and pollen); trace (e.g. glass and fibres); and pattern (e.g. fingermarks and tool marks). Table 1.1 gives some common sources of trace evidence. The primary focus of this text, in terms of physical evidence type, is trace evidence, which is simply anything of investigative or evidential value that is found in

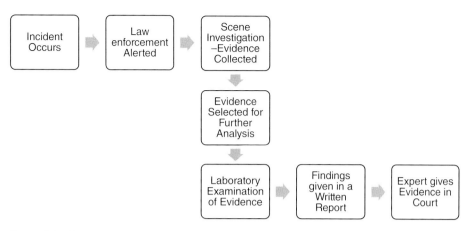

Figure 1.1 Forensic process flow chart. Once the forensic examiners of the crime scene have recovered the evidence a decision will be made about which items to send for further analysis. This should be done in conjunction with the police and the forensic laboratory scientists. After the laboratory analysis the forensic scientist will write a report and may have to go to court to defend their findings.

minute quantities. In order to provide context for the discussion of trace evidence we give a short summary of the basic principles and issues in forensic science and crime scene examination. To be able to correctly interpret their results, the forensic scientist must understand these principles to understand how the evidence came to be at the scene and how it came to be in the form in which it was received in the laboratory.

For any forensic discipline that relies on crime scene examination there is one principle without which the examinations would not be possible. This is *Locard's Exchange Principle*. This states that wherever contact occurs between two objects a transfer from one to the other will take place, commonly stated as 'every contact leaves a trace' (Kirk 1953). This may be a transfer of material or of form (i.e. an indentation). The transfer can occur from only one of the contacting objects to the other (*one-way transfer*) or from both to the other (*two-way transfer*), these are both types of direct transfer. If transfer did not occur there would be no trace evidence, no fingermarks, no trace DNA and a much lower chance of associating a perpetrator with their crime.

However, it must be noted that transfer is not always of practical use. The amount transferred may be too small to be detected or may occur in such a way as to make interpretation impossible, e.g. transfer of a ubiquitous fibre such as blue denim. There is also a third kind of transfer, which is indirect transfer. Indirect transfer is when transfer occurs independently of the 'crime' event, i.e. from a to b and then from b to c. The possibility of indirect transfer and, therefore, no connection to the event, must always be considered when interpreting results. For example, person X's trouser fibre could be transferred from their trousers (a) to a seat cushion (b) on sitting on a chair and then consequently transferred to person Y's trousers (c) when they sat on the same chair. This example could even be extended further by the fibre being transferred again to another chair or other object (d). A forensic examiner must weigh up the likelihood of consecutive transfer in the relevant scenario.

A further transfer related issue is that of *legitimate access*. Transfer may have occurred as a result of normal everyday activities by a person who would normally have access to the scene. For example, at a domestic burglary the homeowner or any person the homeowner has allowed into their home relatively recently would be deemed as having legitimate access. The evidential value (see Section 1.2.3.2) of any evidence associated with a person with legitimate access generally would be low. A perpetrator could exploit this as an explanation for the presence of trace evidence associated with them at a crime scene, which is especially difficult to manage for offences where the perpetrator is often known to the victim, e.g. sexual assault. In some cases, the forensic investigation may be able to corroborate or refute a version of events.

Once the exchange principle is accepted it swiftly follows that *contamination* is a serious problem. The aim of a crime scene examination is to secure, record and recover evidence so that as much information as possible can be gained to accurately reconstruct the events that took place. In most cases a crime scene does not appear clean and fresh with no previous transfer and is not transfer free once the crime has occurred. Contamination occurs when transfer not related to the 'crime' takes place, either by accident or intentionally. It can occur at the scene or at some later date. The scene of crime examiner should not be the source of any contamination, hence the use of suit, gloves, boots and mask and the importance of correct packaging to maintain integrity. Integrity is the preservation of evidence in its state at the time of the crime or more probably at the time of its recovery from the crime scene. Packaging preserves evidence by preventing the addition and loss of material and supplies the chain of custody to identify the keeper of the evidence at all times. An overview of packaging and chain of custody issues follows.

As noted in Section 1.1, most forensic scientists will not visit the scene and so rely on those examining the crime scene not to contaminate samples or to destroy potential evidence. The aim is that the sample analysed in the laboratory resembles the sample as it was at the scene with nothing added or taken away or changed in any other way, i.e. it retains its integrity. Packaging is imperative in protecting samples between recovery at the scene and the point at which the sample is opened for analysis. Scene investigators must use the appropriate packaging and samples must be transported to the laboratory under suitable conditions, including keeping appropriate samples cool and preventing cross-contamination from other samples that may be transported together. In some investigations, the 'scene' may extend beyond the initial place where the potential crime was discovered. This applies to post-mortem examinations (PMEs) where the body may contain potential evidence and is classed as a scene itself. The body must be packaged appropriately for transport for PME and samples taken at PME should be collected in suitable containers, with preservative added if appropriate. They should be stored under conditions that prevent any further changes taking place.

In some instances, there may be no 'scene' to be investigated. Examples include foods investigated for their authenticity or for suspected contamination. These may be samples on the shelves of supermarkets or seized from a shipping container at importation. It will be sufficient for the investigator to note the place of seizure without a photographic record. Similarly, for a sportsperson providing a urine sample to be analysed for potential evidence of a doping offence, the place where the sample is collected is not important in terms of subsequent analysis, although the nature of the collection process is important. (For example, was there an opportunity for the sample to be substituted or adulterated? Was a clean sample vessel used?) In environmental pollution incidents, the first sign that an incident has taken place may be dead wildlife. This may be remote from the original source of the pollution and analysis may be required to identify the cause of death, which may then lead investigators to the source.

The chain of custody supports correct packaging by documenting the history of real evidence (see Section 1.3.4) from the time of collection to the time it is used in court. The chain refers to the paper trail created as the evidence is transferred from one person to another. It ensures that the evidence is in the direct control of the responsible person at all times to ensure the authenticity (and therefore competency) of the evidence.

When evidence is collected at a crime scene, the officer will initiate the chain of custody by photographing in situ, collecting, packaging, sealing and appropriately labelling the evidence. The evidence may be placed in storage by the custodian or may be delivered to the laboratory for analysis. Each of these, and subsequent, transactions must be appropriately documented chronologically on the chain of custody form. A person will sign for the evidence at the time it is received and will sign again when it is handed over to the next person. Each time a forensic scientist examines the evidence, they must also maintain the chain of custody by ensuring that the affixed seals have not been disturbed, and by documenting the condition of the evidence at the time it was received. The form provides a detailed history of the custody of the evidence allowing for any of the custodians to be called as an expert witness to testify about the condition, analyses and storage of the evidence.

If a gap is identified in the chain of custody, even when accidental, this can raise questions in court about the authenticity and integrity of the sample. At the very least it can prevent a case going to court or mean that the evidence is deemed inadmissible, and in the worst case scenario it can result in an acquittal or a guilty verdict being overturned upon appeal. Maintaining the chain of custody is an essential responsibility of any forensic scientist to ensure that the evidence is not adulterated or tampered with in a manner that could affect its probative value.

The chain of custody and correct packaging (and storage) should ensure *continuity* (i.e. that the piece of evidence recovered is the same one as that examined) and ensure that the *integrity* of the evidence is preserved at all times. The forensic examiner must be able to assess whether or not an item received in the laboratory is correctly packaged, which they can begin to do with an understanding of the principles and issues outlined above. However, as there are many methods, which vary between and within different countries, the reader is directed to Horswell (2016) and Gardner (2011) for more in depth discussion.

In addition to recovering items of interest from the scene the examiner must also recover control samples. Whatever type of evidence is recovered or may be recovered from the suspect, the relevant reference samples must also be collected. For example, in an assault scenario if blue fibres have been recovered from the victim that are not from the victim's clothing or possessions, blue fibres should be collected from any suspects' clothing and homes.

While preserving and recovering material from the 'crime' scene, an examiner must also make *contemporaneous notes*. These are a record of everything that they have done at the scene, including the time at which it was done, and any other important observations, who was at the scene, *transient details,* e.g. lights on or off, windows open or closed, phone ringing, etc. Contemporaneous notes also include scene sketches and photographic logs. Some types of analysis may be carried out at the scene. For example, fingerprint evidence will be sought; the scene may be examined with light sources to detect body fluids; and suspected traces of body fluids may be tested in order to make decisions about which samples to collect from the scene and send for more specialised testing. Again, these must be noted and form part of the case documentation. Contemporaneous notes are important in providing evidence of what was done and form part of the chain of custody by noting how items were recovered. They also act as an aide memoire to examiners who carry out hundreds of examinations a year and may not be called to give evidence on a particular scene until months or years later. (Contemporaneous notes/record keeping for the laboratory forensic scientist are discussed in Chapter 2.4.1.)

The issue of contamination is also central in determining the *viability* of a scene examination. The time immediately after the crime is committed, in the context of evidence recovery, is called the 'golden hour'. In this time the maximum amount of evidence is available that retains its integrity and may ultimately be used in court. As time passes the chance of contamination and loss of material increases, particularly if there is lots of activity at the scene, so the chances of recovering evidentially valuable evidence decreases. Once the scene examination is complete and the scene is returned to normal use, the chance for evidence recovery is ended and any further examination would be subject to questions about its viability. It should be noted that the potential for evidence recovery is the same no matter what the severity of the offence, burglars are no more or less likely to leave hairs and fibres, shoe marks etc. than murderers (assuming the same time at a scene and the same level of activity). Whether the evidence is recovered or not will depend, not least, on factors such as the severity of the offence, local policing priorities and budget. Maximising recovery during the golden hour is also important to mitigate against the effects of *attrition,* in which at each stage of an investigation the material that is taken forward diminishes. This could be due to one of many reasons, inconclusive analysis, inadmissibility and, not least, budgetary constraints. The laboratory forensic scientist should always remember that they may not be asked to examine all the evidence recovered and it may be advisable to discuss the selection of items submitted with the investigative team. To maximise evidential value, i.e. the ability of the evidence to assist in proving or disproving a hypothesis, consideration of the issues discussed above, such as the possibility of contamination or indirect transfer, the integrity and potential evidential value, will assist in selecting the most appropriate items for examination.

One measure of evidential value is the type of characteristics available for examination. *Class characteristics* are those that are held by a group of items, e.g. all the glass produced from a particular batch of raw materials or all the shoe soles made from a particular shoe sole mould. Individual characteristics are those that are held only by a single item, e.g. a glass car windscreen that has been scratched by windscreen wipers and the impact of debris or a shoe sole that has been worn down by the wearer's gait and has been scratched by contact with the ground. Clearly, individual characteristics have greater evidential value than class characteristics. A still higher evidential value is gained through a *physical or jigsaw fit*. In this case, two items fit together in such a way that it is apparent that they were originally one object. Evidential value is discussed at further length in Section 1.2.3.2.

Finally, the forensic scientist and the investigative team should always remember that 'absence of evidence is not evidence of absence', i.e. just because you didn't find it doesn't mean it wasn't there. There may be any number of reasons why evidence may not be discovered, recovered or detected, which will need to be conveyed to the *trier of fact*.

1.2.2 Preparatory Issues in Laboratory Analysis

Before beginning an examination in the laboratory, the analyst should take into account case circumstances and they should ask themselves questions such as: What is the purpose of this analysis? What do I hope to show as a result? What evidence may be available from this analysis which may support, or rule out, a prosecution? What evidence may be available from analysis to indicate whether hypothesis A is more likely to be correct than hypothesis B? Allied to these questions are others such as: What technique or combination of techniques that I have available to me will best provide answers to the above questions? (See also Chapter 2.)

Once a strategy has been decided upon the scientist can begin to look at the items submitted. They must first check that the paperwork for the item(s) to be examined is in order, the chain of custody is intact and the packaging is appropriate and has not been compromised. The laboratory should have systems in place for ensuring that the examination environment is clean and there is no opportunity for contamination, such as procedures for cleaning down benchtops, procedures for keeping evidence from different sources apart and installation of air conditioning systems with filters to prevent particles from moving from one laboratory to another.

Examination, whether at a crime scene or in the laboratory, starts with the least destructive or least invasive methods of analysis before progressing to more destructive ones and should start with examination *in situ*. Non-destructive testing is very important in forensic examination to preserve the integrity of the sample but also to ensure that wherever possible the sample is available for re-examination by either legal team. If a test is not repeatable and/or its results are called into question it may lead to a reduction in evidential value or the evidence not being admitted in court.

Visual examination is the starting point, possibly allied with smell and hearing, depending on case circumstances. Depending on the item being examined, the scientist may need to record details of the item as a whole then look for and recover further evidence using tweezers for easily visible evidence followed by other techniques such as tape lifting or similar and swabbing. An example of this would be in a garment examination whereby the condition, colour, shape, size, label details, etc. of the garment would be recorded along with sketches and photographs. Then any trace evidence such as hairs, fibres, glass, paint, etc. would be recovered, packaged and recorded, including where on the garment they were recovered from as this may be important to the case, e.g. in an assault case the position of fibre transfer may indicate where contact has taken place and may support or refute a version of events. The garment may then be tape lifted to recover any evidence not immediately visible and swabs may be taken to test for the presence of body fluids and hence DNA. In this scenario, the original item, the garment, may now be of less interest than the evidence recovered from it in the laboratory.

If not already employed, the visual examination may be followed by investigation using specialist light sources or may proceed directly to physical or chemical analysis or both. Analysis may also involve microscopy to reveal evidence not visible to the naked eye. For most samples, a combination of analyses will be applied. This is because each examination may reveal additional, useful, information; and may highlight additional analyses to be carried out or may indicate that further analysis will not be useful or necessary.

The analyst also needs to think about the nature of the sample and whether it needs to be prepared in some way before analysis can be carried out. For example, if the sample to be examined is urine which may (or may not) contain traces of a drug, there are no analytical techniques which can identify, unequivocally, what drug is present and at what concentration without separating the drug in some way from the urine matrix. Note the use of the word 'unequivocally'. Further discussion of confirmation of the identity of a particular substance can be found in Chapter 3 and further discussion of sample preparation can be found in Chapter 4.

Other questions that need to be considered are: How much sample do I need for analysis and is there sufficient sample? If I only have limited sample, of all the techniques that I could use, which will provide the best evidence?

Another consideration is how many samples need to be analysed. This is particularly pertinent to drug trafficking offences where a seizure may consist of dozens or even hundreds of packages of drugs. Does every package need to be analysed to prove that the suspect is trafficking drugs? What about a seizure consisting of thousands of suspected ecstasy tablets – does every tablet need to be analysed to prove that the tablets contain ecstasy? Is analysis of a single tablet sufficient or is a suitable compromise to analyse a proportion of the tablets? (See also Chapter 2.3)

Issues of sample homogeneity also then arise. If only a portion of the sample is taken for analysis, rather than the entire sample, can the result be taken as typical of the sample as a whole? For a bulk drug seizure, a single package of cocaine may consist of a kilogramme of the drug. A typical analysis using gas chromatography–mass spectrometry (GC-MS) will use 10 mg of sample extracted in 10 ml of methanol. Thus, the analyst requires only 0.0001% of the sample for analysis, a very small proportion. They need to be as certain as possible that the 10 mg taken for analysis is representative of the kilogramme in the package and so they take steps to ensure the kilogramme sample of cocaine is made as homogeneous as possible before the sample taken for analysis (often referred to as a subsample) is withdrawn from the whole sample.

What about situations where the inhomogeneity of a sample may provide crucial evidence? A classic example of this in forensic science is the examination of paint. Motor vehicles are deliberately painted with several layers of paint. The underlying layers are formulated to adhere to the metal substrate and to prevent corrosion of the underlying metal, whereas the surface layers provide the final colour and may also provide gloss and an aesthetically pleasing finish. The colour and thickness of the layers and their chemical composition and the order in which they occur may provide crucial evidence in a case. For this information to be of use, a homogeneous sample is not required. Instead, in the case of paint, the sample is sectioned to provide a profile of the layers with examination by light microscopy; Fourier transform infrared (FTIR) microscopy (see Chapter 7) and possibly by scanning electron microscopy (SEM) combined with energy dispersive X-ray (EDX) analysis or X-ray fluorescence (XRF) (see Chapter 9).

For some samples, the physical form of the sample in combination with chemical information is important and this may influence the technique used for analysis. This is the case with examination of firearm discharge residue (FDR), also known as gunshot residue (GSR). Research has shown that FDR is characterised by the formation of spheroidal particles of the order of a few microns in diameter. For ammunition containing lead,

FDR particles are characterised by their morphology and elemental composition, typically containing barium (Ba), antimony (Sb), and lead (Pb), or sometimes Sb and Ba only (Romolo and Margot 2001). Particles containing Pb, Ba, and Sb have been found on the hands of fireworks technicians but are always found in association with magnesium (Mg) and/or copper (Cu) (Mosher et al. 1998). The classic analysis for FDR is SEM, to show morphology, combined with EDX for elemental analysis. A technique such as inductively coupled plasma mass spectrometry (ICP-MS) (see Chapter 10) could be used to show elemental composition but does not provide information on morphology because (a) samples need to be made fully soluble for ICP-MS analysis and (b) the instrumentation does not involve any sort of imaging. It would be possible to analyse the sample by microscopy and then subject it to ICP-MS but this would be more time consuming than SEM–EDX and would also destroy the sample, whereas SEM–EDX does not.

It should also be borne in mind that forensic analysis is carried out by specialists employed by an organisation. Employing these analysts costs money and the longer they spend analysing a single sample, the more money it costs. Some analyses involve chemical reagents and instrumentation which have to be purchased and, in the case of instrumentation, have a cost associated with continued operation and maintenance. An organisation carrying out forensic analysis will take into account the cost of sample analysis and will put in place processes where analysts use the appropriate combination of methods to minimise costs. An example of this is illustrated by investigation of a white powder seized from a suspect arrested for burglary. The appearance of the white powder and packaging may indicate that it is cocaine. If the potential prosecution is 'possession of a cocaine (as a controlled drug)' the main question to be answered is: Does the white powder contain cocaine? If the sample is being analysed to investigate dealer or trafficking networks, further questions might be: What is the percentage of cocaine in the sample? What other substances are present in the white powder and what is their relative percentage composition in the sample? To answer the more complex questions, an analytical/quantitative technique will be required, however, the examiner may decide to begin with a simple presumptive test. The relatively small cost involved in applying the simple test outweighs the wasted cost of quantitative analysis for a sample that does not, despite well-founded suspicions, contain cocaine.

The forensic scientist will consider all of the questions posed above in combination to devise a plan for testing the evidence that is most likely to yield the information required for meaningful interpretation. This will be straightforward for some samples but not for others. It is imperative that the scientist understands all the techniques available and their relative advantages and disadvantages in order to plan effectively.

1.2.3 Interpretation of Forensic Evidence

Interpretation is at the heart of forensic science; it is what makes a forensic scientist different from an analytical scientist. The results of an analytical test or a physical examination provide information about a piece of evidence but then that has to be considered in terms of the circumstances of a particular case. For example, if a fragment of glass seized at a burglary is analysed and is $x\%$ silica and $y\%$ aluminium oxide, this is an analytical result but it does not tell us the significance of that result nor the inferences that can be drawn (in this case it could be the type of glass, the manufacturer of the glass, the batch of glass, etc. or it could be that since, say, 90% of glass has the same composition it does not tell us very much). The forensic scientist must use many resources, their experience, knowledge, and databases included, to form an opinion on the significance of the result. The competence of a forensic scientist has been expressed as: Ability × Training × Experience × Talent × Motivation × Daily variables (Wertheim 1996). Ability, in the context of trace evidence analysis, includes technical competence, e.g. the ability to choose and use appropriate instrumentation, but, crucially, also competence in interpretation, e.g. the ability to construct an appropriate framework for the interpretation with hypotheses and reference data. Note, whilst experience is important it does not necessarily make for a better result – if you have been doing something badly for 20 years, you have experience but you are still doing it badly. Therefore, continual training and peer review and/or blind testing should always be part of the forensic scientist's job description.

The following sections will look at the role of the expert witness in interpretation, how the value of evidence is assessed and how that value can be expressed in court, including with statistics. The use of statistics in forensic science is problematic and varies greatly with evidence type, therefore only a broad overview of the scientific approach is provided rather than a detailed statistical method. For more in depth discussion of statistics in forensic science see Lucy (2013) and Aitken and Taroni (2004).

1.2.3.1 The Expert Witness and Interpretation

Most witnesses appearing in court are permitted to give factual evidence only. Unless an 'event' has been seen/heard/read first hand by the witness, it cannot be accepted into evidence – that would be 'hearsay'. (Refer to Section 1.3.3 for discussion of the rules of evidence.) An expert witness is a person deemed by the court to have sufficient knowledge and experience in a specific area of expertise to provide evidence that may include opinion. It is the court's responsibility to check the credentials of the witness. Unfortunately, this does not always happen, as was seen in the case of 'Dr' Gene Morrison who posed as a forensic expert in several areas despite having only qualifications purchased online (BBC 2007). The expert's opinion must be based on the facts of the available evidence examined, an accepted method of examination (Section 1.3.5), and, importantly, strictly within the expert's area of expertise (CPS 2019). If the evidence strays into areas outside the expert's field or there is not enough sample to analyse, the expert must make it clear and not provide an opinion on those matters. In particular, an expert should not be asked about guilt or innocence, as this would encroach upon the function of the trier of fact. This will be discussed further in relation to question hierarchies in Section 1.2.3.2. Legally speaking experts carry no more weight than any form of evidence, however, by the very fact that opinion evidence is allowed from an expert witness (and possibly the use of the word expert) more weight may be given to their evidence by the trier of fact. Therefore, the expert must uphold the highest standards at all times and ensure evidence given is clear and balanced and that proper procedures are followed.

In most cases, an expert is employed by someone other than the court, however, their responsibility is to the court and they are bound by the same laws of perjury as other witnesses. Therefore, as well as being an expert scientist they must also be independent and unbiased in their examinations and conclusions. This may be easier in some legal systems than others (e.g. inquisitorial, see Section 1.3.2), however, it is imperative in all.

There are two ways in which expert witness testimony can be presented in court. One is the appearance of the witness in person but the second, and more common way, is by a report detailing the examination, the results and the findings. In practice, the examination of evidence, may or may not have been carried out by the witness (e.g. by a junior examiner or by the opposing counsel's expert), however, the examination should have followed strict protocols and the court reporting officer should have checked and scrutinised the results before coming to any conclusions. A report will be written for all work carried out but it may not be known if a court appearance will be required, therefore, the report must always be such that it can be clearly understood without explanation in person by the author. This is relatively straightforward for the examination and results but is more complex for the interpretation, particularly in assessing the value of the evidence in the case context. Note that under rules of disclosure, findings may need to be provided to the 'opposing' legal teams whether or not they support the position of the team by which a forensic expert is engaged.

Interpretation in the forensic context is not an exact science. As well as all the skills necessary to make a good forensic scientist given earlier, to get reliable/useful results we can also add good external input. This includes high-quality information from the commissioning person/body (which includes correct evidence collection) and access to high-quality and appropriate databases; the better the input the better potential for better output. (Databases will be discussed further in Section 1.2.3.3.) From these required skills and external inputs, we can see that many factors contribute to how an opinion is formed and whether it is valid or not.

1.2.3.2 Evidential Value

As we have mentioned previously, the interpretation of forensic evidence is the most important part of the forensic scientist's job. An analytical scientist familiar with SEM–EDX might be able to follow the protocols specified to carry out the analysis of a glass fragment to determine its composition. This provides some facts (within the constraints of instrument accuracy) about the glass but the forensic expert, and in this case the expert in glass analysis, is needed to bring those facts together, with other information available, to form an opinion on the evidence. Perhaps the glass fragment has the same composition as a reference glass sample but does this mean it has exactly the same origin, can the origin be determined, and was there any possibility of sample mix-up? (See also transfer in Section 1.2.1.) Perhaps there is a difference but it can be explained by a difference in sample preparation or instrument type. Once these kinds of questions have been asked, part of the expert's opinion, given the answers, will consider how informative the evidence is; this is termed its *evidential value*. To give a trivial example, a hair found at a domestic scene of crime that belongs to the home owner is likely to have very little evidential value (excluding some specific circumstances, e.g. found in the hand of a murder victim), whilst a hair from the same person found at a crime scene that they say they have never been to before will have much higher evidential value.

Considering evidential value allows the expert to indicate to the trier of fact how much weight to give it. For this reason, experts must be careful about the terms that they use throughout their report so as not to confuse. In the above glass example, the glass had the same composition as a reference sample and it is tempting to say they 'match', but what if it is due to the capabilities of the instrument? Perhaps it is not sensitive enough to detect differences. What if 90% of all glass has the same composition (as detectable by the instrument), or any small difference is within the expected variation in accuracy of the instrument? 'Match' is a subjective word with potentially different meaning to people of differing expertise. In expert witness reports, subjective phrases such as 'could have', 'consistent with', and 'match' should be avoided and care should be taken with words such as 'unique' and 'identification'. Identification, particularly, may suggest that two pieces of evidence are taken to have the same source and has been used this way in the past. However, an object has been 'identified' when it has been determined to be one of a class of objects, e.g. an object may have been identified as being a book but this does not tell us anything about its source or relevance to the case. As a result, the term 'individual' has become more acceptable in forensic circles to describe two objects/analytical results that are indistinguishably alike. That said, this still may not mean 100% surety and in fact, in practice, it is much easier for a forensic scientist to be definitive about an exclusion of evidence than to be definitive about an individualisation.

The opinion expressed in an expert witness report may be in response to a specific question asked. When submitting evidence, the police or other body will be submitting it with a purpose in mind. They need to know whether it is pertinent to the case and if so why. However, what they ask may be less objective. For example, rather than asking 'is this sample human hair and if so can you determine whose it is?', they may ask 'is this the suspect's hair?', or even 'did the suspect carry out the offence?' This problem, using the viewpoint of the prosecution or defence rather than the investigative body, has been discussed by Cook et al. (1998a). They describe a 'hierarchy of propositions', source, activity, and offence (levels I, II and III, respectively), with propositions that come in pairs, which roughly equate to the prosecution and the defence arguments. In the paper by Cook et al. (1998a), an example of hierarchy of propositions in a burglary case involving glass is given:

Source (I)

The glass fragments came from window X.
They came from some other broken glass object.

Activity (II)

Mr A is the man who smashed window X.
Mr A was not present when window X was smashed.

Offence (III)

Mr A committed the burglary.
Another person committed the burglary.

Forensic scientists are more likely to be able to address source level propositions and it is usually up to the jury to decide on offence level propositions. This is a very important distinction, not understanding it is a common source of confusion (particularly for enthusiastic undergraduate students); the police carry out investigations, forensic scientists examine the evidence, lawyers make their case, and it is only the jury who decide on the ultimate issue of guilt or innocence. The jury's role must not be undermined by a forensic expert's over interpretation, not least because offence level questions require some knowledge or judgement of the defendant's intent, which is not within the scope of forensic expertise.

The 'hierarchy of propositions' approach also supports the idea that the court and the forensic scientist are actually concerned with slightly different questions. The court wants to know about the probability of the defendant committing the crime given the evidence. Whereas, the forensic scientist should ask, if the defendant committed the crime, what is the probability of the evidence and, in order to be balanced, the reverse, if someone else committed the crime, what is the probability of the evidence (Cook et al. 1998b). Bringing the forensic scientist's two questions together, the combined question is posed, given the evidence, how much more or less likely is it that the defendant committed the crime? Asking these questions helps to determine evidential value. We shall come back to this idea when looking at Bayesian analysis in Section 1.2.3.4.

It should be noted here also that to help avoid over interpretation and cross over to level II and level III propositions, there is an argument for the forensic examiner's knowledge of the case details to be limited to what is strictly necessary in order to avoid unintentional/unconscious bias. However, there is also an argument for the forensic examiner to be involved in discussion about which methods of analysis should be carried out, since they have the expertise in what is possible and what is likely to give a return on the investment in time and money that each analysis requires (Cook et al. 1998b). If, as a forensic scientist, someone handed you a man's jacket and trousers and simply said, analyse these and see what you can find, that is a very difficult challenge. It may be far more helpful to know that the jacket may have been worn by someone suspected of a violent sexual assault which included a stabbing, because you have a far better idea of the potential evidence that may be present if the allegation is true and can approach the analysis with that in mind. That said, a forensic scientist should approach all analysis of potential evidence with an open mind and should consider that the information provided by the victim, suspect, and witnesses may not be correct or may be biased. Hence, they should not restrict their investigation or interpretation of findings in a way that rules out other possible scenarios.

The many disparate areas of forensic science mean that it is not equally easy to determine evidential value in all cases. Some forensic disciplines lend themselves to statistical interpretation/calculations, others may in at least some parts, and more still may not, requiring a more intuitive assessment. For example, in glass examination the frequency of a particular chemical composition of glass can be estimated with statistics but not the probability of two glass fragments forming a *physical fit*, i.e. fitting together like jigsaw pieces. A forensic expert will choose the interpretive approach most suitable for their type of examination and must make this approach clear to their audience, which includes drawing attention to any limitations.

Whether or not statistics are used, the forensic scientist must be careful not to overstate the evidence. It is highly unlikely that a proposition can be absolutely proven to be true. There is a gap between what is statistically/intuitively probable and what is actual fact. Evidence suggesting that an individualisation is 99% probable leaves a 1% probability that it is not. Even allowing for the benefit of experience, to declare an individualisation in this case would be to ignore the uncertainty. Ultimately, it is the trier of fact who is most appropriate to make this interpretive leap rather than the forensic scientist. A verbal scale of support can be one way to help with this problem (see Section 1.2.3.4).

1.2.3.3 Statistical Interpretation

Most determinations are made inductively, i.e. from specific to general observations with incomplete knowledge, rather than deductively, i.e. from general observations to specific and logically correct conclusions. In forensic science, particularly, the circumstances surrounding a case and the actions taken are generally too complex to know everything that has occurred precisely, i.e. they are inductive. Statistics and probability are a means to aid inductive reasoning by estimating or predicting the occurrence of something when complete knowledge of the event is either unfeasible or impossible (e.g. it is possible to carry out a fibre examination of all the fibres produced worldwide, but it is not feasible). It must be remembered that statistics offers an estimate rather than an absolute truth. However, the more data we can collect the closer to the 'real' value we hope to be.

When using statistics, the scientist formulates hypotheses and states any assumptions that have been made. Usually, a null hypothesis and an alternative hypothesis are used. In our burglary example these hypotheses may be that the recovered fragment of glass and a control sample have the same chemical composition (null) and that they do not (alternative). These hypotheses should relate to one of the levels in the 'hierarchy of propositions' discussed in Section 1.2.3.2 and the forensic scientist must be sure that they are posing questions that they are able to answer. For example, in many instances it may be possible to determine the composition of a glass fragment but pinpointing the time of breaking is much more problematic (unless there are witnesses, who are not within the forensic scientist's remit in any case). Assumptions will vary widely with the type of evidence and method of analysis, however, some shared examples might be that the evidence being analysed is relevant to the 'crime' and that there is no innocent association to the defendant (Evett 1993) or that events are independent and, therefore, their values can be multiplied together.

Databases are essential to statistical interpretation as without them it would be impossible to give a value to the questions posed. If, to continue the glass example, it is necessary to know how frequently a particular type of glass is used, a database is needed giving the frequencies of use of different types of glass and potentially their use in a specific geographic location, if this may be different.

Another issue that must be taken into consideration in a statistical context is error. This is the discrepancy between the obtained value and the 'true' value. In an analytical context, error comes about in two ways. Systematic error is a bias affecting all values equally, e.g. poor instrument calibration. Random error is unavoidable, unpredictable and affects different observations in different ways, e.g. human error in recording observations, which is why observations should be made multiple times and an average taken (see also Section 2.4.1). Taking into account both types of error, the 'uncertainty of measurement' is the range of values that, with high probability, includes any errors that may arise as the measurement is taken. This uncertainty must be considered when a forensic scientist is evaluating any results, drawing conclusions and conveying them to an audience. In a statistical context, there are also two types of error, Type I is to reject when true and Type II is to accept when false. Again, for any test there should be estimates available for the occurrence of each.

Discussed in Section 1.3.5, the requirements of the Daubert ruling for acceptance of a forensic field in court include an estimate of the error rate. For most fields this is not available but, in the USA, the National Research Council report of 2009 (National Research

Council 2009), which was critical of this aspect of forensic science, has focused thinking onto how this may be achieved.

Currently many fields rely on an intuition, rather than scientific evidence, that no two objects leave identical marks/traces meaning that identical marks/traces imply a common source. In fact, when marks/traces are indistinguishable instead of jumping to the conclusion of identity, there should be information available about how common identical marks/traces from non-common sources are (Saks 2010). This reliance on intuition also does not allow for calculation of error rates, even though blind trial studies (Dror et al. 2006) have proven (as would be expected) that errors do occur. Not only could the sources of error, discussed above, be mitigated against but the error rates could also be estimated using measures such as inter and intra laboratory blind trials, calibration, and quality control.

Until these issues are addressed and error rates can be estimated, court reporting scientists should be clear to their audience about what are the limits of their field, in the respects of knowledge and belief. They should also give an estimate of the 'uncertainty of measurement'.

Once a numerical value for statistical probability has been determined the forensic scientist has to decide how to interpret it. The variety of forensic disciplines and types of evidence within them mean that there is no universal way in which to do this. In addition, they also need to decide how to present that interpretation to the trier of fact. The trier of fact is likely to be someone who has no framework of experience in the relevant field and, therefore, no real perception of how significant is the statistical value. One solution for this is to use a verbal scale of support, which translates a number into a description (see Section 1.2.3.4 for an example). These can use the numerical value directly or, especially when multiple pieces of evidence are combined, a logarithmic scale. The advantages of these scales are that, for the inexperienced, the terms used in a verbal scale will be more easily understood and it is possible to see the relationships between the categories of support, i.e. how much bigger or smaller the numerical values are in each category. However, the major disadvantage is that the category boundaries are arbitrary. For example, if a boundary were set at 0.7 between moderate and strong support, statistically the difference between 0.69 and 0.7 may not be significant but the difference perceived between moderate and strong support may be. In addition, the terms used are subjective and (as previously discussed in relation to the language used in expert witness reports) must be chosen with care, to one person strong support may be definitive whilst to another merely a positive indication.

As mentioned above, the use of an expert witness may provide a certain amount of gravitas to the evidence presented. This skewed perception may also be compounded by the use of statistics to evaluate evidential value. The expert must make it clear that statistical method is simply a way to approximate and does not give a definitive and infallible result, i.e. giving a result a value does not make that value fail-safe.

One way in which evidence with a statistical component can be misunderstood in court is in confusing the meaning of the statistics. The most notorious misinterpretation is the transposed conditional or the 'prosecutor's fallacy', so named by Thompson and Schumann (1987). This erroneously exchanges a small probability of the occurrence of the evidence if it is unrelated to the crime for a large probability that it is related if it has occurred. Thompson and Schumann give the following example: the suspect and the scene sample are the same blood type, which occurs in 1% of the population, so there is a 1%

chance that the blood at the scene came from someone else, which means that there is a 99% chance that the suspect is guilty.

Also to be avoided is the 'defender's fallacy' (Thompson and Schumann 1987), which is used to argue that a piece of evidence does not have value. It uses the probability of the evidence (always less than 1) multiplied by the number of the population and if this is still a 'large' number it is argued that the evidence is of little value (as the defendant remains only one in a large number). However, the size of the whole population has been reduced by the evidence to a smaller set of interest, therefore, the evidence does have value. For example, if the probability of the evidence occurring in a population is 0.04 (4% of the population) and the population is 500 000, then the population of interest is 20 000, which although still large, has been reduced by 480 000.

1.2.3.4 Bayesian Statistics

Bayes' theorem is a statistical theorem widely discussed in forensic science circles because it allows the scientist to employ the approach discussed in Section 1.2.3.2, i.e. given the information available, how much more or less likely is something? The theorem is given as follows:

$$O(A \mid E, I) = \frac{P(E \mid A, I)}{P(E \mid \bar{A}, I)} \times O(A \mid I) \tag{1.1}$$

where, in the forensic context, A is that the suspect is the person who committed the crime, $\bar{A}$ is that some other unknown person committed the crime, I is the initial information, such as assumptions and witness statement information, and E is evidence/new information (includes the results of 'background' information tests/research and scientist's results). Each of the terms in Eq. (1.1) can be considered to evaluate the scenario under investigation: $O(A \mid E, I)$ is the posterior odds, $\frac{P(E \mid A, I)}{P(E \mid \bar{A}, I)}$ is the likelihood ratio (LR), and $O(A \mid I)$ is the prior odds.

The prior odds are the odds that the suspect committed the crime given the initial information (probability that the suspect committed the crime given the initial information divided by the probability that some other person committed the crime given the initial information). The posterior odds are the odds that the suspect committed the crime given the initial information and the evidence (probability that the suspect committed the crime given the initial information and the evidence divided by the probability that some other person committed the crime given the initial information and the evidence). In order to get from one to the other the prior odds are multiplied by the LR. The LR is the probability of the evidence given the suspect committed the crime divided by the probability of the evidence given the suspect did not commit the crime and is what the forensic scientist is primarily concerned with. This is because it acts as a multiplying factor (i.e. how much more or less likely) by considering the evidence both if the suspect is guilty and if the suspect is not guilty. Furthermore, by analysing the sensitivity of the equation we can determine the extent of the contribution of the factors considered, which may help in evaluating their evidential value. For example, if one of the factors in the denominator is very large compared with the others and the numerator is small, the very large denominator factor will make the LR less than one meaning it is less likely that the suspect committed the crime than the prior odds suggest. Even if all the terms in the equation change by the same absolute number, e.g. if

Table 1.2 Example of a verbal scale of support. In this example, verbal strength of support increases with an increase in the order of magnitude of the likelihood ratio.

Verbal scale	Likelihood ratio	
Extremely strong	>10 000	
Very strong	1000–10 000	
Strong	100–1000	Prosecution
Moderate	10–00	
Weak	1–10	
Inconclusive	1	
Weak	0.1–1	
Moderate	0.01–0.1	
Strong	0.001–0.01	Defence
Very strong	0.0001–0.001	
Extremely strong	<0.0001	

they all go up by 2, the large denominator factor will still be the most 'important'. The very large factor has had a greater role in determining the size of the LR than the other small denominator factors, i.e. it has a higher evidential value.

Using the example of fibres recovered from a suspect in a burglary scenario, the questions that may need answering could include: (i) what is the frequency of the fibre in the fibre population (assuming the specific fibre can be identified)? (ii) What is the background number/frequency of these fibres found on an average person unrelated to the crime? (iii) What is the frequency of fibre transfer in the contact scenario in question? (iv) What is the frequency of fibre retention on the surface from which it/they was/were recovered in the time since the crime and given the suspects activities in that time? Some of the answers may be found in the relevant databases, e.g. frequency of fibre in the fibre population, whereas others may require some case specific research, e.g. the frequency of fibres transferred in the contact scenario. Further, the former question is likely to have a relatively accurate value, while the latter is unlikely to, given the many variables involved in transfer of fibres during contact (e.g. force of contact, length of contact, condition of contacting surfaces). Framing the questions correctly, calculating the related values and allowing for the accuracy of the values combine to make the use of Bayesian statistics very complex.

However, one of the benefits of using a LR (or sometimes log (LR)) is that it lends itself to the use of a scale of support (see Section 1.2.3.3). One of the most popular systems increases the verbal strength of support with each order of magnitude of the LR (see Table 1.2).

The LR is a useful model for assessing evidence. However, despite extensive discussion of the application of Bayes' theorem in forensic science, in practice it is hard to apply and is, therefore, controversial. Indeed it has been ruled against in the UK for evidence other than DNA '(and possibly other areas where there is a firm statistical base)' (*R vs. T* 2010).

1.3 Judicial Systems

The definition provided in Section 1.1 demonstrates one of the major differences between forensic science and other natural sciences. Forensic science requires that the scientist not only be well versed in the *scientific* principles but also in the *legal* principles that govern their analyses. A forensic scientist is often tasked with identification, classification and individualisation of evidence, but in addition must be able to present their conclusions to a jury in a format that is understandable to the lay person and impartial (see Section 1.2.3.1). Forensic scientists must therefore be familiar with the judicial system by which the courts are guided in their jurisdiction and must adhere to the rules that govern admissibility of all evidence.

1.3.1 Criminal vs. Civil Law

The main function of a court is to settle disputes whether they be related to a civil or criminal investigation. A civil court case typically involves a dispute between individuals or between an individual and an organisation. A criminal court case involves a dispute between an individual and the state, where the individual's action is considered to be detrimental to society as a whole. Forensic scientists can be called as expert witnesses in both civil and criminal cases but predominantly provide testimony relating to criminal investigations.

The key differences between criminal and civil law are based on the individual, organisation, or entity who files the lawsuit, the standard of proof, the trier of fact, and the action sought. In criminal cases, depending on the jurisdiction, the government or Crown files the litigation on behalf of its people, and subsequently prosecutes the case (i.e. the prosecution). The person accused of the crime, the defendant, is charged in a formal accusation known as an indictment and is defended by a lawyer (i.e. the defence). If the person is charged with a federal offence in the US, the federal attorney's office will act as the prosecution, and if charged with a state crime, the state attorney's office will prosecute. In England and Wales, the Crown Prosecution Service (CPS), on behalf of the state, is the entity responsible for compiling and presenting the prosecution evidence in court. In a non-adversarial system such as those found in Europe, the examining judge determines whether there is a valid case against the accused, and then sends the case for adversarial trial in front of magistrates or by jury (see Section 1.3.2).

In criminal cases, the victim is not responsible for bringing the case to court as this responsibility lies with the government. In certain cases, there may not be a victim, such as the conviction of a driver under the influence of alcohol or drugs. In this instance, the potential harm caused by the defendant is to society as a group, not to one individual. The defendant is charged based on the potential for serious harm that could have been inflicted to individuals as a result of their actions. Since society considers driving while intoxicated a serious criminal offence, a criminal case can be brought against the defendant, even without a victim.

The burden of proof in a criminal case requires the prosecution to prove their case *beyond a reasonable doubt*. The trier of fact in a criminal case is usually either a bench of magistrates for less serious crimes or a jury for more serious crimes. In some jurisdictions the defence may request that there is no jury, in which case the trier of fact may only be the judge.

The burden on the prosecution is therefore to convince the magistrates, jury or judge of the defendant's guilt. This burden is considerably more difficult to prove than in a civil case in order to protect innocent individuals from being convicted of crimes they did not commit. Depending on the jurisdiction, in most cases, the burden of proof lies with the prosecution as it is assumed that the defendant is innocent until proven otherwise. However, there are exceptions whereby the defence holds the burden of proof, such as if a defendant claims self-defence, duress, or insanity. In these cases, the defence may bear the burden of proving the accuracy of the defendant's claims.

Criminal prosecutions are intended to acquit the innocent and convict and subsequently punish an offender for their crime. The convicted defendant may be sentenced to imprisonment, or to pay a monetary penalty to the government and/or victim. In exceptional cases in certain US states, the defendant may be sentenced to the death penalty. In many US jurisdictions, the authorised punishment for conviction of a misdemeanour is a fine and up to 12 months confinement in jail whereas a felony can be punishable by death or imprisonment for more than 1 year. The punishment of a felony relates to the seriousness of the crime committed, and can include crimes such as murder, rape, aggravated assault, burglary, kidnapping, and perjury.

Civil cases differ to criminal cases in each of the factors previously mentioned. A civil case results when an individual or organisation (the plaintiff/claimant) files litigation against another individual or entity (the defendant) for not fulfilling a legal duty or contract owed to the plaintiff/claimant (e.g. breach of contract, divorce, debt collection, house foreclosure). Civil cases can also involve claims for damages resulting from death or personal injury of a person. Civil law involves disputes between private parties and requires the court to resolve the dispute based on the legal rights and obligations of each party.

In civil investigations, the 'victim' is typically the plaintiff who believes they have suffered a legal wrongdoing by the defendant. A civil case is presented before a judge (and sometimes a jury) whereby the plaintiff, or counsel/solicitor for the complainant, argues that the defendant should be held responsible for their actions and should compensate them for any losses or injuries resulting from these actions.

The plaintiff must prove their case by a *preponderance of evidence* or *balance of probability* (the language varies with jurisdiction) meaning that they must convince the judge of the greater weight and value of their evidence. In other words, they must demonstrate that their evidence is more convincing than the other party's and that they should be entitled to a judgement as a matter of law. In cases such as fraud (in jurisdictions where fraud is not a criminal offence) or misrepresentation, a higher standard of evidence is required, and the plaintiff may be required to prove their case by *clear and convincing evidence*.

A forensic scientist called in a civil case may be asked to provide testimony about the apparent truth behind a claim. The assessment of liability therefore becomes important for the expert witness. Occasionally, although less likely, the expert witness may be asked to assess the monetary value of a particular claim. This tends to be outside the expertise of a forensic scientist and is typically only addressed by experts in other fields.

If the defendant loses judgement to the plaintiff, the judge may order them to compensate the plaintiff a monetary value equivalent to their losses at the hand of the defendant. Alternatively, the judge may order them to fulfil the duty of the contract under dispute. Civil suits can be brought in both state and federal courts in the US and in the County or High Courts in the UK. An individual who files litigation against another individual based on breach of contract will typically do so in a state or county court (or possibly High Court if for a higher value claim). However, if an individual or organisation claims a violation against federal statutes or constitutional rights, they will file a civil suit in the federal court.

Occasionally an act can result in both civil and criminal litigation. Continuing with the earlier example of driving under the influence of alcohol, if the driver had caused a vehicular accident as a result of their impairment and subsequently injured a driver or passenger in another vehicle, the injured party may bring a civil suit against the drunk driver. Additionally, the drunk driver would be charged with the criminal offence of driving under the influence of alcohol which involves a criminal lawsuit. Both cases would however be tried separately since the standard of evidence used to judge the case is much higher in criminal investigations than civil investigations.

1.3.2 Adversarial vs. Inquisitorial System

Common law and civil law systems represent the two major systems of law. Most countries that derive their legal systems from the English model are considered common law countries and subsequently use the adversarial system for criminal and civil cases. Other than England, these countries include (but are not limited to): USA, Canada, Australia, New Zealand, Singapore, Hong Kong, South Africa, and Ireland. The adversarial system uses a model whereby two parties assume opposing sides in debating the innocence or guilt of an individual. In a criminal trial, the two parties are represented by the prosecution and the defence, whereas in a civil trial they represent the plaintiff and defendant. The role of the judge in the adversarial system is to act as the unbiased adjudicator between both parties and to ensure that the trial proceeds within the procedural rules and guidelines. The jury is the trier of fact and their role is to determine the innocence or guilt of the individual based on the evidence presented.

As the burden of proof in respect of the offence is on the prosecution, their argument will be based on evidence that indicates that the accused has committed the criminal offence with which they are charged. The defence will then attempt to defend these claims and may call witnesses in an attempt to discredit the prosecution's evidence. The evidence presented and the witnesses called are determined solely by each party (assuming the evidence is deemed admissible) ensuring that the prosecution and defence are in charge of the case, rather than the judge. Although all parties may be involved in case management, the judge does not intervene in the criminal proceedings unless required to clarify the law or maintain procedural impartiality.

The inquisitorial (or non-adversarial) system is the approach employed by many civil law jurisdictions (here we mean constitution based legal systems), which allows the judge to be involved in both the investigation and trial of an individual. The judge acts as both the prosecutor and judge under the inquisitorial system and is responsible for gathering evidence, questioning witnesses, and making a judgement as to whether the case should proceed to trial. In cases involving more serious offences, a tribunal of judges will be employed. The responsibility of the judge is much greater under the inquisitorial system as the judge must conduct the case and play a central role in finding the truth. Once a judgement is made as to whether the accused should stand trial, the case is turned over to a trial court and a different judge and jury will try the accused. The process in this trial is more adversarial but not directly comparable with the adversarial system employed under common law.

There are many arguments both for and against the use of the adversarial and inquisitorial systems and it is recognised that neither system is ideal for every case. Perceived advantages of the adversarial system include the more neutral role of the judge (therefore implying less bias) and the fact that each opposing counsel will presumably work harder than an investigating judge to find facts and evidence which are favourable to their client. This can also be considered a criticism of the adversarial system as each party presents a biased view of the accused, whereas the inquisitorial system considers evidence both favourable and unfavourable to the accused. Additionally, although the seriousness of a case determines the seniority of counsel, less wealthy litigants are at a distinct disadvantage under the adversarial system as they cannot afford the services of expensive lawyers – typically the most successful – and are often defended by overworked or unskilled legal aides whose resources and time are limited. In a forensic context, some argue that the inquisitorial system provides a fairer interpretation of complex scientific evidence, as the judges may have some expertise, or at least more experience than a lay jury, when determining the evidentiary value of forensic evidence presented.

1.3.3 Rules of Evidence

1.3.3.1 Admissibility of Evidence

The US Federal Rules of Evidence (FRE), which were enacted on 1 July 1975, provide a detailed outline of the type of evidence that may be presented to the trier of fact in a US court of law (Bronstein 2016). Any evidence may be deemed admissible in court if it is shown to be both *relevant* and *competent* and this is further defined under Article IV of the rules of evidence (Article IV, Relevance and Its Limits). Although specific to North America, these rules are somewhat similar to the rules of evidence employed in other common law nations including England, Canada, and Australia. While rules of evidence govern admissibility of evidence in court, they do not necessarily govern how that evidence will be used or the weight given to admissible evidence by the judge or jury. They only provide a guideline to ensure the integrity of the evidence presented. Conversely, evidence may be deemed inadmissible and not presented to the trier of fact if it is determined by the judge to be irrelevant and unnecessary (see examples below).

FRE 401 states that 'evidence is relevant if it has any tendency to make a fact more or less probable than it would be without the evidence, and the fact is of consequence in determining the action' (http://www.rulesofevidence.org/article-iv/rule-401). Issues of relevancy question whether the evidence has sufficient probative value to justify allowing it to be admitted as evidence. For example, if an accused purchased a weapon shortly before a fatal homicide with which he or she is charged, is this considered to be relevant and probative evidence to the case? Both the prosecution and defence would argue differently, and such a question warrants further analysis and reasoning. But not all relevant evidence is admissible and even highly relevant evidence could be excluded if it lacks competence.

The second step in determining admissibility is to determine the *competency* of the evidence. FRE 403 states that 'the court may exclude relevant evidence if its probative value is substantially outweighed by a danger of one or more of the following: unfair prejudice, confusing the issues, misleading the jury, undue delay, wasting time, or needlessly presenting cumulative evidence' (http://www.rulesofevidence.org/article-iv/rule-403).

- Prejudicial evidence is any evidence which could be viewed negatively by the trier of fact and subsequently influence their impartiality towards the defendant. Unfair prejudice often results from an emotional response to evidence which causes the trier of fact to view the defendant as guilty before the case has been presented in its entirety. A common example of prejudicial evidence is autopsy photos of a victim. Such photos cause the jury to focus on the graphic nature of the colour images rather than interpreting their value and weight as evidence. In many jurisdictions, autopsy photos are only admissible in black-and-white or if the forensic pathologist can convince the judge of their value in demonstrating probative facts about the case.
- A judge can deem evidence inadmissible if they believe the evidence is not relevant to any of the charges of the case. In such instances, the presentation of the evidence may simply confuse the issues in dispute and mislead the jury, subsequently biasing their opinion towards the defendant.
- Evidence that wastes the time of the court, or is unnecessarily cumulative, may also bias the jury through distraction and confusion of the facts. While the prosecution and defence are typically given sufficient leeway in arguing their case, a judge will not tolerate evidence that is repetitive in nature or lacks relevance. A common example of this type of inadmissible evidence is the request by an attorney to use multiple eyewitnesses to testify to the same event. Typically, a judge would restrict the testimony to one or maybe two witnesses at most so as not to unduly waste the time of the court with repetitive testimony.

Evidence of a person's character or traits is also considered inadmissible if it is attempting to prove that the person acted in accordance with these traits on a particular occasion. Additional rules to determine the relevancy and probative value in the admissibility of evidence can be found in Article IV of the US Federal Rules of Evidence, while guidelines to determine competency of witnesses is outlined in Article VI of the US Federal Rules of Evidence. In other common law countries following the adversarial system, similar rules of evidence are upheld although not always as clearly detailed (Bronstein 2016).

1.3.4 Types of Evidence

Traditionally, there are four types of admissible evidence: real, demonstrative, documentary, and testimonial. Hearsay evidence also may be deemed admissible for a specific piece of evidence in a case. The general rules of admissibility outlined previously apply to all evidence regardless of its type.

Real evidence refers to evidence that is physical and usually involved in some event of the case. In the case of a civil lawsuit, it may refer to the contract that has been breached. In a criminal case, it can include physical evidence such as fingerprints, hairs, weapons, glass, or blood (although classification is by purpose adduced not the nature of the object). To be admissible, real evidence must be relevant and competent as well as being material to the case in dispute. The relevancy and materiality of real evidence is usually evident and straightforward to demonstrate. The competence of real evidence must be demonstrated through authentication, i.e. proving that it is what it purports to be. Real evidence can be authenticated by the testimony of a witness. For example, a person may be able to testify about the authenticity of a signature on a contract if they are familiar with the uniqueness of that particular signature. When a forensic scientist, as an expert witness, is tasked with proving authentication, they do so using the chain of custody (detailed further in Section 1.2.1).

Demonstrative evidence consists of material evidence used to demonstrate the testimony of a witness. It can include diagrams, maps, crime scene sketches, and photos (depending on how the evidence is authenticated). Since demonstrative evidence is presented in order to assist the witness in illustrating their testimony, the witness must authenticate the evidence and testify to its accuracy and factuality. A forensic scientist may use demonstrative evidence to assist in explaining complex analyses such as bloodstain pattern interpretation.

Documentary evidence refers to documents or recordings that detail events. Documentary evidence can also be considered a form of real evidence in the case of contracts under dispute. The authenticity of documentary evidence is paramount in any criminal trial and fortunately for forensic scientists, these standards are already in place in the form of a chain of custody.

Testimonial evidence is provided by competent witnesses under oath or affirmation and requires the statement of fact, and sometimes opinion, as a result of questioning by the prosecution or defence. A lay witness may testify only if they have personal knowledge of the matter. This is commonly referred to as direct evidence because no further analysis or inference is required. For example, eyewitness testimony can be provided by a witness who personally observed an incident or event and can testify to the facts that occurred. Such evidence may be admitted but with the caveat that eyewitness testimony is notoriously inaccurate in situations where the witness was involved in a distressing scenario (e.g. witnessing a fatal vehicle accident). In contrast, expert witnesses are typically called upon to provide an opinion or impression based on their analysis of circumstantial evidence – indirect evidence that is used to infer another fact. For example, if a fingerprint is located at the scene of a crime, an expert witness may testify that the fingerprint at the scene is indistinguishable from the fingerprint of the defendant. The expert cannot state as a direct fact that the defendant was present at the crime scene as they did not witness this occurrence. However, the positive identification of their fingerprint indirectly connects the defendant and the crime scene, and subsequently infers the defendant's guilt.

Hearsay evidence refers to statements made out of court, which includes: second-hand information, such as an overheard comment or evidence provided by someone other than the person who originally acquired the information; death bed and police cell confessions; and reports provided in court without the presence of the author. Second-hand information is generally excluded from criminal trials as the originator of the evidence (whether spoken or written) often cannot be called to testify in court. However, exceptions are made to this rule, for example for expert witness reports when the presence of the expert in court is not necessary.

1.3.5 Opinion and Expert Testimony

Witnesses, including forensic scientists, are called to provide testimony by the issuance of a subpoena or witness summons. A subpoena or summons is a court order to the witness to present themselves on a particular day and time and provide testimony or other evidence. Forensic scientists may receive a *Subpoena Duces Tecum*, a particular type of subpoena which requests their presence in court as well as any documentation in their possession which is considered relevant to the case. Ignoring any type of subpoena or witness summons may cause a witness to be charged with being in contempt of the court and, in extreme circumstances, could result in imprisonment.

Opinion and expert testimony differ in a number of ways from ordinary witness testimony and as a result are often defined separately. Opinion testimony given by a lay witness who is not testifying as an expert, must be valuable to determining a fact in issue, and must not be based on scientific, technical, or other specialised knowledge. For example, a witness who is asked how hot it was on the day of the crime in question may state 'I believe it was about 30 degrees' providing an opinion of the temperature that day without the specific knowledge of the temperature based on a thermometer reading.

Expert witness testimony, on the other hand, is given by witnesses who are qualified as experts by the court in a particular field. A witness can be qualified as an expert based on knowledge, skills, training, experience, and/or education. An expert witness is requested to provide evidence of fact or opinion when their expertise will assist the trier of fact in understanding the admissible evidence or determining a fact in issue. Their testimony must be based on credible facts or data, reliable principles and methods, and the application of both to the facts of the case.

The expert witness is often considered to be the 'interpreter' of complex scientific evidence and is tasked with presenting their opinions and conclusions about such evidence in a manner understandable to the judge and jury. Most judges and jurors do not have sufficient knowledge to interpret the scientific facts of a case and therefore rely on the expert witness to present reliable and valid data and interpretation. However, some measure of assurance is required by the courts before expert testimony can be deemed scientifically valid and thus admissible. Historically, the standards of admissibility of scientific evidence have been founded on two important case rulings in the USA.

1.3.5.1 Admissibility of Scientific and Technical Evidence

In 1923, the District of Columbia Circuit Court heard the case of *Frye vs. United States* (1923) where James Frye was on trial for murder. The case was important in establishing standards of admissibility because the court's ruling established what is commonly known as the 'general acceptance' rule for admissibility of scientific evidence. Essentially, the court ruled that evidence based on scientific analysis is admissible as long as it meets the relevancy standards and is accepted as being valid by the relevant scientific community to which it belongs. Unfortunately, the definition of 'general acceptance' was not provided in the court ruling and the issue has never been clearly resolved. It is commonly assumed to mean that the principles and methods employed by the expert have been rigorously validated and subjected to peer review through publication or equivalent exposure to the field. The limitation to this rule is that many valid scientific methods have never been published and numerous published scientific methods have since been shown to be invalid.

The issue of 'general acceptance' was revisited in 1993 with the US Supreme Court ruling in *Daubert vs. Merrell Dow Pharmaceuticals Inc.* (1993). The plaintiff sued Merrell Dow Pharmaceuticals after taking Bendectin, a commonly prescribed pharmaceutical for nausea, during pregnancy and giving birth to children with birth defects. The plaintiff claimed that Bendectin was responsible for the birth defects and retained a statistician to demonstrate that the instances of birth defects were higher among women who had ingested Bendectin when compared with the instances of birth defects among the general population. In contrast, the defendant's statistician determined that any increases found were not statistically significant and argued that the methods employed by the plaintiff's statistician were not 'generally accepted' and therefore had failed to meet the *Frye* standard for admissibility of statistical evidence relating to birth defects in pregnant women taking Bendectin. The court found in favour of the defendant and Daubert subsequently appealed.

The *Daubert* decision was based on FRE 702, which requires the proponent of the expert witness to demonstrate that the expert is qualified, and that the testimony will be valuable in assisting the trier of fact to interpret the evidence. In the ruling handed down by the Supreme Court following the appeal, the court found that the trial judge should not have used the *Frye* standard for admissibility of evidence, but rather should have used the Federal Rules of Evidence, drawing particular attention to FRE 702 (https://www.rulesofevidence.org/article-vii/rule-702). The decision indicated that the judge should act as the 'gatekeeper' when deciding if scientific evidence is admissible. The ruling provided a list of criteria that judges could use in this determination including error rates, standard and controls, peer review and publication, and general acceptance. The Supreme Court recognised 'general acceptance' as a valuable criterion but determined that additional criteria may be more useful. The following questions could therefore be examined by the courts when determining admissibility:

1. Are there published, peer-reviewed books or articles about the methodology?
2. Is the methodology taught in universities, or discussed and presented at professional scientific meetings?
3. Is there a known error rate and can the methodology be tested for accuracy?
4. Is the methodology generally accepted in the relevant scientific community?

Ultimately, the Supreme Court sent the case back to the trial court and a similar finding resulted in favour of Merrell Dow but this time based on the Federal Rules of Evidence.

Since the *Daubert* decision, courts have found other relevant factors that can assist a judge in determining whether expert testimony is reliable and should be considered by the trier of fact. Factors considered by the judge can include whether the expert has developed their opinions purely for the purpose of testifying, or whether these opinions are founded on scientific research; whether the expert has accounted for alternative explanations; and whether the field of expertise is known to reach reliable results. Two commonly cited cases relevant to the *Daubert* decision and the 'gatekeeper' role of the judge include the case of *Kumho Tire Co. Ltd. vs. Carmichael* (1999) and the case of *General Electric Co. vs. Joiner* (1997).

The *Kumho Tire Co. Ltd. vs. Carmichael* case involved the explosion of faulty tyres on a minivan leading to death and injury. The defendant claimed that the testimony provided by the plaintiff's expert suggesting that the tyre design was defective, was not reliable under the *Daubert* decision. The judge held that the issue in question was whether the gatekeeper role of the judge applied only to scientific evidence or to all types of evidence. The court ruled that the gatekeeper role should cover all expert testimony, in this case technical and engineering evidence, not just scientific evidence. Ultimately, the court agreed with the defendant and concluded that the expert testimony was not reliable under the *Daubert* criteria. The evidence was excluded but the trial did demonstrate the discretionary authority that the judge holds to determine reliability based on the facts and circumstances of each case.

The *General Electric Co. vs. Joiner* case involved a claim by the plaintiff that exposure to transformer fluids containing polychlorinated biphenyl (PCB) compounds promoted the onset of his lung cancer. The judge used his gatekeeper role to exclude the expert testimony relating to the hazards of the fluids, stressing the lack of peer-reviewed epidemiological studies and the fact that the expert testimony was based exclusively on laboratory animal studies. The judge cited the *Daubert* decision and found that the expert had extrapolated from accepted data to an unfounded conclusion.

In other common law countries similar rulings have been handed down, such as in the Sally Clark case in Cheshire, UK (*R vs. Clark* 2003). On 2 July 1998, Sally Clark was charged with the murder of her two baby sons and was sent for trial at Chester Crown Court. While Clark claimed that both children had died as a result of Sudden Infant Death Syndrome (SIDS), the prosecution called an expert in paediatrics to testify that the odds of two children in the same family dying of SIDS was infinitesimally small (1 in 73 million). The expert reached this conclusion by squaring the chance of a single death in the family by SIDS (1 in 8543) without considering environmental or genetic factors that have since been shown to contribute towards SIDS. Clark was sentenced to two terms of life imprisonment.

Clark appealed the decision based on flawed statistical evidence by the prosecution. The appeal judges accepted that the statistical evidence was not scientifically valid but believed that the evidence of her guilt was still overwhelming. In 2002, new evidence was released which demonstrated that the pathologist had overlooked essential medical evidence, which could have been responsible for the death of the second son. As a result of these findings, Clark was freed from her life imprisonment sentences on 29 January 2003.

These court rulings and guidance (e.g. FRE 702 and the CPS Guidance on Expert Evidence 2019), and numerous others like them, have significantly impacted the field of forensic science over the past two decades and have led to the re-examination of certain forensic disciplines with respect to their scientific basis for making qualified conclusions. Several disciplines including fingerprint and document evidence analysis have recently been challenged due to the perceived lack of demonstrated scientific research that proves these techniques as valid and reliable. As technologies advance and new scientific methods are proposed, the courts will no doubt continue to deal with the ongoing issue of the admissibility of scientific evidence in court.

1.4 The Role of Analytical Chemistry in Forensic Science

In the preceding sections we have discussed the background to the work that goes on in the forensic laboratory. In the remainder of the book we will look at the methodology and the methods employed in the analysis of, primarily, trace forensic evidence.

1.4.1 Techniques Used for Chemical Analysis

In Chapter 2 we will discuss the scientific method, what preparations a forensic scientist must make even before they begin their examination of items of evidence, the approach and how they should handle the data once it is obtained. Chapter 3 looks at common presumptive tests that can be employed before expensive analytical techniques in order to help target analysis and Chapter 4 looks at sample preparation techniques that can be used prior to the main analytical technique to ensure the sample is in a suitable form.

Spectroscopic and spectrometric techniques are used widely in forensic science and are covered in Chapters 6–11. The most commonly used are ultraviolet–visible (UV-Vis); spectroscopy infrared (IR) spectroscopy SEM and SEM-EDX mass spectrometry (MS). Other techniques used include fluorescence spectroscopy, near-infrared (NIR) and Raman spectroscopy.

Many of these spectroscopic techniques, particularly IR and NIR, Raman, SEM and SEM-EDX and XRF can be applied to samples with no or minimal sample preparation and are non-destructive to samples, depending on the nature of the sample and how it can be 'presented' to the instrumentation.

Another widely used suite of chemical analytical techniques are the 'chromatographic techniques' which are discussed in Chapters 12–15. Chromatography refers to the separation of the components of the sample being analysed and examples of separation techniques commonly used in forensic science include gas chromatography (GC), high performance liquid chromatography (HPLC), ion chromatography (IC) and capillary electrophoresis (CE). The chromatographic separation of the components alone is insufficient for full analysis in most instances; some form of detecting and identifying the separated components is necessary. Hence many chromatographic techniques are allied with spectroscopic techniques. These are often referred to as hyphenated techniques and the most widely used in forensic science are GC-MS and LC-MS (the abbreviation used for HPLC coupled to MS).

For various reasons, CE has not become a mainstream tool for chemical analysis in forensic science but it is used extensively on the biology side, forming the basis of the final analysis in DNA profiling, where CE is linked with fluorescence detection, providing an extremely efficient and sensitive method of analysis.

Although we recognise that analysis of many samples may also involve analysis by microscopy, because magnification can reveal information not visible to the naked eye, we are deliberately excluding a discussion of light microscopy from this text because, except for some rather specialised analyses, light microscopy alone does not directly give information on the chemical nature of a sample.

As you will see in the chapters that follow, there is a choice of analytical techniques that can be used to provide some types of information. This is particularly true in the case of GC-MS and LC-MS where either technique could equally well be used, for example, for the identification and quantification of cocaine in a white powder or methylenedioxymethamphetamine (MDMA) in suspected ecstasy tablets, i.e. the analysis of low molecular weight organic compounds in complex matrices. Students not familiar with analytical chemistry often find it confusing to have a choice of methods available and, in a given scenario, may not be able to recognise that one technique may be more appropriate for the analysis of a particular sample. This is one of the reasons that an in-depth understanding of the analytical technique in question is so critical – what are the advantages of a particular technique and what are the potential limitations? What samples are suited to analysis by a particular technique and which are not? We aim to cover this aspect in chapters on the individual techniques, taking a look at applications and limitations. In the case of GC-MS and LC-MS, many types of samples can be analysed successfully by either technique to obtain equivalent information. In these situations, the choice of technique may come down to the cost of analysis, the expertise of the analyst or the instrumentation available in a laboratory.

Readers of scientific journals will see a wide variety of analytical techniques used to analyse the sorts of samples that are encountered in forensic science but may be surprised to find that many of these techniques are not used in forensic laboratories. The lifeblood of scientific journals is publishing information new to science, including novel, or more rapid, ways of analysing samples. Laboratories carrying out research involving techniques which may be applicable in forensic science are often not carrying out analyses with a view to presenting evidence in court. Although they will need robust methods of analysis for their research, their methods will not need to undergo the extensive validation required for judicial purposes. A forensic laboratory thinking of adopting such a method will incur significant costs in validating and implementing the method and there is no guarantee that it will be suitable for routine use. The analytical instrumentation used in forensic science can also be costly (often upwards of £100 000). The cost of validation and cost of analytical instrumentation has led to a lag in some forensic laboratories adopting what are considered 'cutting edge' analytical techniques. In the chapters on analytical instrumentation, we will concentrate on techniques routinely used in forensic laboratories but we will also discuss developments taking place in universities and other research organisations, which may become routine in forensic science at some point in the future.

The examples of use described in this book are taken from the myriad of investigations and evidence types that may be encountered. This almost endless possible combination of types of evidence and scenarios to be investigated is what makes forensic science so appealing. We hope that the casework examples included in this book will illustrate, the excitement of forensic science, the sophistication of the techniques used in investigations and show how essential it is for a forensic scientist to have a fundamental understanding of the techniques that they use.

References

Aitken, C. and Taroni, F. (2004). *Statistics and the Evaluation of Evidence for Forensic Scientists*. Wiley.

BBC (2007). Fraudulent forensic expert jailed. http://news.bbc.co.uk/1/hi/england/manchester/6386069.stm (accessed 14 May 2020).

Bronstein, D.A. (2016). *Law for the Expert Witness*. CRC Press.

Cook, R., Evett, I.W., Jackson, G. et al. (1998a). A hierarchy of propositions: deciding which level to address in casework. *Science and Justice* 38 (4): 231–239.

Cook, R., Evett, I.W., Jackson, G. et al. (1998b). A model for case assessment and interpretation. *Science and Justice* 38 (3): 151–156.

CPS (2019). Crown Prosecution Service: Guidance on Expert Evidence. https://www.cps.gov.uk/legal-guidance/expert-evidence#:~:text=The_CPS_Guidance_for_Experts,expert_evidence_and_disclosure_obligations.&text=The_investigator_or_prosecutor_should,his_obligations_as_to_disclosure.

Daubert vs. Merrell Dow Pharmaceuticals Inc. (1993). 509 US 579.

Dror, I.E., Charlton, D., and Péron, A.E. (2006). Contextual information renders experts vulnerable to making erroneous identifications. *Forensic Science International* 156 (1): 74–78.

Evett, I.W. (1993). Establishing the evidential value of a small quantity of material found at a crime scene. *Science and Justice* 33 (2): 83–86.

Frye vs. United States (1923). 293 Fed. 1013, 1014.

Gardner, R.M. (2011). *Practical Crime Scene Processing and Investigation*. CRC Press.

General Electric Co. vs. Joiner (1997). 522 US 136.

Horswell, J. (2016). Crime scene investigation. In: *The Practice of Crime Scene Investigation*, 29–72. CRC Press.

Kirk, P.L. (1953). *Crime Investigation: Physical Evidence and the Police Laboratory*. Interscience.

Kumho Tire Co. Ltd. vs. Carmichael (1999). 526 US 137, 119 S. Ct 1167(1169): 143.

Lucy, D. (2013). *Introduction to Statistics for Forensic Scientists*. Wiley.

Mosher, P., McVicar, M., Randall, E., and Sild, E. (1998). Gunshot residue-similar particles produced by fireworks. *Canadian Society of Forensic Science Journal* 31 (3): 157–168.

National Research Council (2009). *Strengthening Forensic Science in the United States: A Path Forward*. National Academies Press.

R vs. Clark (2003). EWCA Crim 1020.

R vs. T (2010). EWCA Crim 2439.

Romolo, F.S. and Margot, P. (2001). Identification of gunshot residue: a critical review. *Forensic Science International* 119 (2): 195–211.

Saks, M.J. (2010). Forensic identification: from a faith-based "science" to a scientific science. *Forensic Science International* 201 (1–3): 14–17.

Thompson, W.C. and Schumann, E.L. (1987). Interpretation of statistical evidence in criminal trials. *Law and Human Behavior* 11 (3): 167–187.

Wertheim, P. (1996). The ability equation. *Journal of Forensic Identification* 46: 149–159.

2

Analytical Methodology and Experimental Design

Florian Wulfert and Rosalind Wolstenholme

Analysis of a forensic sample and the reporting of a result, such as a concentration, seems at first as straightforward as reporting a simple number. However, getting to that result and being able to trust it, is far from trivial. Analytical science, within a forensic context, needs to apply analytical methods which follow strict standard operating procedures (SOPs) so as to guarantee that the analysis process can be audited and verified, leading, therefore, not only to consistent, reproducible results but also to quantifiable confidence in these results. This chapter will concentrate on the development of these procedures, outlining some of the issues that are important to consider when carrying out an analysis, whether it be analysis of a forensic sample or design of a scenario specific experiment. The latter part of the chapter concentrates on methods of experimental design, particularly relevant to scenario specific experiments, as this is often one of the least well-understood and most complex aspects of scientific studies in many disciplines, including forensic science.

2.1 Scientific Method

The term 'scientific method' came into use in the nineteenth century. Although the methods used in each scientific scenario vary, there are certainly some steps that can be applied to most problems. The first step is the formation of a question, for example, why is the crime scene evidence distributed in this way? There may be some research available that will help answer the question, e.g. fibre analysis research, but, if not, new information is required. Before collecting any new information the next step is to formulate a hypothesis. A hypothesis is a possible explanation for the question posed that can also be tested to see if it is correct. (In statistical analysis, hypotheses come in pairs as a null hypothesis, H_0, which says that there is no effect and the alternative hypothesis, H_1, which says the predicted effect is true.) The consequence of the hypothesis is then predicted based on logical reasoning, usually from existing evidence, and then an experiment is designed and carried out to observe whether the prediction, on the evidence available, is correct or not. The final step is to analyse the data obtained and determine its significance, in some cases using statistics. Where the hypothesis/prediction has been shown to be false/not observed the scientists must go

Analytical Techniques in Forensic Science, First Edition.
Edited by Rosalind Wolstenholme, Sue Jickells and Shari Forbes.
© 2021 John Wiley & Sons Ltd. Published 2021 by John Wiley & Sons Ltd.

through the scientific method steps iteratively, formulating new hypotheses, predictions, and experiments until they are satisfied that they have answered the original question. It should be noted that absolute proof is very hard to find, in most cases, conclusions are drawn from increased confidence in the hypothesis rather than concrete proof.

2.2 What Do We Mean by Analysis?

Analysis is often seen as a very limited task, most will associate the word 'analysis' solely with the act of measuring a certain sample, something that happens in the laboratory on a certain instrument. This view is much too limited and focuses on one (if obviously central) part of the analytical process. The reality is more complex and the quality of the result depends on every single step before the generation of said result and it is, therefore, important to consider the various stages involved in the analytical process. Each single stage needs to be audited and its contribution to errors quantified if we want to understand the uncertainty that is inherent in each result reported. So, whilst developing standardised analysis procedures, quantification of error contribution is important, as it does not make sense to execute the chemical analysis to a very high standard if the sampling or data analysis is not done to equally high standards, the quality depends on the proverbial weakest link (see Section 2.4.1).

2.3 The Stages of Analysis

Within an analytical context, five stages can be identified, namely Sampling, Storage/Transport, Preparation, Instrumental Analysis, and finally Data Analysis. An analysis protocol will have to consider all five stages and define their parameters in order to render analysis not only auditable but also to ensure that the results from different analyses are consistent and comparable. Each stage will contain a number of steps/actions, which will contain in themselves functional/experimental variables. Some practical examples of what these variables could be, for each stage are given below (by no means a complete or exhaustive list):

1. **Sampling**: locations where samples are taken, methods, and their parameters of sample acquirement (including taking subsamples of bulk evidence, see Section 1.2.2 and Section 3.2, for which methods should be included in the SOP).
2. **Storage/Transport**: storage types, temperatures, additives (preservatives), methods, time.
3. **Sample Preparation/Work Up**: extraction methods and their parameters, transfer of samples onto analytical equipment (see Chapter 4).
4. **Instrumental Analysis**: type of instrument, all parameters that affect the working of the apparatus used, including variables of the laboratory environment which might influence the analysis (see Chapters 6–15).
5. **Data Analysis**: all numerical handling, parameters that influence, e.g. the peak integration, internal/external standardisation, baseline corrections, etc.

In the case of forensic evidence examinations, in addition to these five stages, a clear report that a non-scientist, such as a jury member, could understand will be prepared. Further, a defence of the report may also be necessary in court for the benefit of the jury and members of the legal system (see Section 1.2.3.1).

A full inventory of these variables is necessary to get an overview of which of these it could be necessary to include in the method development of the analysis.

In the forensic context, any analysis carried out should be preceded by discussion between the body commissioning the evidence examination and the scientist carrying it out to determine what information is required and whether the evidence can provide that information. Factors affecting the techniques used include the amount of sample available, the techniques available, the forensic question being asked, the severity of the (potential) offence, which will be related to the available budget, and the timescale.

More generally, forensic, and analytical chemistry are typically concerned with one or more of three questions: What is the sample? How much of it is there? What is its specific chemical structure? These are qualitative, quantitative, and structural analysis, respectively. Qualitative analysis will provide information about the characteristics of the sample that can be used for comparison between evidence recovered from, say, a crime scene and a suspect. Here, identification of the sample may not be necessary, rather comparison with other samples. Quantitative analysis will provide information about how much of one or more compounds are present in the sample, which may be important in determining what offence has occurred, e.g. personal drug use vs. supply. There are various ways that quantification can be achieved depending on the sample being analysed (see Section 2.3.1). Structural analysis gives more detailed information on what a sample is and the structure of its molecules, which can be helpful in definitively identifying the sample in cases where the offence relates to a specific compound or group of compounds, e.g. explosives. Note, not all techniques are capable of giving definitive identification of a sample and it may also require the accumulated data from more than one technique to do so.

2.3.1 Quantification

In most cases, quantification cannot be done simply by placing a sample in an instrument and acquiring a measurement. Instruments typically have small variations in response, even for the same sample, and can also give different responses for different analytes at the same concentration. Therefore, additional steps need to be put in place to ensure the data is reliable and can be compared with data from other samples and other instruments.

2.3.1.1 External Standards

Using an external (certified) standard (ES) is a common way to carry out quantification, which most students of science will be familiar with. A standard can be used when (i) the identity of the analyte is known, (ii) a standard for that analyte is available, and (iii) there is minimal effect of the sample matrix on the analytical result, i.e. the matrix does not enhance or supress the signal from the analyte. The standard is used to create a calibration curve. A set of standards of decreasing concentration is prepared. These are analysed, in a random order to reduce error, using the same method as for the sample and the instrument response is recorded. A curve is plotted of concentration vs. response and the equation of the curve is determined (preferably using software rather than manually). This equation can then be used to convert the instrument response from the sample to a concentration. Ideally, the concentration is towards the middle of the range of concentrations covered by the set of standards as this gives more confidence in the value obtained. A further step that can be inserted into the calibration protocol to improve accuracy is to subtract the response of a blank sample, which contains the reagents but no analyte, from calibration standard responses.

Whilst discussing calibration curves two other concepts are relevant, limit of detection (LOD) and limit of quantification (LOQ). The LOD is the smallest amount of analyte that can be reliably detected and the LOQ is the smallest amount that can be reliably quantified. The LOD will be the lowest point on the linear response of an instrument. There are several ways to calculate this parameter but one the simplest is to set it at 2–3 times the magnitude of the noise signal. Similarly, a simple measure of LOQ would be to multiply the noise signal by 10.

2.3.1.2 Internal Standards

Internal standards (ISs) are used when (i) the instrument response can be variable or (ii) when sample preparation methods may lead to the loss of analyte. Quantification in this case is possible as the analyte and the IS have undergone the same sample preparation processes and the ratio of their concentration before sample preparation should be the same as after sample preparation. Unlike in standard addition (Section 2.3.1.3), the IS should not be the same as the analyte. It should be similar and behave in a similar way but should not interfere with the spectrum/chromatograph or the analyte. In order to use the IS in calculations the response factor F should be determined from known concentrations (see Eq. 2.1).

$$\frac{\text{Analyte signal}}{\text{Concentration of analyte}} = F \frac{\text{Standard signal}}{\text{Concentration of signal}} \quad (2.1)$$

F can then be used to calculate the concentration of the analyte in a sample when the other terms in Eq. (2.1) are known.

2.3.1.3 Standard Addition

Standard addition can be used when (i) the composition of the solution containing the analyte is not known and/or (ii) the composition affects the instrument response (matrix effect), and (iii) the instrument has a linear response to the analyte. To use standard addition, samples are prepared with a known amount of the analyte and differing amounts of the standard, which is known to be the same compound as the analyte. The solutions are analysed and the data obtained used to construct a graph, preferably in an electronic spreadsheet. The line of best fit and its equation can be found and then the intercept calculated (by reading from the graph or by using the equation for the line of best fit and $x = 0$). This value is the concentration of the analyte in the solution. The final step is to calculate the dilution that took place when the solutions were made up and apply that to the concentration value obtained, to determine the concentration of the analyte before dilution.

Whichever one of the approaches mentioned earlier, i.e. quantitative, qualitative, and structural identification, is being used, it will contribute to the parameters of the analysis, i.e. choice of technique, sample preparation, data analysis, etc. In most cases, for forensic casework, the choice of technique for a particular type of examination, e.g. analysis of the composition of gunshot residue (GSR), will be straightforward. The examiner will be aware of the most appropriate technique for qualitative, quantitative, or structural analysis of evidence within their expertise and techniques must be validated and accepted by the scientific community (see Section 1.3.5.1) in order to be admissible in a court of law. For example, in the case of analysis of the composition of GSR, scanning electron microscopy (SEM) has been the gold standard for identification for many years. The use of the analytical techniques discussed in this book for particular types of evidence, their advantages, and their limitations can be found in Chapters 6–15.

2.4 Analysis Development

2.4.1 Error Estimation

Ideally, the laboratory analysis of a sample would give a definitive and true answer to whatever question is being posed, with no room for doubt or errors. In reality, each process that a sample undergoes and each technique used comes with inaccuracies and imprecision. Before starting the development of the analysis protocol it is essential to look at every step or handling of the sample and to assess what the influence of those steps is on both accuracy and precision. Accuracy is in this context defined as 'correctness', i.e. how near a measurement is to the real value, whilst precision is defined as 'repeatability', i.e. how near to each other the results of repeated measurements are. Clearly, in order to achieve the most reliable results, data will have both high accuracy and high precision. A scientist must appreciate the capabilities and limitations of the processes/techniques used in order to determine the error involved. Some of this error contribution assessment can be done via a priori knowledge, e.g. through literature searches or instrument specifications, whilst others might need to be quantified experimentally or even estimated from first principles.

As stated above, the occurrence of errors, i.e. deviation from the true value, when acquiring measurements in the laboratory is inevitable in any analysis. However, an understanding of the different types of error and how they occur as well as robust quality control/quality assurance (QC/QA) and experimental design, including statistical analysis, can minimise the effect that they have on the interpretation of data collected.

Errors can be either determinate/systematic or indeterminate/random. Determinate errors are those for which a source can be identified, usually one of: the instrumentation and equipment; the operator; or the method used. Carrying out regular instrument maintenance, calibration using, e.g. certified light sources, etc. and equipment checks should be part of the QC/QA systems in the laboratory and having QC systems will also ensure that the protocols used are designed to minimise errors (see Section 2.4.2). For example, an instrument may be subject to drift in signal over the course of a day. This can be countered by analysing samples in different orders on different days and using software programs to allow for any variation due to the time of analysis. In order to minimise operator errors, training should be thorough and ongoing and cover laboratory systems and use of equipment.

Training in record keeping is also essential so that when errors do occur they can be traced back to their source and prevented in the future. For forensic scientists, notes will make up part of their court going reports and form part of the chain of custody. A forensic scientist's notes should begin with taking custody of the evidence, include how they accessed the evidence (i.e. how any packaging was opened to avoid questions about contamination or seal tampering at a later date), notes on the examinations and tests that were performed using which SOPs (including any actions taken that differ from those set out in the SOP) and how the evidence was resealed. Notes should be signed, dated, and paginated so that there can be no insertions at a later date and they should be kept in a case file.

Indeterminate errors are those arising from random fluctuations in the systems being analysed. These are called noise and will always be present (see Section 2.4.2). Since this type of error/noise is random it can be reduced relative to the signal by adding the data for two measurements together. Random noise will cancel itself out to some extent, whereas sample signal will be additive and, therefore, increase. Again, experimental design can go some way to reducing them but they are very unlikely to be eliminated altogether.

Note that, assuming that errors have occurred in most, if not all, of the measurements when carrying out a particular protocol on a particular day these will contribute to the overall accumulated error for the protocol on this occasion.

Performing this error estimation can give at least a good impression of what type of overall analysis error can be expected. If all variances (= square of standard deviation) of all steps are known, the overall variance and, hence, overall system standard deviation can be estimated as the sum of all the variations of all the steps. Even though this error estimation is seldom very accurate, it can already give an impression of the order of magnitude of the method's overall error as well as identify at a very early stage which steps are the 'weakest link' and therefore lead to improvements, even before the first measurement is performed.

Once a method is developed, the accuracy and precision should be monitored using quality control (QC) samples with known results (for accuracy check) and repeated analysis (for precision).

2.4.2 Quality Assurance and Quality Control

QC and QA are two systems that go hand in hand in the analysis of a sample. QC describes the protocols and policies that are used to carry out laboratory work and that are designed to ensure reliable data is obtained, in which the scientist can have confidence. QA is the scaffolding within the running of the laboratory that ensures QC procedures are followed. Various bodies, such as the International Organization for Standardization (ISO), the American Society for Testing and Materials (ASTM), the International Union of Pure and Applied Chemistry (IUPAC) and the National Institute of Standards and Technology (NIST) publish protocols covering procedures from calibrating instruments to carrying out analysis of a sample. These protocols or SOPs are often the starting point for a laboratory when developing the QA and QC structure for the examination of a particular type of evidence.

SOPs are descriptions of how an examination/procedure should be carried out, regardless of the personnel involved. Clearly, the procedure should provide the same analytical results if a single person repeats the analysis multiple times, called repeatability, as if the analysis is carried out in another laboratory by a second person, called reproducibility. Care should be take here to ensure that multiple analyses are of the right kind and that consideration is given to whether duplicates (samples obtained separately) or replicates (subsamples of the same original sample) are required. A common mistake for students is to use replicates rather than duplicates, which leads to similar analytical results because the replicates have the same sample source but which may not be what the experiment was designed to test. SOPs should also include the use of control samples to test that the procedure is working correctly. These typically include positive and negative controls, which check that a sample containing the analyte gives a positive result and those not containing the analyte do not. Certified standards, controls of known composition, high purity and high stability, are available commercially for use for this purpose as well as to aid in identification and quantification. Sources of error should also be considered in SOPs, for example, signal-to-noise ratio (S/N). A S/N of 3–5 is generally considered to be the lower limit at which the signal is distinguishable from the noise but the SOP for a particular procedure or process should specify the S/N for that procedure/process. Individual software programs may also have built in assessments of acceptable S/N.

SOPs should be available for any process in the laboratory from the most straightforward to the most complex analytical procedures. They may be written for use within a company or may, more likely, be determined by national or international guidelines, as above. The advantages of using pre-existing SOPs is that they will be tested, validated and optimised and their use allows comparison of data obtained in different laboratories. This in turn allows the use of databases of analytical data for comparing and identifying samples. SOPs should be reviewed as part of QA/QC procedures and changed whenever necessary. Where new SOPs are necessary a careful consideration of the design of the experiments is required (see Section 2.4.3).

In addition to SOPs, bodies, such as those mentioned above, also provide certification and accreditation schemes for scientists and laboratories, respectively, which should be renewed regularly. For scientists, certification will look at aspects of their professional skills including competencies in the relevant laboratory skills as well as continuing professional development (CPD). For laboratories, accreditation assesses aspects of the laboratory such as laboratory facilities, compliance with standards and, typically, includes blind tests (in which the scientist/laboratory does not know the expected results of the analysis to be carried out but the accrediting body does).

2.4.3 Method Development and Experimental Designs

As stated above, when developing new analysis protocols, in order to consider all five stages, it is important to design experiments which identify the experimental parameters and variables that have a significant influence on the targeted result (see the list in Section 2.3). Targets for method optimisation have to be carefully chosen, when developing and optimising analysis methods. Typical targets are:

- **Accuracy**: e.g. how near the analysis result is to the known value of a QC sample.
- **Selectivity**: e.g. how much or little interference there is from other compounds or species present in a mixture.

- **Sensitivity**: e.g. how large is the response to a certain input.
- **Detection Limits**: e.g. how low a level can still be seen.
- **Resources**: e.g. parameters such as total analysis time, amount of chemicals used, cost, etc.

In statistics these are called the dependent variables as they depend on the experimental variables and on parameters that can be changed independently from each other (hence independent variables). The independent variables are all the possible instrument settings and experimental choices possible in the five analytical stages mentioned earlier.

At this point it is important not to rush to investigate the influence of each of the independent variables on the dependent (targets) variables in many small experiments. It is only possible to gain a complete picture of which parameters have a significant influence on the result/targets and what the best parameter settings are by performing large integrated experiments that follow a carefully chosen design. The reason is that small isolated experiments would miss the possible interactions between the parameters and one could never be sure of working with ideal settings. For example, one chromatographic column could give superior results with one mobile phase composition, whilst another column would be preferable at a different composition again. This can only be seen with an experiment that changes both column and mobile phase within one experimental design and, hence, in order to develop and optimise an analysis protocol, it is necessary to do it with an overarching design that includes all the experimental variables available in all stages.

There are many experimental designs available, but most can be assigned to either of two classes: factorial and response surface. Generally factorial designs are used to scope out which variables do have significant influence on analysis results, whilst response surface designs are used to make a quantitative model of significant effects to optimise the analysis target(s) (dependent variables). Therefore, to develop an analysis procedure it makes sense to first scope out the important variables with a factorial design and then only optimise for those with a response surface design.

There are other specialist designs, e.g. Latin square designs for locality influences and mixture designs for formulation experiments that will add up to a fixed value (e.g. 100%). These are less common in forensic applications and will not be discussed here but the theory presented hereafter is easy to adapt to those designs as well.

2.4.4 Selecting Critical Variables with Factorial Designs

Experiments planned and run according to factorial designs generally test for each variable's influence by setting it to only two levels, typically the minimum and maximum level expected. By measuring the outcome of all the possible combinations of high and low levels, it is possible to analyse which experimental variables influence the outcome as well as which variables show an interaction. Whilst it is feasible to measure all possible combinations for a low number of variables, it becomes increasingly costly and unwieldy to do that for many, as the total number of possible combinations increases exponentially according to 2^n (where n is the number of variables). So, the table for four variables in Figure 2.1 would contain 16 experimental combinations, and a table for 10 variables would already have 1024 combinations. Considering that one would also want to measure replicates, it becomes clear

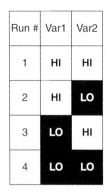

Figure 2.1 Full factorial designs for *n* variables where *n* = 1, 2, or 3.

that applying full factorial designs can be prohibitive. It makes sense, therefore, to decide at this point of the method development to keep certain variables at standard values and, hence, exclude those from any possible optimisation. This will obviously carry the risk that the developed method will be suboptimal but with analytical experience and a priori knowledge it is hoped that such a risk is minimised. Furthermore, it is important to establish that whatever value has been chosen for any now fixed variable it is easy to keep it stable and to ensure that that value is always maintained.

2.4.4.1 Categorical Variables

In the previous paragraphs it has been assumed that each variable can have a low and high value, but this is only true for numerical (any number) or ordinal (first, second, third, etc.) variables. However, some variables will be categorical though, that is, they cannot be captured in a number. This will contribute even further to the inflation of experimental runs, as the numerical part of the design will have to be executed for each level of the categorical variable. If the categorical variable has only two possible levels (e.g. only two chromatographic columns are available/can be used), the number of runs is not increased more than in the numerical case but if the number of possible categories is higher, e.g. 10 columns can be used, the number of runs will be much higher (e.g. 5 times higher for the 2 columns than for the 10 columns example). The total number of possible experimental combinations will, therefore, be given by the product of the number of levels for each variable.

If a categorical variable has many possible levels it might be possible to replace it with a numerical one, on the condition that there is a functional quantity that can characterise the different categorical levels. In the example used previously of 10 different chromatographic columns, one could, for example, consider replacing the column identity with a characteristic such as the polarity of the column's stationary phase. This way the columns are represented by numbers and it is now possible to select just the least and most polar ones as the high and low levels of the column's polarity variable. As a side effect it will, therefore, also be possible to estimate later on an optimal level of polarity and, therefore, easily find the most suitable column, i.e. the column nearest to the optimal polarity.

2.4.4.2 Reduced Designs

Even with only numerical variables, the number of experimental runs given by a full factorial design can become prohibitive, as demonstrated earlier. Hence a reduction of the design is necessary, which means that certain combinations of variables will need to be left out. The choice of the experimental runs to be eliminated is not trivial, after elimination the design still needs to be 'balanced', i.e. each variable and level needs to be represented within the same number of experimental runs as every other one. These reduced designs are known as fractional factorial designs, as opposed to full factorial designs and are best built using specialist statistical software in order to ensure that the final design is balanced.

Reducing designs to fractional factorial ones or, even further, to very sparse specialist designs such as Taguchi designs does affect the quality of the assessment of which factors and interactions are significant. By not providing the full information, the analysis will not always be able to distinguish the higher order interactions from the overall random experimental error. Also, the less experimental runs are performed, the larger the effects will have to be still to be seen as significant effects. Therefore, it can be said that at this point of the method development, when the final experimental design is going to be defined, the quality of the outcome is already fixed but we still need to perform the experiment so as to quantify it.

2.4.4.3 Final Practical Experimental Considerations

If the number of experimental runs demands the experiment to be carried out by various operators or over various days, the experimental design needs a last tweak to include a block variable (e.g. operator or day). Blocked designs include the block variable and will try to correct the results for the block effect (e.g. day to day or operator effects) through the inclusion of replicates that will be measured at each of the blocks. The block variable will not be included in the analysis model but will only serve to correct for block effects.

As a very last consideration, it has to be stressed that the experimental runs should never be performed in a systematic or standard order (as, for example, given in Figure 2.1). Rather, the run sequence should be randomised so as to avoid confusion with temporal drift in the results (e.g. due to ageing of the analytical instrumentation). Often, statistical software will already generate the design and present it in a randomised order. If not (and if not available as an option) then, in standard spreadsheet software, the table with the planned experimental runs can be randomised easily by adding a random number column and sorting the whole table by that random number.

2.4.4.4 Deciding on Significance

Once the experimental run table has been generated, all experimental runs completed and the measured results are filled into their own column(s), the data is ready to be analysed with statistical tools. Balanced designs are very suitable for analysis using generalised linear models (GLMs) (analysis of variance [ANOVA] type models; the difference between GLM and ANOVA is beyond the scope of this chapter and both terms are often used interchangeably) and the choice of design will influence which effects the GLM will be able to 'see' and not able to 'see'.

Whatever the design limitations are with respect to what can be analysed (e.g. a partial factorial design allowing only up to two factor interactions), in the end, the GLM will present a list of the significances (p-values) of the examined variables and interactions. In general, effects with a p-value smaller than a threshold (typically 10 or 5%) are considered significant. In this context it is useful to see the p-value is an expression of the probability that the effects observed are due to random noise only. (*This is a simplification of the meaning of p-values and statistically not correct but is a near enough approximation to the real meaning and is much more intuitive than the strictly correct version.*)

A valid statistical model only includes the significant factors and interactions, so as not to model noise. To eliminate the non-significant factors/interactions (i.e. those that have a p-value larger than 5 or 10%), the Backwards Elimination strategy is most commonly used. This strategy at first builds a model with all possible factors and interactions and then starts eliminating the effect with the highest order (e.g. a two-factor interaction has an order of two) and highest p-value (above the threshold obviously). After each elimination, a new model is built without the eliminated factor, and the next effect with highest order and highest p-value is chosen to be eliminated. This process of eliminations and model building is repeated until the only effects that are left are those that are significant or are necessary due to hierarchy. The latter means that a single (or lower order) factor, which is non-significant, might have to be included if it is part of an interaction (higher order) effect that is significant. For example, if there is a significant interaction between the variables time and temperature, then both single factors (main effects) time and temperature have to be included in the model, irrespective of their p-values. If there is a significant three-factor interaction between variables A, B, and C (denoted as A*B*C), all three main effects (A, B, C) will have to be part of the model as well as all three possible two-factor interactions between the three variables (A*B, A*C, B*C). Once all non-significant effects are eliminated, the final model will present only the significant effect(s) and can be interpreted using the model plots.

2.4.4.5 Interpretation

Model plots will show the effect of significant variables on the measured or dependent variable. It is important to keep in mind that if a main effect is also present in an interaction, that the interaction plot will give the full picture as is shown by the example in Figure 2.2.

In Figure 2.2a the main effect of variable A is depicted as an overall positive contribution to the measurement. By examining the interaction plot (Figure 2.2b), we can see that this is only the case for measurements where variable B is at its high value (level 2) as well, whilst on measurements with B on level 1 the variable A has no effect at all.

Inspection of the model and model plots can now lead to a decision as to which experimental variables are taken further into the analytical method development and which are left:

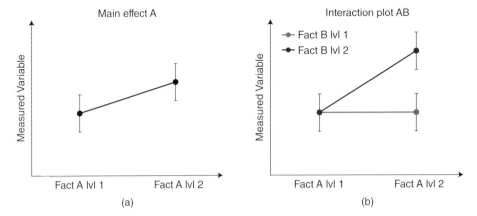

Figure 2.2 Model effect plots: (a) depicting the overall main effect of Factor A; and (b) the interaction effect A*B.

- **Non-significant (eliminated) variables**: These are of no real influence on the measurement within the values/levels used. Therefore, they obviously should not be considered for further method development/optimisation and are best kept constant at a tested level or at least within the range of this test.
- **Significant variables without interaction**: These are important and can be considered for further method development/optimisation. The fact that they do not interact with other experimental variables means that they can be optimised on their own; this can save a great number of experimental runs, time, and resources as can be seen in the following paragraphs. However, if for any external reasons they cannot be used for optimisation, it is essential to note that they need to be kept at a certain level in a very controlled manner, as any fluctuations would significantly affect the measured values.
- **Significant variables with interactions**: Just like the ones without interactions, these variables need to be considered further but the fact that they do exhibit interaction makes them special. They cannot be modelled or optimised separately from their interaction partners and need, therefore, a well-designed modelling experiment (see Section 2.4.3) to do that. In cases when it is decided to keep them constant, this might influence their interaction partners as well. For example, if one was to keep Factor B on level 1 in the Figure 2.2 example, Factor A would become insignificant as it has no effect with B on level 1 (see Figure 2.2b).

It is now the point in the method development where the modelling and optimisation of the method is finalised. From the above optimisation analysis, it is clear which variable(s) need to be optimised and in which context (alone or combined) that should happen.

2.4.5 Modelling the Significant Variables Using Response Surface Designs

With only significant experimental variables left from the factorial designed experiment and a general idea of how these experimental factors influence the measured variable, it is time to fit mathematical models to be able to describe the dependence of the measurement

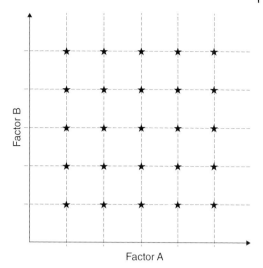

Figure 2.3 Simple response surface design, full grid, for two linear effects (five levels). Stars represent the 25 unique experimental runs possible with this design.

and ultimately to be able to choose the best settings. For this we need more information and, hence, having each experimental variable on only two levels is not enough. The amount of levels needed depends on how complex the effect is on the measured variable. As a rule of thumb, it can be said that a simple linear effect needs at least five levels (hence a linear calibration line should always have a minimum of five concentrations included). A simple curve needs seven levels, a sigmoidal will need nine, a camel hump eleven, and so on. This is due to the fact that the GLM used to fit the mathematical model will use polynomial terms to fit the curvatures and each extra polynomial term needs extra information for a correct fit and to avoid overfitting. (The underlying discussion about degrees of freedom is beyond the scope of this chapter.)

In this context it is easy to see why being able to optimise a single variable in a separate simple one variable design (think calibration line, standard curve) can save quite a few experimental runs. Consider three variables (A, B, C) to be optimised but the factorial design GLM model identified the following significant effects: A, B, C, and A*B. As only A and B show an interaction, C can be modelled separately. If we assume that for A and B there is a simple linear effect, the minimum full grid design for the experimental runs will contain 25 experimental points with unique values for A and B (see Figure 2.3). Hence, even without any repeated measurements, we would have 25 runs as a bare and minimal experiment.

Once we have found the ideal values for A and B, we only need a simple standard line experiment for C with five runs, so in total A, B, and C are optimised in (25 + 5) 30 runs. If we included C in the full grid (only necessary if B*C or additional A*B*C interactions are present), we would need a full experimental cube consisting of five layers of the grid depicted in Figure 2.3. Hence, we would end up with a minimum of (5*5*5) 125 experimental runs, compared with the 30 from before. This shows how important it is to design only for numerical models with the necessary interaction identified during the factorial experiments and model the variables without interactions on their own after the interaction-containing model.

2.4.5.1 Sparse Response Surface Designs

As can be seen from the previous sections, full grid experiments very quickly become prohibitively large with increasing variable numbers and model complexity (number of levels). Hence, just like with the fractional factorial designs, designs also exist that leave out certain experimental combinations (stars in Figure 2.3) and that will estimate those missing experimental runs through interpolation of experiments around them. In this respect, just like they did with factorial designs, categorical variables will make this impossible as one cannot interpolate between non numerical (or at least ordinal) categories. Again, it is therefore very important to try to find a way to express categories into numerical characteristics whenever possible.

There is a plethora of different methods to leave out varying fractions of the experimental combinations to reduce the total number of runs necessary. There are no simple rules for selection of the best candidates to run or leave out, respectively. Some of these reduced designs will concentrate on the middle of the design space, some on the edges or corners, whilst others might try to have an even coverage overall. It goes far beyond the scope of this chapter to discuss the different methods but generally an experimental design software will be able to create different candidates, which can then be inspected for their type of coverage of the design space. Common to all the possible methods to reduce the number of experimental runs is that in the end every reduction comes at a cost. The more points are left out, the more have to be estimated by interpolation (sometimes even extrapolation) and therefore the less accurate a model will be. Complexity may be lost (i.e. higher order terms are not able to be modelled anymore) and the model might become less and less representative of the reality. There is no clear way of assessing when a design becomes too sparse and it is, therefore, important at this point already to plan to run another experiment (validation) with both experimental combinations used during model building as well as those not measured yet in an independent experiment. Only with such an independent set of measurements will it be possible later to assess the accuracy of the model.

2.4.5.2 Analysing Response Surface Models

Once the design has been generated and the experiments run, the data will again be analysed using GLMs, this time as regression models. For a linear, single variable design, this comes down to a simple linear regression, known to many as line of best fit. For a design with two experimental variables, the response is not a line but a surface as shown in Figure 2.4 (hence the name), whilst models with more variables cannot be plotted or represented easily anymore.

Each dimension can now represent more than just one statistical factor. In Figure 2.4 it is easy to see that the experimental Factor B has a curvature, hence the model will include not only the experimental Factor B but also a squared term B^2. As mentioned earlier during the generation of the design, a choice has to be made with respect to the complexity to be modelled. When the GLM is applied to the data, all the possible terms are initially included in the model. Again, during the building of the model, those that are not significant will have to be eliminated, starting with the highest order ones. It is, therefore, very well possible that in the case of a quadratic model design (which allows for simple curvature), the data might exhibit no or no significant curvature and, hence, the squared terms might be eliminated as non-significant. For Figure 2.4 there seems to be very clear curvature for B but very little curvature for Factor A. Unless there is very little noise on the data, one would expect that the A^2 factor would not be significant, hence one would expect only the B^2 term

Figure 2.4 Model plot for a two-variable response surface model. Squares represent the measured values at their experimental settings and the curved grid surface represents the mathematical model fitted to the experimental values.

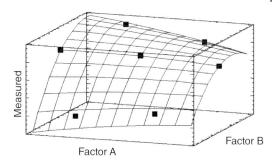

to be in the model. As the curvature for B changes quite unequivocally for different values of A, one would also expect the interaction term to be significant and therefore the final mathematical regression model would be of the type:

$$\text{Measured variable} = c_1 {}^* B^2 + c_2 {}^* AB + c_3 B + c_4 \tag{2.2}$$

where the four c_i denote the different regression coefficients found by the GLM to fit the data.

2.4.5.3 Validation

With mathematical equations now expressing the dependency of the measured variable(s) from the experimental factors, it is easy to see that one can now predict the value of the dependent variable(s) for any combination of values for the independent (experimental) ones. By measuring an independent test set at a variety of new combinations of the experimental values it is now possible to assess whether these new measurements are sufficiently near the prediction given through the regression model established in the previous step. For each experimental run (i) of the validation set, a difference between the measured (y) and predicted value ($\hat{y}$) can be calculated and from those an overall prediction error can be estimated:

$$E_{\text{pred}} = \sqrt{\frac{\sum (y_i - \hat{y}_i)^2}{n}} \tag{2.3}$$

This prediction error or the relative version of it (replace $y_i - \hat{y}_i$ in the brackets by $\frac{y_i - \hat{y}_i}{y_i}$) can now be directly used to asses the validity of the model. If it is (relatively) large, the model is not good enough. Quite likely the design used might have been too sparse/simple or alternatively there is a high noise level on measurements.

2.4.5.4 Optimisation

Assuming that the model has sufficiently low prediction errors, it can now be used to find optimal conditions for the measured variable(s). If the measured variables expressed, for example, the sensitivity, selectivity, and analysis time, one would now have three models one for each of those measured target variables. By applying overall constraints (e.g. maximum sensitivity at a certain analysis time range with a minimum selectivity) and combining the models, optimisation software can now calculate the optimal range for experimental parameters used. Hence, an analytical method can be defined that satisfies all the necessary criteria, is robust and has a clear definition of which experimental variables need monitoring, which need tight control, and which are of influence if the target needs changing.

3

Presumptive Testing

Rosalind Wolstenholme and Shari Forbes

with contributions from Sue Jickells

3.1 Introduction

Presumptive tests are screening tests that give an indication of what a substance might be. They are often chemical spot tests for colour change, which indicates that a reaction has taken place and can be carried out in the field/scene or in a laboratory. They do not give a definitive identification (i.e. they do not confirm the identity of the sample), usually because they are not specific to one substance, rather they detect a chemical reaction that can occur in a range of substances, including the one of interest. For example, the Kastle–Meyer test is often referred to as a test for blood but this is inaccurate as it is a test for peroxidase activity (provided by the haem group in haemoglobin for blood), which means that other substances, such as horseradish, can also give a positive result. This is known as a false positive. Strictly speaking a false positive is not 'false', the reaction should happen for horseradish with the Kastle–Meyer test, but it is not the 'positive' (for blood) that you were interested in. False negatives can also occur when the reaction being tested for should have taken place but did not. This can happen, for example if the reagents are old, the test was not applied correctly, or the substance is present in very small concentrations. For these reasons both a positive and a negative control should be carried out at the same time as the evidence test. The positive control shows that the test is working, how quickly the change is observed, and what colour is produced. The negative control demonstrates that there is no contamination in either the test equipment or reagents. A further issue with colour tests is that colour is a subjective property and may change over time but this can be mitigated by using colour charts and recording the results photographically. Despite these drawbacks, presumptive tests are used because they are generally cheap, quick, easy to carry out (important in fieldwork), can be sensitive, and can give an indication of what confirmatory test it may be sensible to carry out next. However, if sample size is limited presumptive tests may not be carried out and the analyst will move directly to a confirmatory technique, such as mass spectrometry (MS).

3.2 Drugs

The schedule for controlled substances and the penalties that related offences carry, such as illegal possession, use, supply, and production, vary by country but drugs such as marijuana, heroin, cocaine, lysergic acid diethylamide (LSD) and amphetamines are commonly encountered by forensic scientists throughout the world. Presumptive tests exist for all of the drugs listed (and others) and are routinely used to narrow down the possible identity of a sample. However, even before carrying out presumptive tests, the scientist may have to consider how to approach the seizure.

3.2.1 Drugs Seizure Sampling

A drugs seizure can come in many forms, for example, a single wrap, package, or tablet, a haul of similar looking packages/tablets/bottles (units), or a collection of mixed, different sized units with different appearances. For similar looking units it cannot be assumed that the contents are homogeneous or are from the same source. White powder drugs, for example, are often cut with cheaper visibly indistinguishable substances, such as washing powder. The seizure may need to be separated out into subpopulations that are believed to have the same origin. Clearly there is insufficient time and resources to test every unit. Indeed, there may be no need. Once identity and quantity above the legally prescribed weights have been established any further analysis is superfluous. In these cases, a sampling procedure can be used. The simplest methods used in the past are to assay 10% or $\sqrt{n}$ of the total number of units (n) but these numbers are arbitrary. Work has been done suggesting the use of statistical methods, including Bayesian statistics, to calculate a statistically significant number of samples to collect (Aitken 1999; Aitken and Lucy 2002) and European Network of Forensic Science Institutes (ENFSI) and the United Nations (UN) Office on Drugs and Crime have also advocated a statistical approach in their guidelines (United Nations Office on Drugs and Crime 2009a, 2009b). These calculations can be complex and are beyond the scope of this text. However, the forensic scientist must ensure that they consider their approach fully and develop a plan that will be defensible in court.

3.2.2 Major Drug Classes

A brief summary of some of the more commonly encountered drug classes follows to provide some background for the presumptive tests, however, the reader should refer to Gahlinger (2003) for a more in-depth discussion.

3.2.2.1 Marijuana
Marijuana and cannabis are terms commonly used to refer to parts of the plant *Cannabis sativa* L. In particular, the flowering tops of the female cannabis plants are rich in the main psychoactive component, Δ^9-tetrahydrocannabinol (THC). Other cannabinoids found in *C. sativa* include cannabinol, cannabidiol (CBD) and Δ^9-tetrahydrocannabivarin. There is still considerable discussion regarding CBD in terms of its action. It is considered to have antipsychotic activity rather than psychoactive activity. Tetrahydrocannabivarin is generally found in cannabis preparations at low levels (<0.2% w/v) and hence is not considered to contribute strongly to the psychoactive effects of cannabis.

Marijuana is typically supplied dried, known as grass or weed, or in resin form, known as hashish or hash. The presumptive tests for marijuana are the Duquenois–Levine or Corinth IV Salt (see Table 3.1).

Some of the most common presumptive colour tests for drugs are summarised in Table 3.1.

3.2.2.2 Opioids, Cocaine, and Amphetamines

Opioids, both natural (opiates) and synthetic, are central nervous system depressant alkaloid compounds, i.e. they affect brain function and contain an amine group. The opiates most commonly abused are heroin (diamorphine), morphine, and codeine. The synthetic opioids, methadone and buprenorphine, used to treat opiate addiction, are also subject to abuse. More recently, particularly in the United States, addiction arising from the use of opioid analgesics has become a serious problem with opioids such as oxycodone (oxycontin), hydrocodone, fentanyl, hydromorphone, meperidine, and oxymorphone subject to abuse.

Cocaine is a stimulant tertiary amine alkaloid. Amphetamines are stimulants, which as their name suggests, also contain an amine group. Common amphetamines are methylamphetamine, 3,4-methylenedioxyamphetamine (MDA), 3,4-methylenedioxymethamphetamine (MDMA) and 3,4-methylenedioxyethylamphetamine (MDEA). The presence of amines in all groups means that the same presumptive tests can be used, however, different colour changes are observed. The colour changes for the Marquis, Scott, Simon, and Mandelin tests are given in Table 3.1.

Opiates from pharmaceutical origin are commonly supplied as tablets, capsules, suppositories, and liquids to be injected. Heroin produced from the opium poppy is typically supplied as a powder. Amphetamines used for medical purposes are usually white tablets, whereas illegal supplies are more often a white/pinkish powder or in crystal form.

3.2.2.3 Barbiturates and Benzodiazepines

Barbiturates are central nervous system depressants based on barbituric acid (which has a pyrimidine heterocyclic skeleton). Benzodiazepines are also central nervous system depressants with a fused benzene and diazepine ring system. A common example is diazepam (trade name Valium). In medical use barbiturates have largely been replaced by benzodiazepines because they are less likely to cause death in overdose. Both classes of drug are commonly encountered in forensic casework having come originally from legitimate sources, i.e. prescription medication, and/or as adulterants in illegal drugs. The presumptive test for barbiturates is the Dille–Koppanyi test and for benzodiazepines is the Zimmerman test (see Table 3.1).

Both barbiturates and benzodiazepines are usually supplied as tablets, capsules, or liquids to be injected.

3.2.2.4 LSD

LSD is a hallucinogenic indole alkaloid (fused benzene and pyrrole rings) originally of fungal origin but now primarily synthesised as a drug of abuse. LSD is often supplied by soaking blotting paper in a solution of the drug. The paper is divided into 'tabs', which are small squares of paper, each of which typically has a design printed on it. The variety of designs should be carefully recorded as they may link ostensibly separate drug seizures together. A

Table 3.1 Common presumptive colour tests for drugs (Cole 2003; Johns et al. 1979; O'Neal et al. 2000; United Nations Office on Drugs and Crime 2009b).

Target drug	Test	Colour	Notes
Marijuana	Duquenois–Levine	Blue-purple	False +ves can be distinguished by brown colour on addition of further reagent
Opiates, cocaine, and amphetamines	Corinth IV salt	Pink and orange	
	Marquis	Opiates – blue/violet Cocaine – pink/orange Amphetamines – yellow to orange MDA, MDMA, and MDEA – purple/blue	On addition of further reagents psilocybin and psilocin (hallucinogenic mushrooms) result in a yellow/orange color
	Cobalt thiocyanate	Cocaine and diamorphine – blue	Cocaine can be distinguished with the Scott test
	Mandelin	Cocaine – orange Opiates – olive green Heroin – brown Amphetamine – bluish green	
Amphetamines	Simon	Amphetamine (primary amine) – purple Methamphetamine (secondary amine) – blue	Follows +ve result for amphetamines with Marquis test. Different reagents are used to produce the colours stated
Barbiturates	Dille–Koppanyi	Blue	
Benzodiazepine	Zimmerman	Red-pink	
LSD, psilocybin, and psilocin	Ehrlich's *p*-dimethylaminobenzaldehyde	Blue-violet	

Source: Adapted from Heard (2008) and Warlow (2005).

commonly used test for the presence of LSD is Ehrlich's test (see Table 3.1). This can either be carried out directly on a 'tab' or after extraction with methanol.

3.2.2.5 New Psychoactive Substances

In an effort to evade legal penalties, those involved in the manufacture and supply of drugs are continually looking for alternative compounds which mimic the effects of abused drugs but which are not yet subject to legislative control. Since the early 2000s, there has been a dramatic increase in what are referred to as 'new psychoactive substances' (NPS). The most prevalent of these are: synthetic cannabinoids and cathinones, opioids, phenethylamines, piperazines, benzodiazepines, tryptamines, arylamines, and some others, with synthetic cannabinoids and cathinones being the most commonly seized. As this chapter focuses on presumptive testing rather than an in-depth discussion on controlled drugs, only the synthetic cannabinoids will be discussed further here, with emphasis on how these NPS can cause problems for drug control authorities in terms of presumptive testing.

Synthetic Cannabinoids

To overcome legislative penalties associated with trafficking and supplying cannabis, in the mid-2000s, drug suppliers started producing and marketing 'legal highs'. These were herbal materials to be smoked in the same way as cannabis but they did not contain the classic cannabinoids and the herbal material was not generally *C. sativa*, although smoking produced similar psychoactive effects as Δ^9-THC. In 2008, 1-pentyl-3-(1-naphthoyl)indole, commonly referred to as JWH-018, was isolated and identified from a legal high sold under the name Spice. JWH-018 acts on the same cannabinoid receptors as Δ^9-THC and hence is classed as a synthetic cannabinoid receptor agonist (SCRA). Since JWH-018 was identified, many more synthetic cannabinoids have been isolated and identified. As authorities take steps to legislate against the latest compound identified, the chemists synthesising them (based mainly in China), produce a new SCRA. In 2018 the European Monitoring Centre for Drugs and Drug Addiction (EMCDDA) reported that they were monitoring 179 synthetic cannabinoids (EMCDDA 2018, n.d.) divided into seven major structural groups: naphthoylindoles, napthylmethylindoles, napthoylpyrroles, naphthylmethylindenes, phenylacetylindoles, cyclohexylphenols, and classical cannabinoids. This chemical complexity, and the number of compounds potentially in use, makes it difficult to devise a simple presumptive test to detect all synthetic cannabinoids. The United Nations Office on Drugs and Crime (2013) concluded that there was no presumptive test which could cover the whole range of synthetic cannabinoids in use. They recommended the use of thin layer chromatography (TLC) with examination under ultraviolet (UV) light and spraying with Fast Blue RR reagent to detect non-classical cannabinoids. For other synthetic cannabinoids, they proposed use of iodine or iodoplatinate to visualise spots post TLC. These are relatively non-specific reagents but, if suitable standards are run alongside samples, it could give an indication of a presumptive positive. A variety of test kits are available commercially for presumptive testing of synthetic cannabinoids including the tests from MMC International B.V., Sirchie, and EZtestkits. None of these kits test for all known synthetic cannabinoids.

3.2.3 Presumptive Tests for Drugs

3.2.3.1 Colour Tests

Presumptive colour tests for drugs, as discussed in section 3.2.2 for a range of drug classes, detect a drug's functional group(s) and typically have a limit of detection in the region of 1–50 µg depending on the analyte (O'Neal et al. 2000). The drugs most commonly encountered in forensic casework share a limited number of functional groups, e.g. aromatic rings, amines, and phenols, however, they may have more than one functional group. This means that several colour tests will detect more than one drug. In some instances, on addition of further reagents, the test will reduce the presumptive identification from a class of drugs to a specific drug but in other cases a second, separate, test is required to do this.

In addition to being carried out at scenes of crime and in the laboratory, presumptive colour tests for drugs are routinely carried out in airports and other entry ports. Many are available in kit form for ease of use and agencies/government departments, such as the UK Home Office, may issue a list of those that are approved (Home Office 2017).

3.2.3.2 Thin Layer Chromatography

TLC can be used to presumptively identify drugs by comparing spot colour and retention factor values with a positive control, i.e. a drug standard (see Section 12.2.1 for TLC theory). For each class of drugs there are multiple solvent systems that can be used to carry out the separation of components. Visualisation of spots is then carried out by examining under UV light and/or by spraying with a developing reagent which is chosen to react with specific functional groups. Common developing reagents for the drug classes discussed above are (spot colours are given in parentheses): Fast Blue BB for marijuana (red, orange, yellow, brown); Dragendorff's reagent for opiates (orange/orange-red); acidified potassium iodoplatinate reagent for cocaine (blue); Fast Black K for amphetamines (purple or orange-red); mercuric chloride-diphenylcarbazone for barbiturates (blue-violet on pink); sulfuric acid followed by acidified potassium iodoplatinate reagent for benzodiazepines (fluorescent then purple); and Ehrlich's reagent for LSD (purple).

3.2.3.3 Microcrystal Tests

Microcrystal tests use the characteristics of crystals to presumptively identify any drugs present.

A small drop of reagent is added to the sample and once crystallised the sample is viewed using a polarised light microscope. As with other presumptive tests a drug standard, which has undergone the same procedure, should be viewed in the same way.

Table 3.2 summarises common microcrystal tests for drugs.

Whatever the presumptive test method, confirmatory tests are still needed to definitively identify the drug present. In addition, and specifically in drug testing, the quantity of drug present may be important when considering the offence the suspect is to be charged with (possession vs. intent to supply) and this can only be determined with confirmatory techniques that are also quantitative. In drugs testing the quantitative confirmatory techniques are likely to be chromatographic (see Chapters 13 and 14).

Table 3.2 Common presumptive microcrystal tests for drugs.

Target drug	Reagent	Crystal appearance
Cocaine	Platinum chloride with acetic acid	Feathery, spiny, combs
	Gold chloride with acetic acid	X-shaped crystals bisected by a barbed rod
Heroin	Mercuric chloride	Large branched and feathery
Amphetamine	Gold chloride	Long yellow rods
Methamphetamine	Gold chloride	Forked rods
MDMA	Gold chloride	White X-shaped

Source: Adapted from Brinsko et al. (2018) and Smith (2005).

3.3 Firearms Discharge Residue

Firearms discharge residue (FDR), also known as gunshot residue (GSR), is the material that escapes from a firearm upon firing (see also Section 9.3.1). Under high pressure, gases carrying burnt and unburnt primer and propellant particles escape from small gaps in the firearm and the muzzle end. On cooling, particles are distributed in the surrounding area, on the shooter, anyone else in close proximity and, depending on the distance to it, the target. The nature of the explosive event means that not all particles have the same composition or morphology. FDR has both an inorganic and an organic component. The inorganic component is composed of material from the primer and the propellant (detonating explosives, see Section 3.4) and metals from the cartridge case, the projectile and the barrel. More attention has been given to using the inorganic component to determine the presence of FDR in the past. Historically, a combination of lead, barium, and antimony (originating in the primer) in a single spherical particle, was said to be unique to FDR. However, this combination has been seen in other situations, such as motor vehicle brakes and fireworks, and condensate particle morphology rather than spherical particle morphology is now used as an indicator of FDR. Further, lead free primers have been introduced, which would clearly not produce a lead/barium/antimony single particle. Other inorganic constituents include copper, aluminium, iron, silicon, nickel, and zinc and their presence can vary depending on the firearm and cartridge type. The organic component of FDR also comes from the primer and the propellant and can include nitrate and nitrite ions, nitrocellulose, nitroglycerin, diphenylamine, and ethyl and methyl centralite.

In FDR testing, the confirmatory techniques are likely to be scanning electron microscopy (SEM) (see Chapter 9) for inorganic components and chromatographic for organic components (see Chapters 13 and 14).

3.3.1 Firearms Discharge Residue Sampling

FDR is initially deposited without contact, which means that not only the shooter but also other persons present may have FDR deposits on them. FDR can also subsequently be transferred to other surfaces, e.g. by rubbing hands on clothing. Ideally FDR should be collected within three to six hours of the event as it can quickly be lost, depending on the level of activity. Loss will occur rapidly from skin, which may mean the early loss of the FDR with, potentially, the highest evidential value as it may indicate the shooter. Depending on the fabric, clothing can retain FDR for longer than skin, especially in areas that collect material, such as pockets and turn-ups. However, the very fact that these are areas that collect debris could call into question the length of time the FDR has been present, i.e. it may have been there before the incident under investigation. As evidence is evaluated the examiner must be careful to consider where the FDR particles have come from, what has happened since the event and what that may mean in the context of the case.

Common areas for FDR recovery are the hands, the face, hair, clothing, objects in the vicinity of the event and, for distance estimations, the target. Samples are collected from the hands either by swabbing, taping, or by pressing with an adhesive stub (for direct analysis by SEM). Usually more attention is paid to the back of the hands as the palms are to some extent protected by holding the firearm. However, samples should be taken from the palm and back of both hands, including folds of the skin where particles may be retained. If the samples come from the shooter more FDR would be expected on the back of the hand than the palm, whereas this would not necessarily be true for a person standing nearby. The face is sampled particularly in cases where a firearm producing blowback has been used or is suspected to have been used. FDR can persist in hair longer than on skin, particularly if not washed, therefore, hair combings can be a good source of evidence. Collection from clothing should be done by taping and shaking as well as looking for debris in the evidence packaging. Finally, sampling of objects from the scene could be carried out by swabbing or taping depending on the nature of the item.

As well as indicating the shooter and proximity to the shooter, FDR presumptive tests can be used to estimate the distance from the shooter to the target by looking at the FDR distribution. Generally speaking, the further away the target the less material is deposited at the target site. This varies with the firearm used so tests should also be carried out with the specific firearm make involved in the case. In this instance the examiner will use moistened filter paper to recover material from the whole area containing FDR. The material on the filter paper will then be tested and any colour changes used to visualise the pattern of FDR distribution.

In all FDR sample collection, contamination must be very carefully avoided. The suspect must be kept away from areas or items that may have had contact with FDR in the past, such as some police cars, handcuffs, and firearms officers. Examiners who have contact with FDR should also take precautions not to contaminate by showering and wearing protective clothing.

3.3.2 Firearms Discharge Residue Presumptive Tests

Presumptive tests typically test for one or two elements of functional groups. As discussed in Section 3.3, the combination of multiple elements/groups that are not found in other types of residue can be indicative of FDR and there is not one universal formulation for primer and propellant. Therefore, there are several tests that can be used in parallel to presumptively test for FDR. Similar to drug presumptive tests, kits for in-field testing of FDR are available.

Table 3.3 Common presumptive colour tests for firearm discharge residue.

Target elements	Test	Colour	Notes
Barium and lead	Sodium rhodizonate	Magenta indicates lead	On addition of further reagent pink-orange for barium
Antimony	Harrison–Gilroy	Orange	
Nitrites	Modified Griess	Orange-red	Perform after sodium rhodizonate
Nitrates	Diphenylamine	Blue-purple	
Copper	Dithiooxamide	Olive green	Perform before sodium rhodizonate

Environmental sources that give positive results are possible for all of the FDR tests, e.g. fertilisers, amongst others, are sources of nitrates. Therefore, it must always be remembered that presumptive tests are not identification tests and that any indications given must be either confirmed or their interpretation should be conservative and specify any limitations of the test. This was not done in the Birmingham Six case, where the Griess test was used for explosives, as discussed in Section 3.4.2.1.

Some of the most common presumptive colour tests for FDR are summarised in Table 3.3.

3.4 Explosives

Explosives are materials that combine the oxygen and fuel required to produce a rapid release of energy. This usually occurs by build up and release of gas pressure. In this section the emphasis is on chemical explosives rather than mechanical, thermal, or electrical. Chemical explosives can be categorised as either deflagrating (low) or detonating (high), which loosely corresponds to the form of the explosive. Deflagrating explosive blast fronts travel slower than the speed of sound and are used for push applications, e.g. propelling a projectile down the barrel of a gun, whereas detonating blast fronts travel faster than the speed of sound and are used for shatter applications, e.g. bombs. The proximity of the oxygen and fuel is a determining factor in the speed of the reaction and, therefore, the speed of the blast front. Hence, physically mixed oxygen/fuel explosives are generally deflagrating and chemically mixed (i.e. containing oxygen and fuel in each molecule) are detonating. Examples of deflagrating explosives are those commonly found in improvised devices, such as ammonium nitrate/fuel oil (ANFO), and detonating explosives include military explosives such as hexogen (commonly referred to as RDX). In most cases a detonator is needed to start the reaction in the main explosive. These often contain smaller quantities of an explosive chosen for its ease of ignition by spark, flame, or shock wave.

As already mentioned, explosives contain fuel and oxygen and their reaction results in the release of large amounts of energy. The way in which that energy/fuel is stored is in the explosives' chemical bonds, particularly with double and triple bonds. Common constituents of explosives are nitrates, carbon, hydrocarbons, carbohydrates, chlorates, and perchlorates.

3.4.1 Explosive Residue Sampling

When an explosion takes place material from the explosive device and from surrounding structures and objects is carried with the expanding gases produced. Once the initial blast wave has propagated, a second wave moves back towards the seat of the explosion to fill the negative pressure, causing further damage. Pieces of the device, reacted and unreacted explosive particles from the main explosive and any detonators will be distributed throughout the area of damage.

The examination of an explosion site, although extremely complex, follows the same basic principles as any crime scene examination but extra care must be taken with health and safety issues, such as the possibility of further devices being present and the structural safety of buildings. Care must also be taken to collect every fragment of the device and explosive residues, which means casting the cordon wide. Samples should be taken from the crater created by the explosion as well as from device fragments, looking for both reacted and unreacted particles. Unreacted particles provide a better opportunity to determine the initial composition of the explosive.

If a suspect is identified, similar protocols to those used in FDR examination can be carried out in order to recover any explosive particles present on the suspect.

3.4.2 Explosive Residue Presumptive Tests

3.4.2.1 Colour Tests

The nitrates tests used in FDR examination, i.e. the modified Griess test and the diphenylamine (DPA) test, can also be applied to explosive residues, since it is the explosive residue in FDR that is being tested. Trinitrotoluene (TNT), however, does not react in either of these tests. A further test is required to presumptively detect its presence, namely the alcoholic potassium hydroxide test (KOH).

Presumptive tests for explosives have received much criticism in the past due to miscarriages of justice resulting from over interpretation. Infamously, the Birmingham Six were convicted of murder, at least in part, based on a positive Griess test result for nitrites, despite a negative confirmatory gas chromatography–mass spectrometry (GC-MS) result. The scientist in the case claimed that he was sure of the presence of explosives even in the face of the lack of confirmation by GC-MS and the defence argument of contact with playing cards, which contained nitrocellulose.

As with other types of presumptive colour tests commercial kits for in situ testing for explosives are available.

Some of the most common presumptive colour tests for explosives are summarised in Table 3.4.

Table 3.4 Common presumptive colour tests for explosives.

Explosive	Test	Colour
Nitrates/ites (e.g. ANFO)	Modified Griess	Orange-red
	Diphenylamine	Blue-purple
TNT	Alcoholic potassium hydroxide (KOH)	Purple/brown
	Nessler's reagent	Red
RDX	Modified Griess	Pink to red
	Diphenylamine	Blue
PETN	Modified Griess	Pink to red
	Diphenylamine	Blue

PETN, pentaerythritol tetranitrate.
Source: Adapted from Levine and Lewis (2013).

3.4.2.2 Thin Layer Chromatography

As with other types of forensic evidence, TLC can be used for presumptive identification of explosives, against standards, with a variety of solvent systems and developing reagents. Two of the more common developing reagents are DPA and Griess.

3.4.2.3 Portable Instruments

Portable hydrocarbon detectors of the type used in fire investigations (see Section 3.6) can also be used in explosion investigations. Similarly, portable gas chromatography (GC) instruments (see Section 13.6.2) can be used to analyse explosion residues.

If an explosive is presumptively identified by any of the methods discussed above, confirmatory analysis can be carried out using a range of techniques, such as scanning electron microscopy–energy dispersive X-ray (SEM-EDX), GC and high performance liquid chromatography (HPLC) (see Chapters 9, 13, and 14).

3.5 Ethanol (Ethyl Alcohol)

3.5.1 Breath Alcohol Testing

Presumptive tests for alcohol are often referred to as point of collection test (POCT) devices and can be used for pre-employment testing or workplace compliance, diagnostic treatment of medical conditions, and in a law enforcement context – identification of an individual driving under the influence (DUI) of alcohol. Only devices designed for use in cases of suspected DUI will be discussed herein. According to the US National Highway Traffic Safety Administration (NHTSA 2019), in 2018 there were 10 511 fatal vehicle crashes reported that were related to alcohol impaired driving, which represents 29% of all traffic fatalities for the year (most recent data reported). The cost to society of a person DUI of alcohol is considerable and as a result, numerous POCT devices are available to police units to assist in reducing the statistics relating to DUI fatalities.

When ingested orally, alcohol passes from the mouth, into the stomach, and then small intestines where it is absorbed into the bloodstream and distributed throughout the body. Alcohol is metabolised by the liver and eliminated by the kidneys into the urine. The common effect of alcohol is felt when the circulatory system passes it to the brain. A portion of the alcohol will also pass unaltered from the body through the lungs. It is this unaltered alcohol that allows the determination of a breath alcohol concentration (BrAC) that is the commonly utilised sample collection for POCT devices.

Breath alcohol devices were originally developed as a rapid and non-invasive technique for indirectly monitoring alcohol concentration in the blood. Many countries have adopted legislation that allows the collection and testing of a subject's breath sample to determine the alcohol concentration in their body at the time of driving. This method of testing is favoured by most law enforcement organisations due to its ease of use, speed of analysis and reporting of results, and the capability to conduct the testing at the scene of an incident. The electronic instruments currently available for breath alcohol testing can be categorised as either handheld devices (also referred to as preliminary breath testers [PBTs]) or bench-top devices. There is a broad spectrum of breath alcohol testing devices available worldwide and their evidential value varies by country and jurisdiction. Some of these devices are designed for screening purposes only while others are approved for evidential purposes. All devices must conform to the appropriate authority's guidelines and regulations.

3.5.1.1 Electronic Devices

Handheld or PBTs are small units designed for screening an individual's approximate BrAC level at the time of a traffic stop. Most handheld devices utilise an electrochemical fuel cell sensor that oxidises ethanol to acetaldehyde and in doing so produces an electric current that is proportional to the amount of alcohol in the sample. The devices operate on alveolar air (i.e. deep lung air) supplied by an individual blowing into a mouthpiece or tube for a period of time (usually 10–12 seconds). Passive alcohol sensor devices are also available and utilise a fan mechanism to pull expired air into the instrument. The device is held in close proximity to the individual's mouth and a mixture of expired and environmental air passes into a fuel cell sensor where it can detect small concentrations of alcohol. The sensor produces an estimated BrAC level of a driver ranging from 0.02 to 0.12% BrAC. The benefit of this type of instrument is the lack of mouthpieces or other disposable tubes and the ability for police to immediately identify drivers who are suspected to be over the legal alcohol limit.

The majority of PBT instruments will not produce a response in the presence of acetone, an endogenous volatile exhaled in breath, as this ketone is not oxidised at the electrode. This is valuable for reducing false positives resulting from medical conditions such as untreated diabetes or prolonged fasting. However false readings (both high and low) can result from the oxidation of methanol and isopropanol and caution is required when interpreting the test results. For this reason, PBTs should only be employed by trained personnel and in the event of a positive result, a confirmatory (evidentiary) test should follow.

Handheld devices demonstrate several advantages over bench-top devices. Typically, these devices are inexpensive, portable, and require less training and maintenance. However, although PBTs have a degree of sensitivity and specificity, for evidentiary purposes, a bench-top device is generally preferred for BrAC measurements. The instruments used for evidential purposes are larger but more sophisticated and incorporate mechanisms for checking calibration, analysing alcohol-free air in the room the sample is collected, and producing a record of the test result. Many bench-top devices utilise infrared technology coupled with electrochemical sensors to provide a dual-sensor technique that can provide a confirmatory result for the BrAC level of an individual. The result is a quantitative measure of the BrAC, which serves as evidence for prosecuting drivers under the influence.

3.5.1.2 Chemical Test Devices

Screening devices, other than electronic devices, are also available for alcohol detection and typically employ a chemical colour change reaction. Such tests are disposable and only suitable for one use making them useful as self-test devices prior to driving or for workplace compliance where zero-tolerance levels are mandatory (e.g. mines, military, corrections). One example of the chemical reaction incorporates dichromate or permanganate salts in acid-impregnated crystals housed in glass tubes. The individual being tested first crushes the test between their fingers and then blows into the tube for 12 seconds. The yellow crystals change to a blue or blue-green colour in the presence of alcohol and remain unchanged in the absence of alcohol.

Tests often yield a pass/fail result when alcohol is present at or above the cut off level of 0.02% BrAC (i.e. the intoxication level from one standard drink). Other chemically based tests are also available at higher cut-off levels including 0.04 and 0.08% BrAC. An alternative test uses a similar colour change reaction but can approximate the BrAC level at higher concentrations. The breath sample is collected from the individual by inflating a balloon, which is then attached to the test device containing dichromate impregnated crystals. Air is allowed to flow through the tube for one minute before reading the response. In the presence of alcohol, the yellow crystals will change to a green colour and at higher concentrations the green colour will continue to move up the tube. The positive result is read at the highest colour-change level on the tube, which coincides with an approximate BrAC level. Depending on the model, a measurement between 0.02% BrAC and 0.15% BrAC can be estimated.

Regardless of the device employed, the colour change should be determined immediately as other oxidisable components of the breath sample may continue to react with the chromate or permanganate producing false positive readings. These devices are limited to only semiquantitative determinations and suffer from the subjectivity associated with interpreting colour changes.

3.5.2 Saliva-Based Testing

While breath is by far the most common specimen used for POCT devices, saliva may provide an alternative specimen for estimating the blood alcohol concentration (BAC) of an individual. Saliva-based testing is not typically employed for DUI enforcement but holds some promise for future POCT or roadside testing of alcohol and drugs. Its advantage lies in the fact that saliva provides a more direct correlation (i.e. 1:1) with BAC compared with breath, making it a more sensitive testing specimen.

The two main types of tests are the saliva alcohol test and strip-based saliva testers. The saliva alcohol test is a small, portable device similar to a thermometer that provides a quantitative measure of equivalent BAC using a saliva sample. A cotton swab is used to collect the saliva sample from the cheeks, gums, and under the tongue, and then placed in the test device using gentle pressure. Capillary action moves the fluid along the test device and the development of a dark spot indicates that the test has worked. The scale-reading provides a quantitative measure of ethanol concentration. The advantage of this technique over breath alcohol analysis is the reduced cost, simpler collection and testing device, and a high correlation with blood samples when analysed by confirmatory laboratory tests (Jones 1995). Although not regularly employed for DUI testing, it is regularly utilised for workplace enforcement testing.

Strip-based saliva testers do not correlate as well with BAC and only provide a semiquantitative estimation of ethanol concentration in saliva (Penttila et al. 1990). The disposable tester contains a chemically reactive pad, which has been treated with an alcohol oxidase enzyme. The pad is saturated using saliva from the individual's mouth and the enzyme responds to the presence of alcohol by producing a colour proportional to the concentration of alcohol in the sample. The colour change is compared with a calibrated colour chart to estimate the BAC level. Due to the subjective nature of colour determination, strip-based saliva tests do not provide quantitative data and may prove difficult for colour blind or visually impaired users. Additionally, some saliva testers are affected by temperature and will indicate false readings in hot (false high) or cold (false low) temperatures.

The technology and chemical reactions employed in saliva-based tests are currently not as precise or accurate as breath alcohol testing and for this reason are not commonly used in forensic testing. A positive test result using saliva-based tests must be confirmed using an evidential breath or blood test to quantitatively determine an individual's alcohol concentration.

3.6 Ignitable Liquid Residues

The role of a fire investigator is to determine the origin and cause of a fire regardless of whether it was accidentally or intentionally started. To determine the likelihood of an intentionally set fire, samples of fire debris must be collected for confirmatory analysis by GC-MS. Presumptive tests for accelerants or ignitable liquid residues (ILRs) can be conducted at the fire scene to assist the investigator in collecting appropriate samples for laboratory analysis. The importance of correctly locating and identifying ILRs at a fire scene is twofold. Presumptive detection of an ILR can assist investigators to narrow the suspected area of origin, thus reducing the number of samples that need to be collected and submitted for confirmatory analysis. The confirmatory identification of ILRs by the laboratory will further assist the investigator to focus the investigation accordingly. It is important that the investigator be able to place the ILRs in context within the crime scene based on the laboratory results.

One type of presumptive chemical test that may be employed at a fire scene is colorimetric tubes, specifically Dräger tubes, to determine the possible presence of ILRs. Dräger tubes consist of a glass tube containing a powder designed to react with specific chemicals or classes of chemicals to produce a colour change. The ends of the tube are broken, and the tube held directly above the sample. The headspace of the sample is drawn through the tube using a manual or electric pump. If the analyte of interest is present within the sample headspace, it will react with the chemical mixture in the tube providing a colour change. The tube can be calibrated, and a fixed amount of headspace drawn through to allow for an estimate of the concentration of the analyte. However, the specificity and sensitivity of these tubes can be an issue, especially when screening for a range of ignitable liquids.

3.7 Non-Chemical Presumptive Tests

3.7.1 Electronic Detectors

Numerous electronic detectors are available for presumptively locating analytes of interest at a crime scene. Such detectors may also be referred to as electronic sniffers, electronic noses, or portable gas detectors. Electronic sniffers were initially designed for atmospheric analysis of potentially hazardous environments but have made their way into the field analysis of fire, illicit drugs, and explosion investigations. They are often used as a complement to visual observation and canine detection (see Section 3.7.2). Most electronic sniffers use a probe held above the sample or substrate through which vapour samples are drawn using a vacuum pump. The sampled headspace is passed through to a chemical detector and the result is output on a small display or computer system.

3.7.1.1 Electronic Detectors for Fire Investigations

Catalytic combustion detectors represent one type of chemical detector employed in these instruments. The functioning principle is based on a combustible gas that is burnt at the surface of a catalyst-coated resistance element. The heat generated from the combustion reaction causes a resultant increase in resistance that is compared with a reference element and converted to a concentration measurement by the instrument. These electrochemical systems are inexpensive, provide a rapid response, and contain no moving parts making them ideal as portable gas detectors. However, their sensitivity is susceptible to changes in humidity and they lack selectivity, commonly leading to false positive responses. Some commercially available instruments incorporate a needle gauge as well as an audible alarm and demonstrate a high sensitivity for hydrocarbon molecules. Unfortunately, this means that false positive alerts will result from pyrolysis products and fire debris. However, used in conjunction with canines (see Section 3.7.2), this is an efficient method for collecting quality evidence from a fire scene.

A flame ionisation detector (FID) is commonly employed in GC (see Section 13.2.4.1) and is also used in electronic sniffers due to its sensitivity to most combustible compounds. The sample is pumped into a flame and subsequently burned, producing ions and electrons. The electrons are held between two electrodes across which a potential difference is applied. The result is a measurable current whereby intensity is directly proportional to the amount of ions (analytes) in the detector. This detector is particularly valuable for fire debris analysis due to its insensitivity to several interfering compounds including O_2, N_2, CS_2, H_2S, NH_3, CO, CO_2, and H_2O (Stauffer et al. 2008). However, it requires hydrogen and oxygen tanks to produce the flame making it less portable than other systems.

The photoionisation detector can also be employed in electronic sniffers and is based on the same principle as the FID. UV light is used to irradiate a sample, producing ions that are held between plates within a chamber. A potential difference is applied across the two plates producing a measurable current. The intensity of the current is proportional to the amount of ionised molecules that can be recorded. The photoionisation detector responds to volatile organic compounds (VOCs), which may come from the fire debris but may be a result of environmental background. As a result, false positive responses are common with this system making it less valuable to fire investigators.

All of the above detectors operate in a similar manner. The instruments are designed to allow the operator to walk around a scene holding the probe in one hand and the detector in the other hand. Some detectors are designed to produce a response almost instantaneously. The response may be visually displayed on the screen or may be auditory above a certain concentration. The advantage of the latter is that the operator can focus on the scene, rather than the instrument screen. The major drawback to these electronic detectors is that, like most presumptive tests, they lack specificity and, in some instances, sensitivity.

Portable GC-MS has been investigated as a field test for locating and detecting toxic volatiles and accelerants. Details about the theory and operating principle of the portable GC-MS can be found in Chapter 13. One of the first such instruments on the market was the Inficon Hapsite® Field-Portable GC-MS (Inficon, East Syracuse, NY), a miniaturised quadrupole GC-MS weighing approximately 42 lb (~19 kg). The instrument was designed for the analysis of VOCs, toxic industrial chemicals (TICs), toxic industrial materials (TIMs), chemical warfare agents (CWAs), and select semivolatile organic compounds (SVOCs).

The detection limit of the Inficon Hapsite was in the parts per million (ppm) to parts per trillion (ppt) range for most analyses. The mass spectrometer scanned across a mass range of 41–300 amu (1–300 amu in selected ion monitoring mode). The main advantage of the instrument was the ability to produce National Institute of Standards and Technology (NIST) searchable mass spectra allowing for rapid identification and semiquantitation of unknowns in the field. The disadvantages included the high cost and the expertise required to use the instrument and interpret the data. Although the manufacturer promoted the instrument as producing fast, confirmatory results directly comparable with laboratory GC-MS results, a comparative study by Casamento et al. (2005) identified several limitations to the instrument including the reduced chromatographic resolution of the system when applied to ignitable liquids. However, over the past decade the cost of these instruments has decreased, and their capabilities improved with several additional models commercially available including Torion, Griffin, and Gaurdion portable GC-MS technologies. Testing of these newer technologies for the detection and classification of ILRs has provided positive screening results (Leary et al. 2016; Visotin and Lennard 2016) suggesting that portable GC-MS may be increasingly employed for screening fire scenes and locating quality evidence in the future.

3.7.1.2 Electronic Detectors for Explosives and Illicit Drugs

Current global events have demonstrated that explosives present the most prevalent threat to transportation security and particularly to individuals utilising these means of public transport. Terrorists are increasingly using explosive materials, including homemade explosives (HMEs), as weapons of mass destruction and disaster. The detection of these materials is difficult as often only traces of the explosives remain on the suspect. This is further confounded by the short duration available to border security to make a positive identification before allowing the individual to pass. For the same reasons, detection of smuggled contraband such as illegal drugs also poses a considerable problem for customs and border patrol agents.

Over the last two decades, a considerable amount of funding and resources has gone into this area of study with the production of numerous portable instruments for the detection of explosives and illicit drugs. The most commonly employed technique for trace explosives and illicit drugs detection in the transportation security industry is ion mobility spectrometry (IMS). The principle behind IMS technology is based on the separation of ions through a drift tube using a gas phase. The sample is initially ionised, forming negative ions, and accelerated by an electric field into the drift tube through a shutter. As the ions exit the drift tube, they are counted electronically by a Faraday cup. The ions are characterised by their gas phase mobility in the weak electric field based on their charge, mass, and size. Their mobility through a voltage gradient is measured by the time taken to drift the length of the tube. If this drift time (or mobility value) corresponds to a known measurement for explosives, the instrument will provide a positive response. The advantage of the IMS is its portability and specificity of response while its major limitation is poor peak resolution in mixtures.

A variation of the portable GC is also available for trace explosives and illicit drugs detection and employs a surface acoustic wave (SAW) detector. SAW detectors can be used as chemical detectors because of their characteristic acoustic resonant frequencies. When a sample is adsorbed onto the crystal surface, the resonant frequency will change, allowing the signal to be measured. SAW detectors on their own will respond to any adsorbing molecule and as a result they are typically coupled with GC to improve selectivity for trace explosives detection.

In addition to the above-mentioned technologies, there is now a wide variety of portable instrumentation available on the commercial market for the detection of trace explosive and illicit drugs. However, like all portable technologies, their sensitivity and selectivity rarely compares with confirmatory tests and ultimately they can only be used for rapid, positive screening of persons and belongings. Such technologies are not designed for screening large areas or 'clearing' a site (i.e. declaring the site free from explosives). For these instances, canine detection is the preferred screening tool.

3.7.2 Canine Detection

Domestic canines (*Canis lupus familiaris*) have been employed extensively in the military and law enforcement fields for many decades. They can be used for the detection and location of non-biological scents (e.g. narcotics and other illicit drugs, explosives, accelerants and ILRs, currency) and for biological scents (e.g. human remains, human scent tracking, pest species control, wildlife conservation). Their superior scenting ability is a result of their larger olfactory tracts and bulbs, providing a larger proportion of olfactory cells, when compared with humans. The anatomy of their nasal passage also allows for better air flow over the olfactory receptors yielding a more efficient response to odour. Although their levels of sensitivity have not been scientifically validated, it is believed that their sense of smell is tens to hundreds of times more sensitive than our own (Almirall and Furton 2004).

A canine team incorporates a dog and its handler. The team will screen a suspect location, searching for the specific scent the dog has been trained to alert on (e.g. accelerants, explosives, drugs, etc.). If the scent is detected, the dog will positively alert to its handler by sitting or standing still near the indicated area. Other methods for signalling a find may also be used dependent on the training. The dog will then be rewarded with food or playtime if the signal alert is correct.

This method is valuable for screening large areas and narrowing down the source of a particular scent. Samples will be collected from that area and placed in appropriate crime scene containers. At this point, the samples may be checked again by the dog away from the original source and a sufficient distance from each other. If a positive alert is not indicated for one or more of the samples, they may be discarded to reduce the number of samples requiring confirmatory laboratory analysis. Alternatively, the canine team may reanalyse the location from which the samples were collected and if a positive alert is still produced by the dog, additional samples will be collected for analysis.

While canines do not provide confirmatory results, their use to locate valuable evidence at crime scenes can improve the likelihood of a positive confirmation from the laboratory analysis. Their sensitivity is as good as, if not better, than many analytical instruments. Their selectivity is also good and many trained detection dogs can differentiate background interference from analytes of interest in the samples. However, they are not infallible and false positives do occur highlighting the importance of confirmatory analysis by chromatography and MS (see Chapters 10, 13, and 14). It is recommended that canines should only be used as screening tools for locating a particular scent or selecting appropriate samples for further analysis.

References

Aitken, C.G.G. (1999). Sampling – how big a sample? *Journal of Forensic Sciences* 44: 750–760.

Aitken, C.G.G. and Lucy, D. (2002). Estimation of the quantity of a drug in a consignment from measurements on a sample. *Journal of Forensic Sciences* 47: 968–975.

Almirall, J.R. and Furton, K.G. (2004). *Analysis and Interpretation of Fire Scene Evidence*. Boca Raton, FL: CRC Press.

Brinsko, K.M., Golemis, D., King, M.B. et al. (2018). *A Modern Compendium of Microcrystal Tests*. Chicago, IL: McCrone Research Institute.

Casamento, S., Konza, E., Du Pasquier, E., and Royds, D. (2005). Evaluation of a portable gas chromatograph for the detection of ignitable liquids. *Canadian Society of Forensic Science Journal* 38: 191–203.

Cole, M.D. (2003). *The Analysis of Controlled Substances*. Wiley.

EMCDDA (2018). Fentanils and synthetic cannabinoids: driving greater complexity into the drug situation. http://www.emcdda.europa.eu/system/files/publications/8870/2018-2489-td0118414enn.pdf (accessed 22 March 2019).

EMCDDA (n.d.). Synthetic cannabinoids and 'Spice' drug profile. http://www.emcdda.europa.eu/publications/drug-profiles/synthetic-cannabinoids (accessed 22 March 2019).

Gahlinger, P. (2003). *Illegal Drugs*. New York: Plume.

Heard, B.J. (2008). *Handbook of Firearms and Ballistics*, 2e. Chichester: Wiley.

Home Office (2017). Circular 005/2017: testing suspected substances under the Misuse of Drugs Act 1971. https://www.gov.uk/government/publications/circular-0052017-testing-suspected-substances-under-the-misuse-of-drugs-act-1971 (accessed 25 May 2020).

Johns, S.H., Wist, A.A., and Najam, A.R. (1979). Spot tests – a colour chart reference for forensic chemists. *Journal of Forensic Sciences* 24: 631–641.

Jones, A.W. (1995). Measuring ethanol in saliva with the QED enzymatic test device: comparison of results with blood- and breath-alcohol concentrations. *Journal of Analytical Toxicology* 19: 169–174.

Leary, P.E., Dobson, G.S., and Reffner, J.A. (2016). Development and applications of portable gas chromatography-mass spectrometry for emergency responders, the military, and law-enforcement organizations. *Applied Spectroscopy* 70: 888–896.

Levine, B. and Lewis, S.W. (2013). *Presumptive Chemical Tests*, Encyclopedia of Forensic Sciences, 2e. London: Academic Press.

NHTSA (2019). Fatal Crashes and Percent Alcohol-Impaired Driving. https://crashstats.nhtsa.dot.gov/Api/Public/ViewPublication/812864 (accessed 25 July 2020).

O'Neal, C.L., Crouch, D.J., and Fatah, A.A. (2000). Validation of twelve chemical spot tests for the detection of drugs of abuse. *Forensic Science International* 109: 189–201.

Penttila, A., Karhunen, P.J., and Pikkarainen, J. (1990). Alcohol screening with the Alcoscan test strip in forensic praxis. *Forensic Science International* 44: 43–48.

Smith, F.P. (ed.) (2005). *Handbook of Forensic Drug Analysis*. Burlington, MA: Academic Press.

Stauffer, E., Dolan, J.A., and Newman, R. (2008). *Fire Debris Analysis*. Burlington, MA: Academic Press.

United Nations Office on Drugs and Crime (2009a). Guidelines on Representative Drug Sampling. http://www.unodc.org/documents/scientific/Drug_Sampling.pdf (accessed 25 May 2020).

United Nations Office on Drugs and Crime (2009b). Recommended Methods for the Identification and Analysis of Cannabis and Cannabis Products. https://www.unodc.org/documents/scientific/ST-NAR-40-Ebook_1.pdf (accessed 25 May 2020).

United Nations Office on Drugs and Crime (2013). Recommended Methods for the Identification and Analysis of Synthetic Cannabinoid Receptor Agonists in Seized Materials. https://www.unodc.org/documents/scientific/STNAR48_Synthetic_Cannabinoids_ENG.pdf (accessed 3 April 2019).

Visotin, A. and Lennard, C. (2016). Preliminary evaluation of a next-generation portable gas chromatograph mass spectrometer (GC-MS) for the on-site analysis of ignitable liquid residues. *Australian Journal of Forensic Sciences* 48: 203–221.

Warlow, T. (2005). *Firearms, the Law, and Forensic Ballistics*, 2e. Boca Raton, FL: CRC Press.

4

Sample Preparation

Sue Jickells

with contributions from Shari Forbes and Tim Jickells

4.1 Sample Preparation

As noted in Chapter 1, there is often a need to prepare samples in some way before they can be analysed using instrumentation such as gas chromatography (GC) and high performance liquid chromatography (HPLC) (and associated hyphenated mass spectrometry [MS] techniques); isotope ratio mass spectrometry (IRMS); inductively coupled plasma atomic emission spectroscopy (ICP-AES); and inductively coupled plasma mass spectrometry (ICP-MS). There are several reasons for sample preparation, the main ones being:

1. To ensure the portion of a sample analysed is representative of the whole sample.
2. To avoid damage to analytical instrumentation.
3. To separate the analyte from other components in the sample which interfere with the analysis.
4. To increase the concentration of the analyte in the sample extract submitted for instrumental analysis such that the analyte can be detected.

For the purposes of this chapter, sample preparation refers to the steps that follow the sample collection process. Discussion will concentrate on sample preparation procedures associated with the analytical techniques covered in this book and the types of samples encountered in forensic science, i.e. separation of the substances of interest (analytes) from samples so that they can be subjected to instrumental analysis to obtain information which can support police investigations and the judicial process.

In forensic science, it is generally not desirable to destroy the whole sample during analysis because once the sample has gone, so has any further potential evidence. The general rule is to take only as much sample as is required for a valid analytical result to be obtained. For the majority of analyses, the result may be considered invalid if the sample taken for analysis (hereafter referred to as a subsample) is not representative of the whole sample. Analysis then either requires that a large number of subsamples be taken from the whole sample and analysed individually, or that the whole sample is mixed in some way to ensure homogeneity, and then subsampled. The latter approach is generally considered preferable because fewer subsamples need to be analysed, thereby reducing analysis time and cost. There is also greater certainty that the final result is representative of the entire sample. The downside of this approach is that if only one sample is available and all of it is homogenised for subsampling, the original form of the sample is destroyed in the process. If this form is important in terms of the overall analysis, photographs and any accompanying description and other measurements should be made before homogenisation. An example could be a single tablet suspected to be a drug of abuse and where a qualitative and quantitative analysis is to be carried out to identify the presence of the drug and, if present, to quantify it.

The approach of homogenising the whole sample is not appropriate for all samples and situations. An example would be for seizures of tablets suspected to contain a controlled drug. For large scale drug seizures, it would not be unusual for a seizure to consist of many thousands of tablets or even hundreds of thousands. If all tablets are homogenised into one single, powdered sample and one or more subsamples are taken for analysis, the end result will indicate the mean concentration of drug in all tablets. However, the physical form of the tablets is destroyed in the process, which is not desirable from an evidential viewpoint. Furthermore, powdering a large number of tablets is time consuming and may require special equipment, and mixing a large volume of the resultant powder can be difficult and may not result in a fully homogeneous sample. There is also the possibility that a seizure consists of two or more sample populations, each containing a different concentration of a controlled drug, or that each population contains a different drug or one population may contain no drug at all. Homogenising all tablets in a seizure and analysing a subsample of the resultant powder would not reveal these situations. The more usual approach for the analysis of large scale drug seizures is to take a representative portion of tablets for analysis. An examination is first carried out to check whether the tablets appear to be from a single population (examination of packaging, tablet size, colour, logos, mass, size, presumptive testing to identify drug present). A portion of tablets from a sample population is then homogenised and one or more subsamples taken for analysis. In the case of drugs, procedures are specified for representative sampling from large seizures (e.g. United Nations Office on Drugs and Crime, UNODC 2009; Scientific Working Group for the Analysis of Seized Drugs, SWG-DRUG 2016) and sampling of illicit drugs for quantitative analysis (European Network of Forensic Science Institutes, ENFSI 2014).

In the case of liquid samples, shaking the whole sample and then subsampling may be sufficient. This approach is generally considered acceptable for analysis of urine in drug doping testing or workplace drug testing. Blood sampled at post-mortem examination is often clotted, and some form of mechanical blending may be required to obtain a homogeneous sample.

When developing a new method of analysis, evaluating sample preparation must form part of method validation. This may include evaluating the reproducibility of subsampling and the reproducibility of any subsequent extraction procedures and how this influences the precision and accuracy of the overall method. Similarly, when analysing a non-standard sample using an accredited method, some experimentation will be required to check if the sample preparation stages of the standard operating procedure (SOP) give the required precision and accuracy or whether modifications need to be made. Any modification in sample preparation should be validated and documented.

A decision tree scheme for sample preparation is shown in Figure 4.1. The analyst starts with a consideration of the sample to be analysed and the case information or other relevant information. This usually gives an indication as to the compound(s) that are of interest. The analyst then considers whether the compounds of interest are likely to be present in the sample at a high (%) level or at trace levels ($\leq$ppm). The chemical nature of the analytes, their likely level in the sample, the nature of the sample matrix, and the information required guide the analyst towards the most appropriate analytical techniques. The information considered will also indicate to the analyst whether the compounds of interest need to be separated from the rest of the sample matrix in order to provide the required information.

As an example, if the sample is a white powder, and the case information indicates that the information required is whether or not the sample contains cocaine with a view to a charge of trafficking illegal drugs, the analyst concludes that they need to ascertain whether cocaine is present in the sample or not. If the sample is cocaine, and it is being trafficked, cocaine will be present at a high level (%). Analysis by Fourier transform infrared (FTIR) spectroscopy with an attenuated total reflectance (ATR) accessory would be a suitable method of analysis if quantification of the cocaine was not required. This method requires minimal sample preparation and is quick and simple to carry out. If the percentage of cocaine in the sample is required, FTIR is not the most appropriate method and a chromatographic method such as GC or HPLC would be more appropriate. However, GC or HPLC analysis will require separation of the cocaine from the sample matrix because the likelihood is that the powder contains compounds other than cocaine. Solvent extraction is likely to be the most appropriate sample preparation technique.

The scheme set out in Figure 4.1 is not exhaustive in terms of all the sample preparation techniques which might be applicable for all the samples, compounds, and cases which might be encountered in forensic science but serves to illustrate the thinking processes that an analyst adopts when faced with a sample to be analysed. In many areas of forensic science, the analyst will already be familiar with the majority of sample types encountered, the information required, the most appropriate methods of analysis and sample preparation procedures, and will not have to work through a decision tree process. However, forensic science is often not straightforward and the analyst should always consider what approach is most appropriate given the sample, the case, and the information required.

There are many types of sample preparation available to the analytical chemist. It is not possible within this book to cover all the different types and, in any case, not all are used in forensic science; some being used mainly in industrial settings. Hence, we have concentrated on the types of sample preparation most commonly used in forensic science.

Analysis of organic and inorganic analytes often require different approaches in terms of sample preparation. In general, for organic analytes, the analyte is extracted from the sample matrix, leaving the matrix behind, a process often referred to as extraction. For inorganic elements in some types of sample, the sample matrix is digested leaving the elements behind.

Not all samples encountered in forensic science require sample preparation in the sense discussed here, i.e. separating the analyte of interest from the rest of the sample matrix.

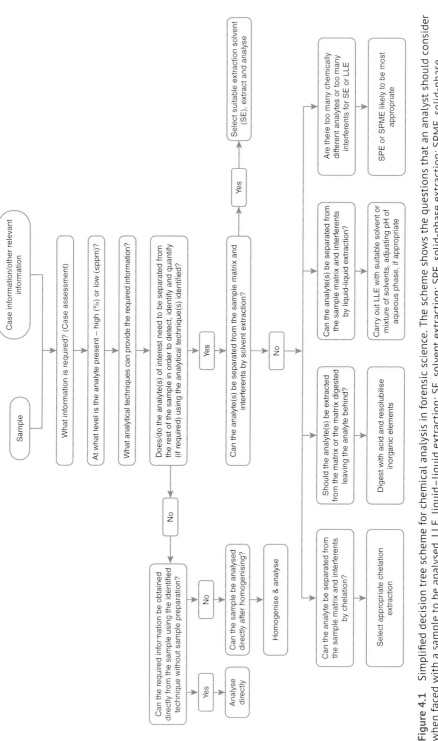

Figure 4.1 Simplified decision tree scheme for chemical analysis in forensic science. The scheme shows the questions that an analyst should consider when faced with a sample to be analysed. LLE, liquid–liquid extraction; SE, solvent extraction; SPE, solid-phase extraction; SPME, solid-phase microextraction.

For example, techniques such as Raman spectroscopy can be used to identify the active component in licit or illicit drugs and excipients or cutting agents. A tablet or sample of powder can be analysed by Raman spectroscopy directly without the need to homogenise or otherwise prepare the sample. Similarly, elemental analysis using scanning electron microscopy–energy dispersive X-ray (SEM-EDX) may be carried out with no or minimal sample preparation for some samples. The classic example is the analysis of firearm discharge residues (FDRs) where a sample of residue can be transferred to an SEM stub and, depending on the type of SEM instrument used, may be analysed with no further sample preparation. Several types of mass spectrometers have been developed which enable direct analysis of samples with no prior sample preparation. Examples include direct electrospray ionisation mass spectrometry (DESI-MS) (Takats et al. 2004) and direct analysis in real time mass spectrometry (DART-MS) (Cody et al. 2005) (see Sections 10.2.3.8 and 10.2.3.9, respectively).

4.2 Extraction

The most commonly employed extraction methods in forensic analytical chemistry are:

- Solvent extraction
- Liquid–liquid extraction (LLE)
- Solid-phase extraction (SPE)
- Solid-phase microextraction (SPME)
- Headspace analysis

These techniques, with the exception of headspace analysis, are discussed in the following sections.

Headspace analysis is discussed in Chapter 13 because it is used exclusively in association with GC. Although SPME is also used for the analysis of volatile substances, it is included here because it can also be used for the extraction of less volatile analytes from liquid samples, including being used in association with HPLC analysis.

Molecularly imprinted polymers (MIPs) are a more recent development and, although not originally designed specifically for sample extraction, can be used for this purpose. An extraction technique called QuEChERS (quick, easy, cheap, effective, rugged, safe) has been widely adopted for sample extraction in the field of pesticides analysis. Like MIPs, QuEChERS has yet to be widely adopted in mainstream forensic science for sample extraction. Both MIPs and QuEChERS are discussed in this section because we anticipate that these techniques will become more widely used for sample preparation in forensic science. Sample preparation methods used in the analysis of inorganic elements or to determine elemental speciation are discussed in Section 4.3.

It is rare that any extraction procedure separates only the analyte of interest from other substances present in the sample. There is usually coextraction of other compounds and incomplete extraction of the analyte. The skill is to engineer the extraction to maximise analyte extraction whilst minimising extraction of other substances, or at least to minimise extraction of substances which interfere with subsequent instrumental analysis or which may damage instrumentation.

An analyst should always be aware of the potential for one or more unexpected compounds to be extracted from a sample, even when using extraction procedures which have been validated. This is because many samples analysed in forensic science come from an uncontrolled and often unknown source. For example, heroin is made from biologically derived starting materials which are subjected to extraction and chemical reaction under poorly controlled conditions. For distribution to users, the drug is typically cut with a variety of substances. Hence, there is the potential for unexpected compounds to be extracted when analysing heroin. Contrast this with the analysis of pharmaceutical drugs where the starting compounds are known, and manufacturing procedures are strictly controlled. It would be far more unusual for an unexpected compound to be present when pharmaceutical drugs are analysed.

This potential for unexpected compounds to be extracted does not mean that new extraction procedures have to be developed for each new sample. It means that an analyst needs to be on the lookout for unexpected signals appearing in the instrument output, indicating that the extraction procedures may have extracted something outside the norm and additional to the analytes of interest. The analyst then needs to consider whether or not the presence of these unexpected substances will have interfered with the identification and quantification of the analytes. Identifying unexpected compounds in a sample may offer an advantage because it may offer evidential value over and above analysis of the target analyte. An example might be in sports doping testing where the presence of an unexpected substance may reveal the presence of a new substance being abused to enhance performance.

4.2.1 Solvent Extraction

Solvent extraction is one of the simplest of the extraction methods in terms of sample handling and cost. Typically, a solvent is selected in which the analyte solves but the bulk of the sample does not. Ideally, the solvent used for extraction is compatible with the instrument to be used for analysis so that the solvent extract can be analysed directly. If this is not the case, the solvent may need to be evaporated and the analyte redissolved in a more suitable solvent at some point during the sample preparation process. The sample, or subsample, is shaken, or otherwise mixed with the selected solvent, to maximise the sample:solvent interaction and to speed up the extraction process. Extraction is likely to be more efficient if solid samples are finely powdered and if semi-solid samples (e.g. foodstuffs) are homogenised to a liquid consistency. Centrifugation or filtering may be required after extraction to separate the solvent containing the analyte (the solvent extract), from non-dissolved material. An example of such an approach is applied in the analysis of drugs of abuse, particularly for drugs such as heroin, ecstasy (3,4-methylene-dioxymethamphetamine, MDMA), cocaine and other drugs which are supplied as powder or tablets. After sample homogenisation, a subsample is shaken with solvent. Non-dissolved material is then separated by centrifugation or filtration. A portion of the solvent extract is then analysed by, e.g. GC-MS, high performance liquid chromatography–ultraviolet (HPLC-UV), and liquid chromatography–mass spectrometry (LC-MS).

Although solvent extraction has distinct advantages in terms of simplicity and equipment required, and can be readily applied to analysis of large numbers of samples, its use is restricted to certain sample types only; analysis of drugs being one of the main areas of application. For many sample types, other compounds in the sample also solve in the selected solvent and interfere with the instrumental analysis. Problems may also be experienced when several, chemically dissimilar analytes are to be analysed. It can be difficult to find a single solvent in which all analytes are readily soluble. In these situations, other extraction procedures must be used which provide more selective analyte extraction. Such procedures include LLE, SPE, or SPME. However, given that it has been estimated that sample preparation accounts for 80% of analysis time (Pawliszyn and Lord 2010), solvent extraction, if applicable, is often a preferred approach because of its simplicity.

4.2.2 Liquid–Liquid Extraction

LLE, more correctly referred to as liquid–liquid distribution (LLD), is defined as 'the process of transferring a dissolved substance from one liquid phase to another (immiscible or partially miscible) liquid phase in contact with it' (McNaught and Wilkinson 1997).

In the simplest form of LLE, where the sample is a liquid, a solvent is chosen for which the analyte has greater affinity and is not miscible with the sample, and which does not chemically react with or otherwise change the nature of the analyte. The two non-miscible liquids are shaken together to maximise physical interaction and the analyte distributes between the two. Normally, in LLE, the sample phase is aqueous and is extracted using an organic solvent. At equilibrium, the concentration ratio of the distribution between the two phases (the partition ratio, K_D, also referred to as the distribution constant) is expressed as:

$$K_D = \frac{[A]_{org}}{[A]_{aq}} \quad (4.1)$$

where $[A]_{org}$ and $[A]_{aq}$ are the concentrations of the analyte in the organic and aqueous phase, respectively. K_D is defined as the 'ratio of the concentration of a substance in a single definite form, A, in the extract to its concentration in the same form in the other phase at equilibrium' (McNaught and Wilkinson 1997). The aspect of 'a single definite form' can be important for analytes which ionise in water. For such analytes, the pH in the aqueous phase is normally adjusted to minimise ionisation. By convention, when calculating K_D using Eq. (4.1), the organic phase is generally the nominator because in LLE, the analyte is normally transferred from an aqueous to an organic phase. Example applications of LLE include workplace drug testing with urine as a test matrix, analysing for potential spiking of drinks in drug-facilitated sexual assault cases, and environmental monitoring of water bodies to check for pesticide contamination.

It is possible to use LLE to extract an analyte from one organic solvent to another immiscible solvent, or to extract an analyte from an organic solvent into water, but this is less common. LLE can also be used for extracting analytes from solid or semi-solid samples but such samples are usually converted to a liquid form for extraction. For example, a subsample of food suspected to be poisoned, or a sample of liver sampled at post-mortem examination for toxicological examination, could be homogenised with distilled water to form a slurry and then a subsample subjected to LLE.

The extracting solvent is selected to maximise K_D, i.e. to maximise partition of analytes from the sample. An example of this would be in the analysis of water suspected to be

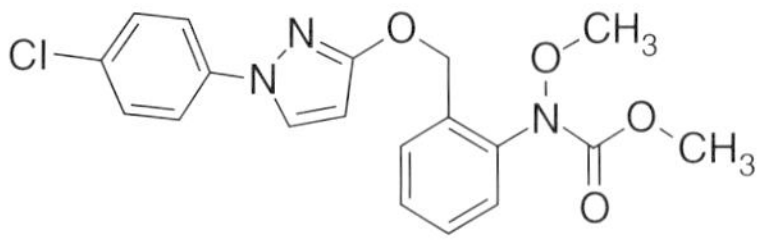

Figure 4.2 Pyraclostrobin.

contaminated with the fungicide pyraclostrobin (Figure 4.2) at a concentration below that at which the fungicide can be directly detected in water. Under these circumstances, the pyraclostrobin needs to be extracted from water and concentrated for analysis. One of the simplest ways to achieve this is by LLE. The solubility of pyraclostrobin in water is 1.9 mg/l at pH 5.8 (FAO 2005). Pyraclostrobin is readily soluble (>500 g/l) in organic solvents including acetone, acetonitrile, dichloromethane, ethyl acetate, and toluene. Of these solvents, dichloromethane, ethyl acetate, and toluene are non-miscible with water. Hence an LLE could be carried out by shaking water suspected to be contaminated with pyraclostrobin with one of these solvents.

Substituting the solubility data into Eq. (4.1), the K_D for pyraclostrobin extracted from water (at pH 5.8) into ethyl acetate is $>2.6 \times 10^5$. This large value for K_D indicates that virtually all the fungicide will partition into ethyl acetate. The fraction extracted (%E) can be calculated as follows:

$$\%E = \frac{100 K_D}{K_D + (V_{aq}/V_{org})} \% \qquad (4.2)$$

where V_{org} and V_{aq} are the volumes of organic solvent and aqueous sample, respectively. For the pyraclostrobin example, if $V_{aq} = V_{org} = 1$ ml, the %E is:

$$\frac{100(2.6 \times 10^5)}{2.6 \times 10^5 + 1} = 99.9996\%$$

i.e. there should be virtually 100% extraction.

Calculations show that for K_D values >100 to >10 000, the predicted %E is >99%. For K_D values >10 to 100, %E varies from 90 to 99%. For K_D values between 1 and 9, %E ranges from 50 to 90%. Thus, for an efficient extraction K_D should be ≥ 100. If solubility data is available for an analyte of interest in water and organic solvents, this forms a good starting point for selecting the most suitable solvent for extraction or for seeing if LLE is an appropriate extraction technique. The solubility of paraquat, a quaternary ammonium herbicide, in water at pH 5.2–9.2 is 620 g/l, in methanol is 143 g/l, and in hexane, dichloromethane, and toluene is <0.1 g/l. Methanol is miscible with water so is not a suitable solvent. K_D for hexane, dichloromethane, and toluene $= 1.6 \times 10^{-4}$, assuming $V_{org}/V_{aq} = 1$; resulting in <0.01% extraction. Hence LLE is not a good sample preparation technique for extracting paraquat from water.

With LLE it is often recommended to repeat the extraction to improve the total amount of analyte extracted. For this, the extracted phase remaining from the first extraction is extracted again with a new portion of solvent and the two solvent extracts combined. The efficiency of repeat extractions depends on K_D and some examples may help to illustrate whether repeat extraction is warranted.

Assume that a Compound X partitions between water and ethyl acetate to give a K_D of 10, and that Compound X is present in an aqueous sample at 10 µg/ml. If 1 ml of sample and 1 ml of ethyl acetate is shaken together, according to Eq. (4.2), 90.09% of Compound X should be extracted into the ethyl acetate, i.e. 9.09 µg, leaving 0.91 µg in the aqueous sample. If this is re-extracted with a further 1 ml of ethyl acetate, a further 0.827 µg will be extracted, i.e. total of 9.91 µg (99.1%) in the combined extracts. Although the total percentage of Compound X extracted is now close to 100%, repeating the extraction takes additional time and adds cost to the overall analysis. Provided the precision of extraction is good (i.e. close to 90.09% of Compound X is extracted from each sample) and the 90.09% extracted results in a concentration that can be measured by the instrumentation used for analysis, a single extraction is probably acceptable, unless laboratory protocols demand a higher percentage extraction. For values of $K_D > 100$, the predicted extraction efficiency is sufficiently high with a single extraction (>99%) such that repeat extraction of the sample produces minimal benefit. For lower values of K_D, replicate extractions may offer significant advantages in terms of the total percentage of analyte extracted. However, if more than two extractions are required to extract sufficient analyte for detection by the instrumental analysis, a more efficient sample preparation method should be used.

Intuitively, there would appear to be an advantage in taking a large volume of sample for analysis relative to the volume of the extraction solvent. However, this assumption is not necessarily valid and depends strongly on K_D. In the case of Compound X above, if 50 ml of a sample containing 10 µg/ml of Compound X was extracted (total mass available for extraction = 500 µg) with 1 ml of ethyl acetate, substituting $K_D = 10$ in Eq. (4.2) and $V_{aq}/V_{org} = 50$, only 16.66% of Compound X would be extracted. This is because there is now a much larger volume of the aqueous phase present relative to the organic phase and hence a greater percentage of Compound X remains in the aqueous phase at equilibrium. Although a total of 83.3 µg would be transferred to the ethyl acetate, and, hence, the concentration of Compound X would be considerably higher than if only 1 ml of sample was extracted, a 16.66% extraction efficiency is poor. With a relatively large volume of sample relative to the volume of extracting solvent, the precision of extraction may be more variable compared with a situation where V_{aq}/V_{org} is close to 1. High extraction efficiency may be important where a quantitative analysis is required but may not be so important for a qualitative analysis where it may be sufficient to confirm the presence of Compound X in a sample without the need to quantify how much is present. Where K_D is large, as in the case of pyraclostrobin for water:ethyl acetate, a large ratio for V_{aq}/V_{org} does not have such a marked effect on %E. For a 100 ml water sample containing pyraclostrobin at 1 µg/ml and extracted with 1 ml of ethyl acetate, 99.96% of the pyraclostrobin will be extracted. This compares with 99.9996% extraction when $V_{aq}/V_{org} = 1$. However, the total mass extracted is increased from ≈1 µg when 1 ml is extracted to ≈100 µg when 100 ml is extracted. This increase in the total mass of analyte extracted may be important in terms of sensitivity of the required analysis.

An approach called liquid-phase microextraction (LPME) was developed to facilitate extraction of organic analytes from aqueous matrices whilst minimising the volume of solvent involved and the need for sample concentration by solvent evaporation. Sarafraz-Yazdi and Amiri (2010) identified three subcategories of LPME: single drop microextraction (SDME); dispersive liquid–liquid microextraction (DLLME); and hollow-fibre liquid-phase microextraction (HF-LPME). In all these approaches, the critical aspect is the use of very low volumes of partitioning solvent – typically of the order of 1–3 µl, relative to the sample volume which would generally be of the order of one to several millilitres. In one form of SDME, a microlitre (1 µl) drop of organic solvent is suspended on the end of a syringe needle into an aqueous sample which is stirred to promote contact between the drop of solvent and the sample, thereby helping to increase the speed of partition of analytes into the solvent. After a suitable time period, the solvent drop is withdrawn into the syringe and the solvent can then be injected directly into a gas chromatograph. Headspace single drop microextraction (HS-SDME) had also been developed where a single drop of solvent is suspended in the headspace of a sample and used to extract volatile compounds.

HF-LPME is more complex than SDME and DDLME but offers more possibilities for extraction. HF-LPME employs a porous hollow fibre which is dipped into an organic solvent. The solvent enters the pores of the fibres where it acts as the extraction solvent when in contact with an aqueous sample. The fibre itself is filled with an acceptor solvent. If the same solvent is used as the acceptor and extracting solvent, the system is a two-phase LLE system, i.e. aqueous/organic. If the acceptor solvent is aqueous, typically pH adjusted, a three-phase extraction system is created (aqueous/organic/aqueous). The advantage of this system is the concentration factor which can be achieved between the aqueous sample and the aqueous acceptor, and the HPLC compatibility of an aqueous acceptor.

It is debatable whether LPME has achieved widespread use in many mainstream forensic laboratories but these approaches are used in toxicology laboratories, e.g. extraction of drugs and metabolites from urine samples, and in laboratories involved in environmental analyses, e.g. analysis of pesticides and heavy metals in water. The reader is referred to Sarafraz-Yazdi and Amiri (2010) for a review of LPME and to Bello-López et al. (2012) for a review of HF-LPME.

Although most of the previous discussion of LLE has indicated the use of a single solvent for extraction, a mixture of two or more solvents can be used for extraction provided they are miscible with one another. This may be advantageous for a particular analyte or for coextraction of several different analytes (multi-residue extraction), i.e. where K_D for a single solvent is not optimal for efficient extraction of all analytes of interest.

Many drugs and their metabolites, pesticides, explosives, and other polar compounds that may be encountered, ionise in aqueous matrices. Such compounds are, by definition, very water soluble and do not, in their ionised state, partition readily into non-polar organic solvents. It is the non-polar organic solvents that are, generally, non-miscible with water and hence suitable for LLE. Therefore, LLE usually requires the analyte to be in its non-ionised form to maximise partition into the organic phase.

The dissociation of a weak acid, HA, is represented by:

$$HA \rightleftharpoons A^- + H^+ \tag{4.3}$$

where A^- is the acid's conjugate base.
For a base, B:

$$HA + B \rightleftharpoons A^- + BH^+ \tag{4.4}$$

The acid ionisation constant, K_a is given by:

$$K_a = \frac{[A^-][H^+]}{[HA]} \tag{4.5}$$

The pH of a solution is defined as −log of $[H^+]$, where $[H^+]$ is the concentration of hydrogen ions in solution. $pK_a = -\log_{10} K_a$. Substituting these terms into the above equation gives:

$$\mathrm{pH} = \mathrm{p}K_a + \log_{10}\left(\frac{[\mathrm{A}^-]}{[\mathrm{HA}]}\right) \qquad (4.6)$$

This is referred to as the Henderson–Hasselbalch equation.
For a base, the equivalent equation is:

$$\mathrm{pH} = \mathrm{p}K_a + \log_{10}\left(\frac{[\mathrm{B}]}{[\mathrm{BH}^+]}\right) \qquad (4.7)$$

Hence the pK_a of an acid is the pH at which it is exactly half dissociated.

At pH > pK_a, an acid exists as A⁻ in water, i.e. it is ionised and is generally soluble in water.

At pH < pK_a, an acid exists mainly as HA, i.e. its non-ionised state and is less soluble in water.

The opposite is the case for a base.

The general rule for extracting an acid from an aqueous solution into an organic solvent is that the pH should be two units lower than the pK_a of the acid for efficient extraction. For a base, the pH should be two units higher than the pK_a.

The Henderson–Hasselbalch equation can be rearranged to find the percentage ionisation at a given pH:

$$\text{For an acid } \%\ \text{ionisation} = \frac{10^{\mathrm{pH}-\mathrm{p}Ka}}{1 + 10^{\mathrm{pH}-\mathrm{p}Ka}} \times 100 \qquad (4.8)$$

$$\text{For a base } \%\ \text{ionisation} = \frac{10^{\mathrm{p}Ka-\mathrm{pH}}}{1 + 10^{\mathrm{p}Ka-\mathrm{pH}}} \times 100 \qquad (4.9)$$

MDMA is a base with a pK_a of 9.9. At pH 9.9, 10.9 and 11.9, the percentage ionisation (calculated using Eq. (4.9)) is 50, 9.09 and 0.99%, respectively, i.e. as the pH is increased above the pK_a, a greater percentage of the MDMA is present in the non-ionised (neutral) form, as indicated by the 2 pH unit guideline noted above. Hence, when using LLE to extract MDMA from an aqueous phase such as urine, the pH of the sample matrix would be adjusted to pH ≥11.9 to favour partition into the organic phase.

For multi-residue analysis, where two or more analytes are to be extracted with differing pK_a, LLE may be carried out by adjusting the pH of the aqueous phase between extractions. The aqueous phase is made basic to promote distribution of bases into the organic phase. Under basic conditions, acidic compounds will be ionised, more water soluble and remain in the aqueous phase. After extraction of basic compounds, the pH of the aqueous phase is lowered to reduce ionisation of the acid compounds, which should then partition more readily into the organic phase. LLE becomes more complex for analytes with several ionisable functional groups.

It is not always a simple matter to select a solvent that fulfils the criteria of extracting the analyte of interest whilst not coextracting potential interferents. If no suitable solvent can be identified, an alternative extraction procedure must be used. If LLE is not suitable, most analysts turn to (SPE).

4.2.3 Solid Phase Extraction

The following definition of SPE is adapted from a definition by Moors et al. (1994):

> A solid phase extraction (SPE) consists of bringing a liquid test portion/solution containing the analyte of interest in contact with a solid or liquid (stationary) phase, whereby the analyte is selectively adsorbed on the surface of the solid phase or distributed into the liquid phase. Other solvents are added in sequence. The first such solvent is usually a wash to remove possible adsorbed matrix components. Eventually, an eluting solvent is brought into contact with the stationary phase to selectively desorb the analyte.

SPE is derived from column chromatography, where the stationary phase is packed into a column and the sample, in liquid form, is passed through the stationary phase, followed by an eluent or a series of eluents. If the stationary phase is chosen correctly, analytes are absorbed onto, or partition into, the stationary phase whilst the other components of the sample matrix are washed through the column. An eluent of higher elutropic strength, or with other properties to favour partition from the stationary phase into the eluent, is then passed through the stationary phase to elute the analytes. The analytes are retained for instrumental analysis. A less common approach is to retain interferents on the stationary phase, allowing the analyte(s) to pass through the phase and to be collected. This approach is less common because it is more difficult to select a stationary phase and eluents which retain all interferents but not the analyte.

Extraction in SPE consists of several stages, as illustrated in Figure 4.3. Four stages are generally recognised:

i. Conditioning
ii. Sample loading
iii. Washing
iv. Elution

i. *Conditioning*
The stationary phase in a SPE cartridge is shipped dry and needs to be conditioned ('wetted') to make the phase ready to interact with the mobile phase and solutes in the sample. Manufacturers provide instructions on the solvent(s) to be used for conditioning and the volume to be passed through the cartridge and the flow rate to be used.

ii. *Sample loading*
The next step after conditioning is loading the sample. The aim of the sample loading step is for the analyte to partition into, or otherwise interact with, the stationary phase, with no loss of analyte from the column. For SPE, the sample is always loaded as a liquid, free of any particulate material which may block the porous nature of the stationary phase material. The manufacturer will typically provide instructions as to how much sample volume can be loaded, guidance on the capacity of the cartridge, and the flow rate to be used for loading.
The aspect of capacity is very important. There will be a limit on the sample volume that can be accommodated by the cartridge and on the mass of solutes that can be accommodated by the stationary phase. The most common limiting factor is the presence of other

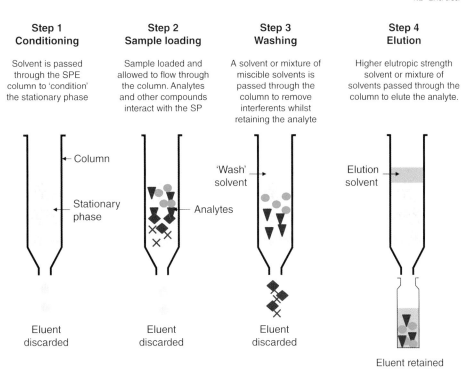

Figure 4.3 Stages of SPE. ▼ and ● represent analytes. ◆ and ✗ represent interferents or other compounds present in the sample that are not of interest in terms of the analysis.

compounds present in the sample which also interact with the stationary phase. If these solutes are present in high concentration and readily interact with the stationary phase, they 'saturate' or 'overload' the column such that the analyte is carried through in the loading solvent, or in the wash solvent. This may result in sample loss and, potentially, a false negative result, i.e. non-detection of the analyte even though it was present in the original sample. Analyte loss will also lead to inaccuracies in quantification. Hence the analyst must consider the capacity of the SPE cartridge relative to the sample being analysed.

The sample may require some form of treatment prior to loading to ensure compatibility with the SPE process. Examples include pre-precipitation and removal of proteins which may otherwise precipitate out on the SPE column and block the flow; filtering or centrifugation to remove particulate material; dilution with a solvent compatible with the conditioning solvent; and adjustment of pH. If quantification of the analyte is required and an internal standard (see Sections 2.3.1.2 and 13.3.3.1) is to be used to aid quantification, it should be added to the sample prior to the sample loading stage.

iii. *Washing*

The sample loading step is followed by a 'washing step' where the aim is to retain the analytes on the column whilst removing unwanted compounds. This is a critical step and experimentation is usually required in method development to find a single solvent or combination of solvents such that analytes are retained but interferents are not. Washing with a sequential series of solvents may be required to achieve this aim.

iv. *Elution*

In the final step, the analyte is eluted from the stationary phase. This requires the use of a solvent, or mixture of solvents, with a higher elutropic strength than was used in the washing stage, i.e. a solvent in which the analyte distributes more readily than it does into the stationary phase. It may be possible to select an eluent that removes the analytes but does not elute potentially interfering compounds that were not removed during the washing step. The eluent from the column is retained for instrumental analysis. Depending on the nature of the solvent used for elution, and the volume, it may be necessary to evaporate the solvent under a stream of inert gas (usually nitrogen) and to redissolve the analytes in a smaller volume of solvent, or in a different solvent. Derivatisation (see Section 4.2.6.2) may also be carried out after the solvent evaporation stage.

The volume of solvent employed through the washing and elution stages may also be critical. Many analytes will have some solubility in the solvents used for the washing steps. If too large a volume of solvent is used for washing, this may elute a significant proportion of the analyte, thereby reducing recovery and leading to problems with analyte detection and quantification.

During method development for sample extraction, it is usual to retain the eluents from the sample loading, washing and elution stages and to analyse these eluents instrumentally. This helps identify if the analyte is retained on the column through the sample loading and washing stages. The total mass of analyte 'recovered' through these stages is calculated. If recovery after the elution stage is not close to 100, a solvent of increased elutropic strength should be applied until analysis indicates that close to 100% of the analyte is recovered. If analyte is recovered during the conditioning or washing stages, the reason for non-retention on the stationary phase must be evaluated and the SPE conditions adjusted.

The fine particle size and close packed nature of the stationary phase in SPE columns restricts the ready flow of solvents. Hence positive pressure must be applied to the head of the column to ensure a flow of mobile phase through the column. Alternately, a vacuum can be applied to the end of the column to draw solvent through. Although eluents can be forced through SPE columns manually using a syringe which forms a seal with the top of the column (i.e. application of positive pressure), this procedure is time consuming because only one sample can be extracted at a time. Also, controlling the flow rate of eluent to obtain a constant flow through the column can be difficult with manually applied pressure. It is more usual to carry out SPE using a vacuum manifold system (Figure 4.4). The SPE columns are held in a container to which a vacuum can be applied. This enables more control over the flow rate of eluents through the columns, simultaneous multi-sample extraction, and also frees the operator to do other tasks. SPE systems are available with pumps and associated switching valves and meters so that the whole SPE procedure can be fully automated. Some of these systems can be directly coupled with HPLC or GC instruments, enabling full automation of sample extraction and analysis once initial sample preparation is complete. Such systems are very effective in terms of analyst time.

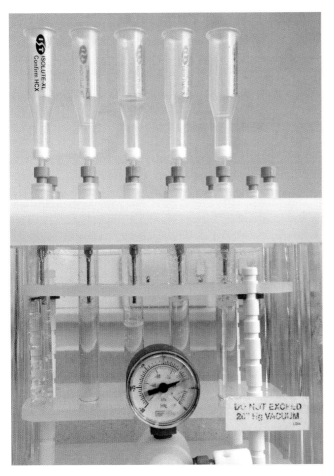

Figure 4.4 Vacuum manifold showing SPE cartridges inserted into the top plate of the manifold. A vacuum is applied to the container such that liquid (sample, wash solvent, or eluent) applied to the top of a cartridge is pulled through the stationary phase. The pressure of the vacuum can be adjusted to control the flow rate. Containers are placed underneath the columns. At the point of conditioning, sample loading and washing, a 'waste' container is placed underneath each column. This is replaced by a sample vial for the elution step when the extracted sample is collected for subsequent analysis, e.g. by GC, GC-MS, HPLC, LC-MS etc. Source: Image courtesy of S. Turfus.

4.2.3.1 Stationary Phases

SPE is a very flexible extraction technique and it is this flexibility that makes it so powerful for separating analytes of varying chemical nature from a wide variety of sample matrices. The flexibility comes from the wide variety of stationary phases that can be used, the wide variety of solvents that can be employed and the variety of physical formats into which the stationary phase can be incorporated. However, this flexibility also makes a discussion of SPE complex because of the many options that can be employed for extraction. For example, stationary phases are available packed into columns, pipette tips and 96-well plates, and formatted as flexible thin circular discs. Numerous stationary phases are available. The earliest types consisted of materials such as silica, alumina, diatomaceous earths, and carbon. These

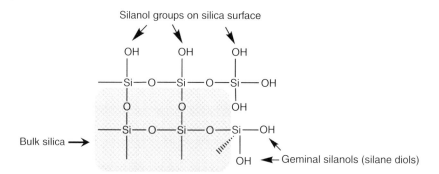

Figure 4.5 Representation of the structure of silica showing silanol groups (−Si−OH) on the surface (the grey area indicates the bulk silica, i.e. the silica network below the surface). Silanol groups on the surface participate in hydrogen bonding interactions and can ionise under certain pH conditions, giving rise to the possibility for ion-exchange interactions. Siloxane groups (Si−O−Si) are also present on the surface of silica but are not shown here. They are far less important in terms of hydrogen bonding interactions and are considered to be 'hydrophobic in nature'. Source: Adapted from Nawrocki (1997).

materials are still used but it is far more common for the surface of these materials, particularly silica, to be modified by chemical reaction to covalently bond functional groups to the surface. The silica is then considered as a 'support' or 'base' material. The structure of silica is shown in Figure 4.5. Covalent bonding of stationary phases takes places through silanol groups on the surface of silica. Table 4.1 shows some of the more common functional groups that are commercially bonded to silica as a support.

Unless the covalent bonding of functional groups is highly efficient, unreacted silanol groups can remain on the support surface. These highly polar silanol groups can also interact with analyte polar functional groups, sometimes giving unwanted or unpredictable interactions between analytes and the stationary phase, often resulting in the analyte or interferent binding more strongly to the stationary phase than anticipated. The chemistry of silica is complex and is affected by the process used to produce the silica; the purity of the silica, in particular the metal content; and, in the case of bonded phases, the type of bonded phase and the reagents used in the production process.

Manufacturers identify several types of silica: Types A, B and C. The earlier types of silica used in chromatography are referred to as Type A. This Type A silica hydrolyses at pH >7 and the Si-O-Si- bond attaching bonded alkyl phases starts to hydrolyse at low pH (pH <2) resulting in loss of the phase. This aspect of hydrolysis at high or low pH is important because sample pH often needs to be manipulated in SPE to induce or supress ionisation, depending on the stationary phase–solute interactions involved in the chromatographic separation process. Type B silica has a lower metal content than Type A and, according to manufacturers, has a wider stability range from about pH 2 to 10. More recently, a Type C silica has been produced where the surface is populated with non-polar Si-H rather than silanol groups. Further developments in silica chemistry include so-called 'hybrid' particles comprised of a 'porous inorganic/organic hybrid monolith material' (Walter et al. 2007) which can be surface modified with bonded phases. This material is claimed to have

Table 4.1 Solid-phase extraction stationary phases, interaction mechanisms with sample components and typical uses.

Phase chemistry	Designation	Support	Main interaction mechanisms[a]	Comments
H₃C—Si—O—Si—C₁₈H₃₇ (with H₃C groups)	C_{18} Octadecyl	Si	Reversed-phase separations van der Waals forces	Strongly hydrophobic/non-polar phase. Used for the extraction of polar to non-polar analytes. Available as monomeric- or polymeric-bonded phases. Monomeric-bonded phases have higher carbon loading than polymeric-bonded C18 phases and hence will retain non-polar analytes more strongly. C_{18} phases may be too strongly interacting for very hydrophobic analytes
H₃C—Si—O—Si—C₈H₁₇ (with H₃C groups)	C_8 Octyl	Si	As for C_{18}	Moderately hydrophobic/non-polar phase. As for C_{18} but with lower carbon loading. If analyte is too strongly retained on a C_{18} phase, try a C_8
	C_4, C_2	Si	Reversed-phase separations Interactions with residual silanols likely to be important, e.g. hydrogen bonding; π interactions	Used when hydrophobic retention mechanism is required but retention is too strong on C_8 or C_{18} phases
H₃C—Si—O—Si—(CH₂)₂—C₆H₅ (with H₃C groups)	Phenyl	Si	Reversed-phase separations π-π interactions	Used for the extraction of non-polar to mid-polar aromatic compounds

Table 4.1 (Continued)

Phase chemistry	Designation	Support	Main interaction mechanisms[a]	Comments
$\sim\sim$Si—O—Si(CH$_3$)$_2$—(CH$_2$)$_3$—NH$_2$	NH$_2$ Aminopropyl	Si	Reversed-phase and normal-phase separations. Hydrogen bonding in normal-phase separations. Ion exchange in reversed-phase separations	Moderately polar. Weakly basic. Extraction of mid to strongly polar analytes. Weak anion exchanger at pH < 8, i.e. when the amine group is in its charged form. For anion exchange, the analyte must also be charged: pH ≥ pK_a + 2
$\sim\sim$Si—O—Si(CH$_3$)$_2$—(CH$_2$)$_3$—CN	CN Cyanopropyl	Si	Can be used in reversed-phase and normal-phase separations. Hydrogen bonding and π-π interactions	Less retentive than silica or diol phases for polar analytes in normal-phase mode. In reversed-phase separations, can be used to extract highly hydrophobic analytes from aqueous solution
$\sim\sim$Si—O—Si(CH$_3$)$_2$—(CH$_2$)$_3$—O—CH$_2$—CH(OH)—CH$_2$—OH	Diol	Si	Can be used in reversed-phase and normal-phase separations. Hydrogen bonding	Strongly polar and strongly hydrophilic. Can be used as an alternative to Si in normal-phase separation or as weakly acting hydrophobic phase in reversed-phase separations
—Si—OH Type A	Si silica	N/A	Normal-phase separations. Hydrogen bonding. Dipole–dipole interaction	Strongly polar and strongly hydrophilic. Used to extract polar analytes from non-polar solvents, e.g. from hydrocarbons. Need to control pH to avoid interactions via ion exchange. Note that Type A and B silicas will undergo stronger hydrogen bonding and will be more prone to ion exchange at higher pH than Type C silica
—Si—H Type C / Al=O, Al—O—Al=O	Alumina	N/A		Highly polar. Can be used as a low capacity ion exchanger

Structure	Sorbent	Base	Mechanism	Description
	Graphitised carbon	C	N/A Reversed-phase separations	Hydrophobic. Used for extracting organic polar to non-polar analytes. Analyte size and shape (rather than interactions with functional groups)
Mg^{2+} ... $O=Si(=O)-O^-$	Magnesium silicate Florisil	N/A	Normal-phase separations	Strongly polar and hydrophilic. For extracting polar compounds from a non-polar matrix
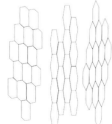 $H_3C-Si(-O\sim\sim)(-CH_3)-(CH_2)_3-N^+(CH_3)_3\ Cl^-$	N^+ Aliphatic quaternary ammonium with Cl^- as counter-ion	Si	Ion exchange	Strong anion exchanger (SAX). $pK_a > 14$ therefore charged at all pHs. Used to extract weak anions, e.g. carboxylic acids. pH of the sample solution needs to be adjusted to ensure analytes are ionised, i.e. $pH \geq pK_a + 2$
$H_3C-Si(-O\sim\sim)(-CH_3)-(CH_2)_2-C_6H_4-SO_3^-\ H^+$	SO_3^- Benzene sulfonic acid	Si	Ion exchange	Strong cation exchanger (SCX). pH of the sample solution needs to be adjusted to ensure analytes are ionised at the load stage, i.e. $pH \leq pK_a - 2$
$H_3C-Si(-O\sim\sim)(-CH_3)-(CH_2)_3-SO_3^-\ Na^+$	SO_3^- Aliphatic sulfonic acid with Na^+ as counter-ion	Si	Ion exchange	Strong cation exchanger (SCX). This is a strong acid phase ($pK_a < 1$) which is charged at all pHs in aqueous solution. Used to extract strong or weak cationic analytes. pH of the sample solution needs to be adjusted to ensure analytes are ionised at the load stage, i.e. $pH \leq pK_a - 2$

Table 4.1 (Continued)

Phase chemistry	Designation	Support	Main interaction mechanisms[a]	Comments
(structure: silica with Si–O–Si(CH₃)₂–(CH₂)₃–COOH)	Carboxylic acid weak cation exchange	Si	Ion exchange	Weak cation exchange
	Mixed mode – ion exchange and C_{18} or C_8	Si	Reversed-phase separations Ion exchange and hydrophobic interactions	Mixture of ion exchange phase (SAX or SCX) and hydrophobic phases (e.g. C_{18} or C_8). pH adjustment required to ensure ionisable analytes are ionised for sample load stage
(structure: polyamide with hydrogen-bonded amide groups)	Polyamide	Polymeric	Reversed phase Hydrogen bonding	Used to extract polar compounds containing multiple —OH or —COOH groups from aqueous solutions
(structure: polystyrene-divinylbenzene polymer backbone with phenyl substituents)	Polystyrene-divinylbenzene (PS-DVB)	Polymeric	Reversed phase Hydrophobic interactions Possible π–π interactions	Hydrophobic. Used for extracting hydrophobic compounds with hydrophilic functional groups.

Structure	Phase	Type	Interactions	Description
benzoic acid (–COOH)	PS-DVB mixed mode weak acidic cation exchange	Polymeric	Ion exchange and hydrophobic interactions. Possible π-π interactions	Mixture of weak cation exchange and hydrophobic phase. Used to extract strong bases. pH usually adjusted during wash and elution stages to ionise and retain analyte during wash step and then neutralise and elute during elution stage
benzenesulfonic acid (–SO$_3^-$ H$^+$)	PS-DVB mixed mode strong acidic cation exchange	Polymeric	Ion exchange and hydrophobic interactions. Possible π-π interactions	Mixture of strong cation exchange and hydrophobic phase. Used to extract weak bases. Strong bases may bind too strongly to be eluted.
benzylpiperazine (NH)	PS-DVB mixed mode weak basic anion exchange	Polymeric	Ion exchange and hydrophobic interactions. Possible π-π interactions	Mixture of weak anion exchange and hydrophobic phase. Used to extract strong acids. pH usually adjusted during wash and elution stages to ionise and retain analyte during wash step and then neutralise and elute during elution stage
benzyltrimethylammonium (N$^+$ OH$^-$)	PS-DVB mixed mode strong basic anion exchange Quarternary ammonium	Polymeric	Ion exchange and hydrophobic interactions. Possible π-π interactions	Mixture of anion exchange and hydrophobic phase. Used to extract acidic and neutral compounds. Strong acids may bind too strongly to be eluted. Operable at pH 1–14.

a) The strongest intermolecular interactions likely to operate are highlighted for each stationary phase. van der Waals forces operate in most separations but, because they are relatively weak forces, they are only mentioned if they are the main intermolecular forces operating in a separation. van der Waals forces include dipole–dipole and dipole-induced dipole interactions and London dispersion forces.

Note that the phase chemistries show the most common functional groups that are available for interaction. Different manufacturers and suppliers of phases have different ways of bonding phases to or within support materials. Where silica is used as a support material, there may be differing degrees of end-capping of the siloxyl groups on the silica support. Most suppliers of SPE phases provide information on solvents to be used for conditioning SPE cartridges and application notes for a wide variety of compounds giving detailed conditions for SPE.

superior stability over a wider pH range than other silicas and reduced problems with silanol groups on the phase surface.

To overcome some of the difficulties with hydrolysis at high pH, and with unwanted interactions with residual silanols, polymeric stationary phases have been developed. The most common polymeric phase is based on styrene-divinylbenzene and can be used between pH 1 and 12. Without modification, polystyrene-divinylbenzene (PSDVB) phases are considered as highly hydrophobic phases. As with silica, polymeric phases can be modified. Typically, more ligands are introduced to offer a greater range of stationary phase–analyte interactions.

Different manufacturers have different ways of modifying phases and use silica of different particle sizes and produced by different processes. All this gives rise to an often bewildering range of stationary phases available for SPE. However, it is this wide range of phases that enables such a wide range of analytes to be extracted from complex sample matrices.

For the purposes of this textbook, discussion concentrates on stationary phases held in columns because this form of SPE is still the most widely used in forensic science. The columns used are referred to as 'cartridges', typically made from poly(propylene). Virtually all forensic science laboratories will purchase ready-made SPE cartridges rather than pack their own columns. Such columns are manufactured to high tolerances, in sterile conditions, guaranteeing reproducible extraction with minimal contamination. Columns can not only be purchased containing different stationary phases but also with a different mass of stationary phase.

We will also concentrate on the stationary phases most commonly used in 'routine' forensic science. Whilst forensic science is not truly routine, most laboratories working to provide evidence in support of judicial purposes establish and validate a suite of analytical methods which are then used for the analysis of the majority of samples encountered. Developing and validating a new method is time consuming and costly and laboratories only change methods if there is a compelling reason. Thus, whilst there are many new developments by manufacturers in terms of stationary phases, most forensic laboratories are likely to be using a limited range of stationary phases for the bulk of their analyses.

Three types of separations are generally recognised in liquid chromatographic techniques such as SPE and HPLC: normal phase (NP), reversed phase (RP) and ion exchange (also referred to as ion chromatography). In NP separations, the stationary phase surface is polar relative to the mobile phase (the eluent). In RP separations, the stationary phase is non-polar relative to the mobile phase. Ion exchange, as the name implies, involves interactions between ionised functional groups on both the stationary phase and the analytes. More recently, a separation mode referred to as hydrophilic interaction liquid chromatography (HILIC) has been developed. This employs polar stationary phases together with the types of mobile phases used in RP-LC, which are generally considered to be polar in nature.

In addition, more analyte specific, polymeric stationary phases have been developed based on MIPs and phases based on immunoaffinity interactions.

4.2.3.2 Normal Phase

Examples of stationary phases used in NP separations are highly polar phases such as silica, including diatomaceous earths such as Kieselguhr; florisil (magnesium silicate);

and alumina (aluminium oxide). Bonded phases include cyano, amino and diol phases (Table 4.1), i.e. strongly polar to moderately polar phases. NP SPE is used for the extraction of strongly polar to moderately polar analytes. These analytes interact with the stationary phase predominantly through hydrogen bonding, π–π interactions, dipole–dipole and dipole-induced dipole interactions. Elution solvents include non-polar and chlorinated solvents (although many laboratories prohibit use of chlorinated solvents for environmental reasons). Strongly polar solvents such as methanol may also be used to overcome very strong bonding interactions. There are limitations in terms of the range of analytes that can be extracted by NP SPE, particularly for non-polar or slightly polar analytes, because these analytes tend to partition more readily into non-polar eluents and hence may not interact sufficiently with highly polar stationary phases to be retained. Very polar analytes may bind so strongly to highly polar stationary phases that it can be difficult to select a solvent of sufficiently high elutropic strength to overcome the analyte–stationary phase interactions. For these reasons, most SPE is carried out by RP chromatography.

4.2.3.3 Reversed Phase

Stationary phases used in RP separations include non-polar bonded hydrocarbon phases such as octadecyl- and octyl- functionalised phases through to bonded phases with some polarity such as phenyl phases (Table 4.1). Elution solvents typically include a mixture of water and a water-miscible solvent such as acetonitrile or methanol. Tetrahydrofuran or isopropanol can be used but is less commonly employed. Although what are generally thought of as highly polar solvents such as acetonitrile and methanol are used in RP separations, these solvents act as 'organic modifiers' because, effectively, they provide an organic property to water, the starting point for the mobile phase. Addition of a higher percentage of organic modifier is considered to make the mobile phase 'less polar' and to promote partition of organic solutes from the stationary phase to the mobile phase.

Silica is still the most commonly employed material as a support for bonded phase columns. It may be particulate in nature or monolithic. The chemistry of production of bonded phases is complex and the reader is referred to Snyder et al. (2011) for a comprehensive discussion. The important aspects are that so-called 'monomeric' or 'polymeric' bonded phases are available; a variety of phase chemistries can be covalently bonded to silica; and the density of the phase bonded per area of silica surface can be varied (often referred to as the 'loading density'). Silica is typically reacted with one or more alkyl silanes, depending on the desired phase chemistry. In monomeric bonded phases, one silanol group reacts with one alkyl silane molecule. In polymeric bonded phases, two adjacent silanol groups react with one alkyl silane molecule to form a polymeric-like structure.

Due to the porous nature of silica, and steric hindrance from the attaching alkyl chains, the resultant bonded phase will contain residual silanol groups on the silica support surface (Figure 4.5). These residual silanol groups can cause unwanted interactions with solutes and manufacturers typically try to minimise residual silanols by 'end capping'. This process involves a reaction with short chain alkyl silanes, leaving a short chain alkyl group on the silica surface. Even with end-capping, some residual silanol groups remain. The pK_a of high purity Type A silica is about 4.8. Thus, at neutral pH (pH 7), residual silanols are largely ionised (SiO^-) and hence can interact via ion exchange mechanisms with basic compounds. At low pH (pH 2.8), residual silanols are non-ionised. High purity Type B silicas have higher pK_a values as do 'hybrid' type particles (pK_a >8). Thus, when dealing with the extraction of ionisable analytes, the analyst may need to consider the type of silica used in the manufacture of a particular stationary phase, whether or not the phase is end-capped, and the pH of eluents used in the separation.

There is considerable debate about the retention mechanisms operating during RP separations and there is still no universally accepted theory. It can be difficult to measure interactions taking place at the molecular scale during chromatographic separations and, until relatively recently, researchers have interpreted these interactions from retention changes seen through varying bonded phase composition, phase coverage, degree of end-capping, mobile phase composition, analyte, etc. during practical experiments. More recently, researchers have turned to the use of computer models and simulations to model molecular interactions and these approaches are starting to reveal finer details of the separation process.

The classical view has long been that the stationary phase in an SPE column, as received from the manufacturer, can be envisaged as a complex network of polymer chains collapsed against the stationary phase support. When conditioned with the appropriate conditioning solvent, the polymer chains were considered to orient at right angles to the stationary phase such that analyte molecules could diffuse between the chains and interact with all functional groups of the phase. Lindsey et al. (2013) reviewed studies using molecular dynamics and Monte Carlo methods to model molecular interactions and concluded that the orientation of chains is more complex and less ordered than previously thought and varies with the loading density of the stationary phase, with the percentage content of organic modifier in the mobile phase, and whether or not the stationary phase contains embedded polar groups. The aspect of chain orientation is important because it may influence penetration of the mobile phase and solutes into the stationary phase region. For good separation to occur, the solute molecules must be able to interact with the stationary phase; in particular with the functional groups which promote intermolecular interactions and bring about separation. Molecular simulations suggest that the phenomenon referred to as 'phase collapse' is not, as was previously supposed, due to reorientation and collapse of the alkyl chains of the stationary phase when highly aqueous mobile phases are used. Rather, the highly aqueous mobile phase does not 'wet' the stationary phase, i.e. it does not penetrate into the stationary phase region. Manufacturers have tried to overcome problems of 'phase collapse' by embedding polar groups into the alkyl chains of the stationary phase. Molecular simulations indicate that embedding polar groups does improve solvent penetration into the bonded phase region and that it can also cause the chains to extend more from the silica surface.

There has also been a debate about whether adsorption or partition is the dominant mechanism in terms of analyte retention in RP separations; adsorption generally being considered to be the situation where an interface region forms between the bulk mobile phase and the stationary phase and where solutes transfer in and out of this interface region but do not fully partition into the stationary phase itself. Where partition is the dominant retention mechanism, solutes are considered to transfer from the bulk liquid phase into the stationary phase itself. Molecular simulations (Lindsey et al. 2013) indicate that for C_{18} alkyl phases, both adsorption and partition operate in RP chromatography for non-polar molecules. Adsorption dominates for polar molecules. For C_8 or shorter chain alkyl phases, adsorption is the dominant retention mechanism both for polar and non-polar analytes.

Simulations also indicate that for C_{18} phases, as the percentage of organic modifier in the mobile phase is increased, solvophilic interactions between solutes and the mobile phase increase. This will act to reduce retention of solutes. Thus far, only relatively simple situations have been modelled, including the predicted molecular interactions of a relatively limited number of solutes only. As computing power increases, it can be expected that more complex situations can be modelled, revealing further information about RP separations.

An understanding of the mechanisms operating is important because it enables the analyst to make predictions about the effect on separation of changes in mobile phase composition; the nature of the organic modifier; pressure; temperature; pH; and stationary phase, including the effect of changing chain length, phase density, use of phases with embedded polar groups, end capping etc. It also enables manufacturers to improve the columns that they produce, to engineer phases more rapidly for difficult separations, and to offer better advice on the suitability of stationary phases for extraction of specified analytes.

4.2.3.4 Ion Exchange

Specialised stationary phases are available for SPE to take advantage of ion exchange chromatography – a form of chromatographic separation that relies on the analyte being in an ionic form and interacting with a stationary phase of opposite charge. Cation and anion exchanger bonded phases are available (Table 4.1) on silica supports or embedded in polymer phases. Polymeric phases have the advantage of operating over a wider pH range than the silica-based bonded phases.

Strong cation exchange phases such as those based on sulfonic acid (SO_3^-) and strong anion exchange phases such as quaternary ammonium (R_4N^+) are ionised at all pHs. Hence, there is no need to adjust pH of the conditioning solvent to ionise the phase itself. Weak cation exchange phases, typically with carboxylic acid ligands, need the pH of the eluent adjusted to ionise the functional group and promote ion exchange. These cation exchange phases are used for the extraction of basic compounds. Amine bonded phases of the type used in NP chromatography for the separation of polar compounds can be used for ion exchange separation if the pH is adjusted to ionise the amine group ($R\text{-}NH_3^+$), resulting in a weak anion exchange phase. This is useful for the extraction of acidic compounds.

It may be necessary to adjust the pH of samples to ensure that analytes are fully ionised at the sample loading stage and interact strongly with the ionised groups on the stationary phase. This ionisation should be maintained through the washing stage but neutralised at the elution stage to convert the analytes to their neutral (non-ionised) form so that they transfer readily to the mobile phase and are eluted from the SPE column.

Mixed mode SPE phases are also available. These consist of a mixture of hydrocarbon phases together with an anion or cation exchange phase. The addition of the hydrocarbon phase extends the range of analytes that can be separated using these phases.

4.2.3.5 Molecularly Imprinted Polymers

MIPs are a more recent development in separation science. The origins of MIPs lie in work carried out to produce silicas for use in chromatography. It was found that silica produced from sodium silicate solutions and dried in the presence of benzene, toluene and xylene retained a molecular imprint of these compounds (Polyakov 1931). At the time, the significance of these findings in terms of sample extraction was not recognised. Although research related to what we now refer to as MIPs continued, it was not until 1972 that the technology of MIPs based on synthetic organic polymers was initiated. Wulff was investigating the possibility of producing synthetic polymers with catalytic sites equivalent to the active sites found in enzymes (Wulff 2013). Since then, the field of MIPs has expanded and the possibility of applying MIPs to sample extraction has been recognised. Several MIPs are now available commercially for sample extraction in SPE form (MI-SPE). Affinisep offer AFFINIMI® for SPE of compounds such as amphetamines (including amphetamine, methamphetamine, MDMA, methylenedioxyamphetamine [MDA], and methylenedioxyethylamphetamine [MDEA]); tetracycline antibiotics; and picolinic herbicides (Clopyralid, Picloram and Aminopyralid). SupelMIP® SPE phases from Sigma-Aldrich are largely targeted towards the analysis of drug residues in foodstuffs but several MIP phases are available for the extraction of biomarkers from urine to detect exposure to secondary smoking. There is also a MIP phase for extracting non-steroidal anti-inflammatory drugs (NSAIDs) from wastewater. These MIPs are available in the traditional column (cartridge) format and the procedure for use involves the same conditioning, loading, washing and elution steps as for more traditional SPE phases. It can be anticipated that the range of commercially available MIPs for SPE will increase and it will be interesting to see if their use is adopted by forensic science providers.

MIPs are produced by the copolymerisation of the target molecule (the analyte of interest) in the presence of a functional monomer, a cross-linking agent, an initiator to start the polymerisation process and a so-called porogenic solvent. The solvent solves the chemicals used to produce the MIP and produces pores or channels in the polymer network to allow access for the extraction or elution solvent when the MIP is used for sample extraction. At the end of the polymerisation process, the target molecule is removed, leaving an insoluble, polymeric, three-dimensional network bearing the molecular imprint of the target molecule as cavities within the network. The imprint should be characteristic in terms of both the shape and size of the target molecule and in terms of the chemical functionality operating between the target and the monomers used in production; enabling molecular recognition of the target in the presence of other compounds. When used for sample extraction, hydrogen bonding, dipole–dipole and ionic interactions are the dominant interactions between the MIP and the target analyte. Those interactions, together with the size and shape aspects of the molecular imprint, provide high specificity and affinity for the target analyte.

A particular problem encountered in the production of MIPs has been the retention of the target molecule within the polymeric matrix despite exhaustive solvent extraction. When the MIP is used for sample extraction, the target molecule may be released. This is clearly not acceptable in forensic science as it may lead to false positive findings, i.e. the identification of an analyte in a sample when, in fact, it is not actually present in the sample. One way to overcome this problem is to use a close structural analogue of the analyte in the production of the MIP. However, it can be difficult to find an analogue with a very similar structure whilst retaining the specificity of extraction which gives MIPs such an advantage. The reader is referred to Vasapollo et al. (2011) for a more detailed discussion of MIPs, including their production and to Martín-Esteban (2016) for more information on the use of MIPs in sample preparation.

4.2.3.6 Immunoaffinity SPE

The specificity offered by antigen–antibody interactions has also been exploited for sample preparation. A wide variety of immunoaffinity SPE phases has been produced for research purposes and used for the extraction of a wide variety of analytes. However, relatively few of these applications seem to have found their way into commercial production; commercial availability generally being a sign that a technology is well accepted and sufficiently robust for routine analytical use. A notable exception is the use of immunoaffinity SPE columns for use in extracting mycotoxins from animal feed and foodstuffs. Monoclonal antibodies are generally preferred because they offer superior specificity for the target antigen and hence reduce the cross-reactivity with other components of the sample matrix that can occur with polyclonal antibodies. Difficulties can be encountered in binding antibodies to surface supports to ensure the correct orientation of the antibody to the antigen (the target analyte in the sample matrix). Immunoaffinity phases are widely used for the extraction of proteins.

4.2.4 Solid-Phase Microextraction

SPME is a sampling technique that uses a fused-silica fibre coated with an appropriate stationary phase (liquid or solid) to extract a range of analytes from different media. The analytes may be volatile or non-volatile and can be extracted from the liquid or gas phase of a sample. SPME was developed by Pawliszyn and co-workers in 1989 to address many of the limitations of LLE and SPE. The technique combines sample collection, extraction, concentration, and sample introduction into a single procedure which is solvent free (Vas and Vekey 2004). The analytes are directly extracted without the need for sample preparation and concentrated onto the fibre before being thermally desorbed directly into the GC inlet.

The technique is used extensively with GC-MS for the analysis of volatile and semi-volatile organic compounds and can also be coupled with LC-MS or capillary electrophoresis–mass spectrometry (CE-MS) to analyse weakly volatile or thermally labile compounds not amenable to GC-MS analysis (see Chapters 13, 14 and 15). Compared with LLE and SPE, the use of SPME saves sample preparation time, removes disposal cost of solvents and can improve detection limits. When it was first developed, the technique was mainly used as a qualitative or semi-quantitative (screening) method. Quantitation using SPME is now available by using internal standards and is analogous to other standard forms of quantitation when carrying out sample preparation and analysis. Its use has been reported in forensic research. Examples include; analysing drugs of abuse in urine (Yashiki et al. 1995; Centini et al. 1996; Huang et al. 2002), blood (Nagasawa et al. 1996), and hair (Lucas et al. 2000; Sporkert and Pragst 2000; Bermejo et al. 2006); illegal trafficking of animal parts (Ueland et al. 2016); explosives (including propellant powders) (Calderara et al. 2003; Kaur et al. 2007; Joshi et al. 2009; Stefanuto et al. 2015); firearm discharge residues (Weyermann et al. 2009); and detection of human remains (Hoffman et al. 2009). The main advantages of SPME include good analytical performance, simplicity and low cost. The process can also be automated for high-volume analysis.

The apparatus used in SPME has the appearance of a modified syringe containing a fibre holder and a fibre assembly. The fibre assembly is a septum-piercing needle containing a 1–2 cm long retractable fibre. The apparatus is assembled by screwing the needle into the fibre holder. The needle is used to pierce the septum of the vial containing the sample

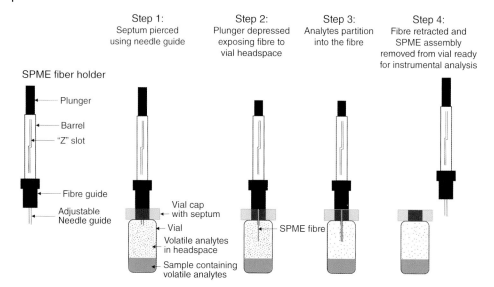

Figure 4.6 Solid-phase microextraction. The process is shown for the extraction of volatile analytes from a sample. The process of transferring the SPME assembly to a gas chromatograph for desorption into the GC injector is not shown but is described in the text.

and the fibre is exposed, by pressing the plunger of the fibre holder to retract the needle (Figure 4.6, Steps 1–2), and left in contact with the sample until the analytes have come to equilibrium between the sample and the fibre matrix (Figure 4.6, Step 3). SPME can be used to sample gases from the headspace of a solid or liquid sample (known as headspace solid-phase microextraction, HS-SPME) or it can be used for direct immersion into the liquid matrix (known as direct immersion solid-phase microextraction, DI-SPME). Once the fibre has been exposed to the sample for a predetermined period of time, the fibre is retracted into the needle for subsequent analysis (Figure 4.6, Step 4).

The fused-silica fibre is coated with a thin polymer film. The coating concentrates the analytes by absorption/adsorption processes (Step 3 in Figure 4.6). Extraction is based on a similar partitioning principle to chromatography, using either gas–liquid (headspace) or liquid–liquid (direct immersion) partitioning. The process begins by adsorption of the analytes onto the extraction phase followed by diffusion of analytes into the bulk of the extraction phase. When the diffusion coefficient is high, the analytes partition fully between the matrix and extraction phase and absorptive extraction occurs. If the diffusion coefficient is low, the analytes remain at the matrix–extraction phase interface and adsorption occurs.

A variety of polymer coatings are commercially available for SPME and are chemically similar to the types of stationary phases used in GC columns. SPME fibres incorporating MIP phases have also been reported but their use does not appear to have moved beyond the research environment into commercial availability.

Analysis of the analytes using GC or GC-MS can occur immediately after extraction. The fibre is placed into the heated GC injection port, which contains an appropriate liner, and the fibre exposed in the hot injector so that the analytes can be thermally desorbed. Injections are carried out in splitless mode to ensure that the analyte is completely transferred to the column to increase the sensitivity of the analysis (see Section 13.2.2). The injection port temperature and desorption times must be optimised for complete transfer of analytes but also to ensure that the sample is not degraded or lost in the injection port.

In SPME, the fibre is reused, i.e. after analysis of one sample, the fibre is used to extract another sample. To avoid analytes from one sample being retained on the fibre such that they are carried over into the next sample, it is recommended that the fibre be desorbed a second time between GC analyses. A fibre blank should always be run prior to any sample collection or analysis to ensure that the coating is not contaminated. For this, the fibre is analysed by GC as if it had been used to extract a sample to check that no substances are present on the fibre. Following long-term storage or before using a new fibre, conditioning is necessary by applying the maximum desorption temperature for a minimum of 30 minutes and a maximum of 4 hours.

For analysis by HPLC or LC-MS, the SPME fibre is placed into a modified valve and the analytes eluted by the mobile phase.

The choice of sample collection (headspace vs. liquid) and analysis (GC vs. HPLC) depends mainly on the volatility and polarity of the analytes. Volatile or semi-volatile samples are best collected using HS-SPME and analysed by GC-MS. Polar and non-volatile samples are most often collected using DI-SPME followed by HPLC or CE analysis. Although the holder and fibres may initially be expensive to purchase, the headspace fibres can be reused up to a hundred times and the direct immersion fibres can be used dozens of times depending on the matrix.

When using SPME for sample collection, several parameters need to be optimised during the method development. The exposure time must be optimised for each sample type to ensure that a concentration equilibrium is established between the sample matrix and the stationary phase. The exposure time may only be a few minutes or may be several hours depending on the properties of the analytes being measured. Exposing the fibre for a longer period than is required to establish equilibrium will not accumulate additional analytes. The time to reach equilibrium can be accelerated with the use of agitation methods such as stirring or sonication, or by gently heating the sample (if safe to do so). These processes improve the transport of analytes from the sample matrix to the fibre phase.

Efficient desorption of the analytes into the GC is dependent on their volatility, the thickness of the fibre coating, the injection depth, and the injector temperature and exposure time. The fibre should be exposed immediately after the needle is placed into the injector port. The exposure depth of the fibre should be adjusted to place it directly in the centre of the injector hot zone. The optimal desorption temperature is typically equal to the boiling point of the least volatile analyte and should be 10–20 °C lower than the upper temperature limit of the fibre. The GC column temperature should start low to allow concentration of the analytes at the head of the GC column and to prevent peak broadening (see Section 12.4.2). Once optimised, all parameters including extraction time, extraction temperature, sample agitation, sample volume, and desorption temperature must be kept constant to ensure reproducibility and precision.

The most important feature influencing the analytical performance of SPME is the type and thickness of the extraction phase (coating). Table 4.2 lists the most common phases that are commercially available. The most commonly used coating is polydimethylsiloxane (PDMS), which is often mixed with other coatings to improve performance. PDMS and

Table 4.2 Commercially available SPME fibres.

Fibre type and phase thickness	Phase thickness (μm)	Phase polarity	Mechanism of analyte/phase interaction (absorption/adsorption)[a]	Maximum operating temperature of phase (°C)[b]	Application
Polydimethylsiloxane (PDMS)	7	Non-polar	Absorption	340	Non-polar high MW compounds (MW 125–600)
PDMS	30	Non-polar	Absorption	280	Non-polar semi-volatile compounds (MW 80–500)
PDMS	100	Non-polar	Absorption	280	Volatile compounds (MW 60–275)
Polyacrylate	85	Polar	Absorption	320	Polar semi-volatile compounds (MW 80–300)
PDMS/divinylbenzene (DVB)	65	Bi-polar	Adsorption	270	Volatiles, amines and nitro-aromatic compounds (MW 50–300)
PDMS/DVB	60	Bi-polar	Adsorption	N/A HPLC use only	Amines and polar compounds
Carboxen[c]/PDMS	75/85	Bi-polar	Adsorption[d]	320	Gases and low MW compounds (MW 30–225)
PDMS/DVB/CAR	50/30	Bi-polar	Adsorption	270	(MW 40–275)
Carbowax[e]	60	Polar	Adsorption	N/A HPLC use only	Alcohols and polar compounds (MW 40–275)

a) Source: Vas and Vékey (2004).
b) The maximum operating temperature is important for GC use, particularly in terms of the GC injector temperature.
c) Carboxen is a carbon molecular sieve.
d) The mechanism of interaction is still not clear. Adsorption processes are involved but there are indications that other processes may also be involved such as capillary condensation (Vas and Vékey 2004).
e) Carbowax is a polyethylene glycol phase.

Information compiled from supplier website (Supelco/Sigma-Aldrich, www.sigmaaldrich.com) and Górecki et al. (1999).

polyacrylate (PA) phases extract the samples via absorption of the analytes into the coating material. All other phases (including mixed phases) extract the sample via adsorption of the analytes onto the surface of the fibre. Following the general principle of 'like attracts like', polar fibres are typically used to extract polar analytes and non-polar fibres are used to extract non-polar compounds. The thickness of the phase can also vary and will impact the equilibrium time and sensitivity of the method. For example, the use of a thicker fibre requires a longer extraction time but increases the recovery of the analytes.

4.2.5 QuEChERS

QuEChERS was developed by Anastassiades and co-workers for the extraction of pesticide residues from fruits and vegetables (Anastassiades et al. 2003). The idea was to use a relatively quick, inexpensive, but effective way of extracting multiple analytes of interest from samples and to minimise sample handling steps and solvent use. The initial multi-residue method (extraction of multiple analytes) consisted of the following steps: (i) extract a sub-sample (10 g) with acetonitrile (10 ml) in the presence of anhydrous $MgSO_4$ (4 g) and NaCl (1 g). (ii) Shake vigorously and then centrifuge to separate the organic and aqueous phase. (iii) To an aliquot of the organic phase, add anhydrous $MgSO_4$ (150 mg) and a small mass of a sorbent, in this case a 'primary secondary amine' (PSA). (iv) Shake and then centrifuge. (v) Analyse a portion of the final organic phase. The second step involving shaking the organic phase from the first extraction step with anhydrous $MgSO_4$ and PSA is referred to as dispersive solid-phase extraction (d-SPE) and removes water, fatty acids, sugars and some polar plant pigments. Hyphenated mass spectrometric techniques are generally required for analysis including GC-MS, LC-MS and MS/MS approaches. These techniques offer additional specificity and sensitivity compared with gas chromatography flame ionisation detector (GC-FID) or HPLC-UV (cf Sections 13.2.4.1 and 14.2.8.2, respectively).

Since the original method was developed, QuEChERS has been adopted by the Association of Official Analytical Chemists (AOAC) and applied to the analysis of a number of other sample matrices, most notably to the extraction of pesticide residues from plants and from soils. Applications in forensic science are also starting to appear with QuEChERS used in the analysis of benzodiazepines in blood (Westland and Dorman 2013) and modified to detect and quantify drugs and metabolites in blood (Matsuta et al. 2013) and for the analysis of opiods, cocaine and cocaethylene in whole blood (Amorim Alves et al. 2017). We anticipate further adoption of the QuEChERS method for applications in forensic science.

4.2.6 Sample Handling Post Extraction

4.2.6.1 Solvent Evaporation

Depending on the concentration of analyte in the sample, the volume of solvent used, the efficiency of extraction, and the method of detection used in the analysis, it may be possible to analyse the solvent extract directly after the extraction process with no further treatment. If the concentration of analyte is below the limit of detection of the analytical method (see Section 2.3.1.1), it can be increased by evaporating some of the solvent. This is usually carried out in the presence of an inert gas such as nitrogen to minimise the possibility of analyte oxidation, often using heat to speed evaporation and to prevent condensation of water in the sample under the cooling that can occur if solvent is evaporated at room temperature. Caution must be exercised because some analytes, such as amphetamine and methamphetamine, are relatively volatile, particularly when heated, and may vaporise. In some instances, the evaporation stage offers the opportunity to switch to a more suitable solvent for the final instrumental analysis. In this case the sample is 'taken to dryness', i.e. all of the extraction solvent is evaporated, and the sample is re-dissolved in a more suitable solvent for analysis.

4.2.6.2 Derivatisation

The analyte may also require chemical modification for instrumental analysis. Where the final analysis is by GC or HPLC, this modification is generally referred to as derivatisation and may be required to maximise chromatographic efficiency (see Chapter 12), or sensitivity of detection (see Section 2.4.3). Derivatisation may also be carried out to produce a more characteristic mass spectrometric fragmentation pattern; to investigate isomeric composition; or to make a compound more amenable to chromatographic analysis, e.g. making it more amenable to GC analysis by replacing polar functionalities in a molecule with non-polar functional groups and thereby reducing the analyte volatility. Provided the derivatisation reagent does not react with the extraction solvent and is soluble in it, it should be possible to carry out the reaction directly in the solvent. If this is not possible, the solvent must be evaporated to dryness before the derivatisation is carried out. Some derivatisation reagents produce by-products which themselves can cause problems in the analysis. An example is fluorinated anhydrides which react to produce acid by-products which can degrade silica GC columns. The acid by-products must be treated in some way to prevent them from reaching the GC column. The reader is referred to Knapp (1979) and Orata (2012) for a more in-depth discussion of derivatisation.

4.3 Sample Preparation for Inorganic Analyses

Analyses of the inorganic composition of samples may be required in some circumstances. Examples include the characterisation of firearm discharge residue; glass and paint analysis; suspect or accidental poisonings involving elements such as arsenic, selenium, thallium, lead, mercury in biological matrices; provenance of foodstuffs (Kelly et al. 2005); and the analysis of geological material (Pye and Croft 2004). In some situations, the analyst will need to determine the total concentration of the element or ion of interest and, in other cases, knowledge of the composition of a particular species (or chemical form) of the element or ion will be required, for instance a particular oxidation state or the metal bound to organic complexes. These issues of total and speciated analyses will be considered separately.

While the total concentration of an element in a particular sample cannot change (unless it is contaminated in some way), the speciation can. Hence samples for speciation require careful and appropriate storage. Elements in solution can also be lost from the sample by adsorption to container walls resulting in analytical results that do not reflect the original sample, even if they pass laboratory analytical quality control procedures. Acidification of a solution is often used as a means to minimise adsorption and retain sample integrity but will alter speciation. Freezing is an alternative that will reduce adsorption and stabilise speciation for liquids. Freezing, drying or a combination of freeze-drying is often used for preservation of solid samples such as soils or biological material. If the concentrations of the analytes are low in a sample, considerable care may be required to avoid contamination during sampling and analysis and hence to ensure that the final analytical concentrations reported reflect the original sample. In particular, soda glass is known to be a source of elemental contamination or may act as a sink, absorbing elements. Rubber stoppers have also been shown to be sources of contamination. Analysis of elemental composition typically employs plastic materials during sample preparation stages, although quartz (silica) glass may be used. Polyethylene (PE) and polytetrafluoroethylene (PTFE) containers are generally recommended but even these materials can be considerable sources of elemental contamination. All containers should be cleaned well before use and kept solely for trace elemental analysis. Soaking with 10% nitric acid is the general recommendation for decontamination followed by copious rinsing with water of low elemental content and protection from contamination by dust and other particulates. Skin can also be a source of contamination as can cosmetics and so direct sample handling should be avoided as far as possible. Use of gloves can minimise contamination but some types of gloves can contaminate samples. If tissue samples are to be analysed for elemental content, suitable precautions should be taken over the materials used for sampling. The evaluation of the integrity of sample storage is a challenging task requiring collaborations between those doing the sampling and laboratory analysts (Howard et al. 1993).

4.3.1 Total Analysis

A key issue to consider in analysis is the matrix to be analysed, which broadly simplifies to a solid or liquid medium. In the case of a sample bound within a liquid medium, the total concentration of many elements can be measured by ICP-AES or ICP-MS. For both of these techniques, the sample is introduced into a high temperature plasma and the concentration of individual elements can be measured (Pye and Croft 2004; Pye et al. 2006; Linge and Jarvis 2009). Older methods based on atomic absorption spectroscopy are now largely superseded by ICP methods, because of improved sensitivity and simultaneous multi-element analysis (see Chapter 11). ICP-MS offers the opportunity to measure isotope ratios of elements which can provide additional sample characterisation (Kelly et al. 2005) and also to measure lower concentrations. However, the trade-off for the higher sensitivity of ICP-MS is a slower analysis time, reduced precision and increased instrument costs compared with ICP-AES. The high temperature of the plasma in ICP instruments reduces or eliminates many matrix interference affects, allowing direct analysis of liquid samples in some situations, particularly if matrix matched standards can be prepared or if standard additions (see Section 2.3.1.3) or internal standards are used for calibration. Matrix matched standards are prepared in the same matrix as the samples but where the matrix is known not to contain the analytes of interest. However, potential interferences remain and the analysis of Certified Reference Materials (CRMs) is recommended (Pye et al. 2006, Linge and Jarvis 2009). CRMs, also called Standard Reference Materials (SRMs), are materials that come with a certification detailing the analytes present and their concentration in the material.

The quantitative value is usually determined by some form of inter-laboratory trial, often using different instrumental methods. An analyst analyses a portion of the CRM alongside samples as part of the analysis. If the CRM gives the certified value, within the degree of precision set for the analysis, the analyst has greater assurance that the quantitative values obtained for samples are correct, at least in terms of the analytical procedure itself.

In situations with liquid samples with low analyte (trace or ultra-trace) concentrations and high concentrations of background ions, such as in seawater, extraction of samples to concentrate the analytes and eliminate matrix effects may be required. The methods of sample extraction are similar to those discussed earlier, broadly solid- or liquid-phase extractions. An alternative approach of co-precipitation to concentrate the analyte can be useful but this usually then requires further analytical steps to eliminate the potential interference from the co-precipitant. SPE for elemental analysis usually employs chelating resins (e.g. Chelex® 100) that have a high affinity for transition metals and a weak affinity for ions of groups 1 and 2 of the periodic table (e.g. sodium and calcium). The affinity of these resins is pH dependent allowing the analytes to be absorbed onto the solid phase at one pH (usually relatively high) and then eluted at a different pH (usually acidic). LLE techniques can involve pH-dependent complexation of the metals of interest by a strong complexing agent such as ammonium pyrrolidine dithiocarbamate (APDC) or 8-hydroxyquinoline to form a metal organic complex which can then be extracted into an organic solvent such as chloroform. Separation and evaporation of the organic layer and redissolution of the metals in acid then allows the analyte to be concentrated and separated from interferences (Howard et al. 1993).

It is possible to analyse solid samples directly using laser ablation coupled to ICP-MS. Glass is particularly amenable to multi-element analysis by this technique. The surface of the glass should be pre-cleaned. Some methods recommend crushing glass between plastic where analysis of the bulk glass is required.

Where solid samples are to be analysed by ICP-AES or ICP-MS techniques, they are usually digested (broken down and then redissolved in acid). Solid samples are typically prepared for total concentration analysis using a strong acid digest; oxidising acids for organic matrices and hydrofluoric acid for aluminosilicate matrices (Howard et al. 1993). For rock samples, a fusion method may be employed, with subsequent dissolution (Pye et al. 2006). The completeness of digestion procedures should be fully evaluated using CRMs. In the case of solid samples, the concentrations of the analyte ions are often relatively high compared with the major ions and hence extraction techniques may not be required to separate analytes from background ions. However, analytical interferences need to be thoroughly evaluated using appropriate CRMs and matrix matching of standards (Howard et al. 1993; Linge and Jarvis 2009).

Solid samples can, in some cases, be analysed directly and potentially non-destructively using a variety of techniques of which the most widely available are X-ray fluorescence (XRF), or X-ray spectroscopy coupled to electron microscopy (e.g. SEM-EDX spectroscopy). The latter technique offers an opportunity for visual characterisation and for the analysis of small samples or small areas of larger samples (Pye and Croft 2007). The accurate quantification of composition using these techniques depends on the availability of suitable standards and reference materials.

4.3.2 Chemical Speciation

In some circumstances, analysis of the concentration of a particular component of the total concentration may be required, for example only the fraction of a metal in a sample that is readily solubilised and hence readily bioavailable, or the fraction of metal in a sample that is in a particularly toxic form, for example chromate (with Cr(VI) rather than Cr(III)). Such analyses are usually referred to as chemical speciation (see [Ure and Davidson 2008] for a comprehensive discussion). Techniques such as ICP or X-ray techniques as discussed above measure the total concentration present. To use such techniques to determine chemical speciation requires pretreatment of the sample to chemically separate the fractions prior to analysis. This can be done offline or, in some cases online, for example, via column chromatography coupled to ICP. In the case of Cr(VI) and Cr(III) this might be achieved by ion exchange resins since Cr(VI) is anionic (chromate, CrO_4^{2-}) under normal aqueous conditions while Cr(III) is cationic (Cr^{3+}).

4.4 DNA Profiling

Although the analysis of DNA is not generally considered as part of chemical analysis in the forensic sciences, several DNA profiling methodologies use capillary electrophoresis (CE) to analyse the short tandem repeat (STR) or single nucleotide polymorphisms (SNPs) products from the polymerase chain reaction (PCR) amplification procedure. In addition, the first stages of DNA profiling require extraction of DNA from samples and purification of the DNA to separate it from components that can interfere with the PCR procedure.

Extraction with phenol/chloroform has long been a staple method of DNA extraction. This is a LLE procedure but with DNA remaining in the aqueous fraction and the organic component used to extract lipids and to help separate DNA from proteins. The addition of isoamyl acetate is incorporated into some phenol/chloroform extraction protocols. In the final stages of the extraction, DNA is precipitated from aqueous solution in the presence of ice-cold ethanol and an acetate salt. On account of environmental and safety concerns over the use of phenol and chloroform, other extraction procedures have been introduced. These include silica in various forms, including as membranes and as silica-coated magnetic particles. Typically, chaotropic salts such as guanidine hydrochloride, guanidine thiocyanate, sodium iodide, and sodium perchlorate are also employed to aid extraction, as is pH. Under acidic conditions, DNA binds readily to silica but can be eluted under alkaline, low salt conditions. Where silica is bound to paramagnetic beads, the magnetic properties of the beads can be used to hold the beads and associated DNA in place during wash stages. This format has enabled automation of DNA extraction and the use of robotic extraction stations for high sample throughput. Silica has also been employed in a more conventional SPE format for DNA extraction, with a silica membrane held in a small column which can fit inside polyethylene micro-sample tubes. This enables sample loading, washing and elution of DNA to be carried out in-tube on the column, with centrifugation used between each stage to draw sample, and wash buffer and eluting solvent through the SPE membrane.

Chelation has also been employed for DNA extraction. Chelex 100 resin binds polyvalent metal ions, including Mg^{2+} ions which are co-factors for nuclease enzymes. Hence the Chelex is used to protect the DNA from enzymatic degradation.

More recently, immunoaffinity methods have been developed for DNA extraction, with monoclonal antibodies bound to magnetic beads.

4.5 Conclusion

As can be seen from the above discussion, there are a variety of techniques that can be used for sample extraction. The skill for the analyst lies in selecting the most appropriate one. As sample extraction can be such a time-consuming step, the trend in analytical chemistry is for the development of new, or redesign of existing, instruments to enable direct analysis of samples without the need for analyte extraction, and for sample extraction techniques which can be automated and directly coupled to instrumental techniques or which otherwise minimise manual sample handling procedures and extraction time.

References

Amorim Alves, E., Sofia Agonia, A., Manuela Cravo, S. et al. (2017). GC-MS method for the analysis of thirteen opioids, cocaine and cocaethylene in whole blood based on a modified quechers extraction. *Current Pharmaceutical Analysis* 13 (3): 215–223.

Anastassiades, M., Lehotay, S.J., Štajnbaher, D., and Schenck, F.J. (2003). Fast and easy multiresidue method employing acetonitrile extraction/partitioning and "dispersive solid-phase extraction" for the determination of pesticide residues in produce. *Journal of AOAC International* 86 (2): 412–431.

Bello-López, M.Á., Ramos-Payán, M., Ocaña-González, J.A. et al. (2012). Analytical applications of hollow fiber liquid phase microextraction (HF-LPME): a review. *Analytical Letters* 45 (8): 804–830.

Bermejo, A.M., López, P., Álvarez, I. et al. (2006). Solid-phase microextraction for the determination of cocaine and cocaethylene in human hair by gas chromatography–mass spectrometry. *Forensic Science International* 156 (1): 2–8.

Calderara, S., Gardebas, D., and Martinez, F. (2003). Solid phase micro extraction coupled with on-column GC/ECD for the post-blast analysis of organic explosives. *Forensic Science International* 137 (1): 6–12.

Centini, F., Masti, A., and Comparini, I.B. (1996). Quantitative and qualitative analysis of MDMA, MDEA, MA and amphetamine in urine by head-space/solid phase micro-extraction (SPME) and GC/MS. *Forensic Science International* 83 (3): 161–166.

Cody, R.B., Laramee, J.A., and Durst, H.D. (2005). Versatile new ion source for the analysis of materials in open air under ambient conditions. *Analytical Chemistry* 77 (8): 2297–2302.

ENFSI (2014). Guidelines on Sampling of Illicit Drugs for Quantitative Analysis. European Network of Forensic Science Institutes Drugs Working Group.

FAO (2005). Pesticide Residues in Food 2004. Evaluations. Part 1: Residues. Paper 182/2.

Górecki, T., Yu, X., and Pawliszyn, J. (1999). Theory of analyte extraction by selected porous polymer SPME fibres. *Analyst* 124 (5): 643–649.

Hoffman, E.M., Curran, A.M., Dulgerian, N. et al. (2009). Characterization of the volatile organic compounds present in the headspace of decomposing human remains. *Forensic Science International* 186 (1–3): 6–13.

Howard, A.G., Statham, P., and Kremling, K. (1993). *Inorganic Trace Analysis: Philosophy and Practice*. Chichester: Wiley.

Huang, M.-K., Liu, C., and Huang, S.-D. (2002). One step and highly sensitive headspace solid-phase microextraction sample preparation approach for the analysis of methamphetamine and amphetamine in human urine. *Analyst* 127 (9): 1203–1206.

Joshi, M., Delgado, Y., Guerra, P. et al. (2009). Detection of odor signatures of smokeless powders using solid phase microextraction coupled to an ion mobility spectrometer. *Forensic Science International* 188 (1–3): 112–118.

Kaur, V., Kumar, A., Malik, A.K., and Rai, P. (2007). SPME-HPLC: a new approach to the analysis of explosives. *Journal of Hazardous Materials* 147 (3): 691–697.

Kelly, S., Heaton, K., and Hoogewerff, J. (2005). Tracing the geographical origin of food: the application of multi-element and multi-isotope analysis. *Trends in Food Science & Technology* 16 (12): 555–567.

Knapp, D.R. (1979). *Handbook of Analytical Derivatization Reactions*. Hoboken, NJ: Wiley.

Lindsey, R.K., Rafferty, J.L., Eggimann, B.L. et al. (2013). Molecular simulation studies of reversed-phase liquid chromatography. *Journal of Chromatography A* 1287: 60–82.

Linge, K.L. and Jarvis, K.E. (2009). Quadrupole ICP-MS: introduction to instrumentation, measurement techniques and analytical capabilities. *Geostandards and Geoanalytical Research* 33 (4): 445–467.

Lucas, A., Bermejo, A., Tabernero, M. et al. (2000). Use of solid-phase microextraction (SPME) for the determination of methadone and EDDP in human hair by GC–MS. *Forensic Science International* 107 (1–3): 225–232.

Martín-Esteban, A. (2016). Recent molecularly imprinted polymer-based methods for sample preparation. In: *Advanced Molecularly Imprinting Materials* (eds. A. Tiwari and L. Uzun), 1–27. Hoboken, NJ: Wiley.

Matsuta, S., Nakanishi, K., Miki, A. et al. (2013). Development of a simple one-pot extraction method for various drugs and metabolites of forensic interest in blood by modifying the QuEChERS method. *Forensic Science International* 232 (1–3): 40–45.

McNaught, A.D. and Wilkinson, A. (1997). *IUPAC. Compendium of Chemical Terminology*, 2378–2385. Oxford, UK: Blackwell Scientific.

Moors, M., Massart, D., and McDowall, R. (1994). Analyte isolation by solid phase extraction (SPE) on silica-bonded phases: classification and recommended practices (technical report). *Pure and Applied Chemistry* 66 (2): 277–304.

Nagasawa, N., Yashiki, M., Iwasaki, Y. et al. (1996). Rapid analysis of amphetamines in blood using head space-solid phase microextraction and selected ion monitoring. *Forensic Science International* 78 (2): 95–102.

Nawrocki, J. (1997). The silanol group and its role in liquid chromatography. *Journal of Chromatography A* 779 (1–2): 29–71.

Orata, F. (2012). Derivatization reactions and reagents for gas chromatography analysis. In: *Advanced Gas Chromatography: Progress in Agricultural, Biomedical and Industrial Applications* (ed. M.A. Mohd) Chapter 5. IntechOpen, pp. 83–108.

Pawliszyn, J. and Lord, H.L. (2010). *Handbook of Sample Preparation*. Hoboken, NJ: Wiley.

Polyakov, M. (1931). Adsorption properties and structure of silica gel. *Zhur Fiz Khim* 2: 799–805.

Pye, K. and Croft, D.J. (2004). Forensic geoscience: introduction and overview. *Geological Society, London, Special Publications* 232 (1): 1–5.

Pye, K. and Croft, D. (2007). Forensic analysis of soil and sediment traces by scanning electron microscopy and energy-dispersive X-ray analysis: an experimental investigation. *Forensic Science International* 165 (1): 52–63.

Pye, K., Blott, S.J., and Wray, D.S. (2006). Elemental analysis of soil samples for forensic purposes by inductively coupled plasma spectrometry – precision considerations. *Forensic Science International* 160 (2–3): 178–192.

Sarafraz-Yazdi, A. and Amiri, A. (2010). Liquid-phase microextraction. *TrAC Trends in Analytical Chemistry* 29 (1): 1–14.

Snyder, L.R., Kirkland, J.J., and Dolan, J.W. (2011). *Introduction to Modern Liquid Chromatography*. Oxford, UK: Wiley.

Sporkert, F. and Pragst, F. (2000). Use of headspace solid-phase microextraction (HS-SPME) in hair analysis for organic compounds. *Forensic Science International* 107 (1–3): 129–148.

Stefanuto, P.-H., Perrault, K., Focant, J.-F., and Forbes, S. (2015). Fast chromatographic method for explosive profiling. *Chromatography* 2 (2): 213–224.

SWGDRUG (2016). Scientific Working Group for the Analysis of Seized Drugs (SWGDRUG) Recommendations. Version 7.1.

Takats, Z., Wiseman, J.M., Gologan, B., and Cooks, R.G. (2004). Mass spectrometry sampling under ambient conditions with desorption electrospray ionization. *Science* 306 (5695): 471–473.

Ueland, M., Ewart, K., Troobnikoff, A.N. et al. (2016). A rapid chemical odour profiling method for the identification of rhinoceros horns. *Forensic Science International* 266: e99–e102.

UNODC (2009). Guidelines on Representative Drug Sampling. ST/NAR/38.

Ure, A.M. and Davidson, C.M. (2008). *Chemical Speciation in the Environment*. Oxford, UK: Wiley.

Vas, G. and Vekey, K. (2004). Solid-phase microextraction: a powerful sample preparation tool prior to mass spectrometric analysis. *Journal of Mass Spectrometry* 39 (3): 233–254.

Vasapollo, G., Sole, R.D., Mergola, L. et al. (2011). Molecularly imprinted polymers: present and future prospective. *International Journal of Molecular Sciences* 12 (9): 5908–5945.

Walter, T. H., J. Ding, M. Kele, et al. (2007). Porous inorganic/hybrid monolith materials for chromatographic separations and process for their preparation. US Patent 7,250,214.

Westland, J.L. and Dorman, F.L. (2013). QuEChERS extraction of benzodiazepines in biological matrices. *Journal of Pharmaceutical Analysis* 3 (6): 509–517.

Weyermann, C., Belaud, V., Riva, F., and Romolo, F.S. (2009). Analysis of organic volatile residues in 9 mm spent cartridges. *Forensic Science International* 186 (1–3): 29–35.

Wulff, G. (2013). Forty years of molecular imprinting in synthetic polymers: origin, features and perspectives. *Microchimica Acta* 180 (15–16): 1359–1370.

Yashiki, M., Kojima, T., Miyazaki, T. et al. (1995). Detection of amphetamines in urine using head space-solid phase microextraction and chemical ionization selected ion monitoring. *Forensic Science International* 76 (3): 169–177.

Part II

Spectroscopic and Spectrometric Techniques

5

The Electromagnetic Spectrum

Rosalind Wolstenholme

Electromagnetic (EM) radiation is described classically as two oscillating, in phase electric and magnetic fields perpendicular to each other, which cause a wave to be propagated, carrying energy. The wave is perpendicular to both fields and travels in a straight line unless it is reflected or refracted (Figure 5.1). Importantly, all EM radiation travels at the speed of light, c, in a vacuum. Alternatively, EM radiation can be described, in terms of quantum mechanics, as a stream of photons or packets of energy. The photons can promote an atom or molecule to an excited state, electronic, vibrational etc., which results in absorption, emission or scattering as the atom/molecule returns to the ground state. For this to happen, the energy gap between the ground state and the excited state must match the wavelength of the incident radiation, i.e. it is quantised. (Fluorescence and Raman scattering are the exceptions that prove the rule, see Chapters 6 and 8.)

Both the classical and quantum models can be useful for understanding the type of radiation incident on a sample and the interactions that occur with an atom or molecule.

The combination of the two models lead to EM radiation being discussed in terms of the interrelated units of wavelength (λ), frequency (ν; number of waves per second), and energy (E), depending on the context. Wavelength is the distance between the peaks of two waves and is typically measured in cm, μm or nm, frequency is the number of waves passing a fixed point in a given amount of time and is measured in Hz (1 wave in 1 second), and energy is measured in J or eV. Equation (5.1) gives the relationship between λ and ν.

$$\nu\lambda = c \text{ or } \lambda = \frac{c}{\nu} \tag{5.1}$$

where c is the speed of light (299 792 458 m/s).

Equation (5.2) gives the relationships between energy and frequency and energy and wavelength.

$$E = h\nu = \frac{hc}{\lambda} \tag{5.2}$$

where h is Plank's constant ($6.626\,070\,15 \times 10^{-34}$ J s)

From Eq. (5.1) it can be seen that ν and λ are inversely proportional to each other. From Eq. (5.2) it can be observed, therefore, that E is proportional to ν and inversely proportional to λ, i.e. as frequency increases energy increases and as wavelength increases energy

---

Analytical Techniques in Forensic Science, First Edition.
Edited by Rosalind Wolstenholme, Sue Jickells and Shari Forbes.
© 2021 John Wiley & Sons Ltd. Published 2021 by John Wiley & Sons Ltd.

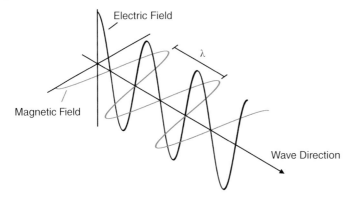

Figure 5.1 Electromagnetic radiation. The classical model describes two perpendicular oscillating electric and magnetic fields causing a wave to be propagated that is perpendicular to both fields.

decreases. This relationship is important when looking at the electromagnetic spectrum (EMS).

A further relationship that is useful to understand, particularly for vibrational spectroscopy, is that between frequency, wavelength and wavenumber, $\tilde{v}$

$$\tilde{v} = \frac{1}{\lambda} \text{ or } v = c\tilde{v} \tag{5.3}$$

$\tilde{v}$ is measured in cm^{-1} and since it is directly proportional to frequency it is also directly proportional to energy.

The EMS is the range (spectrum) of radiation frequencies from around 10^3 to 10^{20} Hz and is often classified into regions of frequencies or wavelengths (the boundaries given can vary slightly in different reference sources since they are not discrete). Figure 5.2 shows the EMS and gives values for the frequencies, wavelengths and energies of the different regions of radiation: radiowave; microwave; infrared (IR); ultraviolet–visible (UV-Vis); X-ray; and γ ray. At the low energy end of the spectrum are radiowaves, which are low frequency and high wavelength, and at the high energy end are γ rays, with high frequency and low wavelength. What we see, i.e. visible light, is in a narrow region in the middle between 380 nm and 760 nm.

Much of our understanding of the structure of atoms has come from research using EM radiation and what are termed *spectroscopic techniques*. Each type of EM radiation interacts with matter or is emitted in a different way and, therefore, can potentially tell us something different about a sample. Consequently, for each region of the EM spectrum a different analytical technique is required to obtain information. Low energy radiation is only capable of causing small changes in a molecule, e.g. nuclear spin direction in nuclear magnetic resonance (NMR) spectroscopy, which uses radio waves. The next region is the microwave region, which can be used to measure the rotational energy of molecules in gases with rotational spectroscopy. IR spectroscopy and Raman spectroscopy are concerned, primarily, with the vibrational energy of a molecule in the IR region (see Chapters 7 and 8). UV-Vis spectroscopy measures molecular or valence electronic excitation (see Chapter 6); X-rays are detected following inner electronic transitions using energy-dispersive X-ray (EDX) spectrometry or wavelength-dispersive X-ray (WDX) spectrometry, which in a forensic

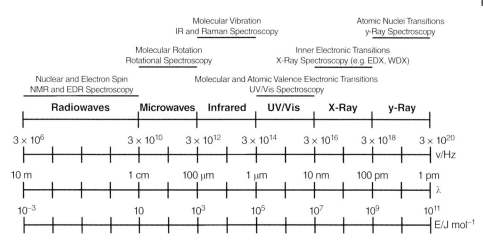

Figure 5.2 The electromagnetic spectrum. Source: Adapted from Banwell and McCash (1994).

context is usually coupled to scanning electron microscopy (SEM) (see Chapter 9). Finally, high energy radiation can cause changes to the configuration of the nucleus as in γ ray spectroscopy. Figure 5.2 relates the frequency/wavelength/energy of the radiation to the corresponding spectroscopic technique.

The output of a spectroscopic analysis is called a spectrum (plural spectra). These take a different form for each technique but are essentially a plot of the intensity of an interaction against the energy of that interaction. They are given on a wavelength or wavenumber scale depending on the specific technique employed. From the position and the magnitude of the features on a spectrum the scientist can infer some of the structural properties of their sample, as is appropriate for the technique used. In addition, the intensity of the peaks in a spectrum can also be used for quantitation since according to the Beer–Lambert law

$$A = \varepsilon l c \tag{5.4}$$

where A is absorbance, ε is the molar absorptivity, l is the pathlength, and c is concentration.

In Forensic Science, spectroscopic analysis tends to concentrate on the range of the EMS from ultraviolet (UV) to X-ray energies, employing UV-Vis spectroscopy, IR spectroscopy, Raman spectroscopy and EDX spectrometry. However, also included within the family of spectroscopic/spectrometric techniques are mass spectrometry (MS) and SEM. The spectroscopic techniques mentioned above use EM radiation to probe the structure of molecules and acquire a spectrum. SEM and MS do not use EM radiation to detect the energy absorbed, scattered or emitted from a molecule.

MS uses an ion source to ionise the sample in order to record its mass to charge ratio (m/z) (see Chapter 10). This information is also presented as a 'spectrum' and can be used to complement and/or confirm structural identifications proposed with spectroscopic techniques. Therefore, MS is usually associated with spectroscopic techniques. MS is used not only in its own right but also can be hyphenated with other techniques, e.g. gas chromatography–mass spectrometry (GC-MS) (see Section 13.2.4.4).

SEM uses a beam of electrons to provide topographical images at high magnification, which can provide definitive evidence, e.g. whether a headlamp bulb was on or off at the time of a collision. It is also often carried out with EDX or WDX to provide information on the elemental composition and distribution of a sample. Again, the topographical image (with information about the distribution of elements) can be complementary to the data acquired with other spectroscopic techniques, and since SEM probes the atomic/molecular structure of the sample it is typically considered with the other spectroscopic/metric techniques.

One of the advantages of spectroscopic/metric techniques for the forensic scientist is that they can be (but are not always) non-destructive (see Section 1.2.2) with little sample preparation. In addition, portable technology is becoming more widely available and much more affordable to forensic agencies so that analyses can be performed quickly and in situ, giving timely analysis and intelligence data to an investigation.

Between them, spectroscopic and spectrometric techniques have been applied to a wide range of forensic evidence types including hairs and fibres, glass, paint, drugs and firearms discharge residue. Whilst there are similarities in instrumentation and/or interpretation between some spectroscopic/metric techniques, there are also many issues specific to each. Chapters 6–11 will discuss in detail the theory of spectroscopic/metric techniques that are commonly used in forensic science (UV-Vis spectroscopy, IR spectroscopy, Raman spectroscopy, SEM, MS, and isotope ratio mass spectrometry [IRMS], respectively) together with how they are applied to forensic samples and the interpretation of the data acquired.

Reference

Banwell, C.N. and McCash, E.M. (1994). *Fundamentals of Molecular Spectroscopy*. New York: McGraw-Hill.

6

Ultraviolet–Visible and Fluorescence Spectroscopy

Rosalind Wolstenholme

6.1 Forensic Introduction

Ultraviolet–visible (UV-Vis) spectroscopy is a well-established technique concerned with electronic transitions within the molecules in a sample. It has been available commercially since the 1940s but in recent years has benefitted from advances in technology, including improved detectors and fibre optic sampling, which have allowed instruments to become smaller, faster, able to analyse increasingly small sample quantities, and to be more portable. UV-Vis spectrometry is commonly used in industry and has been applied, for example, to the analysis of pharmaceuticals, proteins and DNA (Baret et al. 1998; Baldock and Hutchison 2016; Gorog 2018). The forensic applications are various, given the availability not only of standard spectrometers but also of microspectrophotometers (see Section 6.3.4), which can analyse small samples non-destructively, including ink, fibres, and paint (Grieve et al. 1988; Payne et al. 2005; Trzcińska et al. 2013).

In addition to UV-Vis spectroscopy, the UV-Vis region of the electromagnetic (EM) spectrum can be exploited to promote fluorescence, which can be used for filtered light examination (FLE) and fluorescence spectroscopy/fluorometry. As before, the applications are wide ranging from biological applications, such as identifying tumours (Saraswathy et al. 2009) to environmental, e.g. detecting petroleum products in water (Baszanowska and Otremba 2016). Forensic use of fluorescence spectroscopy is not as common as UV-Vis spectroscopy but there are applications, such as detection of human saliva, and it is discussed here for completeness as the 'sister' technique (Denny et al. 2018).

This chapter will look at both techniques and their use in forensic science.

6.2 Theory

6.2.1 Electronic Transitions

UV-Vis spectroscopy analyses the electronic transitions that occur when the EM radiation incident on a sample is in the range 200–750 nm, i.e. the UV-Vis range. Wavelengths from 10

Analytical Techniques in Forensic Science, First Edition.
Edited by Rosalind Wolstenholme, Sue Jickells and Shari Forbes.
© 2021 John Wiley & Sons Ltd. Published 2021 by John Wiley & Sons Ltd.

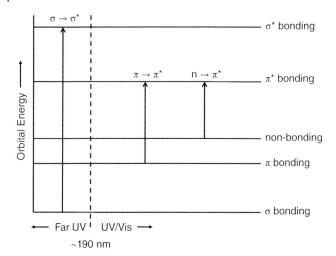

Figure 6.1 Typical electronic transitions for organic molecules.

to 200 nm are classed as vacuum ultraviolet (UV), from 200 to 400 nm as (atmospheric) UV, and 400–750 nm as visible. Over the combined UV and visible ranges, the electronic transitions recorded tend to be molecular rather than atomic so that UV-Vis spectroscopy can provide information about the structure(s) present in a molecule. In a typical excitation, the transition is from the highest occupied molecular orbital (HOMO) to the lowest unoccupied molecular orbital (LUMO) (Figure 6.1). Three types of ground state orbitals are involved in electronic transitions, σ and π bonding orbitals and n non-bonding orbitals, and two excited state orbitals, anti-bonding orbitals σ^* and π^*. Whether a transition is allowed or is 'forbidden' is related to the symmetry and multiplicity of the ground and excited state orbitals and is beyond the scope of this book (Jaffé and Orchin 1962). In practice, we use molar absorptivity, ε, to indicate whether transitions will occur. ε is a measure of how efficiently a molecule absorbs radiation and is constant for any molecule at a given wavelength, it is not dependent on concentration. The higher the value of ε, the more efficient the transition from the ground to the excited state. Theoretically, transitions with ε above approximately 10 000, according to Eq. (6.1), are fully allowed and give rise to high intensity absorptions.

$$\varepsilon = 0.87 \times 10^{20} P \cdot a \tag{6.1}$$

where P is the transition probability and a is the target area of the absorbing system.

Values of ε between 1000 and 10 000 intensity are considered low and transitions with ε below 1000 have low probability and are 'forbidden', although they can be observed if the symmetry of the molecule is disrupted (see Section 6.5.1).

Figure 6.1 shows the electronic transitions that are typical for organic molecules and the wavelength region in which they occur. The favoured transitions are $\pi \rightarrow \pi^*$ at around 200 nm, since they occur in conjugated systems (see below). $\sigma \rightarrow \sigma^*$ transitions are also favoured. However, these are in the low wavelength, vacuum UV region, below 190 nm, making them more specialised to measure and often they are not useful for interpretation. Consequently, they are not typically recorded.

$n \rightarrow \pi^*$ transitions, around 300 nm, are 'forbidden' but molecular vibrations and substitution can disrupt the symmetry that would otherwise result in a low value for ε. Two important transitions of this kind are those of ketones and benzene (see Section 6.5.1).

$n \rightarrow \sigma^*$ transitions, around 200 nm, are also 'forbidden' and are not typically seen in atmospheric UV-Vis spectroscopy.

To promote excitation, the energy of the incident light must be the same as the energy difference between the ground electronic state and the excited electronic state. Therefore, the wavelength (λ) of the absorption indicates the distance between the energy levels of the orbitals, providing information about the structure of the sample molecules. The structures that are responsible for absorption, and colour, are called chromophores (from Greek, meaning to bear colour) and in the UV-Vis region the most efficient are conjugated e.g. aromatic and heteroaromatic rings. Conjugation delocalises electrons and reduces the HOMO/LUMO energy difference requiring lower energy and longer wavelengths to promote excitation. Generally speaking, the more extensive a chromophore within a molecule, the greater the absorption. Delocalisation can also be increased by substitution with groups that have low or no absorption such as —OR, —NO_2, —NH_2 and —CO_2R, where R is an alkyl group. These are known as auxochromes and cause the absorption to be shifted to longer wavelength, i.e. a red or bathochromic shift, as well as increasing the signal intensity. (A shift left is called a hypsochromic or blue shift.)

In addition to the electronic transitions that take place, vibrational and rotational transitions also occur, which can cause small changes in the energy of the transitions. As a consequence, the sharp electronic absorption lines are broadened into peaks. There are also solvent–solute interactions, hydrogen bonding, induced dipole–dipole interactions and London dispersion interactions, which again cause changes to the transition energies that get averaged together. These result in decreasing resolution between peaks the more polar the solvent. This may be countered by analysing the sample at low temperatures. Combined with the other transitions, solvent–solute interactions result in the characteristic UV-Vis curve for samples in solution. In Figure 6.2 the absorption spectra of 1,2,4,5-tetrazine as a vapour, in hexane and in water are given, showing the peak broadening effects of solvent–solute interaction. In the non-polar solvent resolution is better.

Transition metal complexes can also be analysed using UV-Vis spectroscopy. If the complex has an organic ligand, peaks due to the ligand will be present in the spectrum, assuming there is some conjugation, as with any organic molecule. In addition, peaks due to ligand metal charge transfer (LMCT) and metal ligand charge transfer (MLCT) may be present. Here charge is transferred from the molecular orbital of the ligand to the metal resulting in reduction of the metal, LMCT, or from the metal to the ligand resulting in oxidation of the metal, MLCT. LMCT occurs from a bonding or non-bonding π orbital to a σ^* orbital, with $\varepsilon \sim 10\,000$, and results in intense peaks specific to the complex, with peaks for more easily reduced metals occurring at lower energy/higher wavelength. Examples include MnO_4^- and iron (Fe) oxides. MLCT is less common, since metals typically accept rather than donate electrons, but when it occurs it also results in intense peaks. Metals with low oxidation numbers result in low energy/high wavelength peaks and examples include $W(CO)_4$(1,10-phenanthroline) and tris(2,2-bipyridyl)ruthenium(II). It is these intense charge transfer transitions that lead to the characteristic bright colours of transition metal complexes.

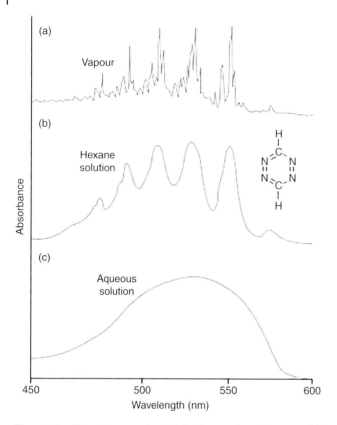

Figure 6.2 Absorption spectra of 1,2,4,5-tetrazine: (a) vapour; (b) hexane solution; and (c) aqueous solution. Source: Reproduced with permission from Cordonnier and Schaep (2008).

Less intense peaks due to the presence of the metal ion are the result of d-d transitions. The ligand electrons and the d orbital electrons repel each other, which causes the d orbital energies to be raised and split into two groups with different energy. If the two groups are far apart more energy and a shorter wavelength is required to promote an electron. The energy gap will vary depending on the metal ion, its oxidation state, and the ligand; therefore, the position of the spectrum peaks will be characteristic for a particular transition metal complex. Typically, ε is low, ~1–100 and, therefore, sensitivity can be low.

Typically, absorbance, A (see Section 6.2.3.1), is plotted against wavelength and the curve is centred on the wavelength where A is maximum, λ_{max}. λ_{max} can be used for characterising the spectrum and A at λ_{max} can be used for quantification. Alternatively, the curve can be plotted as ε vs. wavelength and ε_{max} is ε at λ_{max}. This is useful when comparing spectra from samples at different concentrations as ε does not depend on concentration.

6.2.2 Photoluminescence and Fluorescence

UV and visible radiation can also be used to promote photoluminescence and fluorescence. This can be recorded as excitation or emission spectra (see Section 6.3.2). Here, as before, a sample is illuminated with UV-Vis light, the radiation is absorbed by the sample and this

Figure 6.3 Energy level diagram for fluorescence.

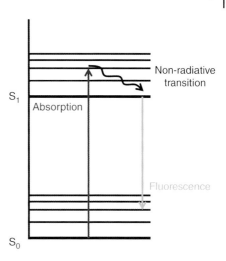

promotes the molecules to an excited electronic state. In fluorescence, the molecule loses energy internally, as vibrational energy, reaching the lowest vibrational level of an electronic excited state. A fluorescence photon is then emitted, in effect instantaneously, at longer wavelength than the excitation wavelength, returning the molecule to the ground state (Figure 6.3). The duration of photon emission is typically in the order of nanoseconds and ceases almost instantaneously when the excitation source is removed. Fluorescence is a type of photoluminescence and in this context, by convention, UV fluorescence refers to emissions stimulated in the UV region and observed in the visible region. Infrared (IR) luminescence refers to emissions stimulated in the visible or near-infrared (NIR) region and emitted in the IR region.

The structure within a molecule that causes fluorescence is known as a fluorophore. The more rigid a molecule the less likely relaxation is to occur by non-radiative means, i.e. the more likely is fluorescence. Aromatic molecules, therefore, are strong fluorophores. As with UV-Vis spectroscopy, substitution can alter the position and the intensity of λ_{max}. —NH_2, —$NHCH_3$ and —OH increase fluorescent intensity, whereas halogens tend to decrease intensity.

Also, as with UV-Vis spectroscopy, the fluorescence spectrum of a sample can be affected by the polarity and pH of the solvent and the temperature at which analysis is carried out. Non-polar solvents and reduced temperatures increase intensity. Further, the proximity of other molecules to the fluorophore can alter the fluorescence signal. This allows fluorescence spectroscopy to be used for monitoring chemical reactions.

Not all molecules are fluorescent. Hence fluorescence spectroscopy is not applicable to as many samples as UV-Vis spectroscopy. This aspect can be exploited because it allows fluorescent molecules to be analysed when mixed with molecules that are non-fluorescent. In addition, the fact that a molecule fluoresces can be used as a differentiating characteristic. Some non-fluorescent molecules may be made to fluoresce simply by changing the solvent due to solvent–solute interactions or by reacting the non-fluorescent molecules to produce a fluorophore, i.e. by derivatisation (see Section 4.2.6.2). Derivatisation can also be used to account for interferences present in a solution by employing difference spectroscopy. Here the molecule of interest is derivatised on the assumption that nothing else in the sample is affected. The difference between the non-derivatised molecule spectrum and the derivatised molecule spectrum can then be calculated and deleted from the non-derivatised spectrum.

The availability of excitation and emission spectra (see Section 6.3.2) also increases selectivity since there are more points for comparison. Samples may produce different spectra with the same excitation or similar spectra with different excitation.

Sensitivity is a further advantage of fluorescence spectroscopy and is increased relative to UV-Vis spectroscopy for two reasons. First, a molecule can be repeatedly excited to promote fluorescence, i.e. many fluorescent photons from a single molecule. However, the structure of a molecule is generally unstable while it is in the excited state and care must be taken not to degrade, i.e. photobleach, the sample. Secondly, although the fluorescent signal is weak relative to the incident light source, the incident light is eliminated from the signal reaching the detector. This results in being able to detect a small fluorescent signal against a low background signal. Whereas, in UV-Vis spectroscopy the measurement is the difference between 100% transmission and 100% minus the absorbed signal. The former (small signal minus ~zero signal) is easier to measure than the latter (large signal minus large signal). Sensitivity can also be increased by using a higher incident light intensity, although this is true of other techniques as well and, again, the possibility of photobleaching and photochemical reactions must be taken into consideration.

The efficiency of fluorescence, i.e. the ratio of the number of molecules that fluoresce relative to the number that absorb (quantum yield), has a value between 0 and 1. A value of 1 indicates that the molecule is 100% efficient, whereas a value of 0 indicates that it does not fluoresce. Further, since the number of molecules that fluoresce is relative to the number of molecules present, and given its sensitivity, fluorescence spectrometry/fluorometry can be useful for quantification (see Section 6.2.3.2).

6.2.3 Quantification

6.2.3.1 UV-Vis Quantification

Ease of quantification is one of the major advantages of UV-Vis spectroscopy over other spectroscopic techniques. Ideally, when quantifying, the area under the curve of the absorbing peak would be used. However, since the typical UV-Vis spectrum has broad overlapping peaks this is not possible. Instead, it can be achieved by using the Beer–Lambert law.

First we must define transmittance, T, and percentage transmittance, $\%T$, which are the fraction of the incident radiation that passes through the sample and the percentage of radiation that passes through, respectively.

$$T = \frac{I}{I_0} \text{ and } \%T = 100\left(\frac{I}{I_0}\right) \tag{6.2}$$

where I_0 is the intensity of the incident light and I is the intensity of the light transmitted through the sample. T can have a value between 0 and 1, i.e. anywhere from none to all of the light can be transmitted.

Absorbance, A, which is dimensionless, is given by

$$A = \log\left(\frac{I_0}{I}\right) = -\log T \tag{6.3}$$

Note, the difference between absorbance and absorption is an important one. Absorption is the process of absorbing the incident radiation, whereas absorbance is the attenuation (reduction in intensity) of the incident radiation, which may be due to processes such as scattering and reflection as well as absorption.

Equation (6.3) describes a logarithmic relationship because the amount of light penetrating the sample decreases with distance into it, therefore absorbance at increasing depths necessarily decreases.

An alternative expression for A is the Beer–Lambert law (Eq. (6.4)), which tells us that A is proportional to c.

$$A = \varepsilon l c \tag{6.4}$$

where ε is molar absorptivity in l/M/cm, although, by convention, it is usually given without dimensions (see Section 6.2.1), l is pathlength though the sample in cm, and c is concentration in M.

Since ε is fixed for a particular molecule at a given λ and l is known, the Beer–Lambert law can be used to calculate the concentration of a solution using the value of A at λ_{max}. The Beer–Lambert law does not hold at high concentrations because interactions between solute molecules can affect the value of ε. The law is considered to be valid when $0 < A < 2$. Ideally, A has a value around 1 to calculate concentration most accurately; this equates to 10% of the incident light being transmitted. An analyst should check their instrument for the values of A with a linear response.

A modified Beer–Lambert law can also be used when a mixture is present in the sample. In this case,

$$A = l(\varepsilon_1 c_1 + \varepsilon_2 c_2 + \ldots + \varepsilon_n c_n) \tag{6.5}$$

where ε_n and c_n refer to the nth compound.

6.2.3.2 Fluorescence Quantification

In the case of fluorometry,

$$I_f = Kc \tag{6.6}$$

where I_f is the power of fluorescent radiation, K is a constant related to quantum efficiency, and c is concentration.

Since fluorescence is proportional to concentration, quantification is typically calculated by plotting a calibration curve using a dilution series of standards in the usual way (see Section 2.3.1) in order to calculate the concentration of the sample, making sure to account for any drift in signal intensity caused by changes in the source intensity over time (usually by calculating a ratio with a reference beam from the same excitation source). Care must be taken to use low concentrations of analyte since self-quenching, where fluorescence is absorbed by sample molecules, can occur at high concentrations resulting in a non-linear relationship between concentration and intensity. Quantification of mixtures can also be carried out provided the emission spectra of the sample molecules do not interfere with each other.

6.3 Instrumentation

6.3.1 UV-Vis Spectrometers

UV-Vis spectrometers consist of the same elements common to most spectroscopic methods: a monochromator, usually with a diffraction grating to disperse the different wavelengths; a detector, such as a photomultiplier, charge-coupled device (CCD) or diode array detector (DAD); and a computer with specialist software. In addition, typical UV-Vis spectrometers are configured with a sample holder that takes a cuvette for samples in solution. Common UV-Vis sources are tungsten filaments or xenon arc lamps to cover the UV and visible region (300–2500 and 160–1100 nm, respectively), deuterium arc lamps to cover the UV region (190–400 nm) or light-emitting diodes (LEDs) for the visible region. Typical UV-Vis spectrometers use a deuterium lamp for the UV that produces light from 170 to 375 nm and a tungsten filament lamp for visible, which produces light from 350 to 2500 nm. Xenon lamps are relatively new and are an example of a pulsed source, which are used because they do not have to be warmed up, they can excite over both the UV and visible regions and they can limit the effects of stray light. Stray light affects the accuracy of the Beer–Lambert law and can be more pronounced at the limits of a radiation source. Therefore, using a single source, such as a xenon lamp, can counter this effect. Pulsed lights also last longer since they are on for less time in total each time a spectrum is captured.

After the light source will be a filter or a monochromator to direct the desired wavelength(s) of light onto the sample. The filter only allows a narrow range through, whereas a monochromator separates the light and positions a slit to allow through the relevant wavelengths. Using a monochromator, the whole wavelength range can be scanned allowing a spectrum to be captured.

Alternatively, instead of the light being separated before the sample, if a photodiode array detector is used, the incident radiation passes through the sample first, is separated using a prism or holographic grating, and is then detected by a linear array of photodiodes. In this case there is no scanning of wavelengths using a slit, rather, the whole spectrum is captured simultaneously, making it faster.

Usually either a UV or a visible spectrum is recorded; however, where a spectrum of UV and visible wavelengths is required, the system will either have a single source for the whole range or two light sources may be needed, which will be switched over to record the entire scan.

Instruments can also be single or double beam. In a single beam system, a reference spectrum of the solvent only is taken first, then the sample cuvette is placed in the beam and the absorbance is recorded. A single wavelength (λ), typically centred on λ_{max}, or a narrow range of wavelengths is selected as the incident radiation so that, coupled with the need to manually swap the reference and sample cuvettes, these systems do not record a full spectrum. In a double beam instrument the beam is split into two equal beams (Figure 6.4). One beam is sent through the reference cuvette and the other through the sample cuvette, then either two detectors measure the absorbance of both at the same time or a single detector measures one at a time. The reference signal is set as 0% and subtracted from the sample–solvent signal, leaving the signal for the sample only.

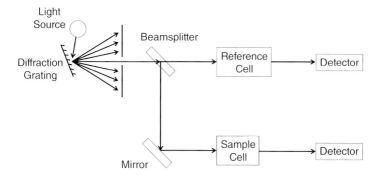

Figure 6.4 Schematic of a double beam UV-Vis spectrometer.

Polarisation analysis can also be carried out to determine if the sample is anisotropic, i.e. has different characteristics at different orientation. This is achieved by placing polarising filters on either side of the sample.

Note that, UV-Vis can also be used as the detection/quantification system for other analytical techniques, principally liquid chromatography (see Sections 6.2.3.1 and 14.2.8).

6.3.2 Fluorescence Spectrometers/Fluorometers

Fluorescence spectroscopy/fluorometry instrumentation is very similar to that of UV-Vis transmission/absorption spectroscopy. The principle difference is that the detector is at 90° to the UV beam. Transmitted light travels in a straight line through the sample and, therefore, is detected at 180° from the source. However, fluorescence is emitted in all directions from the sample, i.e. 360°. This means that the detector can be placed at 90° to record fluorescence rather than the more intense incident/transmitted light that would obscure the fluorescence (Figure 6.5). If the sample is not transparent, the detector can be placed in the same orientation as the source to detect fluorescence from the surface of the sample.

There is also, typically, a second filter or diffraction grating monochromator to remove any stray or scattered light. Stray light is a particular problem for fluorescence spectroscopy

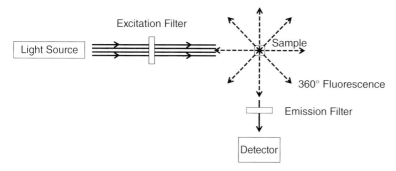

Figure 6.5 Schematic of a fluorescence spectrometer with detector at 90° to the excitation light source.

as the intensity of fluorescent light is typically hundreds to thousands of times less than that of the excitation light. Therefore, any incident light could swamp the fluorescent signal. The low intensity of fluorescent light (but not the presence of stray light) can be counteracted by using a photomultiplier detector to amplify the conversion of photons into electrons. Consequently, photomultipliers are typically used in fluorometry as the detector.

Filter fluorometers are cheaper than spectrofluorometers (instruments with monochromators); however, they are less sensitive and resolution is not as good.

There are two modes for fluorometry measurements, excitation and emission.

In excitation fluorometry, the excitation monochromator allows a range of wavelengths to pass through the sample but the emission monochromator is set at a fixed wavelength (the maximum excitation wavelength should not exceed the fixed wavelength of the emission monochromator). Excitation spectra are typically the same as their corresponding absorption spectra since the greater the number of molecules that absorb and are excited, the greater the number of emissions. Therefore, we will concentrate on emission fluorometry here.

In emission fluorometry, the excitation wavelength is fixed and is chosen to match the excitation maximum/absorption maximum (Ex_{max}/λ_{max}) of the fluorophore as closely as possible, which results in maximum intensity of the emission spectrum. The emission monochromator is scanned to allow wavelengths longer than the excitation wavelength to reach the detector. Where filters are used instead of monochromators, there will typically be a bandpass filter before the sample, as this will only allow a small range of wavelengths to pass through, i.e. a fixed excitation wavelength, and a longpass filter after the sample as this will allow a wide range of wavelengths through, whilst also blocking the incident/stray wavelengths.

The shape of the emission spectrum acquired for a fluorophore is the same no matter the excitation wavelength used. However, the intensity will decrease as the excitation wavelength gets further away from the Ex_{max}. The lamps used in UV-Vis spectroscopy can also be used for fluorometry and the wavelength of choice selected. This is particularly appropriate if the instrument can operate in excitation/UV-Vis absorption and fluorescence modes. However, given that in emission fluorometry the excitation wavelength is fixed, other sources with a narrow range of excitation wavelengths, such as lasers and LEDs, can be used. Lasers produce a very narrow range of wavelengths but are expensive (common wavelengths are 488, 543 and 633 nm), whereas LED sources have a wider excitation range but are low cost, have a long life, and low energy consumption.

Instruments are available that can record both the emission spectrum and the excitation spectrum of a sample. Figure 6.6 gives the excitation spectrum and the emission spectrum of a theoretical molecule on the same axis. Note that the emission spectrum is at longer wavelength, i.e. lower energy. The difference between the wavelengths of the excitation peak and the emission peak is called the Stokes shift (see Section 8.2.1). The larger the Stokes shift the easier it is to block the incident light from the detector as it is further away from the wavelength of fluorescence, reducing background signal. Note also that the spectra are mirror images of each other as the energy differences from the ground state to the excited state are the same as from the excited state to the ground state. Generally, the absorption/excitation spectrum and the emission spectrum of a molecule are near mirror images of each other.

Figure 6.6 Idealised absorption (dashed line) and emission (solid line) spectra of the same 'molecule' showing the mirror image.

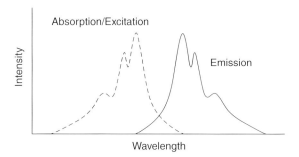

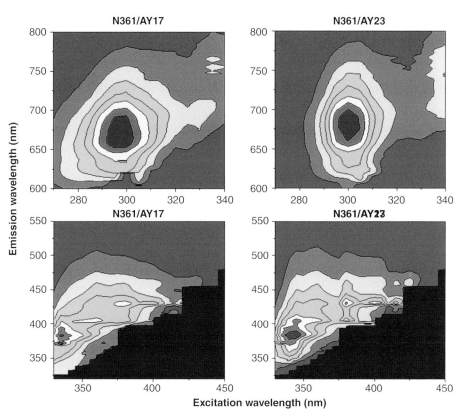

Figure 6.7 Excitation–emission matrix spectra of nylon 361 (N361) cloths pre-dyed with Acid Yellow 17 (AY17) and 23 (AY23) fibre extracts recorded at two different excitation/emission wavelength ranges. Source: Reproduced with permission from Campiglia et al. (2016).

Further, if emission spectra are recorded at regular intervals of excitation wavelength an 'excitation–emission matrix' or 'fluorogram' can be produced (Figure 6.7). This provides more data on the sample and more points of comparison to highlight any subtle characteristics of a sample that may not be captured with a single emission spectrum. In forensic science this may help to highlight, for example, differences in the fluorogram, which cannot be seen in a single emission spectrum, and these differences may allow two samples to be discriminated.

As with UV-Vis spectroscopy, polarisation analysis can also be done using polarising filters.

6.3.3 Coupling Techniques

Both UV-Vis and fluorescence spectroscopy are commonly used as detectors for other analytical techniques, notably high performance liquid chromatography (HPLC) but also capillary electrophoresis (CE). This is because of their general applicability and ease of quantification. With UV-Vis the detection system can be a fixed wavelength, which will simply detect when a molecule is present or not, or if a diode array is used a full spectrum can be recorded giving additional information in addition to the retention time information provided by HPLC (see Section 14.2.8). Where fluorescence is used as a detector, the sample may need to be derivatised in order to give a fluorescent signal. For both techniques, the HPLC solvent must be considered so that unwanted changes in the spectrum and solvent cut-off wavelengths (see Section 6.4.1.1) are avoided, otherwise, comparison with reference data is problematic.

6.3.4 Microspectrophotometers

As the name implies, microspectrophotometry (MSP) uses spectroscopy to acquire spectra from small samples (ng quantities), such as paint chips, ink lines, and single fibres. Microscopic systems can be used with a range of EM regions as excitation, UV-Vis, UV-Vis-NIR, Raman, etc. but the term MSP tends to be used for excitation with UV-Vis radiation. In order to analyse micro samples in this way a spectrometer is coupled with a microscope which, in the case of UV-Vis MSP, is fitted with UV-Vis transparent optics. MSP instruments have a single beam of incident radiation as there is only one path onto the sample via the microscope (and the reverse path) back to the detector. Typically, UV-Vis MSP systems use CCD detectors and, where low signal-to-noise ratio (S/N) is an issue as a result of small sample size, detector cooling can be employed to eliminate electronic noise. Some instruments can be operated in both transmission and reflectance mode. However, in the forensic context, reflectance is particularly useful for solid trace evidence samples. Spot sizes are typically around 0.1 mm^2 meaning that minute traces and points of interest in inhomogeneous samples can be identified and analysed, for example, single fibres (Wąs-Gubala and Starczak 2015). One of the main advantages of MSP is that for samples that look indistinguishable to the naked eye, it can give an objective evaluation of the colour rather than the subjective assessment of a single examiner. For the samples analysed using MSP, spectra can be recorded and interpreted according to λ_{max} and Ex_{max}, or the results can be presented as a numerical colour value.

For further information on MSP systems and analysis, Martin and Eyring (2014) and Adolf and Dunlop (1999) provide a comprehensive discussion.

6.3.5 Hyperspectral Imaging

One of the more recent developments in UV-Vis spectroscopy and fluorometry is the use of hyperspectral imaging (HSI). Here a datacube is acquired with the three dimensions being

Figure 6.8 Hypercube of a blood stain, with two spatial (x,y) and one wavelength (λ) dimension. From the hypercube an image plane is shown for one wavelength (λ_i) and a spectrum is obtained from one pixel (x_j, y_k). Source: Reproduced with permission from Edelman et al. (2012).

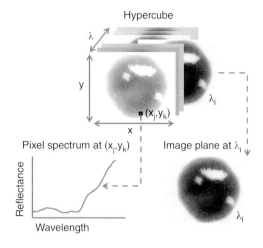

x, y and λ (Figure 6.8). The (x, y) coordinate range is a set of predetermined equally spaced points on the sample and the wavelength range is chosen as appropriate for the sample and technique. Once the (x, y, λ) data cube is acquired, the data can either be interrogated to produce a full spectrum for a particular point on the grid or to produce a map of intensity for a particular wavelength over the whole grid, i.e. a wavelength specific image. The main benefit of HSI is the wealth of data collected and which can be used to plot intensity maps of, in the case of UV-Vis and fluorescence, chromophores and fluorophores. As in other types of UV-Vis and fluorescence spectroscopy, samples can be modified in order to produce a signal/image, for example fluorescent labelling of biological samples. However, collecting the data cube can be time consuming. Note also, that the resolution of the image depends on spatial and spectral resolution. Spatial resolution depends on wavelength (Eq. 8.4) and is better at lower wavelengths, whereas spectral resolution depends on wavelength and instrument specification, e.g. diffraction grating.

6.3.6 Filtered Light Examination

FLE is a method of analysis that uses UV-Vis excitation to view forensic evidence. It is not a spectroscopic technique but is discussed briefly here because it uses UV-Vis light to analyse forensic evidence and because the use of FLE-MSP instruments for spectroscopic measurements is common in forensic analysis (Martyna et al. 2013).

In the forensic context, the technique and associated instrumentation was initially developed for document analysis and the examination of security features on paper currency but it can be used to visualise other types of evidence. It is non-destructive, rapid and low cost and, therefore, can easily be carried out at the beginning of an examination. FLE systems (for example the VSC®8000 manufactured by Foster and Freeman, UK) use a range of light sources, filters and an image capture system to view a sample at various wavelengths. The sources are housed within a single instrument that has an enclosure in which to place the sample. Typically, standard systems are used to view UV fluorescence, IR reflectance and IR luminescence by pairing the incident light with the appropriate filter, as well as ensuring that the image capture system (usually a camera and monitor) is sensitive to the wavelength of light reflected or emitted from the sample. In this way a sample can be viewed in a variety of light conditions affording multiple opportunities to visualise any differences, if they are present, from the rest of the document or any reference samples. As mentioned above, the most advanced systems can incorporate other UV-Vis techniques such as MSP and HSI (see Section 6.3.5) as well as the standard FLE examination. This provides yet more ways in which the forensic examiner can potentially discriminate one sample from another.

6.4 Application to Analyte

UV-Vis is often called a general technique because many samples are coloured and, therefore, absorb in the UV-Vis wavelength range. This is also why UV-Vis is a common choice of detector for other analytical techniques. Further, both reactions and chemical derivatisation can be used to alter the spectrum of a non-absorbing sample so that it falls in the UV-Vis region, increasing the range of samples that can be analysed. Fluorescence in samples is not as common as UV-Vis absorption, although, similarly, molecules can be derivatised to introduce fluorescence. However, fluorescence spectroscopy is more sensitive than UV-Vis due to the excitation of multiple fluorescent photons from a single molecule versus a small amount of light being absorbed over a short pathlength. Both techniques are relatively cheap, fast and, importantly for forensic samples, can be non-destructive.

6.4.1 Transmission Analysis in Solution

6.4.1.1 UV-Vis Solution Analysis

Where a sample is analysed in solution, the usual solvent is 95% ethanol (UV absorbing benzene must be removed), since it is transparent down to 210 nm, has good solvation properties, is inexpensive, and is readily available. Other solvents such as hexane, cyclohexane, chloroform, methanol, and propan-2-ol can be used, as appropriate, for the wavelengths of interest but all solvents have a cut-off wavelength below which $A = 1$ in a 1 cm cell, i.e. they are not transparent. Solvents should be chosen so that they will dissolve the sample, the cut-off wavelength is lower than the λ_{max} of the sample, impurities are minimised, and to be compatible with other coupled techniques if necessary. Some typical solvents and their cut-off wavelengths are given in Table 6.1. Care must also be taken to understand any effects, such as a change in intensity or position, that can be caused by the use of solvents with different polarity (Figure 6.2) and changes in pH.

The sample is usually very dilute, in the order of 1×10^{-6} M, and is poured into a cuvette in sufficient quantity that when placed in the instrument the meniscus is above the beam height. UV-Vis transparent cuvettes are usually made of fused silica (transparent to 165 nm) or quartz glass (transparent to 210 nm) and are used in pairs so that the reference and the sample spectra can be compared. Where volatile solvents are used, caps are available to prevent evaporation. Cuvettes must be cleaned fastidiously before use, making sure not to scratch the internal or external surfaces. Typically, cuvettes have two transparent sides opposite to each other, with the other opposite sides being frosted. The transparent sides are placed into the spectrometer perpendicular to the direction of the beam, so that the radiation can pass through. The cuvette should be handled on the frosted sides to avoid any

Table 6.1 Common UV-Vis solvent cut-off wavelengths.

Solvent	Minimum wavelength/cut-off (nm)
Acetonitrile, CH_3CN	190
Water, H_2O	191
Cyclohexane, C_6H_{12}	195
Hexane, C_6H_{14}	201
Methanol, CH_3OH	203
Ethanol, C_2H_5OH	204
Diethylether, $(C_2H_5)_2O$	215
Dichloromethane, CH_2Cl_2	220

marks or fingerprints transferring to the transparent sides, which might alter the absorption of the sample. Cuvettes often have a pathlength, l, of 1 cm. Alternatively, if sample quantity is an issue, they can retain the 1 cm pathlength, in the direction of the beam, but reduce the volume required by reducing the internal width of the cavity perpendicular to the beam. In addition, and again if concentration is an issue, cuvettes with longer pathlengths are available to increase the absolute value of absorption. In this case the analyst must remember that $l \neq 1$ when doing calculations.

As discussed, UV-Vis spectra, particularly in solution, tend to have a limited number of features (see Section 6.2.1). One way to counter this is to cool the sample, reducing changes to the transition energies, which allows the vibrational fine structures visible in vapour samples to be recorded.

6.4.1.2 Fluorescent Solution Analysis

Solvents can vary more widely in fluorometry than in UV-Vis spectroscopy since whether a molecule fluoresces or not can depend on the solvent and the pH. Fluorescence tends not to occur in water but is more likely in organic solvents, particularly with O_2 removed. As well as affecting the intensity, a change in solvent can affect the position and shape of the spectrum. This is important to remember and check when selecting a solvent and when comparing spectra that have not been collected at the same time on the same instrument e.g. when comparing with reference spectra or databases.

Cuvettes similar to those used in UV-Vis spectroscopy can be used for fluorometry. However, all four sides must be transparent since fluorescence is collected at 90° to the light source. Therefore, care must be taken to prevent and eliminate contamination from the sides before analysis. Typical sample volumes are 2–3 ml but analysis of sample volumes in the region of 100 μl are possible using capillary tubes.

6.4.2 MSP Sample Preparation

Sample preparation for MSP examinations is minimal, given that it is a non-destructive technique. The sample can be placed on a suitable substrate, such as a quartz microscope slide (with quartz coverslip for transmission analysis), which is then placed onto the MSP microscope stage. A mounting medium may be necessary and can be non-destructive, depending on the medium used, as it may be possible to retrieve the sample. If there is no shortage of sample (for example because a whole garment is available for fibre analysis) this is unlikely to be necessary but in cases where a small quantity has been recovered it may be imperative. Therefore, analysis without the mounting medium would be preferred. If a sample is to be mounted, the analyst should take care to select a mounting medium that does not produce a signal, particularly for fluorometry. The analyst must also make sure that the mounting medium used is compatible with all the techniques that are to be carried out on the mounted fibre, e.g. XAM is compatible with UV-Vis analysis but not with IR analysis. Hence, if MSP analysis is likely to be followed by IR analysis, mounting in XAM is not appropriate. If the sample is larger, it can be placed directly onto the stage.

Typically, the stage can be moved in the x, y, and z directions to position the area of interest in the path of the beam spot. The simplicity of sample preparation makes MSP suitable for many types of forensic evidence, particularly trace and coloured evidence, such as textile fibres, paint, ink/documents, and body fluids (Zeichner et al. 1988; Trzcińska et al. 2013; Wąs-Gubala and Starczak 2015; Denny et al. 2018). MSP can also be used in transmission mode for UV-Vis transparent samples. In this case there may be some sample preparation required, such as using a microtome to produce very thin sections so that the light source is able to pass through.

6.4.3 Acquiring a Spectrum

Calibration should be carried out before any sample spectra are captured for quantification. UV-Vis spectroscopy instrument calibration for wavelength accuracy is carried out by looking at the λ_{max} of a source such as a deuterium discharge lamp or a mercury vapour lamp. Calibration for absorbance and stray light are carried out with reference solutions.

In fluorometry, instrument calibration can be carried out using selected wavelengths of the UV-Vis light source e.g. 450.1 nm of a xenon lamp.

6.4.3.1 Capture of Spectra in Solution

The first step in analysing a sample using a spectrometer, where a pulsed source is not used, is to turn on the instrument about 20 minutes prior to analysis so that the lamps can stabilise. The instrument parameters must then be set. This includes selecting the range of wavelengths, or single wavelength, to pass through the sample, the scan interval (how many nm between measurements), the scan speed (how fast to collect the data, typically nm/s) and the scan bandwidth (the width of the monochromator slit, which affects resolution, typically in nm). If the instrument is single beam, calibration of the sample should be carried out by taking a reference spectrum of the solvent only, which can be subtracted from the sample absorbance. For a double beam instrument, the calibration is carried out as the spectrum is acquired by scanning the solvent reference sample at the same time as the dilute sample. The sample, once prepared, is placed in a cuvette, which is placed in the sample holder and the cover of the instrument is closed to prevent ambient/stray light from hitting the detector.

For UV-Vis spectroscopy, the excitation radiation wavelength range should be set to centre on the λ_{max} of the sample molecules if λ_{max} is known, or on another broad peak centre if appropriate, since this is where maximum sensitivity is.

The spectrum is then collected and a method of interpretation employed as appropriate, i.e. identifying λ_{max}, quantification, or spectral interpretation (see Sections 6.2.3 and 6.5.1).

Ideally for a single beam instrument the absorbance for a sample should be between 0.3 and 0.6 absorbance units and for a double beam instrument between 0.6 and 1.2 absorbance units. Keeping within these ranges maximises the precision of the data.

6.4.3.2 MSP and HSI Sample Analysis

Once the sample is prepared and positioned on the microscope stage, MSP and HSI spectrum capture proceeds as with the standard form of UV-Vis or fluorescence spectroscopic analysis. The instrumental parameters are set as appropriate and a spectrum is acquired in a matter of seconds. However, care must be taken with the sample since the high intensity radiation is concentrated in a small area and this may cause photobleaching and degradation if left on the sample for too long.

6.4.4 Forensic Applications

The application of UV-Vis techniques to samples in forensic science varies considerably with the type of technique used. Standard spectroscopic techniques usually require the sample to be in solution for the light source to pass through the sample, i.e. transmission mode. UV-Vis spectroscopy has been used in this way to analyse dilute ink samples and counterfeit whisky samples (Adam et al. 2008; Martins et al. 2017). In the case of some samples, e.g. whisky, this is non-destructive but if applied to ink on a questioned document the ink needs to be extracted from the paper, which is detrimental to the evidence and, therefore, undesirable. In contrast, MSP methods can be carried out on solid samples in reflectance mode and in transmission mode for small transparent samples. Therefore, these methods may be more applicable to forensic samples because they are non-destructive (see Section 1.2.1) and maintain sample integrity. Research and application of MSP, particularly coupled with HSI, has seen an increase in recent years analysing evidence types such as ink, paint, fibres, fingerprints, and firearm propellants (Payne et al. 2005; Edelman et al. 2012; Lin et al. 2016; Starczak and Wąs-Gubała 2016).

Fluorescence spectroscopy/fluorometry is not used as widely in forensic science as UV-Vis spectroscopy. However, it has also been applied to a variety of forensic applications including detecting amylase in dried human saliva collected from drinking glasses, differentiating samples of black ballpoint pen ink, and detecting untreated latent fingermarks (Nakamura et al. 2015; Campiglia et al. 2016; Denny et al. 2018).

Typically, in a forensic context the purpose of an examination is to determine whether or not two pieces of evidence have the same characteristics, bearing in mind the limitations of the techniques used. The implication being that if they do have the same characteristics, they may also have the same source. They may not, but this possibility should be considered when the forensic scientist considers their conclusions (see Section 1.2.3.2). As a consequence, much of the research carried out in forensic science focuses on the ability of techniques to discriminate between samples. Clearly, in the case of UV-Vis-based techniques, it is the ability to discriminate between samples that appear the same colour to the naked eye but do not have the same spectral characteristics, i.e. are metameric, that is of interest. Results are often reported as discriminating power (DP), which calculates the ratio of the number of pairs of e.g. inks differentiated from each other to the total number of pairs of inks compared.

Two types of coloured evidence, inks and fibres, routinely analysed using UV-Vis radiation source methods are discussed at further length in the following two sections.

6.4.4.1 Writing Ink Examination

Typically document examination, whether UV-Vis or other spectroscopic technique, focuses on comparison of the ink component. (See Section 8.3.2.1 for discussion of ink analysis using Raman spectroscopy.) In many cases, the ink in question is ballpoint pen ink, particularly black and blue. Ink is composed of (i) a vehicle that carries (ii) the coloured component, and (iii) other manufacturer specific additives. The portion of interest in UV-Vis analysis is the dye or pigment providing the colour. Common questions are: was a different writing implement, and therefore ink, used in places? Has a portion of writing been obscured or altered? Which ink line was deposited first? Could this pen have been used to write this document? The American Society of Trace Evidence Examiners (ASTEE) Scientific Working Group for Forensic Document Examination (SWGDOC) has published various standards including the Standard for Test Methods for Forensic Writing Ink Comparison (SWGDOC 2013), which reports techniques that have been well tested and that are used routinely by document examiners. Both MSP and spectrofluorometry are included in the standard with reference to published work (Kelly 1973; Zeichner et al. 1988).

Standard UV-Vis analysis has been carried out on ink samples that have been extracted with solvents from a questioned document (Thanasoulias et al. 2003; Adam et al. 2008). This is undesirable, as mentioned above, as it affects the integrity of the document, although, the authors have argued that it allows more techniques to be carried out, including chromatographic techniques, and that it improves resolution.

However, in keeping with the SWGDOC standard, ink analysis research has been carried out in situ with MSP and/or HSI (Pfeffferli 1983; Reed et al. 2014). In these works, the ink was analysed on a substrate with no sample preparation other than to deposit the ink line. In addition, chemometric methods have been applied to help improve differentiation. (See Section 2.4.3 for chemometric experimental design.) For inks of the same colour, particularly dark inks, the spectra are often very similar, and it can be difficult to determine whether any slight differences are due to variation within the same ink or to chemical differences between inks. Principal component analysis (PCA) has been used to objectively separate black and blue ballpoint pen ink spectra, in situ and as extracts, into groups with the same UV-Vis characteristics (Adam et al. 2008; Kumar and Sharma 2017). In addition, preliminary work has been carried out to determine the order of inks deposited in line crossings, where one ink has been deposited on top of the other, using UV-Vis HSI and multi-variate curve resolution–alternating least squares (MCR-ALS) (Martins et al. 2019).

Fluorescence spectrophotometry with PCA has also been used to analyse black ballpoint pen inks (Adam 2008). (The author of the publication refers to the excitation at 590 nm and the emission scan between 620 nm and 770 nm as luminescence but by the definition given in Section 6.2.2 we use fluorescence.) For the 10 inks analysed, a higher DP of 0.6 was achieved using this method rather than 0.2 with FLE.

6.4.4.2 Fibre Examination

The examination of fibre evidence is required in forensic cases where transfer has taken place between two objects with a textile component. This could be, for example, from clothing to car seat upholstery, and vice versa, in a vehicle incident, or from clothing to clothing in an assault or murder case. An examination is more likely to be carried out to associate two fibres and establish the possibility of a common source than it is to determine the type and source of a fibre. The ASTEE Scientific Working Group for Materials Analysis (SWG-MAT) Fiber Subgroup has published various standards and guides for training of experts and examination of fibres, which include the use of UV-Vis spectroscopy as a cheap and quick, widely available, (potentially) non-destructive and reproducible technique (SWG-MAT 2004). However, fibre examination research is moving away from UV-Vis spectroscopy of extractions due to its destructive nature and the availability of MSP systems.

Microscopy should almost always be carried out first on most types of trace forensic evidence but MSP can then be carried out quickly without affecting the outcome of further analytical techniques. UV-Vis spectroscopy is, clearly, well suited to analysing the coloured component of a fibre, i.e. the dye, and for potentially discriminating between two fibres that appear to be the same colour to the naked eye, hence addressing the issue of a common source.

If fibre composition and dye identification are required, further techniques will typically need to be employed to probe different aspects of the fibre, e.g. infrared spectroscopy for fibre composition.

Many different fibre types exist but they can be split into two main classes, natural and man-made/synthetic. Natural fibres include cotton (one of the most ubiquitous fibres), wool, and silk. Synthetic fibres include polyester, polyamide, and acrylic. Fibres also come in different physical forms, i.e. thickness, cross-section, length etc. Further, there are a multitude of different dyes of various types. Whilst UV-Vis spectroscopy is not able to assess all of these characteristics, it is important to know that they can affect a UV-Vis spectrum.

Fibre type is an important consideration for the fibre examiner. For some fibres, the composition of the fibre will contribute to the spectrum at certain wavelengths meaning that, if it is the dye that is of interest, those wavelengths should not be captured. For example, polyester contributes to spectra below 310 nm so that for these fibres only a UV-A and visible spectrum is required (Wąs-Gubala and Starczak 2015). Similarly, the keratin in woollen fibres absorbs below 320 nm so, again, a UV-A and visible spectrum should be captured (Starczak and Wąs-Gubała 2016). Conversely, the cellulose in cotton fibres does not contribute significantly to the UV-Vis spectrum, therefore there is no restriction on the wavelength range that can be used (Wąs-Gubala and Starczak 2015).

Concentration of dye used to colour a fibre has also been investigated. There is agreement that UV-Vis can detect dye on what appear to be colourless fibres but also that as dye concentration increases the spectra change (Suzuki et al. 2001; Wąs-Gubala and Starczak 2015; Reichard et al. 2017). Most papers report that the change is in intensity rather than λ_{max} but the study by Reichard et al. (2017) found that for yellow polyester fibres, λ_{max} had a several nanometre red shift as dye concentration increased. Also, reported in the above literature are differences in spectra between fibres sampled from different areas of the same piece of textile, possibly due to variations in fibre thickness, orientation, or internal scattering. The effects can be more pronounced with natural fibres since they are not uniform like synthetic fibres. Further, where multiple dyes are used it is not always possible to detect the minor component dyes, particularly if there is a high ratio of major to minor dye (Wąs-Gubala and Starczak 2015). Clearly this has implications for analysis of fibres where the dye concentrations and area of origin are not known. Chemometrics can go some way to mitigating against these variations but care must be taken with false negatives.

Wiggins et al. investigated the use of first derivative spectra for ink analysis (2007). They found that in many cases it did not improve discrimination and, due to intra-fibre variation, could lead to false exclusions. However, in some cases where there were fewer features in the spectra it could be useful.

A further development in fibre examination is the use of liquid crystal tunable filter MSP (Markstrom and Mabbott 2011). Here, small lengths of fibre (~1 mm) trapped between two microscope slides were analysed. Spectra of the entire field of view could be captured at a range of wavelengths so that a spectrum of a particular location could be viewed. Advantages of this system are that multiple fibres can be analysed under the same conditions and the areas to investigate selected after the data is captured.

In addition to UV-Vis spectroscopy, fluorescence spectroscopy has also been the subject of recent fibre examination work (Campiglia et al. 2016). Here statistical analysis was used to interpret excitation–emission matrix spectra of fibre dye extracts and to propose fluorescence spectroscopy as a potential tool for the future in fibre examination. Further work by the same group has also tested the level of fluorescence over time in two contrasting climates (Mujumdar et al. 2019).

6.5 Interpretation and Law

In a forensic context, depending on the examination underway, interpretation for most spectroscopic techniques can be carried out on two levels: structural elucidation or comparison of spectra to determine if the samples have the same characteristics or not. This is true for UV-Vis spectroscopy, however, as seen in Figure 6.2, UV-Vis spectra, particularly of solutions, are characterised by a small number of broad peaks and few distinguishing features. (Cooling a sample can counter some of the broadening due to solvent–solute interactions and vibrations.) This is one of the major drawbacks of UV-Vis spectroscopy where interpretation is concerned. It can provide some information about the structure of the sample molecules but it is limited. Due to this, where the structure of the sample molecules is important, UV-Vis is typically used to provide supporting evidence for other analytical techniques, such as IR, Raman, mass spectrometry (MS), nuclear magnetic resonance spectroscopy (NMR), etc. Having said that, data processing methods, such as calculating derivative spectra, can be used to highlight differences where they occur. (Here 'derivative' refers to a mathematical function, differentiation, rather than a chemical alteration of the sample.) Derivative spectra are plotted by calculating the derivative of absorbance with respect to wavelength at each point on the spectrum and can be first, second or higher

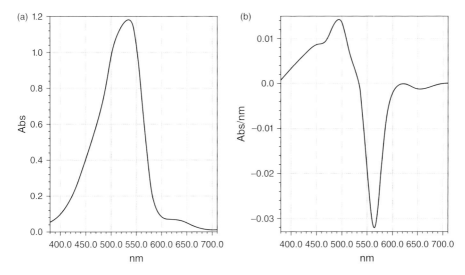

Figure 6.9 Absorbance spectrum (a) and associated first-order derivative spectrum (b) of a red acrylic fibre. Source: Reproduced with permission from Wiggins et al. (2007).

order (Figure 6.9). One advantage of the first-order derivative spectrum is that it identifies any obscured peak maxima. This is because the gradient at a peak maximum in a non-derivative spectrum is zero, therefore, the wavelengths at which a first-order derivative spectrum crosses the x-axis are the wavelengths of peak maxima.

Statistical methods can also be used where there is a contaminant or mixture present to separate the absorption contribution from each, e.g. the Morton and Stubbs correction (1946), which uses ratios of peak intensity, and the Vierordt method, which uses peak intensities and simultaneous equations (Glenn 1960).

Both the structural elucidation and the comparison methods will be discussed in the following two sections.

6.5.1 Interpreting UV-Vis Spectra

As discussed in Section 6.2.1, the chromophores that lead to UV-Vis absorption are typically conjugated systems, where electrons are delocalised. Consequently, single isolated saturated bonds do not appear in UV-Vis spectra. For organic molecules, the absorption wavelengths of basic chromophores can be determined from the literature; examples are given in Table 6.2 (values can slightly vary depending on the source).

In many cases, the analyte of interest will not be a simple molecule and the more complicated a molecule the more the spectrum is affected by the extended structure of the molecule, until for some molecules, the spectrum becomes a fingerprint for that molecule rather than a spectrum that can be broken down for interpretation.

Typically, as conjugation increases the λ_{max} and the intensity of absorbance increase. For example, ethylene has a λ_{max} of 170 nm and ε_{max} of 16 000, but if a molecule has two ethylene groups joined by a C—C bond then λ_{max} increases to 220 nm and ε_{max} to 21 000. In fact, the values for ε_{max} for both $\pi \rightarrow \pi^*$ and $n \rightarrow \pi^*$ transitions increase, which increases

Table 6.2 λ_{max} values for common chromophores.

Chromophore	λ_{max} (nm)
Ether, —O—	185
Ethene, —C=C—	190
Thiol, —SH	195
Amine, —NH$_2$	195
Ketone, >C=O	195, 275
Benzene	254
Azo, —N=O	>290

the intensity of $n \rightarrow \pi^*$ transitions sufficiently for them to be 'allowed' and, consequently, contribute more to the spectrum and to the interpretation. The $n \rightarrow \pi^*$ of ketones and benzenes, mentioned briefly in Section 6.2.1, are two 'forbidden' but important transitions that are observed because of this interaction. The ketone band is around 300 nm ($\varepsilon \sim 10$–100) and the benzene band is around 260 nm ($\varepsilon < 100$).

In practice, UV-Vis spectra are interpreted by identifying potential 'parent' chromophores and then following a set of rules, such as Woodward's rules for dienes, to calculate the effect of further conjugation, substitution, and the presence of auxochromes. The reader is directed to Fleming and Williams (1966) for an in depth discussion but an example is given in Table 6.3 to illustrate the concept. Here, the shift for $\pi \rightarrow \pi^*$ transitions in α,β-unsaturated ketones due to common substitutions are given.

In addition to calculating the effects of conjugation and substitution, the analyst should also take care to note any effect of the solvent on λ_{max} (see Section 6.2.1). Again, data tables are available with corrections for the most common solvents and an example is given in Table 6.4 (Fleming and Williams 1966; Cooper 1980).

Once structural interpretation has been carried out it may be important to carry out quantification. This would be carried out using the Beer–Lambert law as discussed in Section 6.2.3.

If neither structural elucidation nor quantification is important the forensic scientist will typically be trying to determine if two (or more) samples or a sample and reference data have the same characteristics. As discussed above, UV-Vis spectra typically have few distinguishing features (unless the sample is in the vapour phase or cooled) which can make it difficult to discriminate between the spectra of two samples, particularly if they are structurally similar. Spectra where the wavelengths at which absorption occurs are clearly different are the easiest cases to deal with. However, if there are only subtle differences it can be difficult to know if these are due to normal variation or are real differences. Assuming that the differences are reproducible it still may be possible to discriminate between spectra. Increasingly good discrimination has been achieved by employing chemometric methods (see Section 2.4.3) to separate groups of spectra that cannot be distinguished manually (Reichard et al. 2017). Such methods allow the scientist to be more confident in their conclusions. Examples of the use of chemometrics are given in Sections 6.4.4.1 and 6.4.4.2.

Table 6.3 Rules for α,β-unsaturated ketones and aldehyde absorption.

Value assigned to parent α,β-unsaturated six-ring or acyclic ketone	215 nm
Value assigned to parent α,β-unsaturated five-ring ketone	202 nm
Value assigned to parent α,β-unsaturated aldehyde	207 nm
Increments for	
(a) a double bond extending the conjugation	30 nm
(b) each alkyl group or ring residue α	10 nm
β	12 nm
γ and higher	18 nm
(c) auxochromes	
(i) –OH α	35 nm
β	30 nm
δ	50 nm
(ii) –OAc α, β, δ	6 nm
(iii) –OMe α	35 nm
β	30 nm
γ	17 nm
δ	31 nm
(iv) –SAlk β	85 nm
(v) – Cl α	15 nm
β	12 nm
(vi) – Br α	25 nm
β	30 nm
(vii) –NR$_2$ β	95 nm
(d) the exocyclic nature of any double bond	5 nm
(e) homodiene component	39 nm
λ_{calc}^{EtOH}	Total
For λ_{max}^{calc} in other solvents a solvent correction must be subtracted from the above value	

Source: Reproduced with permission from Scott (2013).

6.5.2 Interpreting Fluorescence Spectra

Many of the principles of interpretation in UV-Vis spectroscopy can also be applied to fluorescence spectroscopy, particularly since the emission and excitation spectra are typically mirror images of each other. This enables fluorescence spectra to be analysed for structural information, quantification and for 'pattern matching' comparison. However, fluorescence spectra are more commonly used for monitoring conformational changes, reactions and fluorometry, i.e. as a method of quantification.

Table 6.4 Solvent corrections for α,β-unsaturated ketones.

Solvent	Correction (nm)
Ethanol	0
Methanol	0
Dioxane	+5
Chloroform	+1
Diethylether	+7
Water	−8
Hexane	+11
Cyclohexane	+11

Source: Reproduced with permission from Scott (2013).

6.5.3 UV-Vis and Fluorescence Spectroscopy in Court

UV-Vis spectroscopy and fluorescence spectroscopy have been available for analysis of samples for many years. They are both techniques that are well accepted by the scientific community. However, to be accepted in court they must also be widely accepted as appropriate for the particular type of evidence in question. Forensic laboratories and scientists must ensure that they are abiding by the relevant guidelines for the evidence type being examined and that their choice of technique is included in guidelines. The particular guidelines used will vary with jurisdiction but they are typically published by bodies such as the ASTEE Scientific Working Groups, American Society for Testing and Materials (ASTM) and the European Network of Forensic Science Institutes (ENFSI).

6.6 Case Studies

6.6.1 Case Study 1

In a case in Japan, reported in the literature in 2009, a victim was found dead lying on their front on a bed, covered in a duvet (Suzuki et al. 2009). A suspect was apprehended who admitted entering the victim's house and admitted theft but denied murder. The scene was examined, and numerous fibres were recovered from in and around the bed and from the victim. The numbers of fibres in particular locations were recorded, e.g. on the left upper arm. As a result of information from an eyewitness who was in the area at the time of the incident and reported seeing an unknown man, a dark yellowish-green woollen garment was recovered from the suspect's house. The fibres recovered from the scene included fibres with a similar appearance to the seized woollen garment. They were examined with a microscope and spectra were acquired using a UV-Vis MSP instrument between 240 nm and 780 nm using a scan interval of 2.5 nm. The fibres were found to have the same microscopic and spectral features as those in the suspect's garment.

In this case the suspect did not deny being at the scene, therefore, it might be expected that some fibres from his clothing would be found on items at the scene, giving them low potential evidential value (see Section 1.2.3.2). The suspect gave a full explanation of his actions at the scene, still denying murder, which included finding the victim while carrying out the theft; moving the body to check for life and to try to resuscitate by chest compressions; trying to remove ligatures that were tying the feet and arms together; trying to reposition clothing so that it was correctly on the victim; and covering the victim's body with the duvet again. Taking the suspect's account of his actions at the scene, the scientists in the case reconstructed the incident to determine what the likely distribution of the fibres of interest (target fibres) would be at the scene if the confession was true. Before the simulation, the scene was cleaned to confirm that there were no target fibres present. In addition, the scene examiners were the same examiners that had processed the original scene, ensuring a consistent approach. The finding on repeating the simulation four times showed that the distribution of fibres for the simulation did not match the distribution of the fibres from the scene. Notably, more fibres were found on the lower back half of the victim's body (22 vs. average 1) and front right leg (25 vs. average 1.5) and on the bedcovers under the body (200 vs. average 38). The total target fibres found was also much higher, 314 versus an average of 101. These findings refuted the suspect's statement, suggesting stronger or more prolonged contact and giving more evidential value to the MSP findings. The fibre evidence was admitted in court and the suspect was found guilty of murder, theft, and trespassing.

In this case it was not only the analytical data that helped the court to come to a verdict but also the scenario specific experiments, which give context to the scientific results. It demonstrates one of the most important aspects of a forensic scientist's job when drawing conclusions. This is that they must understand the best way to scientifically analyse the evidence in the laboratory as well as the evidential value issues for the particular evidence type and, crucially, how these two factors affect each other.

6.6.2 Case Study 2

Investigation of an organised crime related double murder case in Italy used UV-Vis MSP and IR spectroscopy followed by Bayesian statistical analysis (see Section 1.2.3.4) to associate the materials in which the two bodies were wrapped with the vehicle in which they were transported (Causin et al. 2004). Fibres were extremely important in the case as there was no DNA or fingerprint evidence.

The male victims had been killed, possibly with a crowbar, and then burnt in a hole in the ground and buried. They were subsequently dug up, wrapped together as one in blankets, which were secured with tape, and then driven in the boot of a vehicle to another location and left in a pond. Crucially, there were tears in the outer wrapping that allowed fibres from the inner wrapping to be transferred on contact.

The examiners determined what fibres were present in the wrapping layers. The outer layer was a beige bedspread with red embroidered flowers and the inner layer was a checked red, green, white and beige blanket. Some of the fibres were not in good condition, particularly the natural fibres, due to having been submerged in water. However, the synthetic fibres were better preserved. Therefore, the analysis focused on the synthetic fibres.

An intelligence source suggested a possible perpetrator to the police so that a forensic examination of the boot of the perpetrator's vehicle was possible. The boot was tape lifted, the location of each tape lifting was noted, and then the examiners examined the tape, using an automated system, looking, in particular, for fibres that might match the synthetic fibres found in the blankets wrapping the bodies. A total of 298 beige, 204 red and 111 green fibres of a total of six different types were flagged as potentially having the same source as the blanket wrappings. Colourless fibres were manually located in the tapings until 20 were recovered. The beige fibres were located all over the boot, whereas the green and red fibres were only on the left-hand side, which is consistent with the beige fibres being transferred from the whole surface of the bundle but the green and red fibres only being able to be transferred from the area with the tears where they were exposed.

A sampling calculation suggested a sample size of 20 fibres per colour/type. UV-Vis MSP analysis and IR spectroscopy were carried out on the fibre samples. It was found that at least 50% and up to 70% of each set of fibres had the same characteristics as the reference fibres from the wrappings. Although the fibres were not rare, the combination, or cumulative evidential value, of six different fibre types recovered from the vehicle and them being compatible with the body wrapping fibres provided strong evidence for contact between the wrapped bodies and the vehicle.

6.7 Forensic Developments

UV-Vis techniques are commonly used in forensic science and well accepted in the courts. Traditionally, this is in the form of standard or MSP UV-Vis spectroscopy or fluorescence spectroscopy to probe the chromophore/fluorophore of a sample or as a FLE to promote fluorescence or luminescence in order to provide an image of a sample. More recent techniques include those discussed in Section 6.3, such as HSI and pulsed light sources. In the future, as with other spectroscopic techniques, the improvements and availability of portable/handheld systems, fibre optic remote sampling and systems with an extended range into the NIR region will further advance the applicability and immediacy of UV-Vis and fluorescence analysis.

References

Adam, C.D. (2008). In situ luminescence spectroscopy with multivariate analysis for the discrimination of black ballpoint pen ink-lines on paper. *Forensic Science International* 182 (1–3): 27–34.

Adam, C.D., Sherratt, S.L., and Zholobenko, V.L. (2008). Classification and individualisation of black ballpoint pen inks using principal component analysis of UV–vis absorption spectra. *Forensic Science International* 174 (1): 16–25.

Adolf, F.-P. and Dunlop, J. (1999). Microspectrophotometry/colour measurement. In: *Forensic Examination of Fibres*, 2e (eds. J. Robertson and M. Grieve), 251–289. Boca Raton: CRC Press.

Baldock, B.L. and Hutchison, J.E. (2016). UV–visible spectroscopy-based quantification of unlabeled DNA bound to gold nanoparticles. *Analytical Chemistry* 88 (24): 12072–12080.

Baret, P., Robert, C., and Cadet, F. (1998). Microquantification of proteins by spectrophotometry. Part II: application procedure for complex mixture containing interfering substances. *Spectroscopy Letters* 31 (3): 487–501.

Baszanowska, E. and Otremba, Z. (2016). Fluorometry in application to fingerprint of petroleum products present in the natural waters. *Journal of the European Optical Society-Rapid Publications* 12 (1): 16.

Campiglia, A.D., Rex, M., de la Peña, A.M., and Goicoechea, H.C. (2016). Excitation–emission matrix fluorescence spectroscopy combined with MCR-ALS as a tool for the forensic analysis of similar and dissimilar sets of textile fiber extracts. *Analytical Methods* 8 (47): 8314–8321.

Causin, V., Schiavone, S., Marigo, A., and Carresi, P. (2004). Bayesian framework for the evaluation of fiber evidence in a double murder – a case report. *Forensic Science International* 141 (2–3): 159–170.

Cooper, J.W. (1980). *Spectroscopic Techniques for Organic Chemists*. Wiley.

Cordonnier, J. and Schaep, J. (2008). Ultraviolet, visible and fluorescence spectrophotometry. In: *Clarke's Analytical Forensic Toxicology* (eds. A. Negrusz and G. Cooper), 393. London: Pharmaceutical Press.

Denny, S.E., Nazeer, S.S., Sivakumar, T. et al. (2018). Forensic application of fluorescence spectroscopy: an efficient technique to predict the presence of human saliva. *Journal of Luminescence* 203: 696–701.

Edelman, G., Gaston, E., Van Leeuwen, T. et al. (2012). Hyperspectral imaging for non-contact analysis of forensic traces. *Forensic Science International* 223 (1–3): 28–39.

Fleming, I. and Williams, D.H. (1966). *Spectroscopic Methods in Organic Chemistry*. New York: McGraw-Hill.

Glenn, A. (1960). The importance of extinction ratios in the spectrophotometric analysis of mixtures of two known absorbing substances. *Journal of Pharmacy and Pharmacology* 12 (1): 595–608.

Gorog, S. (2018). *Ultraviolet-Visible Spectrophotometry in Pharmaceutical Analysis*. CRC Press.

Grieve, M.C., Dunlop, J., and Haddock, P. (1988). An assessment of the value of blue, red, and black cotton fibers as target fibers in forensic science investigations. *Journal of Forensic Science* 33 (6): 1332–1344.

Jaffé, H.H. and Orchin, M. (1962). *Theory and Applications of Ultraviolet Spectroscopy*. Wiley.

Kelly, J. (1973). Spectrofluorometric analysis of ball point ink. *Journal of Police Science and Administration* 1 (2): 175–181.

Kumar, R. and Sharma, V. (2017). A novel combined approach of diffuse reflectance UV–Vis-NIR spectroscopy and multivariate analysis for non-destructive examination of blue ballpoint pen inks in forensic application. *Spectrochimica Acta Part A: Molecular and Biomolecular Spectroscopy* 175: 67–75.

Lin, I.C.P., Hemmings, J., Otieno-Alego, V. et al. (2016). A comparison of conventional microspectrophotometry and hyperspectral imaging for the analysis of blue metallic paint samples. *Journal of Forensic Identification* 66 (5): 429.

Markstrom, L.J. and Mabbott, G.A. (2011). Obtaining absorption spectra from single textile fibers using a liquid crystal tunable filter microspectrophotometer. *Forensic Science International* 209 (1–3): 108–112.

Martin, P.C. and Eyring, M.B. (2014). Microspectrophotometry. In: *Experimental Methods in the Physical Sciences*, vol. 46 (ed. A.K. Shukla), 489–517. Elsevier.

Martins, A.R., Dourado, C.S., Talhavini, M. et al. (2019). Determination of chronological order of crossed lines of ballpoint pens by hyperspectral image in the visible region and multivariate analysis. *Forensic Science International* 296: 91–100.

Martins, A.R., Talhavini, M., Vieira, M.L. et al. (2017). Discrimination of whisky brands and counterfeit identification by UV–Vis spectroscopy and multivariate data analysis. *Food Chemistry* 229: 142–151.

Martyna, A., Lucy, D., Zadora, G. et al. (2013). The evidential value of microspectrophotometry measurements made for pen inks. *Analytical Methods* 5 (23): 6788–6795.

Morton, R. and Stubbs, A.L. (1946). Photoelectric spectrophotometry applied to the analysis of mixtures, and vitamin A oils. *Analyst* 71 (845): 348–356.

Mujumdar, N., de la Peña, A.M., and Campiglia, A.D. (2019). Classification of pre-dyed textile fibers exposed to weathering and photodegradation by non-destructive excitation-emission fluorescence spectroscopy paired with discriminant unfolded-partial least squares. *Forensic Chemistry* 12: 25–32.

Nakamura, A., Okuda, H., Nagaoka, T. et al. (2015). Portable hyperspectral imager with continuous wave green laser for identification and detection of untreated latent fingerprints on walls. *Forensic Science International* 254: 100–105.

Payne, G., Wallace, C., Reedy, B. et al. (2005). Visible and near-infrared chemical imaging methods for the analysis of selected forensic samples. *Talanta* 67 (2): 334–344.

Pfeffferli, P.W. (1983). Application of microspectrophotometry in document examination. *Forensic Science International* 23 (2–3): 129–136.

Reed, G., Savage, K., Edwards, D., and Daeid, N.N. (2014). Hyperspectral imaging of gel pen inks: an emerging tool in document analysis. *Science & Justice* 54 (1): 71–80.

Reichard, E.J., Bartick, E.G., Morgan, S.L., and Goodpaster, J.V. (2017). Microspectrophotometric analysis of yellow polyester fiber dye loadings with chemometric techniques. *Forensic Chemistry* 3: 21–27.

Saraswathy, A., Jayasree, R., Baiju, K. et al. (2009). Optimum wavelength for the differentiation of brain tumor tissue using autofluorescence spectroscopy. *Photomedicine and Laser Surgery* 27 (3): 425–433.

Scott, A.I. (2013). *Interpretation of the Ultraviolet Spectra of Natural Products*. Elsevier.

Starczak, R. and Wąs-Gubała, J. (2016). UV–Vis microspectrophotometric study of wool and polyamide fibres dyed with analogous gryfalan dyes. *Dyes and Pigments* 132: 58–63.

Suzuki, S., Higashikawa, Y., Sugita, R., and Suzuki, Y. (2009). Guilty by his fibers: suspect confession versus textile fibers reconstructed simulation. *Forensic Science International* 189 (1–3): e27–e32.

Suzuki, S., Suzuki, Y., Ohta, H. et al. (2001). Microspectrophotometric discrimination of single fibres dyed by indigo and its derivatives using ultraviolet-visible transmittance spectra. *Science & Justice* 41 (2): 107–111.

SWGDOC (2013) SWGDOC Standard for Test Methods for Forensic Writing Ink Comparison. Scientific Working Group for Forensic Document Examination. https://www.swgdoc.org/documents/SWGDOC%20Standard%20for%20Test%20Methods%20for%20Forensic%20Writing%20Ink%20Comparison.pdf (accessed 26 May 2020).

SWGMAT (2004). A Forensic Fiber Examiner Training Program. Scientific Working Group for Materials Analysis, Fiber Subgroup. https://www.nist.gov/system/files/documents/2016/09/22/fiber_examiner_training_program.pdf (accessed 26 May 2020).

Thanasoulias, N.C., Parisis, N.A., and Evmiridis, N.P. (2003). Multivariate chemometrics for the forensic discrimination of blue ball-point pen inks based on their Vis spectra. *Forensic Science International* 138 (1–3): 75–84.

Trzcińska, B., Zięba-Palus, J., and Kościelniak, P. (2013). Examination of car paint samples using visible microspectrometry for forensic purposes. *Analytical Letters* 46 (8): 1267–1277.

Wąs-Gubala, J. and Starczak, R. (2015). UV–Vis microspectrophotometry as a method of differentiation between cotton fibre evidence coloured with reactive dyes. *Spectrochimica Acta Part A: Molecular and Biomolecular Spectroscopy* 142: 118–125.

Wiggins, K., Palmer, R., Hutchinson, W., and Drummond, P. (2007). An investigation into the use of calculating the first derivative of absorbance spectra as a tool for forensic fibre analysis. *Science & Justice* 47 (1): 9–18.

Zeichner, A., Levin, N., Klein, A., and Novoselsky, Y. (1988). Transmission and reflectance microspectrophotometry of inks. *Journal of Forensic Science* 33 (5): 1171–1184.

7
Infrared Spectroscopy
Barbara Stuart

7.1 Introduction

Infrared (IR) spectroscopy is a technique based on the vibrations within a molecule. An IR spectrum is obtained by passing IR radiation through a sample or reflecting radiation from the sample surface and then determining what fraction of the incident radiation is absorbed at a particular energy. The energy at which a band in an absorption spectrum appears corresponds to the frequency of the vibration of a part of a molecule. For a molecule to show IR absorptions it must possess a specific feature: an electric dipole moment of the molecule must change during the vibration. This is known as the selection rule for IR spectroscopy.

The structural properties of a range of forensic evidence types means that IR spectroscopy can provide fundamental information about the composition of materials collected from a crime scene. An understanding of the molecular composition enables the source of evidence to be determined. The availability of a range of sampling methods enables solid, liquid and gaseous specimens to be investigated. Of particular use for forensic evidence is the availability of microscopic techniques that enable the often small samples to be investigated. The development of portable instrumentation also allows the technique to be used in the field.

7.2 Theory of the Technique

7.2.1 Basis of the Technique

The bands appearing in an IR spectrum can each be assigned to the vibrational modes of the molecular structure under study (Stuart 2004). For a molecule to show IR absorptions it must possess a specific feature, known as a selection rule: an electric dipole moment of the molecule must change during the vibration. The larger this change, the more intense will be the absorption band. A molecule can only absorb radiation when the incoming IR radiation is of the same frequency as one of the fundamental modes of vibration of the molecule. This means that the vibrational motion of a small part of the molecule is increased while the rest of the molecule is left unaffected.

Analytical Techniques in Forensic Science, First Edition.
Edited by Rosalind Wolstenholme, Sue Jickells and Shari Forbes.
© 2021 John Wiley & Sons Ltd. Published 2021 by John Wiley & Sons Ltd.

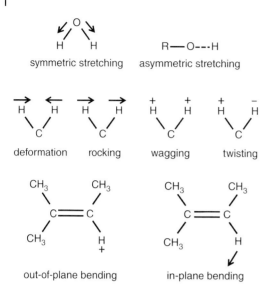

Figure 7.1 Stretching and bending molecular vibrations. Source: Reproduced with permission from Stuart (1996).

Vibrations can involve either a change in bond length (stretching) or bond angle (bending). Some bonds can stretch in-phase (symmetrical stretching) or out-of-phase (asymmetric stretching). Figure 7.1 illustrates examples of the various stretching and bending vibrations observed. If a molecule has different terminal atoms then the two stretching modes are no longer symmetric and asymmetric vibrations of similar bonds, but will have varying proportions of the stretching motion of each group known as coupling. The complexity of an IR spectrum arises from the coupling of vibrations over a large part of or over the complete molecule and such vibrations are called skeletal vibrations. Bands associated with skeletal vibrations produce a pattern or 'fingerprint' for the molecule as a whole, rather than a specific group within the molecule.

7.2.2 Instrumentation

Fourier transform infrared (FTIR) spectrometers are the most regularly used instruments for recording IR spectra (Stuart 2004; Chalmers et al. 2012). The size and cost of such instruments have gradually reduced in recent decades making this technique viable for forensic laboratories. FTIR spectroscopy is based on the idea of the interference of radiation between two beams to yield an interferogram. An interferogram is a signal produced as a function of the change of pathlength between the two beams. The two domains of distance and frequency are interconvertible by the mathematical method of Fourier transformation. In a FTIR spectrometer the radiation emerging from a source is passed through an interferometer to the sample before reaching a detector. The data are converted to a digital form by an analogue-to-digital converter before undergoing Fourier transformation.

The standard interferometer used in FTIR spectrometry is a Michelson interferometer, which consists of two perpendicularly plane mirrors, one of which can travel in a direction perpendicular to the plane (Figure 7.2). A semi-reflecting film, the beam splitter, bisects the planes of these two mirrors. If a collimated beam of monochromatic radiation is passed into an ideal beam splitter, 50% of the incident radiation will be reflected to one of the mirrors

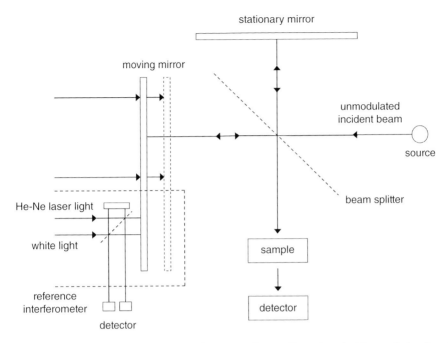

Figure 7.2 Layout of a Michelson interferometer. Source: Reproduced with permission from Stuart (1996).

and 50% will be transmitted to the other mirror. The two beams are reflected from these mirrors, returning to the beam splitter where they recombine and interfere. Fifty per cent of the beam reflected from the fixed mirror is transmitted through the beam splitter and half is reflected back in the direction of the source. The beam that emerges from the interferometer at 90° to the input beam is called the transmitted beam and this is the beam detected in FTIR spectrometry. The moving mirror produces an optical path difference between the two arms of the interferometer.

The moving mirror is a crucial component of the interferometer. It has to be accurately aligned and must be capable of scanning two distances so that the path difference corresponds to a known value. The interferogram is an analogue signal at the detector that has to be digitised in order that the Fourier transformation into a conventional spectrum can be carried out and there are sources of error in transforming the digitised information from the interferogram into a spectrum. The transformation involves an integration stage over a finite displacement rather than over an infinite displacement and the mathematical process of Fourier transformation assumes infinite boundaries. The consequence of this necessary approximation is that the apparent line shape of a spectral line will show a series of negative and positive side lobes (or pods) with diminishing amplitudes. A process of apodisation is used to remove the side lobes by multiplying the interferogram by a suitable function before the Fourier transformation is carried out and spectrometers usually offer a choice of apodisation options (with a general purpose cosine function often used as a default).

In the mid-IR region, FTIR spectrometers commonly use a silicon carbide element heated to about 1200 K as a source. There are two commonly used detectors: a pyroelectric device incorporating deuterium triglycine sulfate (DTGS) and a mercury cadmium telluride (MCT) detector. For routine use a DTGS operates at room temperature and is less sensitive than a MCT detector that requires cooling to liquid nitrogen temperatures. Although a MCT detector is approximately 40 times more sensitive than a DTGS detector, the DTGS has a wider spectral range and can be used to record spectra at lower wavenumbers.

7.2.3 Transmission Spectroscopy

Transmission spectroscopy, where the absorption of radiation is measured after it passes through a sample, is the traditional sampling method in IR spectroscopy (Stuart 2004). Liquids, solids, or gases can be investigated using this approach. Although such sampling approaches have been somewhat superseded in many forensic applications in favour of reflectance and microspectroscopic techniques, these sampling techniques can still be of use where a suitable quantity of sample is available.

Solids can be examined in transmission mode when contained in KBr discs: powdered samples (about 2–3 mg) are mixed with KBr (about 200 mg) and ground in an agate mortar and pestle before being subjected to pressure in an evacuated die to produce a transparent disc. Solids (about 50 mg) may also be mixed and ground with 1–2 drops of a mulling agent (e.g. Nujol – liquid paraffin) to produce a smooth paste known as a mull. Films can be produced for certain sample types (e.g. polymers) by either solvent casting (dissolving in a suitable solvent and allowing to evaporate on an IR window) or melt casting (heating a specimen above the melting temperature and pressing to a thin film). For solid specimens that are too thick as received to measure via transmission, a diamond anvil cell (DAC) can be employed. A DAC uses two diamonds to compress a sample to a thickness suitable for measurement and increases the surface area. Transmission solution cells use alkali halide windows (e.g. KBr, NaCl) and a polytetrafluoroethylene (PTFE) spacer, available in a variety of thicknesses. These cells are filled using a syringe and the syringe ports are sealed with PTFE plugs before sampling. Gases have densities which are several orders of magnitude less than liquids, hence the pathlengths must be correspondingly greater, usually 10 cm or longer. The walls of a transmission gas cell are made of glass or brass with the usual choice of windows and the cells can be filled by flushing.

7.2.4 Reflectance Spectroscopy

Reflectance techniques can be used for samples that are difficult to analyse by the conventional transmission methods (Stuart 2004; Chalmers et al. 2012). The use of reflectance sampling techniques is widespread in forensic applications because these techniques enable non-destructive and field testing to be carried out. The commonly employed reflectance sampling techniques for forensic applications include attenuated total reflectance (ATR), diffuse reflectance and reflection–absorption spectroscopies.

ATR spectroscopy is an internal reflectance sampling technique and involves pressing a crystal on the sample of interest. Figure 7.3 illustrates a micro-ATR accessory. ATR spectroscopy utilises the phenomenon of total internal reflection: a beam of radiation entering a crystal will undergo total internal reflection when the angle of incidence at the interface between the sample and crystal is greater than the critical angle (a function of the refractive

Figure 7.3 A micro-ATR accessory.

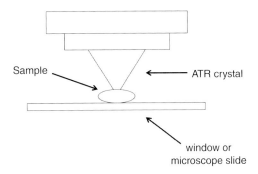

indices of the two surfaces). The beam penetrates a fraction of a wavelength beyond the reflecting surface and when a material that selectively absorbs radiation is in close contact with the reflecting surface, the beam loses energy at the wavelength where the material absorbs. The resultant attenuated radiation is measured and plotted as a function of wavelength by the spectrometer and gives rise to the absorption spectral characteristics of the sample. The depth of penetration in ATR is a function of wavelength, the refractive index of the crystal and the angle of incident radiation. As a consequence, the relative intensities of the IR bands in an ATR spectrum will appear different to those observed in a transmission spectrum; the intensities of bands will be greater at lower wavenumber values. The crystals used in ATR cells are made from materials that possess low water solubility and are of a high refractive index. Such materials include diamond, germanium, and zinc selenide.

An external reflectance technique used for the examination of powdered forensic specimens is diffuse reflectance spectroscopy. Diffuse reflectance results from the energy that penetrates one or more particles and is reflected in all directions. Some samples can be examined as pure samples, but a powdered sample is usually diluted with a non-absorbing alkali halide (e.g. KBr, KCl) powder at a concentration of 1–5 w%. The technique avoids the need to press pellets as the powdered sample can be placed directly in a sampling cup. In order to ensure good quality spectra, the particle size of the sample must be reduced to a fine powder consistency (less than 50 μm) in a homogenous mixture (by employing an agate mortar and pestle or a specialist mixer). Diffusely scattered light can also be collected directly from a sample surface by using an abrasive diamond or graphite sampling pad on a holder that can be directly inserted into the diffuse reflectance accessory. The resulting diffuse reflectance spectra can appear different from the transmission equivalent (e.g. stronger than expected absorption from weak bands), but a Kubelka–Munk function, $f(R_\infty)$, can be applied to compensate for these differences:

$$f(R_\infty) = \frac{(1 - R_\infty)^2}{2R_\infty} = \frac{k}{s} \tag{7.1}$$

where R_∞ is the absolute reflectance of the sampled layer (reflectance when the layer's depth can be said to be infinite), s is the scattering coefficient, and k is the absorption coefficient. For powdered samples diluted in a non-absorbing matrix, the Kubelka–Munk function is linear with concentration.

Another external reflectance sampling technique that is used in forensic analysis is reflection–absorption (or transflection) spectroscopy. Reflection–absorption sampling

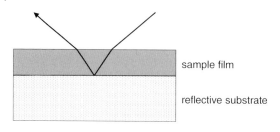

Figure 7.4 Reflection–absorption spectroscopy.

involves measurement of the reflected energy emerging from a sample surface at a given angle of incidence (Figure 7.4). The amount of light reflected depends on the angle of incidence, refractive index, surface roughness and absorption properties of the sample. Reflection–absorption analysis is carried out on a thin film (<10 μm) on a flat reflective metal surface. Where a sample is very reflective, such as crystalline and glassy materials, abnormal dispersion occurs and derivative-like bands result, making interpretation more difficult. However, the reflectance spectrum can be corrected using a Kramers–Kronig transformation (or K-K transformation) to appear in a familiar form.

7.2.5 Infrared Microspectroscopy

An IR spectrometer incorporating a microscope facility in order to study small specimens is regularly encountered in forensic laboratories. Larger laboratory-based instruments have evolved with separate IR microscopes and spectrometer compartments for versatility. The applicability of this technique to forensic applications has been extended with the development of compact integrated instruments that required less user expertise. Samples of the order of 10 μm are able to be investigated.

In FTIR microscopy, the microscope sits above the sample supported on an IR transparent substrate (Chalmers et al. 2012). IR radiation from the spectrometer is focused onto a sample placed on a standard microscope x-y stage. After passing through the sample, the IR beam is collected by a Cassegrain objective that produces an image of the sample within the barrel of the microscope and a variable aperture is placed in this image plane. The radiation is then focused onto a detector using another Cassegrain condenser. The microscope also contains glass objectives to allow visual inspection of the sample. By switching mirrors in the optical set-up, the microscope can be converted from transmission mode to reflectance mode.

IR imaging using FTIR microspectroscopic techniques is emerging as an effective approach to studying complex specimens (Ewing and Kazarian 2017). The technique can be used to produce a two- or three-dimensional picture of the properties of a sample. A large number of focal plane array (FPA) detector elements are read during the acquisition of spectra and this allows thousands of interferograms to be collected simultaneously and then transformed into IR spectra.

If access to an IR microscope facility is not possible, there are special microcells available to allow examination of μg or μl amounts (in addition to the aforementioned DAC). The acquisition of spectra using such cells is accomplished using a beam condenser, a device that ensures as much as possible of the beam passes through the sample.

7.2.6 Handheld and Portable Instruments

In addition to laboratory-based instruments, portable handheld IR spectrometers are commercially available and provide a valuable tool for field work. Handheld and portable IR spectrometers contain similar components to laboratory-based instruments, but are of a more compact and robust design (Chalmers et al. 2012). Advances in optics and light sources have led to the development of such devices that are now employed in the field. Handheld and portable IR devices generally use ATR sampling as this is often the most practical sampling approach in the field. Portable instruments generally contain an ATR device in a rugged container designed for a field laboratory enabling samples collected at a scene to be analysed using the instrument on site as part of a mobile laboratory. Handheld instruments are more compact and are designed to be taken to material requiring direct examination and are applicable when a sample cannot be readily collected, or if a large item is to be examined.

7.3 Application to Analyte

The chemical and physical properties of forensic evidence are broad ranging. The versatility of IR spectroscopy enables solids, liquids and gases to be sampled in a non-destructive manner. The sample preparation method and choice of sampling technique depends on the type of evidence collected. Once spectra are collected, they can be used to determine the composition of the evidence and quantitative analysis can be carried out if required.

7.3.1 Sampling

Several sampling approaches to forensic paint analysis using FTIR spectroscopy are available depending on the size of the specimen and whether individual paint layers are to be examined (Beveridge et al. 2001; SWGMA 2011a; ASTM E1610 2018a, 2018b). Individual layers can be prepared using a scalpel or blade. A DAC is a straightforward means of studying paint samples if individual layers can be separated. Alternatively, the sections may be placed on a salt plate to be examined in transmission mode using an IR microscope. If using a microscopic approach, individual layers of a multi-component paint chip can be sequentially sampled to differentiate the composition of the individual layers if the thickness of the layers is of an appropriate value to be resolved by the instrument. Cross-section samples of paint are prepared by embedding in a suitable media (e.g. acrylic, polyester, or epoxy resin) and microtoming with cutting devices made of materials including diamond or tungsten carbide to produce very thin sections. To produce a good quality spectrum, a paint sample needs to be of the order of 1–10 μm in thickness. IR microscopy can be used to record reflectance spectra of paint, but transmission spectra are preferred as most reference spectra are reported in transmission mode. If coatings or paint smears are required to be examined in situ, ATR sampling using a portable or handheld device is a practical approach.

There is a choice of sampling techniques for fibre evidence depending on the fibre thickness (SWGMA 2011b; Fredericks 2012; ASTM E2224 2018c). If a fibre is not too thick, a transmission approach may be taken. The fibre can be taped at the ends across a hole in a metal disc or by simply laying on an IR window. However, in general fibres are too thick for transmission techniques, with diameters less than 20 μm required to obtain good quality spectra. Fibres should be flattened to reduce the thickness and increase the surface area and a DAC can be used for such an approach. IR microscopy is the most useful technique for obtaining fibre spectra due to their dimensions. IR microscopy can be used to produce transmission or reflectance fibre spectra, where thicker fibres can be potentially examined without modification.

Adhesive tapes may be examined using ATR spectroscopy (SWGMA 2011c; ASTM E3085 2017). Both the adhesive and backing sides of tape can be analysed by targeting uncontaminated sections of the specimen. If separate spectra for the adhesive and backing components are required, a backing sample and/or adhesive can be pressed in a DAC or, alternatively, the adhesive deposited on an alkali halide plate can be examined in transmission mode. For other polymer-based evidence such as packaging (e.g. bags, bottles) or automotive components (e.g. tyres, light enclosures), an ATR technique is usually the most straightforward approach.

Reflectance sampling techniques are generally used when acquiring IR spectra for forensic documents (Bartick 2002; Chalmers et al. 2012; Ferrer 2017). For copy toner analysis, diffuse reflectance or reflection–absorption techniques can be applied. To prepare a film for reflection–absorption sampling, the toner can be transferred from the document to a reflective substrate with heating. ATR spectroscopy is also applicable when examining inks on paper as ink is in relatively low concentrations on the top layer of the paper fibres. Paper itself can be examined using ATR spectroscopy.

When dealing with explosives, the choice of sampling method depends on the nature of the residue (Monts et al. 2000; McNesby and Pesce-Rodriguez 2002; Bartick 2012). Certain particulate explosives can be examined using a DAC by placing the sample on the window without any clean up procedure required. However, where there is an ignition risk for pressure sensitive explosive materials, DAC is not suitable. Given the safety issues involved in explosives analysis, the use of a portable or handheld IR spectrometer in the field is common as the transportation of dangerous materials to a laboratory can be avoided. For plastic explosives, solvent extraction and casting as a film onto an ATR crystal can also be a useful approach.

Transmission sampling techniques, such as KBr discs, have been traditionally used for the analysis of drugs and related chemicals (Bartick 2002; Chalmers et al. 2012). However, benchtop, portable and handheld IR spectrometers using diamond ATR have emerged as a preferred approach to drug identification.

7.3.2 Spectrum Analysis

A spectrum is produced in IR spectroscopy and usually is represented as absorbance, transmittance or reflectance as a function of wavenumber, $\bar{v}$, with units of cm^{-1}. Figure 7.5 provides examples of the ATR spectra collected for a polyethylene storage bag, in terms of absorbance or transmittance. Often it is a matter of personal choice as to how the spectral scale is presented, but absorbance is more applicable for quantitative analysis.

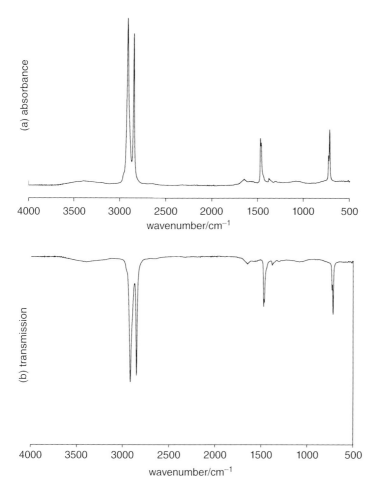

Figure 7.5 Infrared spectra for a polyethylene storage bag. (a) absorbance and (b) transmission.

The bands appearing in an IR spectrum can each be assigned to the vibrational modes of the molecular structure under study (Stuart 2004; Coates 2010). An IR spectrum can be divided into three main regions: the far-IR (<400 cm^{-1}), the mid-IR (4000–400 cm^{-1}), and the near-IR (13000–4000 cm^{-1}); the mid-IR spectrum is the most widely employed for forensic applications. The mid-IR spectrum can be subdivided into the X—H stretching region (4000–2500 cm^{-1}), the triple bond region (2500–2000 cm^{-1}), the double bond region (2000–1500 cm^{-1}), and the fingerprint region (1500–600 cm^{-1}), based on the nature of the functional groups present in the molecular structure. Table 7.1 summarises the principal mid-IR bands that may be encountered when examining materials in forensic applications.

Detailed correlation tables are widely available that allow IR bands to be assigned to a particular vibrational mode and, therefore, the spectrum attributed to a particular molecular structure. A useful strategy for determining the composition of the material under examination is to initially concentrate on the major bands in the higher end of the spectrum

Table 7.1 Common mid-infrared spectral bands.

Wavenumber (cm⁻¹)	Assignment
3700–3600	O—H stretching
3400–3300	N—H stretching
3100–3000	Aromatic C—H stretching
3000–2850	Aliphatic C—H stretching
2300–2050	C≡C stretching
2300–2200	C≡N stretching
1830–1650	C=O stretching
1650	C=C stretching
1500–650	Fingerprint region (bending, rocking)

above 1500 cm⁻¹ (C=O, C—H, O—H and N—H stretching bands). A first step is often to determine if the sample is organic through the identification of C—H stretching bands. Many compound classes produce a characteristic IR pattern due to the presence of the known functional groups in the molecular structures. Thus, the examination of spectral patterns for different compound classes provides an initial diagnostic tool. However, the collection of reference spectra for commonly encountered materials allows the technique to be utilised without necessarily having to assign all the modes in the spectrum.

Spectral library databases are available for materials of forensic interest and allow an unknown to be identified based on a match using the intensities and wavenumber values of bands. The process produces a list of the best matches based on a hit quality index (HQI) on a scale 0–1. The HQI is used as an indicator of the similarity between the unknown spectrum and the reference spectrum and its determination is usually based on a correlation coefficient. Commercial forensic IR databases are available for the identification of materials including drugs, polymers and paint and fibre components.

Quantitative analysis is also feasible using IR spectroscopy (Stuart 2004). If dealing with a single analyte, the Beer–Lambert law is used to relate the amount of light transmitted by a sample to the thickness of the sample. The absorbance is directly proportional to the thickness and the concentration of the sample:

$$A = \varepsilon c l = -\log_{10} T \tag{7.2}$$

where A is the absorbance, ε is the molar absorptivity, c is the analyte concentration, l is the thickness of the sample, and T is the transmittance. A plot of absorbance against concentration will be linear with a gradient of εl and pass through the origin. So to analyse a solution of unknown concentration, for instance, solutions of known concentration need to be prepared, a suitable band chosen, the absorbance at this wavenumber measured, and a calibration graph plotted. The concentration of the compound in solution can be read from the calibration graph, once its absorbance has been measured.

When dealing with more complex multi-component systems, multivariate statistical methods can be applied (Zadora 2010; Kumar and Sharma 2018). Multivariate analytical techniques can be classified as either supervised or unsupervised methods, depending on whether the sample grouping is known in advance. Unsupervised methods, including principal component analysis (PCA) and hierarchical clustering, do not require information for classification and individual samples are clustered based on the similarity amongst the sample data. Supervised methods, which use prior assumptions about the existence of groups, include discriminant analysis (DA) and artificial neural networks (ANNs). These methods are capable of comparing a large number of variables within a data set.

7.4 Interpretation and Law

IR spectroscopy has been adopted as a tool for the forensic analysis of a range of evidence types. Protocols for the sampling and spectral analysis have been developed for the various categories. The technique lends itself to the examination of paint, fibres, tapes, questioned documents, explosives, and drugs.

IR spectroscopy is a valuable technique for the analysis of paint specimens as it can provide structural information about both the inorganic and organic components found in paint (SWGMA 2011a; Ferrer 2017; ASTM E1610 2018a). The technique can be used to simultaneously identify the binder, pigment and additive components of paint specimens. Additional discriminating information can also be provided by the examination of separate paint layers. IR paint analysis is frequently applied to the analysis of automotive paints, with spectral databases available to assist in identifying the make, model and year of a car from which the paint is derived. The Paint Database Query established by the Royal Canadian Mounted Police (RCMP) and the Georgia State Crime Lab Paint FTIR Spectral Library are examples of those used. Guidelines for the IR analysis and reporting of paints include: the American Society for Testing and Materials (ASTM) Standard Guide for Using Infrared Spectroscopy in Forensic Paint Examinations (ASTM E2937 2018b); ASTM Standard Guide for Forensic Paint Analysis and Comparison (ASTM E1610 2018a); and the Scientific Working Group for Materials Analysis (SWGMAT) Standard Guide for Using Infrared Spectroscopy in Forensic Paint Examinations (SWGMA 2011a).

IR spectroscopy may be used to identify and characterise both synthetic and natural fibres (Bartick 2002; SWGMA 2011b; Fredericks 2012; Meleiro and García-Ruiz 2015; ASTM E2224 2018c). Given the variety in composition of synthetic fibres, IR spectral databases containing the spectra of different classes and subclasses of polymer fibres are useful identification tools. The ability to identify the subclass of a synthetic fibre arises from the sensitivity of IR spectroscopy to relatively subtle changes in structure (e.g. different nylon structures can be discriminated). Apart from different polymers used within fibre subclasses, variation in commercial fibres exists due to compositional differences (e.g. different copolymer compositions). Guidelines for the IR analysis and reporting of fibres are provided by the ASTM Standard Guide for Forensic Analysis of Fibres by Infrared Spectroscopy (ASTM E2224 2018c) and the SWGMAT Forensic Fibre Examination Guidelines (SWGMA 2011b).

A detailed compositional analysis of adhesive tapes can be determined using IR spectroscopy based on the identification of the polymeric base materials and the additives (e.g. plasticisers, fillers) used to produce the backing tape (SWGMA 2011c; ASTM E3085 2017). Likewise, the variety of polymeric materials and additives used in an adhesive component can be characterised. Given the common usage of IR spectroscopy for forensic tape analysis, the ASTM E3085 Standard Guide for Fourier Transform Infrared Spectroscopy in Forensic Tape Examinations (ASTM E3085 2017) and the SWGMAT Guideline for Using Fourier Transform Infrared Spectroscopy in Forensic Tape Examinations (SWGMA 2011c) have been established.

IR spectroscopy is a valuable technique for examining many types of polymer-based forensic evidence. Other bulk polymeric evidence types, such as plastic packaging and automotive components, can be discriminated based on the considerable variety in composition (Bartick 2002). For instance, if a plastic bag is produced using the same polymer, but manufactured by different companies, differences in the spectra due to the presence of various additives or the processing conditions can be used to confirm the source of the material.

Different aspects of questioned documents can be investigated using IR spectroscopy, including toners, inks, and paper (Bartick 2002, 2012; Ferrer 2017). For toners and inks, it is possible to differentiate the organic components present. For instance, the ability to identify differences in polymer type, copolymer composition and the presence of additives enables IR spectroscopy to be used to classify toners and identify sources. Likewise, although the cellulose-based components of paper are relatively consistent, the additives (e.g. mineral fillers) vary by manufacturer and the composition can be used to determine where the paper was produced.

IR spectroscopy is an important tool for the identification of explosive materials, particularly in the field (Monts et al. 2000; Bartick 2002; McNesby and Pesce-Rodriquez 2002; Chalmers et al. 2012). The technique particularly lends itself to the identification of organic explosives and precursors and can be used to characterise dark coloured or fluorescent residues (which can be difficult to analyse with a complementary portable Raman spectrometer). The formulations of explosive mixtures can be determined using IR spectroscopy enabling a residue to be linked to a source. The IR analysis of post-blast explosive debris is more problematic due to the quantity of residue available and the presence of contaminants, in which case separation techniques are a more appropriate choice of technique. Spectral explosive libraries have been established by the Federal Bureau of Investigation (FBI) and the Technical Working Group on Fire and Explosives (TWGFEX) have established guidelines for the identification of explosives by IR spectroscopy.

Portable and handheld IR spectrometers are regularly used for the screening of drugs and related chemicals located in clandestine laboratories (Bartick 2002; Chalmers et al. 2012). Complex mixtures are found in such environments and fast identification at the scene can enable appropriate specimens to be collected in a prompt and safe manner for transportation to the laboratory for further analysis. The technique is frequently used in conjunction with handheld Raman spectroscopy in the field for screening of drug evidence. Although Raman spectroscopy has the advantage that it can be used to examine unknown substances within containers, IR spectroscopy has the advantage of being able to identify fluorescent materials. The most established library of drug IR spectra (and related chemicals) was developed by the Georgia Bureau of Investigation. Although the commercially available Georgia State Crime Lab Drug FTIR Spectral Library has been based on spectra recorded for reference materials in KBr discs, ATR spectra are now being incorporated into spectral libraries to enable a more effective match with unknown drug ATR spectra.

7.5 Case Studies – Discrimination of Acrylic Fibres

Although light microscopy is the first method of choice for the identification of individual textile fibres collected as evidence, IR spectroscopy is commonly used to provide confirmation of the fibre type and to provide further information about the subclass of the fibre. IR spectroscopy is particularly useful for the differentiation of acrylic fibres, which easily transfer but can sometimes be difficult to discriminate using microscopic techniques. Acrylic fibres are copolymers of acrylonitrile combined most commonly with either methyl methacrylate, methyl acrylate, or vinyl acetate. As the IR spectra for the different copolymer compositions will be distinctively different, IR spectroscopy can provide more discriminating information about an acrylic fibre.

One of the first approaches to the determination of the composition of acrylic fibre for forensic purposes was to use the ratios of the absorbance bands associated with the acrylonitrile component and ester group of the comonomer (Tungol et al. 1993). The C≡N and C—O stretching bands due to the nitrile and carbonyl groups in the copolymer appear near 2240 and 1730 cm^{-1}, respectively, and are strong and relatively free of overlapping bands. The band height or area can be used to determine the relative composition of the monomers in the copolymer given that a database of suitable fibres is available.

A more discriminating approach was developed by Grieve (1995). Spectra for 200 samples of undyed acrylic fibres collected from a variety of international manufacturers were recorded using FTIR microscopy. The fibres were flattened prior to examination using a small metal roller to enable transmission spectra to be measured. The resulting spectra enabled 20 types of subclasses to be identified based on the various comonomers present and the presence or absence of residual solvent from the manufacturing process.

Subsequent work on the IR spectra of acrylic fibres has focused upon how to derive further discrimination from other regions of the spectra. Miller and Bartick (2001) demonstrated that the 1300–1100 cm^{-1} region of the IR spectra of acrylic fibres can be used to differentiate the acrylic subclasses. Figure 7.6 illustrates the spectra of the major subclasses, acrylonitrile copolymers with methyl acrylate, methyl methacrylate and vinyl acetate, as well as the homopolymer polyacrylonitrile. The presence of the different esters in each class produces different patterns in the C—O stretching region that may be used to identify the subclass.

A quantitative approach to the identification of acrylic fibres has been developed by Causin et al. (2005). In this study, 48 acrylic fibre samples, collected from different sources and known to cover a wide range of polymer compositions, were examined using FTIR microspectroscopy in transmission mode. In addition to the C≡N and C—O stretching bands due to the nitrile and carbonyl groups in the copolymers at 2240 and 1730 cm^{-1}, respectively, an additional band at 1370 cm^{-1} associated with the C—H bonds of the polymer backbone was also considered for the analysis. The areas of these three bands were calculated by integration and the absorbance ratios 1730 cm^{-1}/2240 cm^{-1} and 1730 cm^{-1}/1370 cm^{-1} measured and plotted, allowing a relative

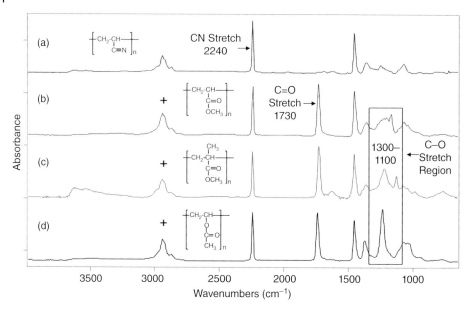

Figure 7.6 Infrared spectra of undyed acrylic fibres: (a) polyacrylonitrile; (b) poly(acrylonitrile-methyl acrylate); (c) poly(acrylonitrile-methyl methacrylate); and (d) poly(acrylonitrile-vinyl acetate). Source: Reproduced with permission from Miller and Bartick (2001).

measure of the comonomer content in each fibre. Such a method provides a straightforward quantitative means of discriminating acrylic fibres.

7.6 Forensic Developments

A significant reason for the widespread adoption of mid-IR spectroscopy for forensic applications has been the development of portable and handheld instruments that are practical for use in the field (Chalmers et al. 2012). A focus has been on producing instruments that are compact, robust and easily transportable to a crime scene. Improvements in the availability of high performance materials and electronics will continue to provide advancements in this area. The quality of the data produced continues to improve and moves closer to matching the quality generated by laboratory-based instruments. At the same time, the cost of instrumentation has reduced making access to the instrumentation affordable for agencies. Benchtop instruments also continue to improve the quality of output and speed of acquisition of data, while becoming more compact and cost-effective.

Software developments will enable more effective identification for forensic applications in the coming years. The introduction of user-friendly software enables the technique to be used in the field with minimal training. The development of suitable spectral databases remains a critical issue for the effective use of IR spectroscopy. Ensuring the unambiguous and correct identification of evidence, often at a crime scene, is an important focus. Spectral libraries continue to expand and must accommodate new forms of materials likely to be encountered. For example, the continuing evolution of illicit synthetic drugs means that the spectra of new chemicals must be included in databases to ensure effective identification.

Currently, the spectral region of IR spectroscopy most commonly used for forensic work has been the mid-IR region. However, in recent years it has been demonstrated that there is also potential for the terahertz region (in the far-IR region 10–200 cm^{-1}) (Burnett et al. 2012). This region of the spectrum is associated with complex vibrational modes arising from crystalline structures. The principal attraction of terahertz spectroscopy for forensic work is the ability to analyse materials contained within packaging as terahertz frequency radiation is transmitted through many non-polar materials such as plastic and paper. The development of commercial terahertz instruments should lead to the expansion of this technique to the safe identification of explosives and drugs.

References

ASTM E3085 (2017). *Standard guide for Fourier transform infrared spectroscopy in forensic tape examinations*. West Conshohocken: American Society for Testing and Materials.

ASTM E1610 (2018a). *Standard guide for forensic paint analysis and comparison*. West Conshohocken: American Society for Testing and Materials.

ASTM E2937 (2018b). *Standard guide for using infrared spectroscopy in forensic paint examinations*. West Conshohocken: American Society for Testing and Materials.

ASTM E2224 (2018c). *Standard guide for forensic analysis of fibres by infrared spectroscopy*. West Conshohocken: American Society for Testing and Materials.

Bartick, E.G. (2002). Applications of vibrational spectroscopy in criminal forensic analysis. *Handbook of Vibrational Spectroscopy* 4: 2993–3004.

Bartick, E. (2012). Criminal forensic analysis. In: *Infrared and Raman Spectroscopy in Forensic Science* (eds. J.M. Chalmers, H.G. Edwards and M.D. Hargreaves), 89–109. Wiley.

Beveridge, A., Fung, T., and MacDougall, D. (2001). Use of infrared spectroscopy for the characterisation of paint fragments. In: *Forensic Examination of Glass and Paint* (ed. B. Caddy), 187–245. CRC Press.

Burnett, A.D., Cunningham, J.E., Davies, A.G. et al. (2012). Terahertz frequency spectroscopy and its potential for security applications. In: *Infrared and Raman Spectroscopy in Forensic Science* (eds. J.M. Chalmers, H.G.M. Edwards and M.D. Hargreaves), 295–314. Wiley.

Causin, V., Marega, C., Schiavone, S., and Marigo, A. (2005). A quantitative differentiation method for acrylic fibers by infrared spectroscopy. *Forensic Science International* 151 (2): 125–131.

Chalmers, J.M., Edwards, H.G.M., and Hargreaves, M.D. (2012). *Infrared and Raman Spectroscopy in Forensic Science*. Wiley.

Coates, J. (2010). Interpretation of infrared spectra, a practical approach. In: *Encyclopedia of Analytical Chemistry* (ed. R. Meyers), 10815–10837. Wiley.

Ewing, A.V. and Kazarian, S.G. (2017). Infrared spectroscopy and spectroscopic imaging in forensic science. *Analyst* 142 (2): 257–272.

Ferrer, N. (2017). Forensic science, Applications of IR spectroscopy. In: *Encyclopedia of Spectroscopy and Spectrometry*, 3e (eds. J.C. Lindon, G.E. Tranter and D.W. Koppenaal), 695–706. Elsevier.

Fredericks, P.M. (2012). Forensic analysis of fibres by vibrational spectroscopy. In: *Infrared and Raman Spectroscopy in Forensic Science* (eds. J.M. Chalmers, H.G.M. Edwards and M.D. Hargreaves), 153–169. Wiley.

Grieve, M.C. (1995). Another look at the classification of acrylic fibres, using FTIR microscopy. *Science and Justice – Journal of the Forensic Science Society* 35 (3): 179–190.

Kumar, R. and Sharma, V. (2018). Chemometrics in forensic science. *Trends in Analytical Chemistry* 105: 191–201.

McNesby, K. and Pesce-Rodriquez, R. (2002). Applications of vibrational spectroscopy in the study of explosives. In: *Handbook of Vibrational Spectroscopy* (eds. J. Chalmers and P. Griffiths), 3152–3168. Wiley.

Meleiro, P.P. and García-Ruiz, C. (2015). Spectroscopic techniques for the forensic analysis of textile fibers. *Applied Spectroscopy Reviews* 51 (4): 258–281.

Miller, J.V. and Bartick, E.G. (2001). Forensic analysis of single fibers by Raman spectroscopy. *Applied Spectroscopy* 55 (12): 1729–1732.

Monts, D., Singh, J., and Boudreaux, G. (2000). Laser- and optical-based techniques for the detection of explosives. In: *Encyclopedia of Analytical Chemistry* (ed. R. Meyers), 2148–2171. Wiley.

Stuart, B. (1996). *Modern Infrared Spectroscopy*, ACOL Series. Chichester: Wiley.

Stuart, B. (2004). *Infrared Spectroscopy: Fundamentals and Applications*. Chichester: Wiley.

SWGMA (2011a). Standard guide for using infrared spectroscopy in forensic paint examinations. *Journal of the American Society of Trace Evidence Examiners* 2: 73–87.

SWGMA (2011b). Forensic fibre examination guidelines: infrared analysis of textile fibres. *Journal of the American Society of Trace Evidence Examiners* 6: 76–82.

SWGMA (2011c). Guideline for using Fourier transform infrared spectroscopy in forensic tape examinations. *Journal of the American Society of Trace Evidence Examiners* 2: 112–121.

Tungol, M.W., Bartick, E.G., and Montaser, A. (1993). Forensic analysis of acrylic copolymer fibers by infrared microscopy. *Applied Spectroscopy* 47 (10): 1655–1658.

Zadora, G. (2010). Chemometrics and statistical considerations in forensic science. In: *Encyclopedia of Analytical Chemistry* (ed. R. Meyers). Wiley https://doi.org/10.1002/9780470027318.a9122.

8

Raman Spectroscopy
Rosalind Wolstenholme

8.1 Forensic Introduction

The Raman effect was predicted in 1923 by Smekal but was first observed by Raman and Krishnan in 1928 (Smekal 1923; Raman and Krishnan 1928). Raman spectroscopy is concerned with interactions induced in the visible region of the electromagnetic spectrum but with magnitude the same as that of radiation in the infrared (IR) region. Therefore, as with IR spectroscopy, it provides information on the vibrational and rotational transitions in a molecule. The peak assigned to a particular bond, e.g. C=O, will appear in approximately the same place, $\sim 1700\,cm^{-1}$, regardless of the molecule in which it is present. Therefore, Raman spectroscopy can be used on its own or to complement IR spectroscopy and other analytical techniques to help elucidate the chemical structure of a sample.

The use of Raman spectroscopy has seen a marked increase in recent years, much of which is due to improvements in instrumentation and a corresponding reduction in cost. Raman scattering is a weak phenomenon, consequently, sensitivity is a factor when acquiring a spectrum. The development of lasers, holographic filters, charge-coupled device (CCD) detectors, Fourier transform (FT) systems and advanced techniques such as surface enhanced Raman spectroscopy (SERS) (Fleischmann et al. 1974) and spatially offset Raman spectroscopy (SORS) (Matousek et al. 2005) have led to improvements in sensitivity and more options for capturing a spectrum (see Section 8.2). Due to these advances, Raman spectroscopy is now used routinely for applications such as in-line monitoring of processes in the pharmaceutical industry and forensic science (Nagy et al. 2017; Fikiet et al. 2018).

8.2 Theory

8.2.1 Raman Scattering

When a molecule is irradiated with electromagnetic radiation, some of the radiation is transmitted, some absorbed, and some scattered. Furthermore, scattering can happen in two ways: elastic, where the wavelength of the scattered radiation is the same as the incident radiation; and inelastic, where it is increased or decreased.

Elastic scattering is termed Rayleigh scattering. It is the result of a molecule in the ground vibrational state being promoted to a higher 'virtual' state and a photon of the same energy as the incident light being scattered as it returns to the ground state.

Raman scattering is inelastic scattering and is a weak phenomenon, about 1 in a million photons. Compared to IR absorption, the occurrence of Raman scattering is 10^{10} times less likely. In Raman scattering, the molecule either begins in the ground state, is excited to a virtual state and returns to an excited state, which is known as Stokes scattering, or it begins in an excited state and returns to the ground state scattering a photon of energy, which is known as anti-Stokes scattering, see Figure 8.1.

As weak as Raman scattering is in general, anti-Stokes scattering is much weaker than Stokes scattering because there is a much lower population of molecules in the excited state than in the ground vibrational state. Stokes scattering results in a decrease in the energy of the scattered radiation relative to the incident radiation, which is recorded as a positive shift in the wavenumber. Anti-Stokes scattering, conversely, results in an increase in the energy of the scattered radiation and a negative shift in wavenumber. A spectrum showing both Stokes and anti-Stokes would have peaks the same distance from and on either side of the incident light frequency (y-axis), but the Stokes peaks would be of much higher intensity, see Figure 8.2. Consequently, in most cases only the Stokes scattering is collected. There are, however, circumstances when the anti-Stokes scattering is important. For example, to counter the impact of fluorescence (see Section 8.2.4) or to measure temperature. (It is possible to use the ratio of anti-Stokes to Stokes to monitor the temperature of a sample as there will be more molecules in an excited state and, therefore, more intense anti-Stokes as temperature increases.)

As mentioned, Raman scattering is measured as a shift from the incident radiation rather than an absolute frequency. This is a result of the way that the incident light, usually a laser (see Section 8.2.4.1), interacts with the molecule in order to cause scattering. The light and the molecule form an unstable, short-lived, virtual state complex with polarised electrons,

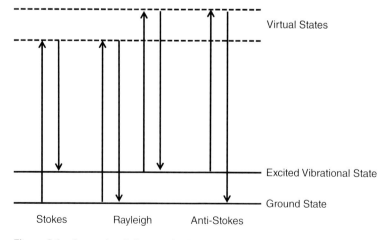

Figure 8.1 Energy level diagram for Raman scattering.

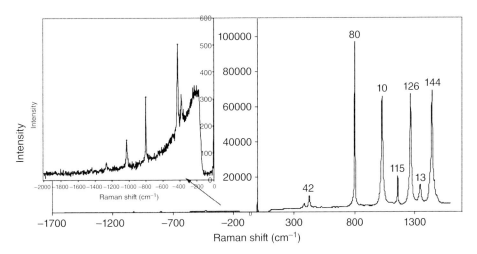

Figure 8.2 Stokes and anti-Stokes scattering for cyclohexane. To show the weak anti-Stokes spectrum, the y-axis has been extended in the inset. Source: Reproduced with permission from Smith and Dent (2019).

the energy of which is dependent on the monochromatic, incident radiation frequency. The molecule does not have time to reconfigure to a stable geometry before the radiation is released with a change in energy corresponding to the molecular vibration and relative to the incident light, i.e. with a Raman shift.

8.2.2 Modes of Vibration

The vibrations that Raman and IR spectroscopy are concerned with are those that are caused by stretching and bending of the bonds in a molecule. Specifically, symmetric and antisymmetric stretching and four types of bending/deformation, namely scissoring, rocking, wagging, and twisting, see Figure 8.3.

Figure 8.3 Modes of vibration for a triatomic molecule.

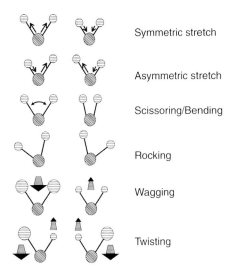

The number of vibrations possible for a molecule is determined by the number of degrees of freedom. Three of these are concerned with translation and three with rotation (two if linear). For any molecule, the number of vibrations is given by $3N - 6$, where N is the number of atoms, or $3N - 5$ for linear molecules. Thus, for a diatomic molecule $3N - 5 = 1$ and there is simply one symmetric stretch where the atoms on either end of the bond move away from and towards each other. Where the atoms at either end of the diatomic bond are the same there is a change in polarisability and a Raman active stretch (see section below for further discussion of polarisability). Where the atoms are not identical there is a change in dipole moment and an IR active stretch. This is a very simple case but it demonstrates what is called the mutual exclusion principle, which is that if a molecule is centrosymmetric, i.e. has symmetry about a centre, a strong Raman active vibration is IR inactive and vice versa. It is as a result of this principle that, although they are concerned with the same ground to excited state energy differences and provide similar chemical information, Raman and IR can be considered as complementary analytical techniques.

As the molecule gets more and more complex more vibrations are possible, and it becomes more difficult to predict whether a vibration is Raman or IR active. Predictions can be made using methods such as density function theory (DFT) calculations but this does not necessarily aid in identification of an unknown given that there are many compounds with very similar structures and, in a forensic context, this data may not be available or within the examiner's expertise. A discussion of DFT is beyond the scope of this book (Nogueira et al. 2003; Dreizler and Gross 2012).

What we can predict is *where* a vibration is likely to occur in the spectrum, which will be discussed in Section 8.4.1.

As mentioned above, the selection rule for IR active molecules is that absorption occurs when there is a change in dipole moment during the vibration. A dipole moment (μ), i.e. a permanent overall difference in charge over distance, is caused where electron density is higher on the more electronegative atoms and lower on the less electronegative atoms and this results in a partial positive charge ($\delta+$) and a partial negative charge ($\delta-$) separated in space. For example, in a linear molecule such as CO_2 there will be a partial negative charge on both O atoms and a partial positive charge on C but, because of the line of symmetry through all atoms of the molecule, the two $\delta-$ cancel each other out. However, in a H_2O molecule, oxygen will be $\delta-$ and both H will be $\delta+$. The additional presence of two lone pairs of electrons causes the molecule to be non-linear and the partial charges will be separated by distance, therefore, there will be a dipole moment. Molecules that have a dipole moment before exposure to a light source and vibrations that do not cause the partial charges to be cancelled out over distance are more likely to result in a change to that dipole moment. Therefore, polar and asymmetric vibrations give the most intense signal in IR spectroscopy as there is a net change in dipole.

The selection rule for Raman active vibrations states that scattering occurs when a change in polarisability of the molecule's electron cloud is induced. Polarisability (α) is the tendency to form an induced dipole in an electric field, i.e. a change in the electron cloud/charge distribution of a molecule, in size, shape or orientation, on exposure to a light source. If a molecule has no dipole moment in its ground state it has more potential to form an induced dipole. It follows that in Raman scattering, symmetric molecules and symmetric vibrations

Figure 8.4 Raman and IR active vibrations for CO_2.

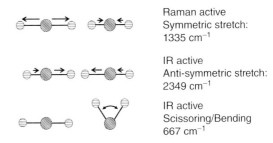

Raman active
Symmetric stretch:
1335 cm^{-1}

IR active
Anti-symmetric stretch:
2349 cm^{-1}

IR active
Scissoring/Bending
667 cm^{-1}

give strong signals. To reiterate, the difference in selection rules is the reason for the complementarity of data provided by Raman and IR spectroscopy.

Figure 8.4 shows the Raman and IR active vibrations for a simple molecule (CO_2). As can be seen, the symmetric stretch in the linear molecule is Raman active but the antisymmetric stretch and the bending modes are IR active.

Further discussion of using group theory to predict which vibrations are Raman active are beyond the scope of this book and can be found in Long (2002), Ferraro et al. (2003), and Smith and Dent 2019)

8.2.3 Raman Shift

Single frequency, or monochromatic, lasers in the ultraviolet–visible (UV-Vis) and near-infrared (NIR) region are used as the incident light source for Raman spectroscopy, commonly between 351 nm and 1064 nm, and a change or shift from the incident frequency is detected. This is an important contrast to IR where energy is absorbed at an absolute wavenumber for a particular bond vibration. In Raman scattering, the incident light does not need to match an energy gap because the shift is relative to the incident light. This has the advantage that instruments with lasers of different wavelengths can be used to promote the same change or shift, i.e. if a sample gives a peak at 320 cm^{-1} with a 785 nm laser it should also give a peak at 320 cm^{-1} with a 514 nm laser (the reasons that a particular laser frequency might be used are discussed in Section 8.2.4.1).

The vibrational transitions seen in Raman spectroscopy occur in the IR region of the electromagnetic spectrum, specifically wavelengths (λ) in the mid-IR region from 2.5 to 25 μm. However, it is convenient to consider spectra in terms of the energy associated with a particular vibration. λ is inversely proportional to energy but since

$$\lambda = \frac{c}{\nu} = \frac{1}{\bar{\nu}} \quad (8.1)$$

where c is the speed of light, ν is frequency, and $\bar{\nu}$ is wavenumber, and

$$E = h\nu \quad (8.2)$$

where E is energy and h is Planck's constant, $\bar{\nu}$ (cm^{-1}) is directly proportional to energy.

Therefore, by convention, both Raman and IR spectra are usually displayed in $\bar{\nu}$ over the range 4000–400 cm^{-1} with $\bar{\nu}$ decreasing from left to right on the x-axis. For clarity, $\bar{\nu}$ is, by definition, a frequency over distance (rather than time) but $\bar{\nu}$ and λ are often used interchangeably to mean the same thing in texts concerning Raman spectroscopy. Strictly speaking, Raman shift (from the incident radiation) should be given as Δ cm^{-1}, however, the

Δ is usually omitted by convention and cm^{-1} is used. The interpretation of Raman spectra is discussed in Section 8.4.1.

8.2.4 Raman Instrumentation

As with most analytical techniques, there is a range of instrumentation options available. As well as the choice of dispersive or FT spectrometer, there is a choice of lasers and attachments, such as microscopes and probes. This section will discuss some of the more commonly used configurations.

8.2.4.1 Lasers, Fluorescence, and Resolution

Monochromatic collimated lasers are used to promote Raman scattering for two reasons: (i) they have a narrow line width, which allows the characteristic narrow bands seen in a Raman spectrum; and (ii) they are an intense source of light (power in mW) which is needed to produce sufficient signal from the relatively weak Raman phenomenon. The lasers used are in the UV-Vis to NIR range of the electromagnetic spectrum, around 351–1064 nm (Ferraro et al. 2003). Those commonly used in Raman are 514 (Ar$^+$), 633 (He-Ne), 785 (Diode), and 1064 (YAG Nd$^+$, InGaAs) nm (green, red, NIR, and NIR, respectively). The choice of excitation depends on several factors, the first being what is available to the spectroscopist but, assuming that there is some choice, the sample being analysed must be considered. The first thing to be aware of is that the intensity of scattering is proportional (i) to the fourth power of laser frequency and (ii) to the square of the laser power, as given by

$$I = Kl\alpha^2 \lambda^4 \tag{8.3}$$

where I is intensity, K is a constant, l is laser power, α is polarisability, and λ is the wavelength of the incident radiation.

Theoretically then, intensity can be maximised by using higher frequency lasers, i.e. UV, and high power. However, the full range of lasers available is used for various reasons (see Table 8.1). At the high frequency end, i.e. UV, intensity will be maximised as well as the interference of fluorescence reduced because fluorescence energy is lower than the Raman scattering at these incident wavelengths and, therefore, fluorescence will not appear in the spectra. However, many compounds absorb UV, which can lead to burning or resonance (see Section 8.2.5.1). Aside from an unrepresentative spectrum, burning is also undesirable in forensic science as the integrity of the sample is an issue (see Section 1.2.1). Additionally,

Table 8.1 Features of the excitation wavelength regions for Raman spectroscopy.

Ultraviolet	Visible	Near-infrared
High frequency/low wavelength		Low frequency/high wavelength
High intensity (can use low power)		Low intensity (can use high power)
Burning		
No fluorescence	Fluorescence	No fluorescence
Better resolution		Poorer resolution

UV spectrometers also require more specialist optics than other systems meaning they are not very common. In the visible region, fluorescence can be a problem and it is greater the higher the frequency. It is also a commonly occurring phenomenon in many forensic samples, such as white paper in questioned document analysis or white powder in drugs analysis. The fluorescence spectrum recorded can swamp the Raman signal from the sample since fluorescence is a much stronger phenomenon relative to Raman scattering. This can be mitigated to some extent by photobleaching the sample, i.e. exposing it to the laser for a period of time to quench the fluorescence, however this can also degrade the sample. In this case it is worth compromising on power in order to reduce fluorescence. It should also be said that in some cases fluorescence may be a discriminatory characteristic of the sample so it is not always necessary to eliminate it (Stewart et al. 2012). Using higher frequency/longer wavelength lasers, i.e. NIR, reduces fluorescence because the laser energy is too low to promote fluorescence. A further way to reduce fluorescence is to record anti-Stokes scattering rather than Stokes. Again, fluorescence is at lower energy than the shifts recorded so it will not be present in the spectra. However, anti-Stokes is not routinely carried out because of its inherent low sensitivity. Finally, SERS can also be used to quench fluorescence; this is discussed further in Section 8.2.5.2.

The choice of laser may also be affected by the required spectral or spatial resolution. Spectral resolution is the capability of the system to distinguish one feature in the spectrum from another along the wavenumber axis. In many cases a resolution of around $4\,\text{cm}^{-1}$ is sufficient. Spectral resolution decreases (i.e. is better) with lower wavelength lasers. However, there are other instrumental parameters (e.g. groove density on a diffraction grating) that determine a system's overall resolution so they should be considered as a whole.

Spatial resolution determines the smallest size feature that can be distinguished when capturing an image. The lower the laser wavelength, the better/smaller the resolution (r) since

$$r = \frac{0.61\lambda}{\text{NA}} \tag{8.4}$$

where NA is the numerical aperture of the microscope objective. This means that for lower laser wavelength the spot size (laser cross section) will be smaller and the instrument, therefore, can potentially distinguish smaller features. Spot sizes are usually in the order of μm. However, care must be taken when reducing spot size as it results in more power per unit area.

8.2.4.2 Dispersive versus FT

There are two types of Raman spectrometer available: dispersive and FT. Dispersive instruments are generally used with visible excitation lasers and FT with NIR lasers. It follows then, given the above discussion, that FT systems offer a better solution if a sample is likely to fluoresce but dispersive instruments allow higher intensity signals, therefore, they are more successfully coupled to microscopes and so are often more suitable for the small samples found in forensic work.

In both instruments the monochromatic laser light is directed onto the sample, using mirrors and lenses. The reflected and scattered light is then collected and focused using the collection optics, which can be at 90° or 180°. The 180° configuration is usual in FT systems and microscopes (see Section 8.2.4.6).

The incident and Rayleigh scattering is much more intense than the Raman scattered light and would saturate the detector and swamp the Raman signal if it was not filtered out at this point. This can be done in one of two ways, namely, monochromators or a notch/edge filter. Multiple monochromators are able to remove the incident light in a very narrow band in the tens of cm^{-1} but they also increase the size of Raman instruments considerably. Holographic notch filters block a band of light in the order of hundreds of cm^{-1}, commonly around 200 cm^{-1}, and are much smaller than monochromators. (Note that in systems where the laser frequency can be changed the notch filter must also be changed to block the appropriate frequencies.) Ideally the wavenumber range removed is as small as possible so that small Raman shifts can be detected, which suggests the use of monochromators. However, notch filters are sufficient for many uses, are smaller and cheaper and, therefore, are used in most modern spectrometers.

8.2.4.3 Dispersive Raman Spectrometers

In a dispersive instrument, a monochromator containing a diffraction grating is used to diffract/disperse the signal. The grating has a series of angled grooves that diffract different wavelengths at different angles allowing them to be separated and then directed to the detector (Figure 8.5). In most cases for dispersive Raman the detector is a CCD array detector, which records a different wavelength range with each pixel.

The signal reaching the detector is linearly distributed by wavelength but this means that it is not linear by wavenumber. Resolution, therefore, is not the same across the wavenumber range of a typical spectrum. There is also a trade-off between resolution and spectral range. Greater resolution is achieved by using a diffraction grating with a higher density of grooves. This spreads the diffracted light over a larger distance but the array is a fixed length with a fixed number of pixels so if the spread of diffracted light is larger than the array

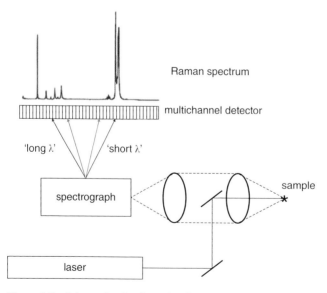

Figure 8.5 Schematic of a dispersive Raman system with a CCD detector. Source: Reproduced with permission from Chalmers et al. (2012).

some of that data will not be detected. Gratings can be interchangeable to provide either very high resolution over a small wavenumber range or lower resolution over the maximum wavenumber range. Alternatively, the range can be scanned in sections and added together.

8.2.4.4 FT-Raman Spectrometers

In a FT instrument rather than use a grating, the frequencies are separated using an interferometer and a mathematical function, a FT. In the interferometer a beam splitter splits the incident light into two paths and directs each onto one of two mirrors. One mirror is stationary, and one moves forwards and backwards increasing and decreasing the pathlength for that beam. When they recombine it causes interference. The wavelengths that 'match' the pathlength of the beam directed to the movable mirror will constructively interfere (they are in-phase), others will destructively interfere (out-of-phase), and most will be between the two. This produces a sinusoidal interferogram that can be detected and then deconvolved, using the Fourier transformation, giving intensity at each wavenumber.

CCD detectors are not sensitive enough at NIR excitation, therefore Ge or InGaAs detectors are used. If the sample is a weak scatterer, Ge is more appropriate.

Generally speaking, dispersion systems are used with visible lasers and FT with NIR but improvements in electron multiplication charge-coupled devices (EMCCDs) and InGaAs arrays have allowed some NIR dispersion systems to be developed (Chalmers et al. 2012).

8.2.4.5 Polarisers

Some Raman systems also have a polariser and an analyser, both polarising filters, before and after the sample, respectively. These are useful if analysis regarding the orientation of molecules in, for example, polymers or crystals is required.

8.2.4.6 Microscopes and Imaging

Raman microscopy systems are particularly applicable to forensic sample analysis, particularly in trace evidence examination. Small samples can be visualised and focused using the microscope in 'normal/visible light' mode. The visible light is then blocked, and the laser light is allowed to reach the sample. The incident light can be sent down the microscope optics via the same path as the visible light (since glass does not give a strong Raman signal) and collected via the return pathway. Before reaching the microscope eye piece the incident and Raman light are directed towards the detector via the filter which blocks the incident light so that the Raman scattering shifts can be detected.

In addition to being able to handle small samples, further advantages to Raman microscopy are related to the fact that the laser spot sizes are usually in single figure μm. This allows the analyst to select the precise location on the sample to analyse, potentially removing interferences from an inhomogenous sample, such as fluorescence. Although, this must be weighed against the increased power per unit area, particularly for samples that are prone to decomposition or burning.

As with other modern analytical techniques, imaging and mapping are also a possibility using Raman microscopy. In imaging, rather than a grating, a rotating carousel of filters, each allowing $\sim 20\,\text{cm}^{-1}$ of light to pass to the detector, is used allowing an image to be produced showing scattering intensity over the sample. This is often carried out over a small range of wavenumbers of interest and the width of the filters results in poor spectral resolution. In mapping, a full spectrum is captured at regular intervals in a predetermined area. Closer sampling points allow better spatial resolution up to the limits of resolution for a particular Raman instrument (which are determined by the choice of laser and system optics, see Section 8.2.4.1). Once the data has been collected it can be interrogated to plot the intensity of the bands of interest in the sample. Multiple images can be produced since a complete spectrum is available for each point. Mapping is very time consuming, given that each individual spectrum can take ~10 seconds to capture but it has the advantage of having all the spectral information for each sampled point. In contrast, imaging is faster but will only record the data for the selected frequency.

Raman spectroscopy is essentially a surface technique; however, confocal Raman spectroscopy is a further feature of some Raman microscopes. Here the laser can be focused at varying depths of the sample either producing a series of spectra or, with an imaging system, virtually slicing the sample and producing an image at a specific depth. This is useful for samples with layers, which in a forensic context might be crossed lines of ink on a document, where the question is, which was deposited first (Borba et al. 2017).

8.2.4.7 Portable Instruments and Probes

In forensic science particularly, there any many applications, such as drugs and explosives, where it would be desirable to carry out analysis in situ at the scene. Portable Raman systems are now available and can be around the size of a small suitcase with a computer attached. In addition, since glass is largely inactive in Raman spectroscopy, fibre optic probes can be used for 'stand-off' applications where the sample to be analysed is either inaccessible or carries a high potential risk of harm.

There are limitations to these systems, most significantly reduced resolution relative to a benchtop system. Also, the interpretation of the scene data needs to be considered. Is it to be done by someone with experience of Raman interpretation at the scene or perhaps remotely or will the data simply be compared against a suitable database? This will have an impact on the level of certainty in the conclusions.

Further, in the use of probes, the silica used in the glass production can contribute to the spectra due to the repeated reflection off the internal surface, therefore, extra care should be taken with interpretation.

8.2.4.8 Quantitation

Quantitation is possible with Raman spectroscopy, however, for many of the samples occurring in forensic science it would not be appropriate or necessary. It is more likely that forensic samples are being analysed using Raman spectroscopy for chemical information rather than for quantitation. Theoretically, there is a linear relationship between intensity and concentration, however, variations in intensity can be found between spectra. Raman intensity is routinely reported on a verbal scale (weak, medium, strong), rather than with a numerical magnitude. That being the case, quantitation is usually carried out using a ratio of peak intensity and is relative rather than absolute. Since forensic scientists are unlikely to use Raman for quantification on a routine basis, quantification will not be discussed further here. See Larkin (2011) for discussion on quantitation with Raman spectroscopy.

8.2.5 Advanced Techniques

There are an increasing number of advanced techniques for collecting Raman data, although many of these are not routinely used in a forensic context. The most commonly seen are resonance Raman spectroscopy (RR), surface enhanced Raman spectroscopy (SERS), surface enhanced resonance Raman spectroscopy (SERRS) and SORS, which are discussed here. The reader is directed to Chalmers et al. (2012) for further discussion of techniques, such as tip enhanced Raman spectroscopy (TERS), in a forensic context.

8.2.5.1 Resonance Raman Spectroscopy

As well as reducing unwanted effects (see Section 8.2.4.1), the laser choice may be used to promote positive effects, for example, resonance. Here a laser wavelength is chosen to be close to an electronic absorption band of the relevant group in the molecule. This causes resonance, thereby, increasing the signal intensity by up to 10^6. Care needs to be taken with resonance as it can cause burning. It should be noted that it can also change the relative intensities of the peaks present, which should be taken into account when comparing with other spectra from the laboratory, the literature and databases to ensure that the sampling conditions are the same or that differences are taken into consideration.

8.2.5.2 SERS/SERRS

SERS is accomplished by placing the sample in contact with a roughened metal surface and yields an increase in sensitivity over Raman spectroscopy of up to 10^6 (Fleischmann et al. 1974). This can be carried out either by applying a colloid or by using a commercially available prepared surface. The mechanism by which the enhancement is achieved is not entirely known, although it is likely that there are two types of enhancement occurring. Field enhancement results from an increase in electric field at the surface and probably accounts for a 10^4 increase in sensitivity. Chemical enhancement results from interactions between the sample and the metal, such as charge transfer, and accounts for the remaining 100-fold increase in sensitivity. Spectra of single molecules have been reported using this method (Kneipp et al. 1997). However, reproducibility can be an issue, particularly for colloids.

As well as the increase in sensitivity, SERS can be used to quench fluorescence in order to analyse samples that would otherwise produce a spectrum with a fluorescence background and no discernible peaks from the sample. An example of SERS using silver colloid on the dye Acid Red 52 is shown in Figure 8.6 (Geiman et al. 2009). Normal Raman (NR) spectra were obtained from solid dye samples, while SERS spectra were obtained using a solution of the dye. The NR spectra at 633 and 785 nm have strong fluorescence backgrounds. Fluorescence is reduced, as expected, at 1064 nm and when using SERS with 633 nm excitation, despite the lower concentration of dye present. However, the SERS spectrum at 785 nm does not provide any improvement on the NR 785 nm spectrum. This demonstrates the variety of options available to the analyst in order acquire a spectrum of sufficient resolution and signal-to-noise ratio (S/N), depending on the instrumentation available.

Further, RR and SERS can be combined creating SERRS. The increase in sensitivity obtained using this technique can be up to 10^{10}. As a result of this increase the use of

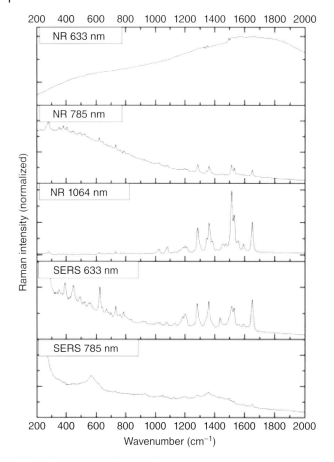

Figure 8.6 Normal Raman (NR) and surface-enhanced Raman spectra (SERS) of Acid Red 52. Source: Reproduced with permission from Geiman et al. (2009).

lower laser powers is possible, which is desirable to minimise sample degradation and fluorescence. The technique is also able to discriminate structurally similar molecules because spectral bands are sharper.

The application of a colloid to a sample is ostensibly a destructive or damaging technique, however, it is possible to apply the colloid in such a manner that it cannot be seen unless viewed under a microscope because the colloid area dimensions are in the micrometre range. SER(R)S is, therefore, considered to be a non-destructive or quasi-non-destructive technique depending on the way in which the proximity to the metal surface is achieved.

8.2.5.3 SORS

SORS is a technique that was developed to deal with samples for which sampling at depth is desired but which have a diffuse scattering or fluorescent surface layer, for example coloured glass (Matousek et al. 2005). Here scattering is detected from the site of illumination and also from a site laterally offset, usually within mm. The scattering at the excitation point will be dominated by scattering from the surface layer, whilst the offset scattering with include scattering from deeper into the sample. A (weighted) subtraction is carried out to remove the contribution from the surface layer leaving the spectra from the sample at depth. Increasing the offset increases the depth from which scattering is detected, up to a point, so that different depths can be sampled to determine composition.

8.2.6 Advantages and Disadvantages of Raman Spectroscopy

Many of the features and challenges of Raman spectroscopy are discussed throughout this chapter. However, it is useful, particularly to those new to the technique, to briefly consider the most significant in one place.

One of the main advantages of Raman spectroscopy, in a forensic context, is that in many cases it is non-destructive with minimal sample preparation. As discussed in Section 1.2.2, non-destructive techniques are important in forensic science to preserve the integrity of a sample and to retain some sample for analysis by the defence's scientific team, if necessary. If using a Raman microscope, the sample simply can be placed on the stage and the microscope focused; when using a probe it needs only to be directed at the sample. In addition, Raman microscopy employs a laser beam with a diameter of approximately 2 μm so it can target a very small sample area, which is important when sample size is small as is often the case with forensic samples.

The need for only limited sample preparation and the availability of portable systems allows for rapid analysis and analysis in situ at a crime scene. To give an example of potential use in a HAZMAT (hazardous material) scenario; analysis of an unknown white powder could be undertaken at a scene with a portable system and fibre optic probe to quickly indicate the structure of the compound. If it is found to be hazardous, steps could be taken to minimise harm by carrying out HAZMAT protocols, including closing off the scene and using protective clothing or, if found to be benign, by standing down the HAZMAT team and continuing with a standard scene examination.

Another advantage to Raman spectroscopy for forensic science is that, in contrast to IR spectroscopy, neither glass nor water give a strong Raman signal. This gives one advantage in that IR transparent optics are not necessary and a second advantage as, for example, in the HAZMAT scenario above, the sample can be analysed without opening its glass container, which reduces the risk of harm still more. Further, aqueous samples can be analysed without contribution from water to the spectrum (possibly in glass containers as well).

In addition, the availability of different wavelength lasers allows different species within a mixture to be probed producing different spectra via resonance, generating more features for comparison and offering potentially greater discrimination.

There are disadvantages inherent to Raman spectroscopy, however, which should also be considered when selecting the analytical method. Amongst these, the laser source can cause heating over the sample area and, consequently, decomposition or photodecomposition of the sample. Also, sample fluorescence can be strong, and since fluorescence is a more photon efficient phenomenon than Raman scattering, it can easily mask the weak Raman scattering signal. Furthermore, as it is essentially a surface technique, there can be interference from the substrate when the sample is present in thin layers.

8.3 Application to Analyte

8.3.1 Acquiring a Spectrum

Sample preparation for many forensic evidence types is minimal consisting of placing the sample on a slide so that it can be put on a microscope stage. However, the properties of the material should always be taken into account when planning any analysis and the effects of sample preparation on the spectrum acquired, i.e. any artefacts, should also be considered in the interpretation stage.

In addition to the simplicity of sample preparation, the calibration of a Raman instrument on a day-to-day basis is straightforward. Typically, a spectrum is acquired of a silicon standard with standard instrument settings and the shape and position of the peak, which should be at $520\,cm^{-1}$, are checked for consistency with previous acquisitions.

In addition to daily calibration, checks of the optics and laser should also be carried out periodically to ensure the instrument is optimised and correctly aligned.

Depending on the specific instrument used, spectra can be acquired in various modes. Renishaw plc instruments allow spectra to be acquired in static or extended mode. In static mode the user specifies the wavenumber to be at the centre of a scan and a few hundred wavenumbers on either side are also scanned. Therefore, a static scan is quick but scans a narrow range. This is the mode used for silicon calibration as only a small range is needed to check the silicon peak position. In extended mode the user can input the wavenumber range to be scanned, which can be much larger than in static mode, allowing a full spectrum to be acquired. This is more time consuming, however, a single extended scan is usually still acquired in seconds rather than minutes. Extended mode, or its equivalent, is the norm for most Raman analysis. Once acquisition mode has been selected the main instrument parameters that are optimised for a particular sample are the time taken to acquire a spectrum, the laser power and the number of accumulations. Laser power and time will be optimised to reduce issues such as burning and to increase signal (since signal strength is proportional to power, see Eq. 8.3). Multiple accumulations are used to acquire multiple spectra of the same sample consecutively and sum them together. Where there is signal from the sample the count will be summed to create peaks with higher intensity, where there is noise, since noise is random, it will be decreased. Some instruments allow the sample to be continually scanned until a predefined S/N threshold is met. The overall effect is to increase the contribution from the sample and reduce noise. This is particularly useful where sensitivity is an issue.

Once the spectrum of a sample has been acquired, mathematical manipulation of the data may take place such as baseline correction, background removal etc. Examples of the types of functions that can be applied to spectra include first or second derivatives to correct the baseline, deconvolution to separate out peaks that overlap or are wholly contained within other peaks, and smoothing to reduce signal to noise. However, any manipulation ultimately alters the data and if not done correctly can obscure or give a false impression of the sample's characteristics. It should be done only by personnel with expertise in the area and must be logged so that comparison of spectra is done on a like for like basis.

8.3.2 Forensic Applications

Raman spectroscopy can be used on samples in any state (solid, liquid, or gas). However, in forensic analyses it is most often used as a surface analysis technique on solid samples. It has been applied to a wide variety of forensic evidence types, including ink and paint, fibres, drugs, explosives and condom lubricants (Lewis et al. 1995; White et al. 1996; Claybourn and Ansell 2000; Day et al. 2004; Buzzini et al. 2006; Bradshaw et al. 2013). As with all forensic examinations the aim may be to identify an unknown sample, or it may be to determine if two samples may have the same source. Three evidence types have been selected for discussion in the following three sections (pen ink, paint, and drugs of abuse).

8.3.2.1 Pen Ink

Pen ink is made up of three main parts: the coloured component/colourant – a pigment or dye; the vehicle – for example, water or glycol; and other additives that give the ink the specific properties that the manufacturer wants it to have, for example corrosion inhibitors to prevent damage to the pen itself and plasticisers to improve the flow of the ink. The majority of the work carried out on inks using Raman has focused on the coloured component since many pigments/dyes have a high degree of symmetry and, therefore, give a strong Raman signal.

The analysis of inks can be carried out for several reasons: to establish if there is an association between inks from two different sources, for example, a suspect note and a pen seized from a suspect; to determine the age of a sample; or to determine the sequence of deposition of crossing lines (Borba et al. 2017). Common documents received by the forensic laboratory include letters, cheques, ID documents, wills, and contracts. Typically, there is no sample preparation required, unless SER(R)S is being used. However, care must be taken when selecting the laser power as samples can be prone to burning. For some colours fluorescence needs to be considered, for others the issue is that there is little variation in the dyes or pigments present in different inks and/or the resonance effect causes one coloured component to dominate the spectrum, for example, crystal violet tends to be the major dye component in the spectra of blue ballpoint pen inks. Raman, including SER(R)S, has been used to discriminate between pens with the same colour of ballpoint inks (glycol and dye based) (White 2000) and between gel pen inks (water and pigment based) (Mazzella and Khanmy-Vital 2003), amongst others. Claybourn and Ansell (2000) found fluorescence to be an issue for analysis of black ballpoint inks with a 514 nm excitation laser but successfully used a 782 nm laser to reduce the fluorescence contribution to the spectra. SE(R)RS has also successfully been used to quench fluorescence allowing differentiation between black ballpoint pen inks (Seifar et al. 2001) and blue ballpoint inks (Borba et al. 2017).

8.3.2.2 Paint

There are many different types of paint that could be examined as evidence in a forensic laboratory, including artist's paint, household paint, and automotive paint. Paint consists of a coloured component, i.e. the pigment, a resin, a solvent or liquid component and manufacturer specific additives.

The capability of Raman spectroscopy to analyse with no sample preparation or damage to a sample is very important if analysis is being carried out on valuable artefacts or art works. However, if the sequence of paint layers is important, some preparation may be necessary. Single layer paint samples can be analysed as small chips, whereas multiple layers may need to be cross-sectioned with a microtome before analysis. As with pen inks, fluorescence and the dominance of particular pigments can affect analysis.

For household and automotive paints, the objective might be to determine if the combination of layers of paint are the same or not (see case studies, Section 8.5) (Buzzini et al. 2006; Zięba-Palus and Trzcińska 2013). For artworks often the question is about the authenticity of the work and Raman has been used to check whether the composition of the paint is appropriate for the period in which the piece is supposed to have been made, i.e. are there any anachronisms in the paint available at the supposed time of painting and the paint used in the work being examined. A case in 2002 proved that works supposedly by Jackson Pollock could not have been made by him as pigments present, including Pigment Red 254 and Pigment Yellow 151, were not in production until after the artist's death (Khandekar et al. 2010).

In paint analysis Raman has typically been used for identification of or differentiation by pigments, without the interference of binder peaks since these are weak relative to the pigments, whereas IR has been used to gain information on the binder content. Raman has been successfully used in studies to differentiate large numbers of light coloured (white or yellow) paint samples that look similar to the naked eye (Kuptsov 1994; Bell et al. 2005b) as well as other colours, such as red (Massonnet and Stoecklein 1999). However, Bell et al. (2005a) argued that Raman spectroscopy could also be a better analysis method for binder analysis since they achieved the same discriminating power for Raman as they did for IR and Raman has the advantage that it is more rapid.

8.3.2.3 Drugs of Abuse

Drugs of abuse are a common evidence type for which portable Raman spectroscopy is applicable. Rapid analysis at the scene of an incident is desirable; it could, for example, potentially save lives in cases of overdose. Again, fluorescence can be a problem, although in the case of drugs it is the cutting agents rather than the drugs themselves that cause the majority of the fluorescence problems. However, using longer wavelength lasers, particularly NIR, can counter the effect to a large extent. There are two possibilities for forensic drug analysis with Raman. The first is to identify the drug present, which may not be trivial for drugs with similar structures and/or in mixtures. The second is to determine whether or not two samples have come from the same source. In this case, the non-drug components can provide useful points of difference.

The identification of drugs of abuse in many different forensic scenarios, where trace amounts are present, has been investigated by several groups: Day et al. (2004) analysed five drugs of abuse in doped fingerprints and found that they could be identified and discriminated from a selection of common adulterants; West and Went (2009) successfully acquired spectra from particles of banned substances trapped between fibres on tape lifts; and Eliasson et al. (2008) used SORS to obtain spectra of cocaine in alcohol solutions through glass containers.

There are many more applications of Raman spectroscopy to forensic samples. The reader is directed to Chalmers et al. (2012) for more extensive discussion.

8.4 Interpretation and Law

8.4.1 Interpreting Raman Spectra

The interpretation of Raman spectroscopy can be done on two levels depending on the question to be answered. The first is a simple pattern matching of the spectra or to put it another way a spot the difference/similarity to identify a sample either as a particular substance or as having the same characteristics as another. Of course, this has its limitations since it will only tell you what is in a sample if you have an appropriate reference spectrum to compare it with (criteria for appropriateness including factors such as the sample preparation method, instrumentation and acquisition parameters, and any data manipulation such as baseline correction, i.e. comparing like with like). As with many techniques, databases of reference spectra, or spectral libraries, are available to compare spectra acquired with those of known compounds. These are useful in the laboratory and, especially, at a crime scene where operators may not always have expertise in Raman spectroscopy. Since there are many, many compounds that could be included in a library, the libraries that tend to be available include a selection of spectra appropriate for a specific application, such as pharmaceuticals or forensic applications.

The second type of approach is to use the peak position and shape/intensity to work out the functional groups present and, from these, part or all of the structure present. This is not always straightforward if the sample is a mixture and can often only give an indication of the class of compound rather than the specific structure if, for example, the sample is a long chain aliphatic since Raman spectroscopy cannot tell you how many of each type of functional group is present and, therefore, cannot distinguish between say pentane and heptane. Often, this method is carried out in conjunction with interpretation of an IR spectrum, with each providing information that can confirm or clarify the other.

Using a classical model, if we imagine that a bond is two balls (atoms) on the end of a spring (bond) and assume that the spring obeys Hooke's law, the frequency of the harmonic oscillation will given by

$$\bar{v} = \frac{1}{2\pi}\sqrt{\frac{k}{\mu}} \qquad (8.5)$$

where k is a constant for the spring and μ is the reduced mass given by

$$\mu = \frac{m_1.m_2}{m_1 + m_2} \qquad (8.6)$$

where m_1 and m_2 are the masses of the atoms.

We can see that as the mass of the atoms decreases, the frequency (or wavenumber by convention in a Raman spectrum) increases. Similarly, as bond strength increases, frequency tends to increase; a strong bond requires more energy to vibrate it, and since frequency is proportional to energy, a strong bond will vibrate at a higher $\bar{v}$. So, in general, the lighter the atoms and the stronger the bond, the bigger/higher the Raman shift/wavenumber. Clearly,

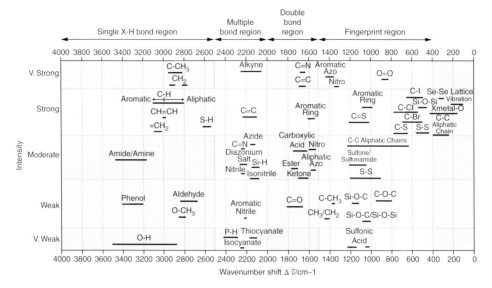

Figure 8.7 Characteristic Raman frequencies. (Note, there is no significance to the vertical position of a bond within a horizontal band.)

the converse is also generally true; the heavier the atoms and weaker the bond, the smaller the Raman shift. However, since there are many combinations of light/heavy/weak/strong bonds it is not the case that all strong bonds occur at higher shift than all weak bonds. It is safer to say that, for example, a stronger double bond C=C peak occurs at ~1600 cm^{-1} (higher shift), whereas, a weaker single bond C—C peak occurs at ~800 cm^{-1} (lower shift). In addition, the position of a bond within a molecule can affect its vibrational frequency (see C—H example below), which can give additional information on a possible structure/functional group.

Again considering the spring model, it is easier (requires less energy) to bend a spring than it is to stretch it, so the bending vibrations for a particular molecule tend to appear at lower frequencies/wavenumbers than the corresponding stretching vibrations.

The above rules of thumb are important to remember when we look at assigning peak frequencies to particular bonds. In order to help identify a particular sample there are many reference tables available giving the position, strength and shape (i.e. broad, sharp) for functional groups, an example, is given in Figure 8.7. More comprehensive tables are available in Socrates (2004).

The first indication of the type of functional group can be obtained by using the following, very general, bond strength region classifications: 4000–2500 cm^{-1} is the single X—H bond region (low bond strength but high shift because H is light); 2500–2000 cm^{-1} multiple/cumulative bond region; 2000–1500 cm^{-1} double bond region; and below 650 cm^{-1} are inorganic groups, metal–organic groups, or lattice vibrations. The fingerprint region, between 1500 cm^{-1} and 650 cm^{-1}, does include specific bond vibrations, however, given that the structures giving rise to them can produce complex profiles and that there may be overtones of higher/lower energy bonds present this region is so difficult to interpret that it is not usually done. However, it can give a characteristic peak pattern that identifies a particular molecule, hence the term 'fingerprint' region.

Ideally, the process of identifying a structure will identify the functional groups peak by peak following a methodical order. This is best done by identifying peaks with a high wavenumber shift, rather than in the fingerprint region. The fingerprint region cannot necessarily be relied upon for identifying peaks, even if they are intense, however, they can provide supporting evidence for peak identification in another region. The most reliable peaks are ones that consistently appear when the relevant group is present in the sample and give a high intensity, sharp Raman signal without overlapping other peaks. In addition, functional groups with more than one characteristic band can be helpful for confirming the presence of a particular group.

When looking at a spectrum, check the high wavenumber shift for groups such as amines and alcohols. Determine whether there are aliphatic and/or aromatic groups present by looking at $3000\,\mathrm{cm}^{-1}$; strong peaks just above indicate aromatic C—H bonds and those just below aliphatic C—H bonds. Note the presence or absence of the strong aromatic ring breathing vibration at $1600\text{–}1550\,\mathrm{cm}^{-1}$ and the presence or absence of methyl groups at around $1380\,\mathrm{cm}^{-1}$. Continue on through the bond regions, following a methodical process to identify, if possible, the remaining peaks, not forgetting that the absence of peaks can also be used to help confirm or rule out identifications.

Once the analyst has identified a possible structure they should always check that the identification makes sense for the sample they have in front of them, e.g. if the sample is a white powder, is the proposed structure consistent with a white powder?

For both approaches to interpretation, pattern matching and peak assignment, it becomes increasingly difficult the more constituents a mixture has, not just because of the mixture but also because some constituents will, inevitably, contribute more to the spectrum than others. Statistical methods such as those discussed in Section 2.4.3 can be used to separate out the contributions of different components to the spectrum. In many cases the analyst will want to confirm an identification by means of another confirmatory technique, such as mass spectrometry.

8.4.2 Raman Spectroscopy in Court

Raman spectroscopy is still an emerging technique in the forensic sciences. To be accepted as a method used in court as probative evidence a technique must be accepted by the scientific community as being applicable to a particular type of evidence. Although Raman has been applied to many forensic problems this has been more often in research rather than in casework. Consequently, there are a limited number of accepted protocols for analysis of trace evidence types. However, Raman spectroscopy is beginning to be incorporated into workflows. For example, it is included in the Federal Bureau of Investigation (FBI) Scientific Working Group for Materials Analysis (SWGMAT) Fibre Subgroup training programme (2004) and is listed as one of the available techniques in the European Network of Forensic Science Institutes (ENFSI) Best Practice Manual for the Forensic Recovery, Identification and Analysis of Explosives Traces (2015). In addition, case work has been carried out by various Raman groups working with local law enforcement agencies (Buzzini et al.

2006; Lepot et al. 2008). As the technique is more widely adopted Raman will, no doubt, be seen increasingly in the courtroom.

8.5 Case Studies

8.5.1 Case Study 1

Buzzini et al. (2006) describe a case study that involves a classic example of the transfer of trace evidence; transfer of paint from a door to an object used to force it open. In the case, a burglary had taken place in an industrial setting, during which a crowbar had been used to force open a door into a room containing a safe. No trace evidence was recovered from the scene; however, a suspect car was detained in the local area and a red crowbar was recovered from the boot. A reference sample of paint from the door at the crime scene and the crowbar were sent to the laboratory for examination. The paint from the door was observed to have two layers (one blue and one pink). Paint traces, due to transfer, were observed on the crowbar blade. These were white, blue, and pink. Consequently, the blue and pink paints from both the reference and crowbar samples were examined. The pink paint sample spectra were dominated by fluorescence and were not able to be analysed by Raman spectroscopy successfully. However, the blue trace evidence sample was analysed with a 514 nm laser and found to have Pigment Blue 15 (a phthalocyanine) and the rutile form of titanium dioxide present. Comparison with the spectrum from the door reference sample allowed the conclusion that the samples were indistinguishable. Following the chemical analysis of the samples the data was considered in the context of the case under investigation. First, the significance of the paint colours being found in combination on the crowbar was considered. In a database available to the investigating team, blue paint had only once been observed as a trace on a crowbar seized by the police and pink had not been observed at all. Secondly, using a further database, comprising the details of samples of household paints from domestic burglaries, and statistical sampling of the database, only 5.4% of the samples were blue and none were blue and pink. Further, the spectra of the blue paints in the database did not have the same features as the samples in the case. Therefore, due to the rarity of the paint layer colour combination, the rarity of the paint spectrum and the fact that the trace evidence and reference sample were indistinguishable, the team concluded that there was strong evidence for the crowbar paint trace evidence having come from the door at the scene.

8.5.2 Case Study 2

Zięba-Palus and Trzcińska (2013) describe the analysis of evidence in a hit-and-run incident. A young man's body was found on a side road of a village. There were no indications of a collision and the only trace evidence collected was small green paint and black plastic samples from the hood of the man's coat. The police had information that two identified vehicles had been seen nearby. They collected reference paint samples from both vehicles and submitted them to the laboratory for comparison with the trace evidence found on the hood. The hood samples were observed to have four layers (two green, one grey, and one yellow). IR was used to analyse the binder component of the paints and Raman for the colourants. The top green layer was found to have an acrylic urethane styrene resin and to contain chrome yellow and chlorinated pthalocyanine. The lower green layer was an alkyd melamine. When the reference samples from the two vehicles were analysed one was a different colour paint but the other had the same paint layer combination, i.e. green, green, grey, and yellow. Further, this second vehicle reference sample, when analysed, had the same chemical composition as the paint traces recovered from the man's hood. The laboratory team concluded that the trace evidence and the reference sample from the second vehicle were indistinguishable, that the presence of two chemically different green layers increased the evidential value of the comparison (it may indicate a re-spray, which is individual to a single vehicle), and that the vehicle may have been involved in the incident.

8.6 Forensic Developments

Raman spectroscopy has great potential for use in the forensic sciences. It has been demonstrated to be applicable to a wide variety of forensic evidence types but much of the work carried out is still in the realms of research. Although there are exciting technological developments in Raman spectroscopy, such as TERS (Stöckle et al. 2000), what is more pertinent for Raman spectroscopy to become a mainstay of forensic analysis is work on integrating it into existing work flows and adoption into protocols. The availability of portable instrumentation could be one of the ways in which to achieve this. Portable systems can be very easy to use, allowing acquisition of spectra in a matter of seconds, making them very attractive for in situ analysis by non-experts. However, work needs to be done on creating databases containing relevant reference samples that allow an unambiguous interpretation of the spectra for the non-expert.

One of the major hurdles for Raman spectroscopy is that it is not taught as extensively as other analytical techniques in educational institutions. Consequently, there is less expertise in the area and less awareness of what the technique is capable of than for more mainstream/established techniques. Is it anticipated that this will change in the near future.

References

Bell, S.E., Fido, L.A., Speers, S.J. et al. (2005a). Forensic analysis of architectural finishes using Fourier transform infrared and Raman spectroscopy, part I: The resin bases. *Applied Spectroscopy* 59 (11): 1333–1339.

Bell, S.E., Fido, L.A., Speers, S.J. et al. (2005b). Forensic analysis of architectural finishes using Fourier transform infrared and Raman spectroscopy, part II: White paint. *Applied Spectroscopy* 59 (11): 1340–1346.

Borba, F.D.S.L., Jawhari, T., Honorato, R.S., and de Juan, A. (2017). Confocal Raman imaging and chemometrics applied to solve forensic document examination involving crossed lines and obliteration cases by a depth profiling study. *Analyst* 142 (7): 1106–1118.

Bradshaw, R., Wolstenholme, R., Ferguson, L.S. et al. (2013). Spectroscopic imaging based approach for condom identification in condom contaminated fingermarks. *Analyst* 138 (9): 2546–2557.

Buzzini, P., Massonnet, G., and Monard Sermier, F. (2006). The micro Raman analysis of paint evidence in criminalistics: case studies. *Journal of Raman Spectroscopy* 37 (9): 922–931.

Chalmers, J.M., Edwards, H.G., and Hargreaves, M.D. (2012). *Infrared and Raman Spectroscopy in Forensic Science*. Wiley.

Claybourn, M. and Ansell, M. (2000). Using Raman spectroscopy to solve crime: inks, questioned documents and fraud. *Science & Justice: Journal of the Forensic Science Society* 40 (4): 261–271.

Day, J.S., Edwards, H.G., Dobrowski, S.A., and Voice, A.M. (2004). The detection of drugs of abuse in fingerprints using Raman spectroscopy I: latent fingerprints. *Spectrochimica Acta Part A: Molecular and Biomolecular Spectroscopy* 60 (3): 563–568.

Dreizler, R.M. and Gross, E.K. (2012). *Density Functional Theory: An Approach to the Quantum Many-Body Problem*. Springer Science & Business Media.

Eliasson, C., Macleod, N., and Matousek, P. (2008). Non-invasive detection of cocaine dissolved in beverages using displaced Raman spectroscopy. *Analytica Chimica Acta* 607 (1): 50–53.

ENFSI (2015). Best Practice Manual for the Forensic Recovery, Identification and Analysis of Explosives Traces. ENFSI-BPM-EXP-01.

Ferraro, J., Nakamoto, K., and Brown, C.W. (2003). *Introductory Raman Spectroscopy*. Academic Press.

Fikiet, M.A., Khandasammy, S.R., Mistek, E. et al. (2018). Surface enhanced Raman spectroscopy: a review of recent applications in forensic science. *Spectrochimica Acta Part A: Molecular and Biomolecular Spectroscopy* 197: 255–260.

Fleischmann, M., Hendra, P.J., and McQuillan, A.J. (1974). Raman spectra of pyridine adsorbed at a silver electrode. *Chemical Physics Letters* 26 (2): 163–166.

Geiman, I., Leona, M., and Lombardi, J.R. (2009). Application of Raman spectroscopy and surface-enhanced Raman scattering to the analysis of synthetic dyes found in ballpoint pen inks. *Journal of Forensic Sciences* 54 (4): 947–952.

Khandekar, N., Mancusi-Ungaro, C., Cooper, H. et al. (2010). A technical analysis of three paintings attributed to Jackson Pollock. *Studies in Conservation* 55 (3): 204–215.

Kneipp, K., Wang, Y., Kneipp, H. et al. (1997). Single molecule detection using surface-enhanced Raman scattering (SERS). *Physical Review Letters* 78 (9): 1667.

Kuptsov, A.H. (1994). Applications of Fourier transform Raman spectroscopy in forensic science. *Journal of Forensic Science* 39 (2): 305–318.

Larkin, P. (2011). *Infrared and Raman Spectroscopy: Principles and Spectral Interpretation*. Elsevier.

Lepot, L., De Wael, K., Gason, F., and Gilbert, B. (2008). Application of Raman spectroscopy to forensic fibre cases. *Science & Justice* 48 (3): 109–117.

Lewis, I., Daniel, N. Jr., Chaffin, N. et al. (1995). Raman spectroscopic studies of explosive materials: towards a fieldable explosives detector. *Spectrochimica Acta Part A: Molecular and Biomolecular Spectroscopy* 51 (12): 1985–2000.

Long, D.A. (2002). *The Raman Effect*. Wiley.

Massonnet, G. and Stoecklein, W. (1999). Identification of organic pigments in coatings: applications to red automotive topcoats. Part III: Raman spectroscopy (NIR FT-Raman). *Science & Justice: Journal of the Forensic Science Society* 39 (3): 181.

Matousek, P., Clark, I., Draper, E. et al. (2005). Subsurface probing in diffusely scattering media using spatially offset Raman spectroscopy. *Applied Spectroscopy* 59 (4): 393–400.

Mazzella, W.D. and Khanmy-Vital, A. (2003). A study to investigate the evidential value of blue gel pen inks. *Journal of Forensic Sciences* 48 (2): 419–424.

Nagy, B., Farkas, A., Gyürkés, M. et al. (2017). In-line Raman spectroscopic monitoring and feedback control of a continuous twin-screw pharmaceutical powder blending and tableting process. *International Journal of Pharmaceutics* 530 (1–2): 21–29.

Nogueira, F., Castro, A., and Marques, M.A. (2003). *A Tutorial on Density Functional Theory. A Primer in Density Functional Theory*, 218–256. Springer.

Raman, C.V. and Krishnan, K.S. (1928). A new type of secondary radiation. *Nature* 121 (3048): 501.

Seifar, R.M., Verheul, J.M., Ariese, F. et al. (2001). Applicability of surface-enhanced resonance Raman scattering for the direct discrimination of ballpoint pen inks. *Analyst* 126 (8): 1418–1422.

Smekal, A. (1923). Zur Quantentheorie der Dispersion. *Naturwissenschaften* 11 (43): 873–875.

Smith, E. and Dent, G. (2019). *Modern Raman Spectroscopy: A Practical Approach*. Wiley.

Socrates, G. (2004). *Infrared and Raman Characteristic Group Frequencies: Tables and Charts*. Wiley.

Stewart, S., Bell, S.E., Armstrong, W.J. et al. (2012). Forensic examination of multilayer white paint by lateral scanning Raman spectroscopy. *Journal of Raman Spectroscopy* 43 (1): 131–137.

Stöckle, R.M., Suh, Y.D., Deckert, V., and Zenobi, R. (2000). Nanoscale chemical analysis by tip-enhanced Raman spectroscopy. *Chemical Physics Letters* 318 (1–3): 131–136.

SWGMAT (2004). A Forensic Fiber Examiner Training Program. https://www.nist.gov/system/files/documents/2016/09/22/fiber_examiner_training_program.pdf (accessed 26/7/2020).

West, M.J. and Went, M.J. (2009). The spectroscopic detection of drugs of abuse on textile fibres after recovery with adhesive lifters. *Forensic Science International* 189 (1–3): 100–103.

White, P. (2000). SERRS spectroscopy – a new technique for forensic science? *Science & Justice* 2 (40): 113–119.

White, P., Munro, C., and Smith, W. (1996). In situ surface enhanced resonance Raman scattering analysis of a reactive dye covalently bound to cotton. *Analyst* 121 (6): 835–838.

Zięba-Palus, J. and Trzcińska, B.M. (2013). Application of infrared and Raman spectroscopy in paint trace examination. *Journal of Forensic Sciences* 58 (5): 1359–1363.

9

Scanning Electron Microscopy

Grzegorz Zadora and Aleksandra Michalska

9.1 Introduction

In general, a scanning electron microscope is a type of microscope that produces images of a sample's surface by scanning it with the use of a focused beam of electrons (Lawes 1987; Newbury et al. 1986; Watt 1997). The first scanning transmission electron microscope was built by von Ardenne in 1938 (von Ardenne 1938a, b) but the first 'modern' scanning electron microscope was built by a Cambridge group (Oatley et al. 1965; Smith and Oatley 1955). Their work led to the production of the first commercial instrument 'Stereoscan' by the Cambridge Scientific Instrument Company in 1965 (delivered to DuPont). Further details concerning the history of scanning electron microscopy (SEM) can be found in McMullan (2006).

Various signals are produced when high energy electrons interact with atoms in the sample. These signals can be detected and they contain information about the sample's surface (see Section 9.2.1). This allows the analysis of morphology (e.g. that a feature suspected to be a gunshot residue [GSR] particle does or does not have a spherical shape or exhibits features of having been molten), topography and composition of the analysed objects, including homogeneity of sample, size of inclusions, defects, etc. The electron beam is generally scanned in a raster scan pattern, and the beam's position is combined with the detected signal to produce an image (resolution can be better than 1 nm). Samples can now be observed with the following parameters: in high vacuum, in low vacuum, in wet conditions (environmental scanning electron microscopy [ESEM]), and at a wide range of cryogenic or elevated temperatures.

When it is combined with X-ray analysis (e.g. a scanning electron microscope coupled with an energy dispersive X-ray [EDX] detector for SEM-EDX) then elemental analysis is possible (see Section 9.2.2). This option is the most popularly applied in forensic laboratories and is broadly discussed in this chapter.

Analytical Techniques in Forensic Science, First Edition.
Edited by Rosalind Wolstenholme, Sue Jickells and Shari Forbes.
© 2021 John Wiley & Sons Ltd. Published 2021 by John Wiley & Sons Ltd.

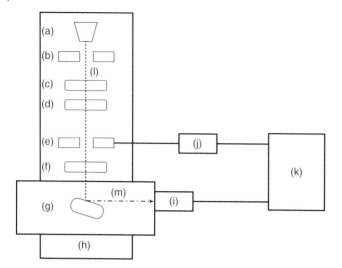

Figure 9.1 Diagram of a scanning electron microscope: (a) electron gun, (b) anodic disc, (c, d) condensers, (e) scan coils, (f) objective lens, (g) specimen stage, (h) vacuum pump, (i) detectors (e.g. energy dispersive X-ray detector), (j) magnification control including waveform generator, (k) cathode ray tube brightness control, scan coil and display screen, (l) electron beam, and (m) beams of secondary electrons, and/or backscattered electrons, and/or X-rays.

9.2 Theory of the Technique

9.2.1 Scanning Electron Microscope

A schematic diagram of a typical scanning electron microscope is presented in Figure 9.1 (Lawes 1987; McCarthy 2002; Newbury et al. 1986; Watt 1997). The electron gun (a in Figure 9.1) produces an electron beam by the process of thermionic emission, caused by heating a filament (e.g. made from tungsten or LaB_6). A bias voltage of between 2 kV and 50 kV is applied between the filament and the earthed anodic disc (b in Figure 9.1) in order to accelerate the electrons away from the charged filament. This is the basis for the design of an electron gun system, which typically consists of a filament in an enclosed metal assembly. The tip of the filament will emit high energy electrons that are accelerated through an electron field, thus acquiring kinetic energy. This forms a concentrated electron beam that can be directed down towards the sample. To avoid collision of the electrons with gas molecules (air), and thus scattering of the beam, a vacuum system (h in Figure 9.1) is required. Hence the electron gun is housed inside a sealed column which is attached above the sample chamber. The condensers (c and d in Figure 9.1) reduce the diameter of the beam, e.g. from 50 to 0.005 µm (i.e. 10 000 times) and the beam is focused using a series of electromagnets.

The thermionic electron gun is the cheapest and the most often used; however, for the production of electron beams other types of electron guns based on field emission (e.g. Schottky emission gun) can be used.

When the electron beam, generated by the electron gun, hits the specimen (g in Figure 9.1), several different events can take place (m in Figure 9.1, and Figure 9.2), from

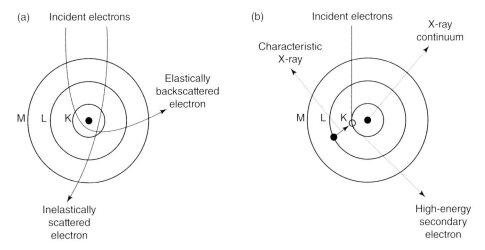

Figure 9.2 Schematic showing selected electron interactions in an atom used commonly in SEM-EDX analysis: (a) elastically backscattered electron, inelastically scattered electron; and (b) X-ray continuum, high-energy secondary electron, characteristic X-ray.

which secondary electrons (SE), backscattered electrons (BSE) and characteristic X-ray radiation are used and detected by suitable detectors (i in Figure 9.1).

Elastic scattering takes place over a range of beam energies of circa 20 kV (Figure 9.2a). The electron beam bounces off the specimen, changing direction, but maintaining most of its energy (with a loss of less than 1 eV). Among these elastic scatterings are BSEs, which are formed when the primary electron beam interacts with the nucleus of a sample atom and may be deflected in any direction with little loss of energy. These deflections can continue and hence some of these electrons (from the primary beam) work their way back out from the sample allowing them to be detected. These BSEs possess high energies and may escape from a greater depth within the sample. This results in very little surface (topographical) information regarding the sample. However, BSEs can provide good compositional information as the intensity of the signal is proportional to the mean atomic number of the sample. This relates to the fact that the higher the atomic number of an atom, the greater the positive charge of its nucleus and hence the more likely an interaction that produces a BSE. Hence, the higher the mean atomic number of the sample, the brighter the image obtained.

Inelastic scattering results in a large amount of energy transfer between the electron beam and target atoms. In this instance information can be obtained regarding elemental composition of the specimen surface. For example SEs (Figure 9.2a) are formed when electrons are knocked out of their orbitals by the incident electron beam. These ejected electrons from the sample exhibit some amount of kinetic energy. If they are created near to the surface of the specimen, they may escape and be detected as SEs. The SEs have typical energies below 50 eV and therefore are sensitive to the topography of the sample. Thus, edges within samples can emit more SEs than flat surfaces whilst holes will produce considerably less, if any. Therefore, SEs are important for obtaining information relating to the three-dimensional topography of the surface.

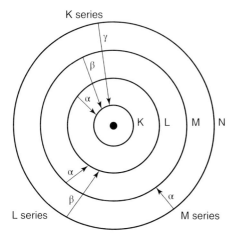

Figure 9.3 Naming convention for X-rays generated by specific electron jumps.

Characteristic X-rays are obtained if an inner shell electron (e.g. K, L, M in Figure 9.2b, and Figure 9.3) is knocked out of its orbital by the incident beam thus ionising it. An electron from an outer shell drops down to replace it, thus stabilising the atom once again (Figure 9.2b). The excess energy of this electron transition is released as electromagnetic radiation in the form of a characteristic X-ray (10^3–10^5 eV, Figure 9.2b), and detected independently from detection of SEs and BSEs. They provide an additional source of information about the elemental composition of an analysed sample and are used for qualitative and quantitative analysis of solid samples. Characteristic X-rays are produced since the energy released as an electron transition occurs is unique to each element because the orbitals around the nucleus of each atom have distinct energies which are dependent on the positive charge, and hence atomic number, of that atom. X-rays are designated double-barrelled names based on the shell name of the vacating electron ('K' is equivalent to the electron shell described by the principal quantum number (n) equal to 1 ($n = 1$), 'L' is equivalent to $n = 2$, etc.) and the origin of the 'replacing' electron ('α' if from one shell higher, 'β' if from two shells higher, etc.). For example, if the incident beam knocked out of an L shell electron and an N shell electron dropped down to replace it (Figure 9.3), the X-ray produced from the release of the excess energy would be called an L_β X-ray.

If the energy or wavelength of the X-ray is known, then it is possible to identify the element that was responsible for producing it. This is because the energy of the X-rays produced is determined by the structure of the atom. Moseley's law (Eq. (9.1)) describes the relationship between atomic number Z and wavelength λ of a spectral line.

$$\lambda = \frac{K}{(Z - \sigma)^2} \tag{9.1}$$

where K and σ are constants for a given spectral line. Alternatively, a quantum of electromagnetic radiation could also be described by its wavelength (λ) or energy (E) according to the equation:

$$E = \frac{hc}{\lambda} \tag{9.2}$$

where h is Planck's constant and c is the speed of light in a vacuum.

Therefore when a spectrum is obtained (see Section 9.3.2) peak localisation allows information on the presence of particular elements in a sample to be obtained in a relatively easy way.

Quantitative analysis is possible because there is a linear dependency between the intensity of the characteristic X-ray of a particular element and its concentration (c) in the analysed sample (Eq. (9.3)). It is a relative method as it is based on determination of the intensity of a characteristic X-ray measured with a standard I_w (i.e. a sample of pure element or pure chemical compound) and then a determination of the intensity of a characteristic X-ray of a particular element measured on a sample I:

$$c = \frac{I}{I_w} \tag{9.3}$$

It should be mentioned that it is difficult to accurately translate X-rays detected into exact quantities of elements present. This is due to:

- Differences in X-ray behaviour which are dependent on the atomic number of the element (Z).
- The number of X-rays absorbed before escaping the sample (A).
- The number of X-rays resulting from secondary fluorescence (F) within the sample.

Most software allows quantitative or semi-quantitative data for each element, made possible by mathematical correction algorithms (e.g. 'ZAF correction'), which are used to rectify the problems associated with atomic number, absorption and fluorescence. Nevertheless, results obtained on the basis of Eq. (9.3) are semi-quantitative. If aiming for quantitative analysis, it is necessary to make a calibration curve, when suitable standards are available.

Moreover, when quantitative or semi-quantitative analysis is made then, according to theory, the surface of the sample should be smooth and flat (see Section 9.3.2) in order to record a strong signal. When the analysed surface is rough (see Section 9.3.2) then a weaker signal could be recorded. This is because if a sample is smooth then all signal can go to a detector (Figure 9.4a). If it is rough then it could meet a barrier/obstacle in the form of another part of the sample and could be absorbed, which could further affect a reduction of the X-ray beam intensity in comparison with a signal intensity measured for a smooth and flat sample (Figure 9.4b; see also Section 9.2.4).

The area of influence of the electron beam within a solid sample depends on the energy of the electrons, i.e. it depends on the accelerating voltage/probe current, on the angle of

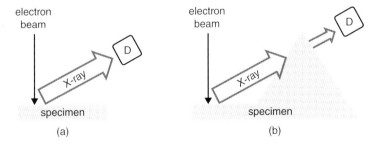

Figure 9.4 An illustration of the difference in X-ray intensity reaching a detector (D) when a sample surface is smooth (a) or rough (b).

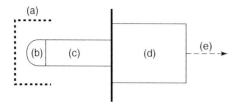

Figure 9.5 Diagram of a scintillator/photomultiplier detector used for measuring backscattered and secondary electrons: (a) wire cage, (b) scintillator tip, (c) light guide, (d) photomultiplier tube, and (e) connection to display screen (k in Figure 9.1).

incidence on a sample surface as well as on the average atomic number of the elements which create the solid sample. The depth of penetration of electrons in a sample increases with the energy of the electron beam (kV) and decreases when the average atomic number of the elements that create the solid sample increases. The depth of penetration of the most frequently analysed signals is typically as follows:

(a) Tens of nanometres (normally not exceeding 80 nm) for SEs.
(b) From hundreds to thousands of nanometres for BSEs.
(c) From hundreds of nanometres to several micrometres for X-rays.

With the aim of obtaining an image of a sample, e.g. a subject of interest for forensic analysis, two pairs of scan coils (e in Figure 9.1) are arranged perpendicular to one another and to the electron beam (l in Figure 9.1). They are supplied with current waveforms produced by the waveform generator (j in Figure 9.1). A raster scan is performed in which the beam is moved through a grid, which might contain, for example, one million frames (one thousand horizontal by one thousand vertical; k in Figure 9.1). BSEs and SEs can be detected by one scintillator/photomultiplier detector (Figure 9.4), called the Everthart–Thornley detector. To detect SEs, the wire cage (a in Figure 9.5) is set to a voltage of +250 eV. This stabilises the beam and improves collection of SEs. The scintillator tip (b in Figure 9.5) has a positive charge and, therefore, attracts the low energy SEs from the specimen and converts them into photons using materials such as plastic, glass and calcium fluoride doped with europium impurities. The light guide (c in Figure 9.5) sends the photons to the photomultiplier tube (d in Figure 9.5) by total internal reflection. The photons then strike an electrode with a high electrical potential, thus producing many photoelectrons. These, in turn, strike another electrode producing many more. This process continues until a large number of electrons have been produced (usually 10^5–10^6 electrons per photon introduced). This electronic signal is then sent to the display screen (DS) via the connection (e in Figure 9.5) and is then forwarded on to the DS brightness control (k in Figure 9.1) where it is converted into a spot of variable brightness (bright if the signal is strong and dark if the signal is weak). Finally, an instant image of the specimen is presented by viewing the entire grid, again, for example, one million frames in a one thousand horizontal by one thousand vertical grid. The image can be enhanced using the magnification control (j in Figure 9.1). In order to detect BSEs, the wire cage (a in Figure 9.5) is held at a potential of between −50 eV and 0 eV. SEs are repelled but high energy BSEs, travelling towards the detector, enter the scintillator tip (b in Figure 9.5), and are amplified in the same way as discussed above.

Nowadays for detection of BSEs and SEs usually separate detectors are used. The Everthart–Thornley detector is applied for detection of SEs while a scintillator/photomultiplier detector called the Robinson detector, or a p-n junction is used for BSEs.

9.2.2 X-Ray Detection

Detection of characteristic X-rays allows the elements present in the sample to be determined (Lawes 1987). It can be done by analysis of wave features of X-rays by application of a wave dispersion detector (wavelength dispersive spectroscopy [WDS] or wavelength dispersive X-ray spectroscopy [WDXS]) or an energy dispersive detector (energy dispersive spectroscopy [EDS] or energy dispersive X-ray spectroscopy [EDXS]).

In a WDS detector diffraction crystals are used. They work like a monochromator, which aims to disperse and separate X-rays having different wavelengths. According to Braggs' equation (Eq. (9.4)) a diffraction is observed only for a wave having length n (positive integer);

$$n\lambda = 2d_{hkl} \sin \Theta \tag{9.4}$$

To cover the whole range of wavelengths of characteristic X-ray for elements from boron to uranium, it is necessary to have several crystals on which the electron beam falls (under Θ angle), and which are characterised by different between-surface distances (d_{hkl}). This is a disadvantage of this detection technique as usually only one or sometimes several peaks can be scanned.

Therefore, detection of a range of X-ray energies is usually carried out by dispersion of its energy (EDS, EDXS), rather than wavelength, where semiconductive detectors are used, e.g. p-n junction (Figure 9.6), in the form of silica crystals or crystals of silica doped with lithium Si (Li). The atmospheric thin window (ATW) (a in Figure 9.6), acts as an electron filter, preventing BSEs and SEs from entering the detector, as well as X-ray energies below that of Be (i.e. Li, He, and H). Thus, these elements could not be detected. When an X-ray (b in Figure 9.6) passes through, it strikes the first gold film (c in Figure 9.6) on the front of the silicon-lithium (SiLi) semiconductor crystal (d in Figure 9.6), releasing a free electron that scatters throughout the crystal, producing more free electrons and electron holes.

When all of the energy is absorbed within the crystal, a 1 keV field is applied to the gold contacts, separating the electrons and electron holes, thus creating a current pulse. The current produced is very small and must be amplified by a factor of $>10^{10}$. The field effect transistor (FET; e in Figure 9.6) is a detector pre-amplifier that limits the electronic noise from the crystal, thus improving the signal.

It should be mentioned that this kind of detector requires cooling, in order to prevent lithium atoms from migrating in the semiconductor, and it greatly improves the signal-to-noise ratio (S/N) by keeping the FET cool. One of the ways in which to achieve

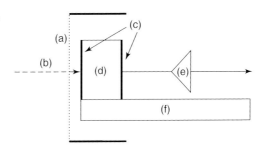

Figure 9.6 Diagram of an energy dispersive X-ray detector: (a) atmospheric thin window, (b) X-ray, (c) gold film, (d) silicon-lithium semiconductor crystal, (e) field effect transistor, and (f) liquid nitrogen.

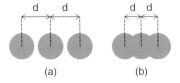

Figure 9.7 Illustration of resolution, i.e. capability to distinguish three particles separated from each other by a narrow gap: (a) sufficient resolution as three particles separated from each other by a narrow gap, d; and (b) insufficient resolution as three particles are not separated from each other by a narrow gap, d.

this is the use of liquid nitrogen in the cooling process (f in Figure 9.6), which is cumbersome, or more commonly, by devices working under a Peltier cooling process.

9.2.3 Operating Conditions

There are some operating principles that allow good quality images to be obtained, but there are also advantages and disadvantages related to the increase or decrease of the value of a particular parameter.

In general SEM is used when the resolution (i.e. capability to distinguish two particles separated from each other by a narrow gap; Figure 9.7a) of an optical microscope is not sufficient for a particular type of sample. If the particles presented in Figure 9.7a move closer together then, in one moment, only a single, somewhat fuzzy, shape will be seen (Figure 9.7b). In such a situation the optical microscope is now incapable of resolving the distance, d between these centres.

In SEM the resolution is mainly limited by the diameter of the probing beam of electrons, the so-called 'spot size'. When, a large diameter beam is used for particles presented in Figure 9.7a then an indistinct signal would be produced and as a result a fuzzy shape would be obtained. On the other hand, a smaller beam would produce a signal showing them as two separate particles. Nevertheless, operating with the smallest available spot size, although providing the highest possible resolution, also produces very 'noisy' images, i.e. poor S/N.

It is assumed that the 'noise' inherent in the system is unchanged. Therefore, if the beam is left at every point on the specimen for a longer time, then a larger signal from that point would be collected. In other words scanning speed has an influence on the S/N. A fast scanning speed gives poorer S/N than a slow scanning speed. Scan speed is the speed at which the electron beam scans over the specimen (and display) and is variable over quite a range. Fast scan speeds mean that the specimen can be examined in real time, because the displayed image is being updated many times a second. It is very useful to work with fast scan speeds especially when elemental analysis is performed as it helps to orientate the specimen and locate areas of interest (e.g. for elemental analysis), after making some adjustment to the specimen position for example. However, slow scan speeds could be very useful as well, especially if someone would like to take a photo of a specimen. In such a situation image quality is improved as it allows a reduction in S/N, as described earlier.

Magnification is closely linked to resolution and is calculated as a ratio of screen resolution to beam diameter (spot size). Manufacturers sometimes provide much higher magnifications than those one would expect from their specified resolution capabilities. These over-magnified images contain no more information (details), because of the limit of resolution.

Thus, if fast scan speeds are used but high resolution is not required (e.g. at low magnifications) then increasing the spot size greatly improves the S/N, and image quality. In such situations, bombarding the specimen with larger numbers of primary electrons would, predictably, increase the yield of secondaries, X-rays, etc. and in this case the specimen absorbs much more energy, and is, therefore, more likely to charge up and could be damaged by overheating.

Accelerating voltage (from a few hundreds of eV to dozens of keV) plays a very important role in determining microscope performance as the highest voltage should be used for best resolution. When the highest voltage is used then the electron beam produced is obviously much more energetic and penetrates much more deeply into the surface of the specimen. It is especially important in quantitative X-ray microanalysis. Additionally, because the sample is absorbing much more energy at high kV, specimen charging and damage is more common.

Another parameter is beam current. Reducing the beam current, reduces charging and thermal damage to the specimen. Of course, bombarding the specimen with fewer primary electrons does produce lower signals from it, and inevitably poorer S/Ns.

The working distance (WD) of the SEM is the distance from the final lens of the microscope, to the surface of the specimen. It can be varied, from about 5 to 40 mm, by moving the sample vertically up or down, using the specimen stage Z-axis shift. Long WDs provide images with good depth of field but poorer resolution whereas short WDs produces the opposite (better resolution, poorer depth of field).

Another parameter which could affect resolution or S/N is aperture size. In general, scanning electron microscopes are provided with mechanical apertures, which can be altered by the operator. These are small metal plates with very small (25–1000/in diameter) holes (apertures) through them. They define the effective diameter of the lenses and reduce some of the inherent aberrations. In general, large apertures give lower resolution but better S/Ns.

Despite the rules described existing, experience in manipulating particular SEM equipment is the best method to learn how to obtain the best image for a particular sample.

9.2.4 Specimen Preparation

In the forensic field most of the samples that are analysed with the use of the SEM technique (or SEM-EDX) are solid and most of them are electrically non-conductive. When non-conductive specimens are examined by SEM they tend to collect a surplus of electrons on their surface, in other words they become negatively charged and this causes image defects (charging). This effect could be reduced by using very low accelerating voltages (<1 keV), and as a result sacrificing resolution, which is not a desired effect.

Therefore, instead of changing the operating conditions of the microscope it is better to change the condition of the specimen to be conductive. This could be done by coating the specimen surface with a very thin, uniform conductive layer. The material used for coating should have the following properties: good conduction of heat and electrons; chemical inertness (resistant to oxidation and tarnishing layers); good secondary electron emission (large signal); smallest possible grain size (to avoid 'decorating' the specimen); easy to produce thin films. These conditions are met by gold, gold/palladium alloy, and carbon. Vacuum evaporation is popularly used for coating in forensic laboratories (see Section 9.2.4.1).

Another restriction in sample preparation is related to the fact that in order to produce a beam, the microscope operates with a high vacuum inside it. Any volatile material would be vaporised at these low pressures. This is one of the most severe restrictions on the type of specimen that can be examined and, therefore, samples containing water cannot be analysed (unless ESEM is used), as water molecules may be ionised near the high voltage gun. The very reactive hydroxyl ions resulting from this ionisation, can damage the gun's hot filament. One solution could be an environmental scanning electron microscope. The disadvantage of using this technique in the forensic field is mainly related to X-ray detection (e.g. gas surrounding a sample generates X-rays). This effect could be reduced by the application of dedicated algorithms to deduct the effects of gas on the information extracted during analysis. Moreover, the presence of gas in a specimen chamber also has an effect on the distance between the electron gun and a specimen. The higher the pressure, the shorter the distance – in other words the distance varies for low vacuum from ca. 10 mm to a fraction of a millimetre when the pressure is close to one atmosphere.

Moreover, some chemical samples may be totally volatile at these low pressures. Therefore, it is good practice, in the case of unknown samples, to expose them to a similar vacuum (e.g. in a coating unit), before risking contamination of the microscope. Additionally, volatile samples can cause the vacuum to become poorer, and can lead to a general loss of resolution. In addition, some of the beam electrons collide with the freed vapour molecules causing beam instability.

Moreover, the electron beam's energy could be converted to heat and this can produce quite significant increases in temperature, enough to damage delicate structures of samples with low melting points.

The energy actually absorbed could be reduced by changing some of the SEM operating conditions (e.g. spot size). Alternatively, much of the heat can be carried away by ensuring that there is good contact between the sample and the microscope, by using conductive adhesives for instance. Coating the surface of the specimen with a thin metal layer would also help conduct the heat away. Nevertheless, in the case of very sensitive specimens (e.g. some low melting point waxes), the only real solution is to use a cooled specimen stage.

9.2.4.1 Vacuum Evaporation

Vacuum evaporation is the simplest of two techniques that can be used to produce a wide variety of metal coatings and, unlike sputtering, it can also be used for depositing carbon on the surface of the sample.

The process is carried out inside a vacuum chamber. A small amount of the metal to be evaporated, in the form of a fine wire, is suspended from the tungsten filament. The dry, mounted specimen is placed at the bottom of the vacuum chamber (several cm below the heater). The chamber is then sealed and evacuated. When a suitable vacuum has been reached the current to the heater is switched on, and steadily increased. As the heater temperature is raised the fine metal wire melts and particles are driven off. These radiate outwards in straight lines, coating any surface directly exposed to them.

The same equipment can be used for vacuum evaporation by carbon. In this case, carbon rods or a few strands of carbon fibre are used. Once the required vacuum has been attained, the carbon is heated, again using a high current (low voltage) supply. Carbon is evaporated and, as before, coats anything inside the work chamber, including the specimens. It is advisable to use spectroscopically pure carbon, especially if the elemental composition of the specimens will be under investigation.

When vacuum evaporating is applied, this process can cause damage to delicate specimens. One type of damage, thermal damage, is caused because of radiant heat from the filament during evaporation. It can be greatly reduced by keeping the heater/specimen distance as large as possible (e.g. less than a few cm) and keeping the metal source as small as possible. During sputtering these effects can raise the temperature of the specimens by as much as 40 °C, enough to cause pitting, melting and, in extreme cases, destruction of the specimens.

With microscopes capable of resolving finer and finer detail, SEM is now revealing that conventional coating techniques can produce coatings made up of distinguishable particles or grains. These agglomerations or clumps of deposited material are often called 'decoration' artefacts.

Moreover, different samples can require specific treatment before they can be analysed by SEM or SEM-EDX. For example, when quantitative analysis is performed then it is required that a sample surface is smooth and flat. Therefore, in the case when samples of minerals, rocks or glass are subject to analysis, a common way of sample preparation is embedding them in a resin (e.g. epoxy resin, acrylic resin, Bakelite) and polishing the surface (e.g. using cerium oxide powder, alumina or diamond paste) with the use of a flat grinding machine. However, due to the fact that this approach is rather impractical e.g. for very small glass fragments (linear dimensions less than 0.4 mm), which are most frequently the subject of forensic examination, rather than embedding, the procedure described in Michalska et al. (2016) could be applied (see also Sections 9.5.2 and 9.5.3). The authors propose using an optical microscope (magnification 40×) for selecting very small glass fragments as smooth and flat as possible and directly depositing them on an SEM stub.

9.3 Application to Analyte(s)

SEM has a much greater depth of field than optical microscopy, and can, hence, provide much clearer images of fibres, hairs, bullets, cartridge cases and other items that do not have a flat surface. A combination of SEM with X-ray detection (SEM-EDX) additionally allows the chemical composition from any area that is observed to be determined. Three basic analytical techniques could be applied in X-ray microanalysis, i.e. an analysis of the surface, linear analysis, and point analysis. A map of the distribution of a particular element is obtained, in the form of bright points/regions (e.g. application of BSEs; see Section 9.2.1), when an analysis of the surface is made. It allows the homogeneity of the sample to be established, which could be further used in solving a comparison problem (see Section 9.4). Changes of concentration of a particular element along a determined line can be obtained when a linear analysis is made. A point analysis is applied especially when precise quantitative analysis is required (see Section 9.3.2).

Another important advantage of SEM-EDX is the possibility of analysing elemental composition of traces much smaller than those that could be analysed by other techniques commonly applied in forensic laboratories, i.e. laser ablation inductively coupled plasma mass spectrometry (LA-ICP-MS), laser-induced breakdown spectroscopy (LIBS), micro-X-ray fluorescence (µ-XRF). All the above-mentioned reasons in combination with the non-destructive nature of the method mean that SEM-EDX is widely used in the forensic sphere.

Two popular applications of the method relate to analysis of GSR (Section 9.3.1) and examination of glass microtraces (Section 9.3.2). These types of evidence are broadly discussed in the following.

9.3.1 Gunshot Residue

GSR known also as cartridge discharge residue (CDR) (Wallace and McQuillan 1984), firearm discharge residue (FDR) (Flynn et al. 1998), primer discharge residue (PDR) (Bergman et al. 1991) as well as potential firearm discharge residue (P-FDR) or full GSR (Lebiedzik and Johnson 2000) are metallic particles under investigation in shooting events (European Network of Forensic Science Institutes [ENFSI] 2008; Scientific Working Group on Gunshot Residue [SWGGSR] 2011). Their 'uniqueness' arises from the combination of highly specific chemical composition and untypical morphology, both affected by the way in which they are formed (Basu 1982; Wolten and Nesbitt 1980).

When a gun is fired, the firing pin strikes the back of the cartridge case and activates the shock sensitive primer. The heat of combustion melts the primer and causes the ingredients to vaporise. Due to supersaturation, the primer vapours recondense into small droplets, some of which may grow by coalescence. Formed GSR droplets as well as other gases, ignite the propellant (gunpowder) and force the bullet down the barrel and on its path (Basu 1982). Together with the bullet a cloud of organic and inorganic vapours consistent with burned and unburned propellant, GSR droplets as well as other residues are released from the gun (Meng and Caddy 1997). GSR droplets are ejected from the barrel as well as from any small gaps in the gun and become deposited onto the hands, face, hair and clothing of the firer as well as on other surfaces nearby (Michalska 2007). As a result of rapid cooling GSR droplets solidify in their existing usually spherical form with sizes typically between 0.1 µm and 10 µm. Some of them may take irregular shape, and occasionally, be up to several dozen micrometres (e.g. 55 µm in size reported by Basu [1982]). Regardless of their size and shape they exhibit features of forming in extremely high temperature and pressure, i.e. they have the appearance of having been molten (Figure 9.8). The chemical composition of GSR is determined mainly by the elemental composition of the primer (Basu 1982; Wolten and Nesbitt 1980). There are several types of primers (a component of ammunition), which could have an effect on GSR composition (Table 9.1) (Collins et al. 2003; Harris 1995; Michalska 2007). However, the most popular primers are still those based on compounds of lead, antimony and barium, therefore GSR particles are usually comprised of these elements (ENFSI 2008; SWGGSR 2011). GSR residue may also contain other elements such as tin, copper, iron, aluminium, calcium, and silicon from the bullet, the bullet jacket, the cartridge casing, or the lubricant (Basu 1982; Meng and Caddy 1997; Romolo 2002; Wolten and Nesbitt 1980).

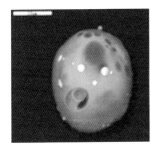 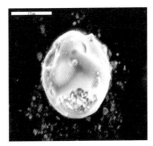 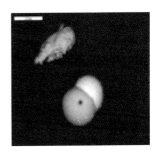

Figure 9.8 The morphology of three-component PbSbBa particles found on the suspect's jacket. Source: Reproduced with permission from Michalska (2007).

Table 9.1 Elemental composition of GSR versus types of ammunition (Michalska 2007).

Ammunition	Ammunition name	Elemental composition of GSR
Traditional	9-mm TZZ	Pb, Sb, Ba
	Geco Sintox™	Pb, Sb, Ba
	7.65 mm LV Geco	Pb, Sb, Ba (Al, Si, Fe, Cu, Zn)^a
	5.56 mm Winchester	Pb, Sb, Ba, Al
	0.22 long-rifle Eley	Pb, Ba
	Sellier Bellot of Prague	Pb, Ba, Ca, Si, Sn
		Pb, Ba, Ca, Sn
	6.35 Hirtenberg	Pb, Ba, Ca, Si
	0.38 Kynoch	Pb, Ba, Ca, Si
	9-mm perabellum Israel	Pb, Sb, Cl, K
	7.62 × 25 mm China	Pb, Sb, Sn, Cl, S
	7.62 mm Russian Federation	Sb, Sn, Hg, S, K, Cl, Cu (Al, Si, Zn, Fe)^a
	Eastern Germany	Sb, S, Cl, K
Modern	9 mm Sintox	Ti, Zn
	'X Lead Free'	Mn, K
	Hirtenberg Lead Free™	Sr
	CCI Blazer®	Sr (Sr, Ba)^b

a) Rarely detected elements.
b) Barium presence is caused by strontium natural contamination by barium.

Identification of GSR particles is based on both the chemical composition and the morphology of the analysed residue. The ability to examine these two features in one instrument led to the SEM-EDX technique being preferred in their recognition (ENFSI 2008; SWGGSR 2011).

When a person is suspected of firing a gun, police investigators collect the potential GSR by dabbing the surface of interest (around 100 times) using double-sided self-adhesive

Table 9.2 Classification scheme of GSR particles.

Category of GSR	Chemical composition of particles related to:		Evidential value[a]
	Traditional ammunition	Modern ammunition	
Characteristic	PbSbBa	GdTiZn	Very strong support
	PbBaCaSiSn	GaCuSn	
Consistent	PbBaCaSi,		Strong support
	PbBa, PbSb,	TiZn	
	SbBa, BaCaSi,	Sr	
	BaAl		
Commonly associated	Sb, Pb, Ba (in the absence of S)	Not mentioned	Limited support

a) Level of support for hypothesis that detected particle is GSR.
Source: Data from SWGGSR (2011).

carbon tape deposited onto a 12 mm aluminium SEM stub (see Figure 9.11) (Trimpe 2011). When particles are seized from the cartridge case, e.g. found at the scene of crime, GSR particles from the cartridge interior are beat out onto a SEM stub surface. The SEM stub is usually fixed into the cup of a plastic container, which prevents contamination. In the case of a lack of an SEM stub, potential GSR may be collected by the use of a small piece of clear adhesive tape. In this situation an additional clear piece of tape should be sent to the forensic laboratory in order to check if the selected material masks any heavy elements typical for GSR. Moreover, using common tape demands an extra coating before analysis. Carbon is the recommended coating material, as it provides a negligible backscattered signal and does not emit interfering X-ray lines (ENFSI 2008; SWGGSR 2011; Trimpe 2011). The coating should be uniform and of appropriate thickness in order to eliminate charging of the samples during analysis. Potential GSR particles are searched for manually or, as is recommended, with the use of an automated system mode (ENFSI 2008; SWGGSR 2011). When automated searching mode is used, the operator defines the sets of parameters e.g. position of searching stages, location of reference material, minimum particle size which should be detected as well as chemical composition of expected residues. The SEM stub is then scanned in a systematic, defined sequence during which the analysed surface is divided by software into smaller fields. The number and size of the scanned field depends on prearranged magnification. During analysis GSR particles, having high mean atomic number, are recognised by their BSE image (see Section 9.2.1). Once a particle with a bright BSE image is detected it is analysed by EDX (Nesbitt et al. 1976) (see Section 9.2.2). The automated system classifies detected residues according to their chemical composition, and saves the coordinates describing the particles' position. After the programme search is complete, the particles recognised by the automated system as GSR must be verified by the expert, who reacquires the X-ray spectrum and documents the particle's morphology (ENFSI 2008; SWGGSR 2011).

Only residue which meets specific criteria may be identified as GSR (ENFSI 2008; Romolo 2002; SWGGSR 2011). The most crucial is elemental composition. When composition is considered the expert compares the qualitative SEM-EDX spectrum obtained for a particular particle with a classification scheme created by a group of experts (e.g. Scientific Working Group for Gunshot residue; www.swggsr.org). The scheme is based on longstanding examination of GSR particles as well as other residues from non-GSR sources. The scheme describes in detail typical GSR composition produced by particular ammunition and shows the type and level of other elements that might be found in these GSR particles. It shows also the gradation of evidential value of particles based on their chemical composition. According to this scheme (Table 9.2), GSR particles can be assigned into three categories: characteristic, consistent, and commonly associated. The highest evidential value is given to particles classified as characteristic, because their composition is typical for primer explosions, and is very rarely found in other sources. High evidential value is associated with particles classified as consistent with GSR. These particles are generally formed during firearm discharge, however single particles of similar composition are occasionally detected in other sources. Particles classified as commonly associated with GSR have a composition commonly found in numerous sources, therefore, alone their evidential value is rather low. However, when they are found in addition to characteristic or consistent particles, they increase the likelihood that the analysed particles belong to GSR (Giacalone 2002; Michalska 2007; Romolo 2002). The proposed classification scheme does not include all types of GSR composition, especially those originating from lead free ammunition. In the case of ammunition with primers based on mercury fulminate (Zeichner and Levin 1992), it gives only general guidance. Evaluation of the evidential value of particles in these cases, as well as in a situation where particle composition from a particular fired ammunition becomes changed as a result of interaction between GSR particles originating from this and previously discharged ammunition (Zeichner et al. 1991), depends on the expert's knowledge and experience.

It should be also noted that SEM-EDX easily deals with GSR originating from traditional ammunition with primers based on components containing heavy metals. Some novel ammunition, in which primers are divested of heavy elements, generates particles which do not contain high atomic number elements. Such particles will not be readily detected using the same parameters as in the case of detection of traditional heavy-metal GSR (SWGGSR 2011).

As already mentioned, as well as specific elemental composition, the particles should also possess characteristic morphology, e.g. having spherical shape or exhibit features of having been molten. Any particle that has sharp edges, corners or crystalline structure should be excluded because its morphology is inconsistent with the mechanism of GSR formulation (ENFSI 2008; SWGGSR 2011). The expert should also focus on the presence of other particles or elements that are not typically found in GSR. Some studies have shown that in a few examples such as detonation of fireworks (Kosanke 2003; Mosher et al. 1998; Trimpe 2003), friction of brake pads (Ingo et al. 2004; Torre et al. 2002), or air bag explosion (Berk 2009a, 2009b) single particles similar in both elemental composition and morphology to GSR residue could be occasionally formed. However, each time samples from these sources contained additional elements, e.g. high concentration of magnesium and/or other elements such as chlorine or potassium in the case of fireworks or an elevated level of iron in the case of brake pads. These elements are not typically found in GSR, therefore when found should exclude particles as originating from gun discharge (ENFSI 2008; SWGGSR 2011).

9.3.2 Glass

Glass (used here to mean silica glasses) is a material that is used in many areas of human activity and it appears most frequently as window glass, car windows and windscreens, car headlamps, car mirrors, light bulbs, bottles, jars, tableware, and decorative items. Very small fragments of glass can be formed during events such as car accidents, burglaries and fights. These fragments may be recovered from the scene of the incident, as well as from the clothes and bodies of people present (Caddy 2001; Curran et al. 2000; Zadora et al. 2014).

During the production of glass (Caddy 2001), many different elements are incorporated into the molten mixture. Certain elements are crucial for glass production and are always present. These major components are silica, sodium, calcium, and magnesium oxides. Silica, calcium, and sodium oxides are the basic ingredients that form glass. Sodium oxide reduces the softening point of silica, while calcium oxide and magnesium oxide make glass more physically or chemically resistant (e.g. avoids the recrystallisation process).

Minor components are also present, such as potassium, aluminium, and iron oxides. Their role is to improve the physical properties of glass samples (e.g. K_2O is added in order to improved optical properties while Al_2O_3 is added in order to avoid recrystallisation). Iron oxides are cheap additives used in order to produce a coloured glass (green or brown). In contrast to coloured glass, in ordinary colourless windowpanes the presence of iron oxide is revealed only as a green shade, e.g. in a cross-section. The concentration of iron oxides in coloured glass is sufficient to be detected by the SEM-EDX technique, whereas in non-coloured glass objects their amount is below the detection limits of the method (e.g. 0.1 [wt%] for iron).

Due to the fact that in glass analysis both qualitative and quantitative SEM-EDX results are used, the preparation of glass objects requires their surface to be smooth and flat (see Sections 9.2.1 and 9.2.4; Michalska et al. 2016). Therefore the embedding procedure is recommended for glass sample preparation (Figure 9.9). If this procedure is applied, the analysed sample is placed on a flat surface, mounted in appropriate resin, mechanically ground and finally polished. In the embedding process epoxy resin (Croveri et al. 2010; Verità et al. 1994), acrylic resin (Croveri et al. 2010; Echlin 2009; Falcone et al. 2006; Verità et al. 1994), or conductive Bakelite (Becker et al. 2001) can be used. Silicon carbide paper or diamond abrasive paper is applied for grinding, whereas cerium oxide powder (Verità et al. 1994), alumina (Becker et al. 2001), or diamond paste of different grades (Becker et al. 2001; Croveri et al. 2010; Falcone et al. 2006; Verità et al. 1994) are proposed for polishing. Preparation of the samples in this way guarantees a well-polished surface maintaining the sample's flatness and minimising relief and edge rounding of the object (Verità et al. 1994). Moreover, the grinding and polishing stage of embedding allows the removal of the original glass surface (e.g. metal coatings covering optical glasses or computer screens or tin layer used in the manufacturing process of float glass surfaces), which is extremely important in determination of the bulk elemental content of the analysed glass samples. Nevertheless, the embedding process is time-consuming and is impractical for very small samples, e.g. glass fragments with linear dimensions less than 0.4 mm, which are frequently found at the scene of events such as car accidents, burglaries, and street fights (Curran et al. 2000; Zadora et al. 2014), and hence the most often analysed in the forensic field. Therefore, in such cases glass objects can be selected, under the stereomicroscope,

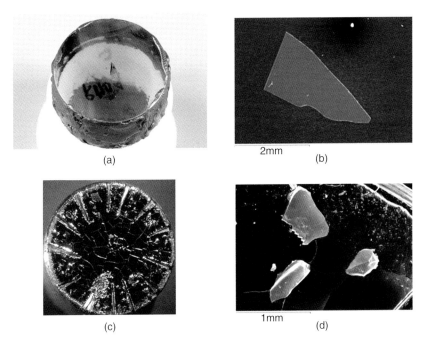

Figure 9.9 Type of glass fragment preparation: (a) embedded technique; (c) non-embedded technique; and (b, d) glass surface view in SEM-EDX microscopy related to the chosen method of preparation, i.e. respectively, embedding and non-embedding.

with surfaces as smooth and flat as possible (Figure 9.9) and then directly located onto an SEM stub (Michalska et al. 2016). It has been verified that a non-embedding procedure of glass sample preparation also allows reliable quantitative SEM-EDX results to be obtained (Zadora et al. 2014; Michalska et al. 2016).

The identification of the samples as a glass by the SEM-EDX technique is based on qualitative SEM-EDX results. Spectra recorded for glass are usually easily recognised as different from other samples such as aluminosilicate or silica (Figure 9.10). However, ascertaining that a seized sample is a glass object is insufficient for fact finders (judges, prosecutors, police, etc.) They usually want to gain information about the type of glass from which the analysed fragments originate, e.g. windows, containers (i.e. they want to solve a classification problem) or receive proof that known and questioned glass fragments originate from the same source (a comparison problem). In order to solve these problems, quantitative elemental composition of glass samples has to be used. Taking into account that the SEM-EDX technique delivers information only about major and minor elements found in glass, which usually differ only slightly in amount, and bearing in mind that this technique does not allow detection of trace elements, which possess the highest discrimination power in order to solve these problems, quantitative SEM-EDX results should always be considered with the application of appropriate statistical approaches (see Section 9.4). In most cases the combination of quantitative elemental results with statistical tools allows detection of any significant differences in the amounts of major and minor elements (Zadora et al. 2014) so a classification or comparison problem related to glass analysis can be solved.

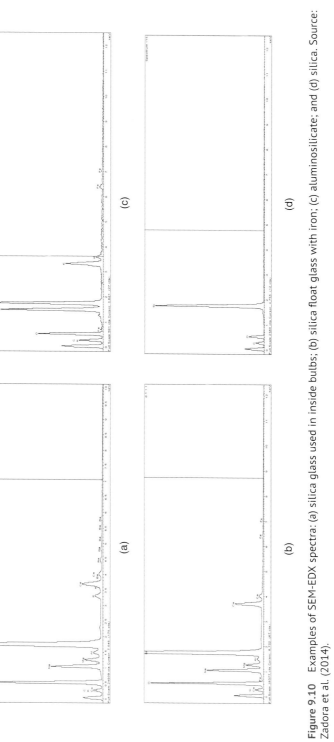

Figure 9.10 Examples of SEM-EDX spectra: (a) silica glass used in inside bulbs; (b) silica float glass with iron; (c) aluminosilicate; and (d) silica. Source: Zadora et al. (2014).

9.3.3 Other Samples

SEM can be used in the examination of handwritten documents in order to detect forgeries (Krüsemann 2001). It can help to determine which of two intersecting lines was drawn first. Paper is non-conductive and therefore in the past it was coated with a metal layer to eliminate charging at high kV. After coating (see also Section 9.2.4), written lines can only be imaged if the imprint has changed the topography of the paper. Today, instead of using a high kV for imaging, an extreme low voltage (e.g. 2.0 keV) is used and this allows imaging of paper in its natural state.

Due to the ability to observe a fibre's surface at high magnification SEM is used in identification and discrimination of animal fibres based on cuticle scale edge height (Tucker 1998), in characterisation of fibre cross-section (Scientific Working Group for Materials Analysis [SWGMAT] (1999), and analysis of fibre damage (Pelton 1995). The latter may sometimes contribute to establishing the instrument type that caused the fibre damage, e.g. scissors or knife (Krüsemann 2001). The SEM-EDX technique is also applied to identification of inorganic pigments and delustrants present in fibres (SWGMAT 1999).

In addition to standard footwear analysis, SEM-EDX can be used in the analysis of shoe surfaces. The method is usually applied in order to reveal the presence of flower or tree pollens, which may stick to the shoe. The identification of such evidence may additionally confirm that a shoe has been at the scene of the crime (Krüsemann 2001). The main difficulty with such analysis is the size of the shoe relative to the instrument chamber, hence the scanning electron microscope must be able to handle this.

Nowadays, SEM-EDX is also used in identification of inorganic ingredients present in automotive paints as well as in the examination of their morphological features such as thickness of particular layers as well as their sequence in paint chips (Caddy 2001; Zadora and Brożek-Mucha 2003).

Due to the fact that SEM-EDX allows analysis of very small objects as well as the ability to gain information on elemental analysis the method is applied in analysis of unknown substances, e.g. powders or pills. If it is suspected (e.g. after analysis by Fourier transform infrared [FTIR] spectroscopy) that the sample is an inorganic substance then SEM-EDX analysis is usually performed, especially when other analysis of elemental composition cannot be done due to the limitation of sample quantity.

Other fields in which SEM or SEM-EDX techniques are used are forensic pathology, taphonomy, and anthropology (Milani et al. 2012).

9.4 Interpretation and Law

9.4.1 Evidence Evaluation on Source Level

The application of numerous analytical methods, including SEM or SEM-EDX, to the analysis of evidence samples returns various kinds of data:

- Morphological data – e.g. shape of residues suspected to be GSR particles, the number and thicknesses of layers in a cross-section of car paint.

- Qualitative data – e.g. elements present in a sample.
- Quantitative data – e.g. the concentration of elements in a glass fragment.

In general, representatives of the administration of justice, who are not specialists in chemistry, are not interested in details such as the composition of the analysed objects, except in a situation such as the concentration of alcohol in a driver's blood sample or the content of illegal substances in a consignment of tablets or body fluids. Therefore, results of analyses should be presented in a form that can be understood by non-specialists, but at the same time the applied method of data evaluation should express the role of a forensic expert in the administration of justice. This role is to evaluate physicochemical data (evidence, E) in the context of two alternative hypotheses – the so-called prosecution proposition (H_1) and a defence proposition (H_2) – that is, to estimate, if possible in a numerical way, the conditional probabilities $Pr(E|H_1)$ and $Pr(E|H_2)$ (see Section 1.2.3.4).

One of the problems solved by the expert on the basis of obtained data is an identification/classification problem, e.g.:

- In GSR cases it is related to answering the question: do the recovered particles originate from discharge of a gun (H_1) or do they arise from human activity other than shooting (H_2)?
  In order to answer this question, information on both morphological and qualitative elemental composition are used (see Sections 9.3.1 and 9.5.1).
- In glass cases it is related to the questions: is a microtrace found in debris a glass object (H_1) or not (H_2), and if yes, then does it originate from container (H_1) or float glass fragment (H_2)?
  In order to answer this question qualitative and quantitative elemental composition is used (see Sections 9.3.2 and 9.5.2).

The other forensic problem which could be solved on the basis of obtained data, e.g. from analysis by SEM or SEM-EDX, is a comparison problem. This task is related to the question – could two samples, recovered (e.g. traces recovered from the suspect) and control (e.g. samples collected at the scene of crime), have originated from the same object (H_1) or are they are originated from two different objects (H_2)? For example, could the glass fragment revealed on a suspect's clothes have originated from a broken car windscreen, or in GSR cases do GSR particles detected on a suspect's hands originate from a cartridge case found at a scene of crime?

When solving a comparison problem, especially in a case when similarities are found between compared samples, the evaluation process should include analysis of observed similarity in order to establish whether they are observed by chance. Thus, some knowledge about the rarity of the measured physicochemical properties in a population representative of the analysed casework, called the relevant population (e.g. the population of car windows in the case of a hit-and-run accident) and the between-object variability is required. For instance, one would expect elemental composition from different locations of the same glass object to be very similar. However, equally similar elemental composition values could also be observed for different glass items. Therefore, information about the rarity of a determined elemental composition has to be taken into account. The value of the evidence in support of the proposition that the recovered glass fragments and the control sample have a common origin is greater when the determined values are similar and rare in the relevant population, than when the physicochemical values are equally similar but common in the same population (Zadora et al. 2014). The information about the rarity of the physicochemical data can be obtained from relevant databases available in forensic laboratories. In the case of a classification problem, the process of evaluation of evidential value of physicochemical data should also take into account the fact that the same features could occur in samples from various categories.

The information about the rarity as well as, if possible, about within- and between-object variability, included in one calculation run can be applied when the likelihood ratio (LR) approach is used. The law of likelihood (Edwards 1990) tells us that: if H_1 implies that the probability that a random variable X takes the value x is $Pr(X = x|H_1)$, while hypothesis H_2 implies that the probability is $Pr(X = x|H_2)$, then the observation $E = x$ is evidence supporting H_1 over H_2 if and only if $Pr(X = x|H_1) > Pr(X = x|H_2)$, and the likelihood ratio $LR = Pr(X = x|H_1)/Pr(X = x|H_2)$, measures the strength of that evidence. The LR is not a probability, but a ratio of probabilities, and it takes values between 0 and infinity (∞). In the case of analytical chemistry most commonly continuous measurements are considered, therefore the probabilities are replaced by probability density functions $f(E|H_1)$ and $f(E|H_2)$. The decision rules are that values of LR above 1 support H_1, and values of LR below 1 support H_2. A value of LR close to 1 provides little support for either proposition, while a value equal to 1 supports neither of the hypotheses. Also, the larger the value of the LR, the stronger the support of E for H_1, and conversely the lower the value the stronger the support for H_2. For example, a LR equal to 100, in the case of solving a comparison problem, means that the observed similarity of analysed physicochemical data (E) within compared samples provides 100 times more support to the hypothesis that an evidence sample and control sample originate from the same object (H_1) than that they originate from different objects (H_2).

As already mentioned, the estimation of rarity of characteristic features determined for compared samples is a crucial part of evaluation of the evidential value of evidence. The best method to estimate it is to use information collected in a particular database, e.g. a database of elemental composition of glass fragments analysed by the SEM-EDX technique (Aitken et al. 2007). If a suitable database is available then the exact value of $Pr(E|H_1)$ and $Pr(E|H_2)$ and LR could be calculated (Zadora et al. 2014) (see also Sections 9.5.2 and 9.5.3).

Details concerning equations that could be used for evaluation of physicochemical data (e.g. elemental composition of glass by SEM-EDX) as well as a description of suitable software can be found in Zadora et al. (2014) and for GSR analysis can be found in Damary et al. (2016).

However, creation of a database and using it in evaluation of evidential value of evidence in some cases can be very difficult or even impossible. Nevertheless, in such a situation, data obtained from physicochemical analysis by SEM or SEM-EDX also can be evaluated within an LR approach. In such cases experts do not perform calculations concerning an exact value of LR (ENFSI 2015), but use their subjective estimation (based on their experience and knowledge). This still allows which probability – $Pr(E|H_1)$ and

$Pr(E|H_2)$ – is larger to be established. Suitable examples, can be found in Sections 9.5.1 and 9.5.3.[1]

9.4.2 Evidence Evaluation on Activity Level

It should be pointed out that the fact finder, frequently, would also like to know what kind of activity caused the microtraces which bear similarities to the control sample, to be transferred, to persist and then recovered from the body, clothes, shoes, etc. (Aitken and Taroni 2004; Taroni 2014) e.g.:

(a) Did the person who wore the clothes on which the glass microtraces were found with the same composition as a window in a car from which a GPS was stolen, break the car window (H_1) or not (H_2)?
(b) Did the person on whose palms GSR particles were found fire the gun (H_1) or not (H_2)?

This is an analysis on so-called activity level (ENFSI 2015; Taroni et al. 2014). A difference between analysis on source and activity levels can be explained by the following example. A burglary to a car took place and a car window was broken. A sample of glass from the broken window was collected (control sample) and a suspect was arrested after several hours. The trousers and a jacket of the suspect were collected for analysis. Eight glass fragments that revealed similarities of elemental composition determined by SEM-EDX to the control glass sample were found during analysis of debris collected from the clothes. In such a situation, a fact finder would like to know which of two hypotheses is more likely to be supported by this evidence:

H_1 – the suspect broke the window.
H_2 – the suspect did not break the window.

In order to answer this problem, other factors, in addition, to those included in an evaluation on a source level (rarity of elemental composition determined in compared glass fragments by SEM-EDX, within- and between-object variability) should be taken into account (Curran et al. 2000; Taroni et al. 2014), i.e. probabilities of the following events:

(a) Primary transfer – an event by which, during the breaking of a glass window, a particular number of glass fragments, which reveal similarities to the control sample, are transferred, persist and could be recovered, e.g. from clothes, after a particular time.
(b) Glass found by chance – an event when glass fragment(s) which by chance reveal similarities of physicochemical features to the control sample is(are) transferred to a suspect's clothes from other glass objects broken during this person's ordinary daily activities or they were in contact with broken glass by chance.
(c) Secondary transfer – an event when a glass fragment(s) present on the clothes of person A is(are) transferred to the clothes of person B during their direct contact, or a glass fragment(s) is(are) transferred from an object (e.g. car seat) onto another person's clothes.

---

1 It should be highlighted that the LR approach is used not only for data obtained by SEM-EDX (Aitken et al. 2007; Zadora et al. 2014) but it is also used for evaluation of other kinds of data delivered by different techniques (Aitken and Lucy 2004; Alladio et al. 2017; Bolck and Alberink 2010; Martyna et al. 2013a, 2013b, 2014, 2015, 2016; Michalska et al. 2015; Pierrini et al. 2007; Własiuk et al. 2015; Zieba-Palus et al. 2008).

GSR evidence interpretation on an activity level is rather difficult and demands broad knowledge and experience of the expert (see Section 9.3.1 and 9.5.1). In fact, both positive and negative identification of GSR particles do not allow formulation of categorical conclusions that a suspected person discharged (H_1) or did not discharge (H_2) a firearm (Maitre et al. 2017; SWGGSR 2011). Positive GSR results (source level analysis) could result from such circumstances (activity level) as firing a gun, being in a proximity to a discharged firearm, or being in contact with a surface or person who has GSR particles on it/them (a contamination problem). Conversely, the absence of GSR in examined material may result either from the suspect not having used a gun or be the effect of inappropriate or too late examination of the evidence, incidental or deliberate removing of GSR by the suspected person as well as using ammunition producing non characteristic particles that are not readily detected by the SEM-EDX technique (SWGGSR 2011).

Nevertheless, the strength of conclusions (support to H_1 or H_2) in GSR cases always relates to the type of identified particles (e.g. detection of *characteristic* and/or *consistent* with GSR residues; see Section 9.5.1), their number and arises also from the circumstances of the event (Michalska 2007; SWGGSR 2011). When multiple *characteristic* GSR particles are found with a large number of *consistent* with GSR or *commonly associated* with GSR (Table 9.1) particles, these findings strongly support the hypothesis of having used a gun or having been in close proximity to a gun when it was fired (H_1) rather than their occurrence as an effect of contamination (H_2). When only a few particles are identified, the formulation of the findings becomes more complicated. In such cases the obtained analytical results should always be strictly considered with the circumstances of the event, which usually strongly affects the interpretation of the results (see Section 9.5.1 or examples in ENFSI [2015]). In some laboratories the expert uses an appropriate scale in which the number and classes of GSR determines the strength of conclusions (ENFSI 2015). Some laboratories have established a minimal number of GSR particles (Trimpe 2011) necessary for formulating a positive conclusion. Usually the threshold is two or three three-component PbSbBa particles (Table 9.2, traditional ammunition) of characteristic morphology. This number arises from studies concerning contamination problems and results from examinations of workers using a gun, e.g. police officers, hunters, etc. However, there is still no definitive guildeline concerning the number of particles, which should be detected in order to give positive findings (Trimpe 2011).

More details on evaluation of various kinds of evidence on an activity level can be found in the literature (Aitken and Taroni 2004; Aitken et al. 2012; ENFSI 2015; Jackson et al. 2015; Puch-Solis et al. 2012; Roberts and Aitken 2012; Taroni et al. 2014) as well as in Section 9.5.

9.5 Case Study

Examples of application of SEM-EDX in real casework are presented in this section.

9.5.1 GSR – Case Study

There was a bank robbery. The robber threatened the bank cashier with a gun. The frightened women tried to press the security buzzer when the assailant shot her and escaped.

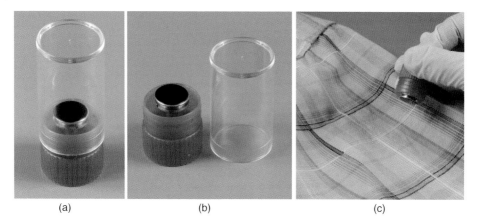

Figure 9.11 (a, b) Professional SEM stub used for seizing of GSR particles; and (c) the way in which GSR is collected from the suspected cloths.

Three hours later police investigators located a suspect. Before he was handcuffed the police officers collected microtraces from both the suspect's hands (evidence 1 – left hand, evidence 2 – right hand). They used an SEM stub (Figure 9.11), which was pressed around 100 times onto each hand, mainly in the region between the thumb and forefinger. They also seized the shirt (evidence 3) and the jacket (evidence 4), which the suspect probably wore during the shooting event. All evidence was immediately delivered to a forensic laboratory. Together with the evidence, the police officers also prepared an interrogation report in which the suspect denied having shot a gun in the 48 hours before being arrested.

The forensic expert was asked to examine the evidence in order to answer the following questions:

- Are GSR particles present in microtraces collected from both hands of the suspect?
- Are GSR particles present on the suspect's shirt or jacket?
- Does the submitted evidence point at the suspect as a person who has fired a gun?

With the aim of answering these questions, the following examination was performed. Potential GSR microtraces were collected from the suspect's shirt by pressing an SEM stub around 100 times on the external surface of the evidence. The stub was placed mainly in the region of turn-ups, sleeves, and pockets. The same operation was repeated with a jacket using a new SEM stub (Figure 9.11). All stubs (those with microtraces from the suspect's clothes and from both hands) were carbon coated (see Sections 9.2.4 and 9.2.4.1) using an SCD sputter coater (Bal-Tech, Switzerland) and mounted in the chamber of a scanning electron microscope. A Jeol JSM-5800 scanning electron microscope (Jeol, Japan) equipped with a Link ISIS 300 EDX spectrometer (Oxford Instruments Ltd, UK) was used in examination of both chemical composition and morphology of potential GSR residue. Searching for these particles was carried out using an automated system mode. The chemical composition and morphology of all particles identified by the automated program as GSR particles were manually confirmed by the expert.

The results of examination concerning the number and the type of GSR particles based on their chemical composition are listed in Table 9.3 while examples of morphology for

Table 9.3 Type and number of particles identified during case examination.

GSR class	Evidence Chemical composition	No. 1 left hand	No. 2 right hand	No. 3 blouse	No. 4 jacket
Characteristic	Pb Sb Ba	—	—	—	5
Consistent	Sb Ba	2	1	—	1
	Pb Ba	8	4	2	1
	Pb Sb	18	12	45	10
Commonly associated	Pb	82	53	211	120
	Sb	52	80	10	45

selected three- and two-component GSR residues are shown in Figure 9.8. Five characteristic particles containing lead, antimony and barium were found among microtraces collected from the suspect's jacket. Such particles are residue from a detonated primer of a discharged gun (Section 9.3.1). At the same time, multiple particles containing antimony/barium, lead/barium, or lead/antimony were found on the SEM stubs collected from both the hands and clothes of the suspect. Such particles are found mainly in primer residues, although they may originate from other sources (Section 9.3.1). All particles classified as characteristic or consistent with GSR were spherical or exhibited features of having been molten. The size of the recovered particles was approximately between 0.5 µm and 5 µm. Together with the particles classified as characteristic or consistent with GSR, numerous particles commonly associated with GSR were found from each of the analysed pieces of evidence. These particles can be found in numerous environmental sources, however their presence with characteristic and consistent residue confirmed that the examined particles belong to GSR.

GSR particles can be deposited on clothes or hands in circumstances such as firing a gun, handling a gun, being very close to a gun when fired or by way of contamination for example by having contact with a person who had fired a gun. The number of GSR particles identified depends on the time that has elapsed after firing and on the physical activity of the shooter between firing and sampling, e.g. washing the hands or rubbing of the hands.

Taking into account the analytical results as well as the circumstances of the event, the identification of so many GSR particles in all the evidence allows the conclusion that the hypothesis of the suspect's contact with a gun (firing or being close to a gun on firing) is more likely to be supported than the hypothesis that these particles originate from environmental contamination.

9.5.2 Glass – Comparison and Classification Problem

On an evening in December, a dead man was found on the side of the road. The police investigators suspected a hit-and-run accident. The victim's body was transported to a medical department and during post-mortem examination, his clothes were packed into separate bags, which were described as follows:

- Evidence 1 – jeans trousers
- Evidence 2 – t-shirt
- Evidence 3 – briefs
- Evidence 4 – socks
- Evidence 5 – shoes.

A few days later, police investigators located the car suspected of being involved and from its broken windscreen the control glass sample was collected (evidence 6).

All evidence was delivered to a forensic laboratory in order to answer the following questions:

(a) Are glass fragments present on the victim's clothes and shoes?
(b) If yes, could they originate from the same object from which a control glass sample was collected?

During examination, the surface of each of the victim's items of clothing (evidence 1–5) were brushed. Recovered debris collected for each piece of evidence was put separately into plastic Petri dishes. Collected microtraces were then checked under a stereomicroscope (MZ 16 Leica Company, magnification 40×) in order to find possible glass fragments. As a result of microscope examination two, three and three 'glass fragments' were found in debris collected from the jeans trousers (evidence 1), socks (evidence 4), and shoes (evidence 5), respectively. Their linear dimensions were in the range of 0.1–0.5 mm. All potential glass fragments were placed directly onto an SEM stub. Additionally, three glass fragments of linear dimension ca. 0.1 mm originating from the broken windscreen of the suspected car (evidence 6) were put on a second SEM stub.

The prepared stubs were coated using an SCD sputter coater (Bal-Tech, Switzerland). An elemental analysis Jeol JSM-5800 scanning electron microscope (Jeol, Japan) equipped with a Link ISIS 300 EDX spectrometer (Oxford Instruments Ltd, UK) was used. During SEM-EDX examination the following measurement conditions were applied: accelerating voltage, 20 kV; count time, 50 seconds; magnification, 1000–5000 ×; and calibration element, cobalt. The Quant option was used to calculate the weight percentages of selected elements (and element oxides) in each sample. The selected analytical conditions allowed all elements to be determined except lithium and boron.

SEM-EDX analysis confirmed that all the selected fragments were glass. The mean elemental composition (expressed in [wt%]) for the recovered glass fragments is presented in Table 9.4. The mean elemental composition (expressed in [wt%]) of the control sample (evidence 6) was as follows: O, 47.55 ± 0.72; Na, 10.30 ± 0.31; Mg, 0.16 ± 0.02; Al, 0.20 ± 0.03; Si, 33.92 ± 0.64; and Ca, 7.83 ± 0.35.

In order to answer the question whether the glass fragments found on the victim's clothes and shoes (evidence 1–5) could have a common origin with the control sample collected from the suspected car's windscreen (evidence 6), their elemental composition was compared using the LR test (see also Section 9.4) . The model used in the calculation was published in Zadora et al. (2014) – see eqs. (4.5) and (4.6). The calculation was performed in R software using routines written by the expert.

Table 9.4 Mean elemental composition of glass fragments found in debris collected from jeans trousers (evidence 1), socks (evidence 4) and shoes (evidence 5), and the results of the likelihood ratio (LR) calculations.

		Mean elemental content [wt%] ± standard deviation						
Evidence	Trace	O	Na	Mg	Al	Si	Ca	LR
1	1	47.46	9.98	0.13	0.19	34.19	7.98	30 170
		0.69	0.14	0.04	0.02	0.53	0.24	
	2	44.84	9.37	0.15	0.21	36.22	9.13	1314
		0.89	0.22	0.02	0.02	0.61	0.45	
4	1	44.97	9.35	0.15	0.20	36.08	9.15	1109
		0.11	0.14	0.01	0.02	0.17	0.08	
	2	45.78	9.77	0.14	0.19	35.49	8.59	19 200
		0.18	0.04	0.02	0.01	0.08	0.08	
	3	48.30	9.71	2.45	0.73	33.73	4.60	2×10^{-3}
		0.82	0.05	0.01	0.01	0.59	0.18	
5	1	46.02	8.98	2.18	0.79	36.00	5.93	7×10^{-3}
		0.55	0.20	0.05	0.05	0.55	0.18	
	2	47.33	10.09	0.16	0.20	34.31	7.87	55 800
		0.21	0.04	0.01	0.01	0.19	0.10	
	3	52.57	10.20	2.18	0.64	30.32	4.06	1×10^{-4}
		0.11	0.04	0.03	0.01	0.15	0.01	

During comparative analysis the following hypotheses were considered:

- H_1 – ith glass fragment found in debris of evidence 1, 4, or 5 originates from the same windscreen from which a control sample was collected (evidence 6).
- H_2 – ith glass fragment found in debris of evidence 1, 4, or 5 does not originate from the same windscreen from which a control sample was collected (evidence 6).

The results of the LR test are interpreted as follows: if LR value is above 1, the evidence (elemental composition of glass object) provides more support for hypothesis H_1, while a value below 1 provides more support for the alternative hypothesis H_2. A LR value equal to one supports neither of the hypotheses. Moreover, in LR interpretation the simple rule is applied that the higher than 1 (lower than 1) the value of the LR is, the stronger the support for hypothesis H_1 (H_2).

On the basis of performed calculations (Table 9.4) it was concluded that the elemental composition of five glass fragments recovered from the victim's clothing provided more support for the hypothesis that they did have a common origin with the control sample collected from the broken windscreen of the suspected car (H_1) than the alternative one (H_2). In the case of the three glass fragments recovered in debris from the victim's clothing, the

Table 9.5 Results of likelihood ratio (LR) calculations within a classification problem, which do not reveal similarity to a control sample (evidence 6).

Evidence	Trace	LR category
4	3	339
		p
5	1	606
		p
	2	1×10^{19}
		p

Category: p, glass containers.

elemental composition provided more support for the hypothesis that they did not originate from the car's broken windscreen (H_2) than the opposite (H_1).

For the three glass fragments, which turned out to have a different origin than the control sample, an additional LR calculation was carried out (as per a classification problem). The calculation was performed in order to determine from which category these fragments could originate, i.e. to establish if they originate from another car windscreen. In this experiment, the following hypotheses were considered:

- H_p – a particular glass fragment originates from category *glass containers* (p).
- H_{cw} – a particular glass fragment originates from category *car and building windows* (cw),

On the basis of the LR results presented in Table 9.5, it could be concluded that the elemental composition of these three fragments more likely supports the hypothesis (H_p) that they originate from category *glass containers* (p) than the alternative hypothesis (H_{cw}) that they originate from category *car and building windows* (cw).

9.5.3 Glass – Was the Car Bulb Switched on During the Accident?

On a dark evening in November, two cars collided on a crossroads. The driver of car A which drove into car B explained that he did not see the second car due to the fact that car B did not have their lamps switched on. The driver of car B denied this. In order to determine whose testimony was true, police investigators collected the broken halogen lamp of car B (evidence 1) and delivered it to a forensic laboratory. The aim of the examination was to answer the question – whether the bulb of car B was switched on during the accident.

Forming a conclusion as to whether a bulb was switched on or off during a car accident is possible when the glass bulb becomes broken during the accident event. In such cases the expert searches for a white or yellow coating fixed to the filament. The coating is associated with particular oxides forming as a result of a reaction between the hot filament and oxygen present in air. The presence of such oxides, e.g. tungsten(IV) oxide, on a filament made from tungsten provide more support for the hypothesis that the bulb was switched on during a collision (H_1) than the alternative (H_2). It can also happen that when a bulb that is on is broken the glass fragments of broken bulb melt into the still warm filament. Revealing such evidence confirms the bulb was lit during the accident.

The filament of the broken bulb in the case was examined under a stereomicroscope (MZ 16, Leica Company), however no coating was revealed on the filament surface. Nevertheless, in microscope examination an area of molten substance fixed to the filament surface was observed.

The filament was detached from the bulb and placed directly onto an SEM stub. Its surface was carbon coated (see Sections 9.2.4 and 9.2.4.1) using an SCD sputter coater (Bal-Tech, Switzerland). For filament surface observation as well as analysis of elemental composition a Jeol JSM-5800 scanning electron microscope (Jeol, Japan) equipped with a Link ISIS 300 EDX spectrometer (Oxford Instruments Ltd, UK) was used. The applied conditions were the same as in the example described in Section 9.5.2.

The elemental analysis of a clear surface of the filament allowed it to be established that this particular filament was made of tungsten. Due to the fact that the expert did not have reference material concerning the type and amount of coating arising from reaction between tungsten and oxygen contained in air, the examination was focused only on analysis of the molten substance fixed to the filament (Figure 9.12). Elemental analysis of this substance confirmed that these were molten fragments of glass, with mean elemental composition (expressed in [wt%]) as follows: O, 49.86 ± 0.67; Na, 6.63 ± 0.42; Mg, 1.85 ± 0.14; Al, 1.55 ± 0.67; Si, 34.84 ± 0.86; K, 0.79 ± 0.07; Ca, 3.21 ± 0.15; and Ba, 1.19 ± 0.25.

This elemental composition – especially relatively large amounts of potassium and barium – is characteristic for glass having special optical properties (Figure 9.12).

Although the presence of a coating was not observed on the filament surface, the confirmation of molten glass fixed to the filament is sufficient evidence to conclude that the hypothesis that the bulb (evidence 1) was switched on during the accident (H_1) was more likely than the alternative one (H_2).

Other examples of cases where glass fragments were the subject of analysis can be found in Michalska et al. (2015).

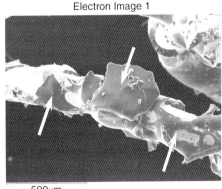

Figure 9.12 Filament of car bulb (evidence 1) observed under SEM-EDX with melted glass traces (indicated by arrows).

References

Aitken, C.G.G. and Lucy, D. (2004). Evaluation of trace evidence in the form of multivariate data. *Applied Statistics* 53: 109–122.

Aitken, C.G.G. and Taroni, F. (2004). *Statistics and the Evaluation of Evidence for Forensic Scientists*. Chichester: Wiley.

Aitken, C.G.G., Zadora, G., and Lucy, D. (2007). A two-level model for evidence evaluation. *Journal of Forensic Sciences* 52: 412–419.

Aitken, C.G.G., Roberts, P., and Jackson, G. (2012). *Fundamentals of Probability and Statistical Evidence in Criminal Proceedings: Guidance for Judges, Lawyers, Forensic Scientists and Expert Witnesses, Practitioner Guide 1*. London: Royal Statistical Society.

Alladio, E., Martyna, A., Salomone, A. et al. (2017). Evaluation of direct and indirect ethanol biomarkers using a likelihood ratio approach to identify chronic alcohol abusers for forensic purposes. *Forensic Science International* 271: 13–22.

von Ardenne, M. (1938a). Das elektronen-rastermikroskop. Praktische ausführung. *Zeitschrift für Technische Physik* 19: 407–416.

von Ardenne, M. (1938b). Das Elektronen-Rastermikroskop. Theoretische Grundlagen. *Zeitschrift für Physik* 109: 553–572.

Basu, S. (1982). Formation of gunshot residues. *Journal of Forensic Sciences* 27: 72–91.

Becker, S., Gunaratnam, L., Hicks, T. et al. (2001). The differentiation of float glass using refractive index and elemental analysis: comparisons of techniques. *Problems of Forensic Sciences* 47: 80–92.

Bergman, P., Springer, E., and Levin, N. (1991). Hand grenades and primer discharge residues. *Journal of Forensic Sciences* 36: 1044–1052.

Berk, R.E. (2009a). Automated SEM/EDS analysis of airbag residue. I: particle identification. *Journal of Forensic Sciences* 54: 60–68.

Berk, R.E. (2009b). Automated SEM/EDS analysis of airbag residue. II: airbag residue as a source of percussion primer residue particles. *Journal of Forensic Sciences* 54: 69–76.

Bolck, A. and Alberink, I. (2010). Variation in likelihood ratios for forensic evidence evaluation of XTC tablets comparison. *Journal of Chemometrics* 25: 41–49.

Caddy, B. (2001). *Forensic Examination of Glass and Paint Analysis and Interpretation*. London: Taylor & Francis.

Collins, P., Coumbaros, J., Horsley, G. et al. (2003). Glass-containing gunshot residue particles: a new type of highly characteristic particle? *Journal of Forensic Sciences* 48: 538–553.

Croveri, P., Fragalà, I., and Ciliberto, E. (2010). Analysis of glass tesserae from the mosaics of the 'Villa del Casale' near Piazza Armerina (Enna, Italy). Chemical composition, state of preservation and production technology. *Applied Physics A* 100: 927–935.

Curran, J.M., Hicks, T.N., and Buckleton, J.S. (2000). *Forensic Interpretation of Glass Evidence*. Boca Raton: CRC Press.

Damary, N.K., Mandel, M., Levin, N., and Izraeli, E. (2016). Calculation of likelihood ratios for gunshot residue evidence – statistical aspects. *Law, Probability and Risk* 15: 107–125.

Echlin, P. (2009). *Handbook of Sample Preparation for Scanning Electron Microscopy and X-Ray Microanalysis*. New York: Springer.

Edwards, A.W.F. (1990). *Likelihood*. Baltimore: The Johns Hopkins University Press.

ENFSI (2008). ENFSI Guide for Gunshot Residue Analysis by Scanning Electron Microscopy/Energy-Dispersive X-ray Spectrometry, Version 2. European Network of Forensic Science Institutes.

ENFSI (2015). ENFSI Guideline for Evaluative Reporting in Forensic Science: A Primer for Legal Practitioners. European Network of Forensic Science Institutes.

Falcone, R., Sommariva, G., and Verità, M. (2006). WDXRF, EPMA and SEM/EDX quantitative chemical analyses of small glass samples. *Microchimica Acta* 155: 137–140.

Flynn, J., Stoilovic, M., Lennard, C. et al. (1998). Evaluation of X-ray microfluorescence spectrometry for the elemental analysis of firearm discharge residues. *Forensic Science International* 97: 21–36.

Giacalone, J.R. (2002). Distinguishing characteristic for the determination of forensic microtrace particles. *International Association for MicroAnalysis* 3 (2): 8–10.

Harris, A. (1995). Analysis of primer residue from CCI blazer lead free ammunition by scanning electron microscopy/energy dispersive x-ray. *Journal of Forensic Sciences* 40: 27–30.

Ingo, G.M., D'uffizi, M., Falso, G. et al. (2004). Thermal and microchemical investigation of automotive brake pad wear residues. *Thermochimica Acta* 418: 61–68.

Jackson, G., Aitken, C.G.G. & Roberts, P. (2015). Communicating and Interpreting Statistical Evidence in the Administration of Criminal Justice: 4. Case Assessment and Interpretation of Expert Evidence. London: Royal Statistical Society.

Kosanke, K.L. (2003). Identification of pyrotechnic reaction residue particles. *International Association for Micro Analysis* 4: 2–7.

Krüsemann, H. (2001). SEMs and forensic science. *Problems in Forensic Science* 47: 110–121.

Lawes, G. (1987). *Scanning Electron Microscopy and X-Ray Microanalysis*. Wiley for ACOL.

Lebiedzik, J. and Johnson, D.L. (2000). Rapid search and quantitative analysis of gunshot residue particles in the SEM. *Journal of Forensic Sciences* 45: 83–92.

Maitre, M., Kirkbride, K.P., Horder, M. et al. (2017). Current perspectives in the interpretation of gunshot residues in forensic science: a review. *Forensic Science International* 270: 1–11.

Martyna, A., Lucy, D., Zadora, G. et al. (2013a). The evidential value of microspectrophotometry measurements made for pen inks. *Analytical Methods* 5: 6788–6795.

Martyna, A., Sjastad, K.-E., Zadora, G., and Ramos, D. (2013b). Analysis of lead isotopic ratios of glass objects with the aim of comparing them for forensic purposes. *Talanta* 105: 158–166.

Martyna, A., Zadora, G., Stanimirova, I., and Ramos, D. (2014). Wine authenticity verification as a forensic problem. An application of likelihood ratio approach to label verification. *Food Chemistry* 150: 287–295.

Martyna, A., Michalska, A., and Zadora, G. (2015). Interpretation of FTIR spectra of polymers and Raman spectra of car paints by means of likelihood ratio approach supported by wavelet transform for reducing data dimensionality. *Analytical and Bioanalytical Chemistry* 407: 3357–3376.

Martyna, A., Zadora, G., Neocleous, T. et al. (2016). Hybrid approach combining chemometrics and likelihood ratio framework for reporting the evidential value of spectra. *Analytica Chimica Acta* 931: 34–46.

McCarthy, C. (2002). SEM-EDX examination of GSR originating from selected types of ammunition. Glasgow: Forensic Science Unit, University of Strathclyde.

McMullan, D. (2006). Scanning electron microscopy 1928–1965. *Scanning* 17: 175–185.

Meng, H.H. and Caddy, B. (1997). Gunshot residue analysis – a review. *Journal of Forensic Sciences* 42: 553–570.

Michalska, A. (2007). Badanie morfologii, struktury i składu chemicznego charakterystycznych cząstek powystrzałowych metodami fizykochemicznymi. MSc thesis. Jagiellonian University.

Michalska, A., Martyna, A., Zięba-Palus, J., and Zadora, G. (2015). Application of a likelihood ratio approach in solving a comparison problem of Raman spectra recorded for blue automotive paints. *Journal of Raman Spectroscopy* 46: 772–783.

Michalska, A., Zadora, G., and Martyna, A. (2016). Optimized sample preparation for glass fragments by scanning electron microscopy-energy dispersive X-ray spectrometry. *Analytical Letters* 49 (12): 1884–1895.

Milani, M., Gottardi, R., Savoia, C., and Cattaneo, C. (2012). FIB/SEM/EDS complementary analysis for proper forensic interpretation. In: *Current Microscopy Contributions to Advances in Science and Technology* (ed. A. Méndez-Vilas), 179–185. Formatex Research Center.

Mosher, P.V., McVicar, M.J., Randall, E.D., and Sild, E.H. (1998). Gunshot residue – similar particles produced by fireworks. *Canadian Society of Forensic Science Journal* 31: 157–168.

Nesbitt, R.S., Wessel, J.E., and Jones, P.F. (1976). Detection of gunshot residue by use of the scanning electron microscope. *Journal of Forensic Sciences* 21: 595–610.

Newbury, D., Joy, D., Echlin, P. et al. (1986). *Advanced Scanning Electron Microscopy and x-Ray Microanalysis*. New York: Springer.

Oatley, C., Nixon, W., and Pease, R. (1965). Scanning electron microscopy. *Advances in Electronics and Electron Physics* 21: 181–247.

Pelton, W. (1995). Distinguishing the cause of textile fiber damage using the scanning electron microscope (SEM). *Journal of Forensic Sciences* 40: 874–882.

Pierrini, G., Doyle, S., Champod, C. et al. (2007). Evaluation of preliminary isotopic analysis (^{13}C and ^{15}N) of explosives. A likelihood ratio approach to assess the links between Semtex samples. *Forensic Science International* 167: 43–48.

Puch-Solis, R., Roberts, P., Pope, S., and Aitken, C. (2012). *Assessing the Probative Value of DNA Evidence: Guidance for Judges, Lawyers, Forensic Scientists and Expert Witnesses, Practitioner Guide 2*. London: Royal Statistical Society.

Roberts, P. & Aitken, C.G.G. (2012). The Logic of Forensic Proof: Inferential Reasoning in Criminal Evidence and Forensic Science. Guidance for Judges, Lawyers, Forensic Scientists and Expert Witnesses, Practitioner Guide 3. London: Royal Statistical Society.

Romolo, F.S. (2002). Standardization in gunshot residue analysis by scanning electron microscopy/energy-dispersive spectroscopy. *International Association for Micro Analysis* 3: 1–3.

Smith, K.C.A. and Oatley, C.W. (1955). The scanning electron microscope and its fields of application. *British Journal of Applied Physics* 6: 391–399.

SWGGSR (2011). Guide for Primer Gunshot Residue Analysis by Scanning Electron Microscopy/Energy Dispersive X-ray Spectrometry. Scientific Working Group on Materials Analysis Gunshot Residue Subgroup.

SWGMAT (1999). Forensic Fiber Examination Guidelines. Scientific Working Group on Materials Analysis Fiber Subgroup.

Taroni, F., Aitken, C.G.G., Garbolino, P., and Biedermann, A. (2014). *Bayesian Networks for Probabilistic Inference and Decision Analysis in Forensic Science*, 2 edn. Wiley.

Torre, C., Mattutino, G., Vasino, V., and Robino, C. (2002). Brake linings: a source of non-GSR particles containing lead, barium, and antimony. *Journal of Forensic Sciences* 47: 494–504.

Trimpe, M. (2003). Analysis of fireworks for particles of the type found in primer residue (GSR). *International Association for Micro Analysis* 4: 1–8.

Trimpe, M. (2011). The current status of GSR examinations. *FBI Law Enforcement Bulletin* 80: 24–32.

Tucker, P.A. (1998). Scale Heights of chemically treated wool and hair fibers. *Textile Research Journal* 68: 229–230.

Verità, M., Basso, R., Wypyski, M.T., and Koestler, R. (1994). X-ray microanalysis of ancient glassy materials: a comparative study of wavelength dispersive and energy dispersive techniques. *Archaeometry* 36: 241–251.

Wallace, J.S. and McQuillan, J. (1984). Discharge residues from cartridge-operated industrial tools. *Journal of the Forensic Science Society* 24: 495–508.

Watt, I.M. (1997). *The Principles and Practice of Electron Microscopy*. Cambridge University Press.

Własiuk, P., Martyna, A., and Zadora, G. (2015). A likelihood ratio model for the determination of the geographical origin of olive oil. *Analytica Chimica Acta* 853: 187–199.

Wolten, G.M. and Nesbitt, R.S. (1980). On the mechanism of gunshot residue particle formation. *Journal of Forensic Sciences* 25: 533–545.

Zadora, G. and Brożek-Mucha, Z. (2003). SEM–EDX – a useful tool for forensic examinations. *Materials Chemistry and Physics* 81: 345–348.

Zadora, G., Martyna, A., Ramos, D., and Aitken, C. (2014). *Statistical Analysis in Forensic Science: Evidential Values of Multivariate Physicochemical Data*. Wiley.

Zeichner, A. and Levin, N. (1992). Gunshot residue particles formed by using ammunitions that have mercury fulminate based primers. *Journal of Forensic Sciences* 37: 1567–1573.

Zeichner, A., Levin, N., and Springer, E. (1991). Gunshot residues particles formed by using different types of ammunition in the same firearm. *Journal of Forensic Sciences* 36: 1020–1026.

Zieba-Palus, J., Zadora, G., and Milczarek, J.M. (2008). Differentiation and evaluation of evidence value of styrene acrylic urethane topcoat car paints analysed by pyrolysis-gas chromatography. *Journal of Chromatography A* 1179: 47–58.

10

Mass Spectrometry

Mark C. Parkin and Alan Brailsford

10.1 Introduction

Mass spectrometry (MS) is one of the most powerful analytical tools available to the forensic scientist. The instrumentation enables the selective measurement of mass-to-charge ratio (m/z) from which we can determine a compound's molecular weight and subsequently information as to its identity. The technique can be used to determine both elemental composition and molecular structure. Upon calibration of the instrumentation, it can also be used to quantify species. Analysis is undertaken by converting individual atoms or molecules into ions and measuring their motion and subsequent separation in a magnetic or electric field. MS has its origins in atomic physics and the rapid development of nuclear physics. The technique was originally pioneered by the physicist Sir Joseph John Thomson at the Cavendish laboratory in Cambridge where he was interrogating the composition of cathode rays (Thomson 1897). Thompson's work led to the eventual discovery of the electron and he soon turned his attention to so called 'positive rays'. These travelled in the opposite direction to the electron in his discharge tubes and he therefore thought there must be a unit of 'positive electricity'. Thomson showed that these positively charged particles could be deflected by a magnetic field and in doing so he developed a 'mass spectrograph' to measure the atomic weights of elements (Thomson 1911). Francis William Aston, who was invited to the Cavendish by Thomson, further refined the instrumentation and used it to separate two forms of neon, proving the existence of stable isotopes (Figure 10.1) (Aston 1920).

Around the same time and independently, Arthur Jeffery Dempster developed an instrument at the University of Chicago that was able to provide accurate measurements on the relative isotopic abundances as opposed to the mass accuracy of Aston and Thomson's instruments (Dempster 1918). During the Second World War, the technique became more applied and had a pivotal role in the preparation of fissionable uranium and weapon-grade plutonium. There was also an increasing need for higher octane petroleum fuels to power aircraft and this paved the way for the adoption of MS as a major analytical tool used by the petroleum industry for quality control in refining the oil. The method was beginning to leave the domain of the physicist and the determination of masses and relative abundances and

Analytical Techniques in Forensic Science, First Edition.
Edited by Rosalind Wolstenholme, Sue Jickells and Shari Forbes.
© 2021 John Wiley & Sons Ltd. Published 2021 by John Wiley & Sons Ltd.

220 | 10 Mass Spectrometry

Figure 10.1 Aston's first mass spectrometer. The coils in the background make up the magnetic component for separating the isotopes. The bulb in the foreground is the gas discharge tube that was used to generate ions prior to their separation. Source: Reproduced with permission from Cavendish Lab UoC.

was beginning to become a tool for the chemist to undertake qualitative structural analysis. Not long after the war, many in the petroleum and chemical industries realised that the power of the technique could be used to identify and characterise organic compounds. This led to further advances with the technique such that it could be used to ionise and handle more polar and non-volatile species. Development of so-called 'double-focusing high resolution' mass spectrometers in the 1950s introduced the concept of using accurate mass measurements to help identify compounds. John Herbert Beynon realised that with sufficient mass accuracy it would be possible to obtain sufficient information on the elemental composition of a molecule such that its identity could be elucidated from a limited number of choices of chemical formula (Beynon 1959). Further developments in the 1950s by Wolfgang Paul and others led to the development of the quadrupole mass analyser and the quadrupole ion-trap (Paul and Steinwedel 1953), advancing the technology beyond the use of magnetic fields for ion separation. The 1960s saw the development of gas chromatography (GC) and the subsequent coupling to mass spectrometers (gas chromatography–mass spectrometry, GC-MS). This heralded a new era as GC could be used to separate highly complex samples with the constituents eluting subsequently analysed by MS. The use of GC-MS made it an indispensable tool for the petrochemical, medical, environmental, agricultural, food and forensics industries. Further, the coupling of GC to Wolfgang Paul's Nobel Prize winning transmission quadrupoles provided a small and fast analyser that took up much less space in the laboratory and required much less expertise to operate. The 1970s saw a number of attempts to couple high performance liquid chromatography (HPLC) with mass spectrometers and commercialise the approach. This would bring benefits to biologists as it would allow for the analysis of more polar species without the need for lengthy derivatisation processes to enhance volatility for GC separations. However, the hyphenation of HPLC to a mass spectrometer was difficult to achieve largely because of the high flow rates of HPLC and the production of huge volumes of gas when the liquid phase expands in the vacuum of the mass spectrometer. This was eventually overcome in the 1980s by another Nobel Prize winning concept (along with those awarded to Aston and Paul) known as 'electrospray ionisation' developed by John Fenn (Whitehouse et al. 1985). Also during the 1970s, Fourier transform ion cyclotron resonance (FT-ICR) MS was developed by Alan Marshall and Mel Comisarow (Comisarow and Marshall 1974). The approach involves the measurement of the frequency of an ion's rotation in a magnetic field and still to this day provides the highest degree of mass accuracy and resolving power. The late 1990s saw the second generation of high mass accuracy instruments, this time coupled to high resolution chromatography in the form of capillary GC columns and small, sub 2 µM, HPLC particles. Alexander Makarov developed the Orbitrap mass analyser, producing a benchtop accurate mass instrumentation that could measure mass to four or five decimal places (Makarov 2000). Unlike the accurate mass instruments of the 1950s, 1960s and 1970s, small amounts of complex biological material could now be probed with femtomole levels of sensitivity.

10.1.1 Forensic Application of Mass Spectrometry

Over the last 50 years, MS has become an integral part of the forensic sciences, with instruments now ubiquitous in forensic laboratories around the world. This popularity has been driven by the ability of MS to identify, and if required, quantify, unknown substances with the high degree of certainty required by the legal system (Hoffmann and Jackson 2015). In common with its application to other areas of analytical and bioanalytical science, MS in forensic science is frequently hyphenated to a chromatographic technique, usually GC (see Chapter 13) or liquid chromatography (LC) (see Chapter 14), which add a second dimension of separation, facilitating the analysis of more complex sample types. Developments in chromatographic techniques, in conjunction with the ever-increasing sensitivity and sophistication of MS, continue to increase the range of forensic applications for which MS can be applied. Further, developments in automated analysis have reduced the need for labour intensive sample preparation while allowing hundreds of samples to be analysed in a day.

While much of the early work in forensic science involved GC-MS due to the earlier commercially available instrumentation and relative ease of hyphenation, modern laboratories use a variety of MS-based approaches taking advantage of the differing characteristics of each instrument. This may include LC-MS, tandem mass spectrometry (MS/MS) (in triple quadrupole or ion trap configurations), high resolution mass spectrometry (HRMS), and isotope ratio mass spectrometry (IRMS) (see Chapter 11). Within the forensic sciences, MS has likely found its widest application to the area of drug detection, either in biological samples (drug metabolites) or as bulk powders. Prior to the late 1960s, even though MS instrumentation was capable of detecting organic compounds in mixtures, it was not used to provide evidence for courts. The reason for this is thought to be due to the costs and difficulty in operating instruments as well as the capability in interpreting the data. Initially, MS was

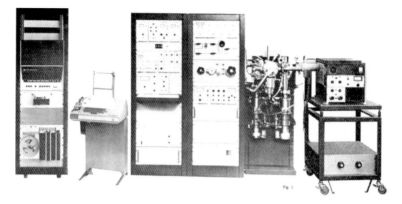

Figure 10.2 Atlas MAT CH-5 magnetic sector mass spectrometer coupled to a gas chromatograph as used at the 1972 Olympic Games in Munich. Source: Reproduced with permission from Hemmersbach (2008).

used for bulk drug analysis to identify seized material, but hyphenation to chromatographic MS instrumentation facilitated the analysis of more complex sample types, particularly those of biological origin, such as urine, blood, and more recently so-called 'alternative matrices' such as hair (Vincenti et al. 2013), nails (Cappelle et al. 2015), fingermarks (Huynh and Halamek 2016), and oral fluid (saliva) (Moore and Crouch 2013). These are matrices that are used to determine drug exposure in a wide range of forensic scenarios ranging from determining causes of death, to drug driving convictions and workplace drug testing. MS has also played a pivotal evidential role in doping control in both human and equine sport being first introduced at the Olympic games in Munich in 1972 (Hemmersbach 2008) (Figure 10.2). At the 1972 Games, MS was used as a tool complementary to immunoassay to identify doping substances from athletes' biological samples. In recent years, the analysis of endogenous substances within these biological matrices has also been developed. In particular, small molecules such as amino acids and macromolecules such as proteins and lipids have been investigated in the forensic context. This has led to a diversity of applications from enhancing fingermark identification (Wolstenholme et al. 2009) through to determining the sex of the sample's origin (Ferguson et al. 2012).

MS, then, is not solely associated with drug related crime and offences. It has also been used extensively for the analysis of explosives and combustibles to aid in applications related to monitoring terrorist activities and arson. As far back as 1959 MS was used by the Chicago police crime laboratory to identify small quantities of volatile liquid found at suspicious fires (Nicol 1959). The use of improvised explosive devices has become a highly effective weapon in the terrorist arsenal and so the detection of explosives and energetic materials has become a major research area in forensics. Finally, another forensic area where MS is often deployed is trace evidence analysis, particularly to paints, inks, and dyes. This application area facilitates the linking of objects, people and places associated with criminal activity and can potentially provide information on the source of the materials used. Vehicle accidents can be investigated by analysing glass and paints by use of elemental MS. This approach involves the use of mass spectrometers that can generate ions from atoms rather than molecules and help build elemental profiles that are a 'fingerprint' of the material under investigation (Hobbs and Almirall 2003). Similarly, documents and paintings can be analysed for fraudulent activity by analysing the dyes used in the inks and paints and building up the chemical profile for probable matches (Matthews et al. 2011).

MS has a long and rich history associated with the forensic sciences and the legal system. It continues to be one of the gold standards for evidential analysis and is still undergoing development at an unprecedented speed. The cost of instrumentation has come down significantly in recent years and lower levels of sensitivity now mean that many more exhibits from a crime scene can be analysed. Further exciting developments include the use of non-destructive approaches for ionising molecules. These include ambient ionisation methods such as desorption electrospray ionisation, limiting the amount of sample preparation needed and helping to preserve evidence. This allows for further testing by complementary approaches such as detecting drugs in fingermarks and their subsequent development for suspect identification.

10.2 Theory of the Technique

10.2.1 Principles of Mass Spectrometry

Compounds are detected and identified in MS by the generation, separation and detection of gas-phase ions based on their m/z in the mass spectrometer. This means that the main components of the instrument are the *ion source*, the *mass analyser* and the *detector*, as shown in Figure 10.3. Once formed in the ion source, ions are transmitted around the mass spectrometer either by electric or magnetic fields from one location to another in the gas phase. In order to prevent the ions from having their trajectory deflected, becoming neutralised or fragmented by their collision with atmospheric molecules, the instrument is operated at low pressure under high vacuum. The mass analyser component is usually operated at around 10^{-4}–10^{-7} Pa. This high vacuum is achieved in two stages. First using a mechanical rough

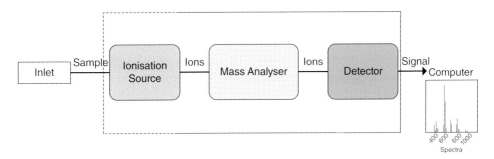

Figure 10.3 The main components of a mass spectrometer. The inlet provides a means for sample introduction either as a syringe pump or the outlet of a chromatographic/separations system (GC, LC, or CE). The ionisation source generates ions for their subsequent separation in the mass analyser and in some modes is either at vacuum or atmospheric pressure. The detector amplifies the signal obtained from the ions on exiting the mass analyser and this is recorded as a signal (spectra/chromatogram) in the processing software of the computer. Upon generation of ions, the mass spectrometer is held under vacuum to prevent ion losses from bombardment with atoms/molecules from the atmosphere.

pump, which takes the vacuum down to around 10^{-1} Pa, and then a turbomolecular pump to high vacuum, $<10^{-2}$ Pa, in the mass analyser region. Ion losses can also occur between the major components of the instrument, from the ionisation source to the mass analyser and from the mass analyser to the detector. Depending on the type of mass analyser, ions are moved from one location to another as a beam or a packet and have a tendency to diverge their path leading to losses. An ideal instrument would suffer no ion losses leading to very high levels of sensitivity. Ion *guides*, *lenses* and *funnels* (also known collectively as the *ion optics*) have been developed by various instrument manufacturers to try to limit these losses and increase ion transmission. Once the ions reach the mass analyser, separation of the ions can occur using either electric or magnetic fields or in some instruments a combination of the two. Upon exiting the mass analyser, ions are sent to the detector where amplification of the ion's signal occurs by the secondary emission of electrons or photons generated from the striking of the ion beam into a reactive surface.

A typical procedure for MS analysis is as follows:

- Sample introduction as either a solid, liquid or gas.
- Sample ionisation,
- Sample ions separated according to their *m/z*.
- Sample ions detected using a device to measure charged particles.
- Resulting mass spectrum displayed on a computer.
- Spectrum interpreted either by using intact *m/z* values or fragmentation data.

Upon separating the ions, a mass spectrum is generated showing the *m/z* values on the x-axis of the graph against their relative abundance (signal intensity) on the y-axis. Later in this chapter, the features of the spectrum will be discussed in more detail (Section 10.2.6). In forensic science, there are two ways in which the mass spectral data is used: (i) for identification; and (ii) for determining the amount (quantification) present. Identification involves carefully interpreting the spectra and often comparing with reference compounds also run through the instrumentation. In some cases, the spectra can be compared with a reference library of spectra. Usually a combination of these approaches is used. For quantification, calibrants of known concentrations are prepared and a calibration line is produced as a plot of signal intensity against concentration. The analyte's concentration can then be measured and read from this line.

10.2.2 Sample Introduction

On modern mass spectrometers, the introduction of sample to the instrument is now commonly undertaken by hyphenation to a chromatographic system and in some cases following separation by capillary electrophoresis. There also remains the option on most instruments to introduce sample, unseparated, by means of a simple syringe pump and the filling of a syringe. The type of hyphenation and therefore sample introduction determines the mode of ionisation. An exception to this setup is the matrix-assisted laser desorption/ionisation (MALDI) source (discussed in detail later), which allows for the spotting and crystallisation of sample onto a metal target plate. In this case, there is no preceding analytical separation step and relatively pure sample is introduced to the mass spectrometer ionisation stage directly.

The use of chromatography, to separate compounds prior to introduction to the mass spectrometer, allows for those components to be analysed individually if there is sufficient chromatographic resolution. This brings benefits to the subsequent data acquired since the mass analyser cannot necessarily separate those compounds that have the same molecular weight, which would complicate the spectra produced. Furthermore, the ionisation step can also be affected by sample complexity. Mixtures can cause the ion signal to be either enhanced or more usually suppressed thus negatively impacting on sensitivity and quantification.

10.2.3 Modes of Sample Ionisation

The measurements undertaken by the mass spectrometer involve ions, rather than neutral molecules or atoms, as their motion can be manipulated readily by applying electric or magnetic forces. Their energy and velocity can be controlled to aid in their separation and detection. This could not be achieved if the species was neutral, since molecular motion would be random. As the analytes enter the mass spectrometer, they must be ionised prior to separation in the mass analyser. Ions (either positive or negative) can easily be manipulated by the electric (or magnetic) field facilitating separation based on their mass relative to the overall charge state (m/z). Ionisation is therefore the first crucial step in the analysis process of the mass spectrometer. There is now a variety of ionisation techniques and in principle they relate to the mode of hyphenation or coupling that is used by the mass spectrometer. The component of the mass spectrometer in which ionisation takes place is known as the *ion source* and, as will be discussed, some sources are operated at vacuum pressures and others at atmospheric pressure. We will look in detail at the most common approaches used by the types of mass spectrometers forensic scientists would use and describe how they influence the type of spectra that can be obtained. The different modes of ionisation encountered on instruments in the forensic laboratory can be loosely categorised as *gas-phase methods*, *atmospheric pressure methods*, and *desorption and ambient methods*. These are summarised in Table 10.1.

10.2.3.1 Electron Ionisation

Electron ionisation (EI) is a gas-phase method and one of the oldest and most widely used modes of ionisation in mass spectrometers. It was first developed by Dempster back in 1918 (Dempster 1918) and remained the principal mode of ionisation through to the 1960s until

Table 10.1 Modes of ionisation and their categories.

	Modes of ionisation
Gas-phase methods	Electron ionisation, chemical ionisation
Atmospheric pressure methods	Electrospray ionisation, atmospheric pressure chemical ionisation
Desorption and ambient methods	Matrix-assisted laser desorption ionisation, secondary ion mass spectrometry, desorption electrospray ionisation, direct analysis in real time

chemical ionisation was introduced. The energy for ionisation to occur is provided by the bombardment of the sample with 70 eV electrons emitted from a filament. The ionisation chamber inside the source is maintained at 10^{-1} Pa to minimise ion/molecule collisions and is compatible with the operating pressures of the ion transfer region of the mass spectrometer. A thin filament of rhenium wire acts as a cathode and is heated to an incandescent temperature in order to emit electrons. These electrons then enter the ionisation chamber through an entrance slit and an applied potential difference of 70 V between the filament and the ionisation chamber accelerates them. An electron trap acting as the anode is placed outside the chamber and is held at a small positive potential to the chamber. Any unconsumed electrons are collected by the trap. To focus the electrons and increase the probability of ionisation, they are made to travel in a tight helical path of increased length. This is formed by applying a weak magnetic field parallel to their direction of travel as they traverse the 15–20 mm gap between the entrance slit and the trap. The analyte sample stream, typically from a GC outlet, enters the ionisation chamber through a small hole at the top and perpendicular to the electron beam. As the molecules in the sample stream pass through the beam they interact with the electrons' 70 eV of kinetic energy and typically absorb around 14 eV as internal energy. This causes the ejection of one of the electrons from the analyte, usually an electron that is not tightly bound within the molecule such as those in non-bonding orbitals of heteroatoms. In Eq. (10.1), an electron from the analyte molecule (M) is expelled during this collision process to convert the molecule to a positive ion with an odd number of electrons. This positive ion is called a radical cation and is indicated as the *molecular ion* in the spectrum. It is represented by the symbol M^+ in the following equation:

$$M + e^- \rightarrow M^+ + 2e^- \tag{10.1}$$

This process is frequently referred to as 'knocking out' an electron, however the term is slightly disingenuous as the chance of accelerated electrons actually colliding with those in the analyte and 'knocking them out' is incredibly low. Previously EI was known as 'electron impact' ionisation but this name implies that an elastic collision between an ionising electron and the molecule occurs. The probability of a head-on collision occurring between the ionising electron and the nuclei or electrons of one of the molecule's atoms would be infinitesimally small. The ions are pushed through an exit slit into the accelerating and focusing region of the mass spectrometer's ion optics. The entire chamber is also heated to around 300 °C to avoid condensation of the sample and maintain it in the gas phase. Figure 10.4 is a schematic of the EI ion source.

Many classes of compounds can be ionised by EI, but it should be noted that its use is limited to thermally stable and volatile compounds, typically those below a molecular weight of 1000 Da and that are non-polar in nature. As such, whilst many analytes are volatile under the reduced pressure of the mass spectrometer's vacuum, many must be vaporised at the sample introduction stage. This aspect makes the coupling to a GC an ideal match since GC involves the separation of compounds based, in part, on their volatility. However, those compounds that are not particularly volatile can be derivatised to alter their molecular structure and hence polarity to make them more volatile. These days, this additional step is considered a hindrance and, as we shall see later, the use of ionisation processes that work well with polar as well as non-polar species are preferred. In general, all GC-MS

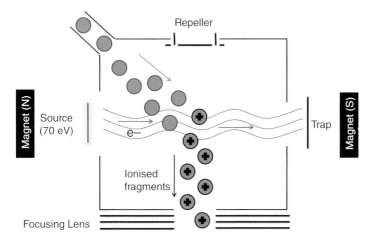

Figure 10.4 Schematic diagram of an EI source. The main body of the EI source is a metal block with entrance and exit holes drilled into it. Molecules enter the ionisation chamber and pass through a beam of electrons generated from a cathode filament held at a potential difference between the ion source block at 70 V. The electrons form a beam and travel in an extended path focused between the poles of a magnet and drawn towards the anode trap. Analyte molecules (green dots) enter the ion block through a hole and travel towards the electron beam. Ionisation occurs as an electron is lost from the molecules interacting with the electron beam resulting in the formation of positively charged ions (cations – positive green dots). The cations are then drawn towards the accelerating and focusing region (ion optics) of the mass spectrometer assisted by a repeller electrode and holding the exit slit at ground potential.

instruments come with an EI source. An important concept to understand is the influence this type of ionisation has on the subsequent spectra obtained. Due to the relatively high amount of energy involved, 70 eV, which is typically transferred to the molecules at around 10–20 eV and is in excess of the 10 eV required to ionise most organic molecules, the excess energy leads to extensive fragmentation of the molecule upon ion formation. EI is therefore typically referred to as a 'hard ionisation' technique. However, the large degree of fragmentation has proved very useful, as it produces highly diagnostic mass spectra with multiple m/z signals linked to specific fragments (Figure 10.5). The diagnostic power of these spectra has led to them often being referred to as 'fingerprints', and has enabled the development of large spectral libraries against which acquired spectra can be searched and matched to identify unknowns. As shown in Figure 10.5, very little of the molecular ion remains following ionisation and interpreting the spectra is analogous to building a jigsaw where the pieces (fragments) are placed together to visualise (interpret) the whole. The abundance of the molecular ion is largely dependent on its stability and the amount of energy required to ionise it. Some structural features tend to provide greater stability than others. For example, those that are aromatic will provide greater molecular ion abundances than those that contain branched alkanes.

10.2.3.2 Chemical Ionisation

Chemical ionisation (CI), not to be confused with atmospheric pressure chemical ionisation (APCI), which will be discussed later in this section, is another technique frequently used in GC-MS. It is also a gas-phase method and is a relatively less energetic mode of ionisation

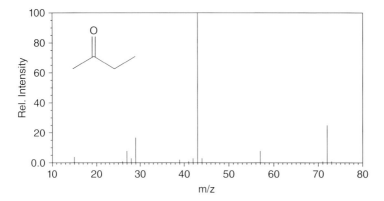

Figure 10.5 Example EI spectra of butanone. Note the last two peaks furthest to the right in the spectra – the peak at m/z 72 is the molecular ion together with the corresponding ^{13}C isotope at m/z 73. The most dominant peak at m/z 43 is the most stable fragment that is formed from the excess energy that is put into the EI process. The mass spectrum is normalised to the relative abundance of the most dominant peak.

that is used to generate a signal for the molecular ion for those species that fail to yield one by EI. This is particularly important for the analysis of unknowns where spectral interpretation could prove to be very difficult by not knowing the m/z value for the molecular ion. As such it is a technique that is complementary to EI and was first developed by Munson and Field (1966).

As with EI, a heated filament is used to generate a beam of electrons, which are accelerated into the source. However, in addition to the sample stream a reagent gas, typically ammonia or methane, is also added into the source in large excess relative to the sample stream. As the reagent gas is in excess, electrons produced from the filament preferentially ionise the reagent gas rather than the analyte directly. These reagent ions then collide with each other within the source forming a plasma through a series of reactions which subsequently ionises the analyte. Most commonly CI is used to create positive ions through the donation of a proton to create fragments with a mass of +1 Da, $[M + H]^+$. However, negative CI can be used if the analyte is capable of stabilising a negative charge. As not all analytes are capable of this, negative CI may add a degree of selectivity to the analysis, aiding in its identification.

Conceptually, the CI process is quite distinct from EI. CI is achieved through *gas-phase acid–base reactions* between the sample molecules and the reagent gas ions. Three steps tend to be involved in the CI process. In the first step, the reagent gas is ionised by bombardment with a beam of 200–500 eV electrons. In the second step, one or more stable reagent ions are produced by ion–molecule reactions. Finally, the sample molecules are ionised by gas-phase ion–molecule reactions with the stable reagent ions (Figure 10.6).

We can look at this in a bit more detail by considering methane as the CI reagent gas, here ionisation of the sample molecule (M) proceeds through the following reactions:

Step 1: ionisation of the reagent gas

$$CH_4 + e^- \rightarrow CH_4^+ + 2e^- (+CH_3^+ + \text{other ions})$$

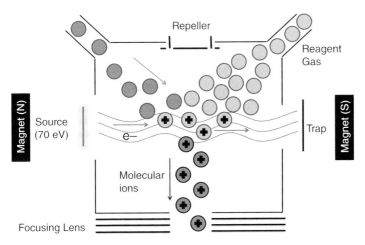

Figure 10.6 Schematic diagram of a chemical ionisation source. The analyte (green dots) along with the reagent gas (blue dots) enter the ionisation chamber. The reagent gas (which is in vast excess) is ionised (positive blue dots) by the stream of electrons, usually creating a positively charged species. In-source collisions between the ionised reagent gas molecules, create a ionisation plasma, which subsequently ionises the analyte, most commonly by protonation (positive green dots), resulting in $[M+H]^+$ adduct ions. The positive ions and any fragments resulting from the ionisation that carry a positive charge are then deflected by the repeller towards the exit of the source.

Step 2: formation of stable reagent ions

$$CH_4 + CH_4^+ \rightarrow CH_5^+ + CH_3$$

or

$$CH_4 + CH_3^+ \rightarrow C_2H_5^+ + H_2$$

Step 3: ionisation of the sample molecules

Proton transfer : $M + CH_5^+ \rightarrow [M+H]^+ + CH_4$

Adduct formation : $M + CH_5^+ \rightarrow [M+CH_5]^+$

Adduct formation : $M + C_2H_5^+ \rightarrow [M+C_2H_5]^+$

Hydride ion abstraction : $M + C_2H_5^+ \rightarrow [M-H]^+ + C_2H_6$

Unlike the ions formed in EI, the ions formed by CI are *even-electron ions* (no electron is removed from the neutral molecule). The molecular mass of the analyte can be deduced from the m/z values of those adduct ions after accounting for the mass of attached or abstracted species. Compared with EI, the ions produced in CI involve much lower energy processes, which results in a decrease in fragmentation, although it is still typically referred to as a hard ionisation technique. While the lack of fragmentation limits the structural information that can be gained, it instead leads to an increase in the molecular ion peak, facilitating the determination of the molecular mass, which may not be possible from a highly fragmented EI spectrum.

10.2.3.3 Electrospray Ionisation

EI and CI are modes of ionisation that function in the gas phase making them readily suitable for ionisation in GC-MS instruments. When LC is coupled to a mass spectrometer (LC-MS) we are faced with dealing with the liquid phase and the mainly hydrophilic species that are dissolved in solution. Many drugs are difficult to vaporise and are much more amenable to chromatographic separation in the liquid phase. This is particularly true of metabolites of drugs that may be present in blood or urine. The same is also true for larger biomolecules such as peptides, proteins and DNA, all of which are of forensic interest and can be analysed by MS. Indeed, large biomolecules tend to degrade and break apart upon heating (think about frying an egg), making them even more difficult to get into the gas phase of the mass spectrometer. In yet another Nobel Prize for the field of MS, the 2001 award was given to John Fenn for his development of the *electrospray* process (Fenn et al. 1989). It elegantly solved two major problems in coupling MS to LC. Electrospray ionisation (ESI) is likely to be the most common type of ionisation encountered in the forensic MS laboratory. This is especially true if the trend for using LC-MS continues to be more popular than GC-MS.

ESI has become the most popular ionisation source used with LC as it not only ionises the sample but also acts as an interface between the LC and MS, evaporating off the liquid eluting from the LC and forcing analyte molecules out of the liquid phase and into the gas phase as ions. Simplistically, the mechanism of ESI involves the production of a fine aerosol or mist, at atmospheric pressure, that is rapidly evaporated due to the increase in surface area created by aerosol formation. This aerosol is formed by passing the liquid through a metal needle that is held at a high voltage (typically 3–5 kV). The aerosol is formed by the surface of the liquid picking up charge. As the liquid surface becomes charged, repulsion forces work against the surface tension of the solution and the liquid begins to balloon out into a cone-like shape known as a *Taylor cone*. Eventually the liquid cone ruptures as the repulsive electrostatic charge overcomes surface tension. When this happens, the liquid breaks up into droplets and these droplets begin to shrink and as they do the surface charge builds up again and they further explode into even smaller droplets. This process is known as the *coulombic explosion* (or fission) and it repeats itself until most of the resulting mist has evaporated away. To aid in this desolvation process the source is heated and frequently an additional flow of inert gas such as nitrogen is added concentrically around the needle. Any analytes dissolved inside the droplets will be squeezed out by the shrinking liquid droplets and they will either pick up charge as they exit the droplet through its charged surface or they have been pre-charged by placing a small amount of acid or alkali into the liquid prior to the electrospray step. In other words, if we acidify or basify our LC mobile phase (or the injection/infusion solvent if LC is not used), we can encourage protonation ($[M + nH]^{n+}$) and deprotonation ($[M - nH]^{n-}$) to take place. It is these protonated (if operated in positive mode) and deprotonated (if operated in negative mode) species that emerge from the droplet already ionised. A schematic of the evaporative process is shown in Figure 10.7.

So, in a way, we consider the process more of a desolvation method rather than an ionisation technique. The ions themselves are produced prior to solvent removal in the liquid sample stream as the analytes flow through the high voltage needle. To encourage ionisation chemical modifiers, such as acids (e.g. formic acid) acting as a source of protons, may

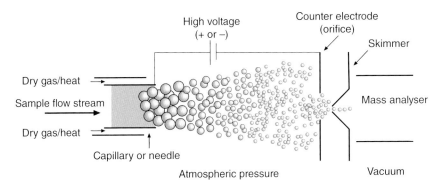

Figure 10.7 Schematic diagram of the ESI process. Droplets are formed by the application of a high positive or negative voltage to a metal needle through which the liquid flow stream of the sample passes. The high voltage causes nebulisation of the liquid flow and increasingly smaller droplets are formed through which the protonated or deprotonated species emerges, stripped from its surrounding liquid and in the gas phase. Ionisation occurs either from the addition of an acid or base to the liquid phase or by transfer mechanisms as the analyte leaves the droplets.

be added. Two theories currently exist as to how a singly charged ion is released into the gas phase. The simplest, termed the charge residual model (CRM), theorises that the droplets continue to undergo cycles of evaporation and coulombic fission until the droplets contain only a single charged molecule. Alternatively, the ion evaporation model (IEM) states that as the size of the droplet decreases, single charged ions can be desorbed from the droplet. It is currently thought that both processes occur, with the IEM model predominating in the ionisation of small molecules while the CRM model is more likely in the ionisation of larger molecules.

Further advances in recent years, such as microspray or nanospray (Wilm and Mann 1996), are essentially low flow (100–10 nl/min) versions of ESI. These methods offer increases in sensitivity as the ratio of analyte to solvent is increased and the droplets formed are smaller, promoting both desolvation and ionisation efficiency. However, these methods are not commonly used routinely in forensic laboratories as they are less robust.

As well as being very useful for the analysis of small molecules, ESI has proved particularly suitable for the analysis of large molecules such as proteins. As a soft ionisation technique it causes little fragmentation producing relatively simple mass spectra. In addition, ESI frequently results in the production of multiply charged species. While this may complicate the spectra, as multiply protonated/deprotonated species will be present in different charge states, it does have the advantage of effectively extending the mass range of the instrument (the m/z of larger molecules is reduced by the multiple charges). The lack of fragmentation does, however, create a problem as the amount of structural information contained within the spectra is limited, this though can be overcome by the use of tandem MS as described later in this chapter (Section 10.2.4).

10.2.3.4 Atmospheric Pressure Chemical Ionisation

APCI is another atmospheric pressure ionisation technique frequently hyphenated to LC. The principle of ionisation in APCI is identical to that described for conventional CI. The difference is that APCI is performed at atmospheric pressure at which many more ion–molecule collisions can occur between the sample molecules and the reagent ions. This improves ionisation efficiency and detection sensitivity.

Like ESI, it can be performed in positive and negative mode; however, rather than applying a voltage to the spray itself, the voltage is applied to a needle placed in the aerosol region of the source. The high voltage needle creates ions by a process known as corona discharge. Like CI, rather than ionising the analyte directly, it is either gaseous solvent molecules or specifically added nebulising gas, which are ionised, due to their presence in excess within the source. Similar to ESI, APCI results in the formation of a quasi-molecular ion created by the addition or removal of hydrogen ($[M + H]^+$, $[M - H]^-$).

The APCI technique consist of three main parts – a removable nebuliser probe (300–500 °C), an ionisation region, and an intermediate-pressure ion transfer region. The LC effluent flows through a fused-silica capillary tube and the nebuliser gas flow coaxially. A mist of fine droplets emerges from the central tube. The combination of heat and gas flow converts the droplets and analytes into a stream of gas. The ionisation region contains a discharge electrode (a corona pin), which is held at 2–3 kV potential with respect to the exit aperture of the counter electrode.

10.2.3.5 Desorption and Ambient Methods

The ionisation methods described above can be coupled to chromatographic techniques or in the case of ESI, a coupling can be made with capillary electrophoresis. There are a number of other ionisation modes that the forensic scientist might encounter that use so-called *offline* sources where the sample is processed without any chromatographic separation and/or sample preparation. These approaches are usually desorption methods that allow for the direct analysis of the sample, usually in situ and at the surface, bringing a number of forensic advantages. These include preserving the sample, limiting the amount of sample preparation required, and the ability to 'image' or visualise the sample in the case of fingermarks and fingerprints.

10.2.3.6 Matrix-Assisted Laser Desorption/Ionisation

MALDI involves dissolving the analyte in a solvent that also contains a solution of a small organic compound called the matrix and forming a solid crystal. MALDI was developed by a German group, Franz Hillenkamp and Michael Karas (Karas and Hillenkamp 1988), and is a variant of another desorption method using a liquid matrix developed simultaneously by Koichi Tanaka (Tanaka et al. 1988). The matrix is present in large molar excess (typically 1:5000 M ratio of analyte to matrix). This large ratio separates the analyte molecules and prevents analyte–analyte interactions (clustering and ion-pairing) during the ionisation/desorption process. The resulting mixture forms a solid solution of analyte-doped crystals. The solvent containing the mixture of analyte and matrix is deposited onto a MALDI sample plate. The solvent is evaporated and then the sample–matrix crystals are irradiated with a laser beam under vacuum conditions. The laser beam is of high irradiance power and with short pulse widths (a few nanoseconds). One of the key features of the matrix is to absorb a large amount of energy from the laser irradiation and then relay that to the sample molecules permitting their desorption. The matrix minimises sample damage from the laser pulse by absorbing most of the incident energy and increasing the efficiency of energy transfer from the laser to the analyte. A high energy density at the laser spot effectively

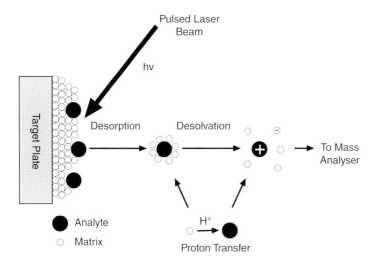

Figure 10.8 Schematic diagram of the MALDI process. The analyte and matrix (small organic acid) are mixed together in a solvent that evaporates leaving small crystals. A laser is fired at the crystals on a target plate and this causes a rapid desorption and expansion from the solid phase to the gas phase. Within a gas cloud that is formed, proton transfer occurs causing ionisation of the analyte.

causes a phase transition from a solid to a gas (sublimation) and in this explosive expansion of the localised matrix, neutrals are mostly dispersed along with ions with high initial velocities (400–800 ms^{-1}). There is very little increase in internal energy of the analyte leading to little or no fragmentation. This makes MALDI a soft ionisation technique. Figure 10.8 is a schematic of the MALDI process.

The actual ionisation process that occurs in MALDI is not fully understood. The most widely accepted ion formation mechanism involves proton transfer in the solid phase before desorption or gas-phase proton transfer in the expanding plume from photoionised matrix molecules. MALDI produces a protonated molecule $[M + H]^+$ or a deprotonated molecule $[M - H]^-$ if operated in negative mode. MALDI is a very soft ionisation technique and is ideal for the analysis of larger molecules such as DNA, proteins and sugars, particularly when linked to time of flight (TOF) MS, due to the large mass range of these instruments. Unlike ESI, it is less likely to result in the formation of multiply charged ions simplifying the spectra produced. In MALDI, the sample is contained within the matrix and is bombarded with a pulsed laser at 337 nm, with pulses every 5–20 ns. The energy of the laser is mainly absorbed by the matrix, which breaks up causing both the matrix and the analyte molecules contained within to become gaseous. There have been two models proposed for ion formation in the MALDI process. The older model proposes that ionisation occurs by the initial photoionisation of the matrix molecules followed by charge transfer to the analyte molecules in the subsequent plume. The more recent explanation, known as the 'lucky survivor' model assumes that the analyte is incorporated into the matrix as a charged species (Karas et al. 2000). Most of these charged species become re-neutralised in the desorbed clusters of matrix and analyte but a few survive and are attracted into the mass spectrometer.

10.2.3.7 Secondary Ion Mass Spectrometry

Secondary ion mass spectrometry (SIMS) is a desorption ionisation technique that involves the generation of secondary ions released from a surface when it is bombarded by an energetic primary ion beam. The primary source of ions can have a low current and are known as *static sources* as they do not cause any damage to the sample's surface. In contrast, so-called *dynamic sources* can produce surface erosion. The primary ion beam is used to sputter the surface to determine elemental, isotopic or molecular composition to depths of 1–2 nm. There are three types of ion sources or *ion guns* as they are sometimes known. One type uses gaseous elements such as Ar^+, Xe^+, O_2^+ with primary ionisation occurring within the energetics of a plasma or by electron ionisation. The second type involves the generation of caesium ions where caesium atoms are vaporised through a porous tungsten plug and then ionised during evaporation. A third type is known as a liquid metal ion gun (LMIG). The LMIG uses metallic alloys, which are liquid at room temperature. The liquid metal covers a tungsten tip and emits ions under influence of an intense electric field. SIMS can be regarded as being the most sensitive surface analysis technique with detection limits ranging from parts per million to parts per billion. A caveat, however, is that they are highly specialised and expensive instruments requiring significant expertise to operate. Whilst the forensic scientist should be aware of their utility, it is unlikely that a general forensic laboratory would require one.

10.2.3.8 Desorption Electrospray Ionisation

Desorption electrospray ionisation (DESI) is another desorption-based ionisation technique developed by Zoltan Takas and Graham Cooks. It involves either depositing the sample on to an insulating surface or plate or directly placing the sample into the DESI spray (Takats et al. 2004). It is undertaken under ambient conditions, where the ionisation takes place outside of the mass spectrometer and without sample preparation or separation. The sample is directly subjected to the nebulised beam of electrosprayed droplets. Essentially, DESI is a modified ESI source in which only the ESI solvents are passing through the needle. The solvents used are typically a mixture of 1:1 methanol and water or acetonitrile and water with a small amount of either acetic acid or ammonium hydroxide to encourage ionisation. The flow rate of this solvent mixture through the needle is typically 1–20 μl/min and like ESI is pneumatically assisted with a stream of heated nebulising gas (nitrogen). Analytes on the surface of either the plate or the sample are rapidly desorbed by the electrosprayed droplets splashing the sample and carrying the analyte away towards the inlet of the mass spectrometer. Figure 10.9 shows a schematic of the process and later in the chapter we will look at some potential forensic applications of this ionisation method and how its ability to analyse samples under ambient conditions, without the need for any pretreatment, can be highly beneficial.

10.2.3.9 Direct Analysis in Real Time

Direct analysis in real time (DART) is also, like DESI, an ambient ionisation technique. It was developed by Robert Cody and James Laramée in 2005 as an ionisation source that could be placed onto the front of conventional mass spectrometers (Cody et al. 2005). Excited gas molecules of helium or nitrogen are used as a primary source of ionisation particles. These species are formed by the use of an electrical discharge similar to the

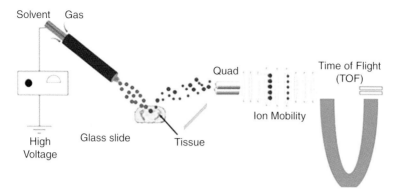

Figure 10.9 Schematic diagram of the DESI ion source and process. Source: Reproduced with permission from Waters Corporation (2020).

corona discharge used in APCI. This beam is focused towards the entrance orifice of the mass spectrometer and protonated water clusters are formed with atmospheric water when the source is operated in positive mode. The protonated water species, H_3O^+, then go on to transfer the proton to the analyte molecule thus forming the ions. In negative mode anions, $[M-H]^-$, are formed from ion–molecule reactions between negative ion clusters of oxygen and water. Due to the ambient nature of the technique, samples can be placed directly into the DART source without any further preparation. The technique has proven to be useful in a number of forensic applications ranging from detecting drugs on bank notes used by drug smugglers through to detecting counterfeit pharmaceutical products. Later in the chapter (Section 10.3) we will look at some of these examples.

10.2.4 Ion Separation – Mass Analysers

Following the formation of ions in the source, the mass analyser's (also termed ion separator) function is to separate the ions produced in the source either in space or time based on their m/z. All mass analysers operate under a high vacuum to reduce unwanted collisions involving the ions analysed. Usually a turbo molecular pump is used to reduce the vacuum from the rough pumps further. Typically, the mass analyser region might be $<10^{-2}$ Pa. The mass analyser is the heart of the instrument and it exploits the concept that a mobile ion can be distinguished from another ion on the basis of the differences in their kinetic energy, momentum and velocity. As we shall see, the mass analyser will use one or more of these properties to separate or *resolve* the ions. Some of the most common mass analysers used in forensic science are discussed below. In order to understand the performance of the different types of mass analyser we need to understand some important theoretical concepts of MS and how they play out in real world instruments. In particular, we need to consider the accuracy of the instruments and how close the m/z measurement can be to the true 'exact mass' and how well we can 'resolve' these masses from each other.

10.2.4.1 Mass Range, Resolution and Accuracy

The *mass range* of the mass analyser is the range between the maximum and minimum m/z values that the instrument can resolve. It is usually determined by the physical property of the ion that is being exploited to provide the mass separation. The *resolution* of the mass analyser is its ability to separate two neighbouring mass ions. It is a highly desirable figure of merit for instruments as it helps in determining the instrument's mass accuracy. There would be no benefit if an instrument that was capable of measuring mass to four decimal places was not able to separate masses with differences of 0.0001 from each other. The *mass resolving* power of an instrument is a performance characteristic and is the ability of a mass spectrometer to provide a specified value of mass resolution.

The *nominal mass* of an ion or molecule is the mass of the most abundant isotope of each element rounded to the nearest integer. The *exact mass* is the calculated mass of an ion or molecule determined by knowing the isotopic composition and charge state, in other words it is the *theoretical mass*. It is important to consider the charge state for some instruments. The mass of the electron (0.00055 Da) may not be negligible depending on the accuracy of the instrument. *Mass accuracy* is the measured, experimental mass of the ion or molecule and is determined by the accuracy and precision of the mass analyser. The mass spectrum obtained should be annotated with either the nominal or accurate mass through the appropriate use of significant figures relevant to the degree of accuracy of the mass analyser.

10.2.4.2 Magnetic Sector

Magnetic sector mass analysers are the oldest type of mass analyser and were the type first used by J.J. Thomson for his analysis of positive rays. They remained one of the most popular types of mass analyser until the quadrupole and ion-trap instruments were commercialised in the 1960s. They are rarely encountered these days, as more sensitive, smaller and accurate mass analysers now exist. It is unlikely that these types of instruments would be encountered in the forensic laboratory but they are important to consider from a historical perspective.

A magnet sector mass analyser uses a magnetic or electric sector to separate ions. Ions from the source are accelerated into a flight tube where they are separated based on the m/z. As the ion enters the magnetic field, the charge causes the ion to be deflected into a circular path by the magnetic field following the curvature of the flight tube. The radius of the analyte's flight path is dependent on the velocity, mass and charge of the ion. Therefore, if the acceleration is constant and collisions are minimised, separation occurs due the m/z, with ions with a lower m/z being deflected to a greater degree than those with a higher m/z. Only ions with the specific m/z relative to the applied magnetic field that allows them to follow the correct flight path to reach the detector will be detected. However, by varying the strength of the magnetic field different m/z can be brought into focus. A typical magnetic sector instrument was capable of high resolution and of delivering mass accuracy to three decimal places. This is sufficient to calculate the elemental composition of the ions detected based on their isotopic composition.

10.2.4.3 Quadrupoles – Quadrupole Mass Filter

The quadrupole is probably the most widely used mass analyser today and is certainly one of the cheapest, most robust and most compact. It consists of four mutually parallel circular metal rods such that the electric field between them is hyperbolic (quadrupolar). A dynamic electrical field is created by applying a simultaneous radio frequency (RF), which is 180° out of phase on the opposite pairs of rods. Each pair of opposing rods is also supplied with

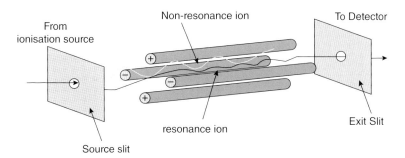

Figure 10.10 Schematic diagram of a quadrupole mass filter. Resonant ions will travel down through the poles towards the detector in a corkscrew trajectory. Non-resonant ions will strike the poles and become neutralised.

direct current (DC) of equal magnitude but opposite polarities with respect to ground. The electrical field created in the mass analyser acts as a filter in that it only allows ions of certain masses ('resonant ions') to pass down the gap between the four rods, i.e. the quadrupole is a mass-to-charge *filter*. By controlling the ratio of the RF to DC voltages, a *narrow band* filter can be produced allowing ions of one m/z value to take a stable trajectory between the rods to the detector, a simple analogy being that of a corkscrew-like path. Ions with different m/z values (non-resonant ions) will collide with the rods and become neutralised. By ramping (increasing) the DC and RF voltages, stable trajectories for ions of different m/z values pass through to the detector (Figure 10.10). This whole process is extremely rapid and typically after less than 1/6th of a second, ions over a wide m/z range may be analysed or *scanned*. The mass range of the analyser is governed by the size of the rods, the distance between the rods, and the frequency of the radio wave, producing an upper limit between m/z 2000 and m/z 4000 (depending on the instrument manufacturer). If an increase in sensitivity (lower limit of detection) is required, selected ions can be filtered to allow the detector to integrate signals of relevant ions for a longer time. Quadrupoles have a mass accuracy and resolution that is lower than sector instruments. Typically, the m/z value is only accurate to one decimal place and the instruments struggle to resolve to unit mass resolution.

10.2.4.4 Quadrupole Ion Trap

The quadrupole ion trap was first introduced by Wolfgang Paul as a type of mass spectrometer (Paul et al. 1958) and it earned him the 1989 Nobel Prize for Physics.

Effectively it is a three-dimensional (3D) analogue of the linear quadrupole mass analyser described above. Rather than being a beam-type instrument like the quadrupole where an uninterrupted ion beam passes through the mass analyser to the detector, it holds ions in a discrete region of space. These days there are now two basic designs, the 3D ion trap or *Paul trap* and the linear or two-dimensional (2D) ion trap. The 3D trap consists of a ring electrode sandwiched between two end-capped electrodes. Holes in the end caps allow ion injection into the analyser from an external source and ion ejection from the analyser to the detector. It is called a trap because it enables ions to be trapped in an electric field by applying an RF voltage to the ring electrode (when the source is positive ESI, a small negative direct current offset is also applied to all three electrodes). Ions formed from sample molecules, such as by ESI, are *stored* for a finite time (1–30 ms), the ions being trapped in a Lissajous 'figure of 8' trajectory. Increasing the RF potential amplitude destabilises the ion trajectories and resonant ejects ions sequentially from the ion-trap region through the end cap electrodes. The ions can be ejected and detected according to their m/z to yield a full scan mass spectrum. A small amount of 'damping gas' (helium) improves resolution and also enables multiple MS experiments (MS^n) experiments. The linear or 2D trap has four electrode rods and two end cap electrodes making a box-like shape. The RF and DC voltages are applied to the four rods, which induces ion confinement in the radial direction only. A simple repulsive voltage set on the end caps provides axial trapping. Scanning of the RF and DC voltages on the rods allows for the ions to be radially ejected through large slits between the rods on either side of the 'box' towards a pair of detectors. Whilst 3D traps are still quite commonly used on GC-MS instruments, due to their compact size, they are limited by their sensitivity. The small space inside the 3D trap means that fewer ions can be stored in comparison with the 2D design. This has led to the 2D trap being a popular mass analyser for LC-MS instruments. The mass accuracy and resolving power of these instruments is similar to the single quadrupole instruments. However, significantly higher mass resolution can be achieved if a narrow m/z range is scanned slowly.

10.2.4.5 Time of Flight

TOF MS is a form of mass analyser that separates ions based on their m/z using time rather than a spatial-based distinction. The basic principle was first described by William Stephens in 1946 (Stephens 1946). Some years later (in 1955), the first commercial instrument was developed by Wiley and McLaren (Wiley and McLaren 1955). The principle of the TOF mass analyser is one of the simplest. Following their initial acceleration in an electric field, ions are separated in the TOF analyser according to their velocities when they drift in a field-free region called a flight tube. Ions of the same kinetic energy are accelerated towards the detector and with a constant energy. Due to the relationship between kinetic energy and mass ($E_k = \frac{1}{2}mv^2$), ions of larger mass will have a lower velocity. Therefore the time an ion takes to reach the detector at a known distance from the entrance to the mass analyser can be used to determine its m/z. Figure 10.11 shows a schematic of a linear TOF instrument.

Initially, TOFs were linked to pulsed laser MALDI sources as the time from each pulse to detection could be used to calculate the m/z, however since these early days, delayed orthogonal extraction, where ions are essentially delayed in their entrance to the TOF, has been utilised. While delaying their entrance into the TOF, the kinetic energy of the molecules post ionisation is equilibrated, which therefore results in improved mass resolution. Orthogonal extraction also allows the use of ionisation techniques other than MALDI with TOF analysers. Early TOF instruments were so called 'linear' TOFs, where the ions would travel down the flight tube in one direction. Further improvements to TOF-MS resolution were provided by the use of a *reflectron*. The reflectron, as the name suggests, alters the flight path of ions by reflecting them from their initial direction of travel back towards a detector using an electrostatic field. This further compensates for any variation in kinetic energy as ions of the same mass with greater energies (and hence velocities) penetrate deeper into the reflectron and therefore have a slightly longer flight path compared with lower energy ions. The distribution in the time taken for ions of the same m/z to reach the detector is therefore minimised. Due to the field-free nature of the TOF, theoretically there is no upper limit to the mass range measured. A further distinction is that as no electric fields need

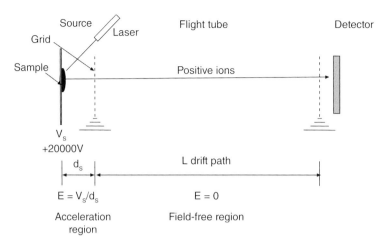

Figure 10.11 Schematic diagram of a TOF mass analyser following production of ions usually from a MALDI source. Following ion formation, during a laser pulse the ions are then subjected to an applied electric field. The ions are continuously accelerated and drift in a field-free region travelling through this region with a velocity dependant on their m/z ratios. Ions are, therefore, dispersed in time. L, length.

to be applied, the instruments do not undergo a scan to achieve the mass range required. This makes their operation (duty cycle) much quicker to acquire spectra than their scanning counterparts. Depending on the size of the flight tube, TOF instruments can have very high resolving powers and this allows for accurate mass measurements similar to those of a sector instrument. This means that like the sector instruments, elemental composition can be determined.

10.2.4.6 Fourier Transform Instruments – Ion Cyclotron Resonance

The concept of ion cyclotron resonance (ICR) was first described by Lawrence and Edlefsen in 1930 (Lawrence and Edlefsen 1939) and developed as a mass spectrometer in 1949 by Sommer et al. (1951). These mass analysers are another form of trapping instrument and function by trapping ions in a cell, or *Penning* trap. Ions are radially trapped by the application of a magnetic field and trapped axially by a static repulsive electric potential on the front and back trapping plates. Once trapped, the ions are then excited by the application of an RF pulse causing them to move in a coherent orbital motion known as their *cyclotron frequency*. The ions induce a charge in the walls of receiver plates and the frequency of their motion can then be measured. By measuring this frequency, the m/z values can be calculated by applying a Fourier transformation and converting the time domain into m/z. The signal is very complicated due to each m/z value creating its own characteristic frequency and necessitating the need for a fast Fourier transform algorithm. Frequencies can be measured with great accuracy this way and so very high resolving powers can be achieved. The high resolving power provides FT-ICRs with the ability to measure mass to five decimal places. Also the nature of the applied electric and magnetic field means that the mass range of the instrument can extend to around m/z 10 000. This makes the instrument particularly useful for determining the elemental composition of both large and small molecules. The cells of FT-ICR instruments require ultra high vacuum to limit the chances of any collisions with air molecules and as such require a number of large turbomolecular pumps. This together with the need for a strong superconducting magnet and accompanying helium cooling makes them large, costly and expensive to run instruments. It is unlikely that the forensic scientist would use these types of instruments. Their context is important as a similar instrument, the Orbitrap, is described next and these instruments can achieve similar resolution and accuracy and at a much cheaper cost.

10.2.4.7 Fourier Transform Instruments – Orbitrap

The Orbitrap is the most recent type of mass analyser to be developed. It is also, as its name suggests, a trapping mass analyser. It was invented by Alexander Makarov and later commercialised by ThermoFisher Scientific in 2005 (Makarov 2000; Hardman and Makarov 2003). The operating principle is based on electrostatic trapping first proposed by Kingdon in 1923 (Kingdon 1923). The Orbitrap is made from a pair of cup-shaped outer electrodes that when placed together have a barrel shape and contain a central inner electrode in the shape of a spindle (Figure 10.12).

Ions are injected into the space between the outer and inner electrodes and begin to orbit the spindle due to a balance between the electrostatic and centrifugal forces. Ions are injected into the Orbitrap at right angles to the z-axis of the spindle electrode. They then follow a circular orbit around the spindle electrode and between the barrel outer electrode. The radius of the ion's circular orbit is a balance between the attractive electrostatic centripetal force with the centrifugal force obtained from its tangential velocity upon injection into the Orbitrap. This radius has no mass dependency and is the same for all ions regardless of their mass. The shape of the barrel outer electrode and the field induced causes the ions to oscillate axially along the x-axis from left to right analogous to a pendulum bob. The frequency of this oscillation is mass dependent and is detected as a time domain signal and transformed into a mass spectrum using a Fourier transform similar to the FT-ICR.

Figure 10.12 Cutaway picture of the partial cross-section of an Orbitrap electrostatic mass analyser. The central spindle electrode can be seen inside the barrel-like outer electrodes. The width of the barrel is approximately the diameter of a one-euro coin. Source: Reproduced with permission from https://commons.wikimedia.org/wiki/File:Orbitrap_mass_analyzer_-_partial_cross-section.JPG.

The Orbitrap can achieve a similar performance in terms of mass accuracy and resolution to the ICR instrument allowing for accurate mass measurements of four to five decimal places. It is also smaller and cheaper to run, requiring no magnet and cryogen. These features have made the Orbitrap a popular instrument for the unambiguous identification of molecules with very high levels of sensitivity. It is quite likely that the forensic scientist using mass spectrometers will encounter this type of instrument along with the quadrupole time of flight (Q-TOF) instruments described in Section 10.2.4.11. These types of 'affordable' high sensitivity, high mass accuracy instruments are becoming more popular and later in the chapter we will see how they are being used to help identify unknown compounds in a forensic context where reference materials may not be available (Section 10.6.2).

10.2.4.8 Tandem Mass Spectrometry – Ion Fragmentation by Collision Induced Dissociation

Some of the mass analysers described above are low mass accuracy instruments. If you couple these types of instruments to soft ionisation techniques such as ESI, then very little structural information is obtained in the mass spectrum. This can make identifying unknowns difficult and in the forensic context leaves the laboratory with data that could be ambiguous. If the mass analyser is only capable of providing unit mass accuracy, then it is conceivable that a large number of possible molecules may share the same nominal mass. This makes the analytical approach less selective and specific. With hyphenated instruments, retention times can be used to provide a degree of selectivity since these can be matched against standards put through the instrument made from certified reference materials. An ion's bond will break in a particular way that can be understood by following a set of rules. Therefore, breaking the intact ion into smaller fragments can allow us to determine the structure of the original species. This is analogous to having the pieces of a jigsaw and placing them together to see the complete picture. EI spectra by their very nature contain these fragment ions. Soft ionisation methods on the other hand produce spectra that do not contain or have very little (at low abundance) fragmentation.

The coupling of mass analysers with low mass accuracy to ESI or APCI usually involve a second stage of MS to produce and separate fragment ions. To produce structurally specific information for soft ionisation methods it is essential to perform experiments that yield ion fragments. This is either undertaken within the same mass analyser as in the case of quadrupole ion traps (both 3D and 2D) or an additional mass analyser as part of the mass spectrometer and is known as *tandem mass spectrometry* usually abbreviated to MS/MS. We can conceive tandem MS as occurring in two ways: the first by the coupling of two distinct mass analysers known as *tandem in space*, or if the same mass analyser is used in the case of ion trapping then this is known as *tandem in time*. In both cases the first stage is to isolate the so-called *precursor ion* and then the second stage is to produce *product ions* usually by an energetic activation process with an inert gas known as *collision induced dissociation* (CID). MS/MS experiments using this approach have been undertaken since the late 1960s (Jennings 1968). Tandem MS does not need to be limited to two stages of mass analysis and it can be undertaken multiple times known as *multistage MS* and is abbreviated as MS^n. Multistage MS experiments are usually undertaken in ion-trap instruments and are typically tandem in time experiments. The multistage nature makes the use of beam-type instruments difficult as additional mass analysers would be required.

10.2.4.9 Tandem Mass Analysers – Ion Traps

The quadrupole ion trap in both 3D and 2D configurations is highly suited to tandem MS experiments and they have evolved into highly suitable MS/MS instruments. MS/MS experiments can be undertaken in the trap by using three time-separated steps: ion isolation, ion excitation and dissociation, and product ion isolation. Selecting its m/z value for confinement initially traps the precursor ion. All other ions are ejected by the resonance–ejection procedure described in Section 10.2.4.4. The selected ion packet inside the trap then undergoes an excitation pulse of a low enough amplitude not to displace any precursor ions but gives them enough energy to interact with a bath of helium that is introduced to the inside of the trap. The precursor ion is then fragmented by CID with helium and product ions are formed that can be scanned out of the trap by ramping the RF amplitude producing the product ion spectra. A notable feature of ion trap is its ability to undertake MS^n experiments by further dissociation of specific product ions that can be targeted by ejecting all other ions from the first tandem stage out of the trap. The ion excitation and collision step can then be repeated for the new precursor and this process can be further repeated until there is insufficient ion population to generate any meaningful signal from the instrument.

10.2.4.10 Tandem Mass Analysers – Triple Quadrupoles

Another instrument that is popular for undertaking MS/MS experiments is the triple-quadrupole (QQQ) instrument that consists of three quadrupoles arranged sequentially. Richard Yost and Chris Enke developed it in the late 1970s (Yost and Enke 1978, 1979) and these instruments remain very much the workhorse mass spectrometer commonly found in forensic laboratories. The mass analyser region consists of the first (Q_1) and last (Q_3) quadrupoles functioning as normal mass filters. The middle (Q_2) quadrupole is operated with only the RF potential. This allows for all ions to pass through it from Q_1 towards Q_3 and the detector and as a transient containment region in which ions can undergo CID. Q_2 is commonly referred to as a collision cell and in reality, to improve ion transmission efficiency, is often an octopole or hexapole. CID is induced by applying a low amount of energy to the ions as they pass through the Q_2 field and bathing them in gas, usually argon. This tandem in space configuration allows for modes of the instrument's operation that are unique to QQQs. Another configuration of the QQQ was created by the company Sciex and consisted of a linear ion trap in place of Q_3. Known as a 'Q-Trap' this instrument is really a type of hybrid instrument consisting of multiple types of mass analysers.

10.2.4.11 Tandem Mass Analysers – Hybrid Instruments

The ion-separation nature of a linear TOF instrument means it is inherently unsuitable for MS/MS experiments. Any product ions formed in a collision cell would have to be refocused and accelerated again down a further field-free region making the architecture difficult to conceive and build. These types of TOF/TOF instruments do exist but the forensic scientist is unlikely to encounter them. A more popular tandem MS instrument based on a TOF that may be encountered is the Q-TOF. First developed by Morris and colleagues and commercialised by Waters Corp in the late 1990s (Morris et al. 1996), this hybrid instrument consists of a normal mass resolving quadrupole and an RF-only quadrupole serving as a collision cell similar to the QQQ above. The ion beam can be focused out of the second quadrupole and into a TOF that is placed orthogonally to the direction of the ion beam. Precursor ions can be selected by the filtering nature of the first quadrupole and then fragmented in the second quadrupole collision cell. Product ions can then be separated and measured in the TOF mass analyser. The orthogonal arrangement allows for the ion beam to be accelerated down the flight tube by pulsation of an accelerating potential orthogonal to its direction exiting the second quadrupole. These instruments have very high ion transmission efficiencies and as such have very high sensitivity as well as good mass accuracy similar to the Orbitrap. Additionally, the TOF section can be operated as a standalone mass analyser by putting the first quadrupole into RF only mode. This allows the higher degree of mass accuracy of the TOF to be applied to precursor ions.

Further types of hybrid instruments that facilitate MS/MS are the Orbitrap configurations. These instruments can be supplied either with a 2D ion trap ahead of the Orbitrap mass analyser or with a quadrupole mass filter followed by a collision cell ahead of the Orbitrap. This allows for MS^n experiments with the former and MS^2 experiments with the latter. The quadrupole-Orbitrap or Q-Exactive configuration is proving to be a popular instrument in forensic laboratories as it can provide quantitative information and high degrees of sensitivity similar to that of a QQQ instrument but with much greater mass accuracy.

10.2.5 Ion Detection

The detector on a mass spectrometer provides information on the ion abundances after their exit from the mass analyser. The beam of ions is converted into an electrical signal that can be amplified, stored and later displayed by software on a computer attached to the mass spectrometer. Usually the signal from the ions generated by the ion source is too weak to be detected directly and some form of amplification is required. Another important feature of the detectors used in mass spectrometers is that they should generate a signal, as an electric current, that is proportional to the abundance of the ions. A number of approaches are used but detection is always based on the ion's charge, mass or velocity. A notable exception to this are the detectors used in FT instruments. In FT-ICR and Orbitrap instruments, ions are detected by the image current they produce in a circuit connecting a pair of metal plates held close to where the ion trajectories take place.

10.2.5.1 Electron Multipliers

The most common type of ion detector used is an electron multiplier. It functions by a process known as *secondary electron emission* (secondary electron multiplier, SEM). (In this chapter SEM refers to secondary electron multiplier, elsewhere in this book it refers to scanning electron microscopy.) When the ion beam strikes the specially coated dynode, several secondary electrons are emitted. These electrons go on to strike the surface of the dynode on the opposite side and several more electrodes are released. There are a variety of designs that are used. A *discrete dynode electron multiplier* is made up of a series (typically 12–20) of copper-beryllium dynodes. A high voltage (at opposite potential to the ion's charge) is applied to the first dynode known as the conversion dynode and the ions slam into it with a high kinetic energy. A small number of electrons are then emitted from the conversion dynode and are accelerated towards the second dynode where several more electrons are released. This process is repeated at the subsequent dynodes causing amplification at each successive stage. This cascade of electron current (excess of 10^7 electrons for each striking ion) arrives at the end of the electron multiplier at the anode and is then further amplified by conventional electronic amplification.

Another common design of the SEM is a horn-shaped assembly known as a *continuous dynode* or *continuous electron multiplier* (CEM). Here the discrete dynodes are replaced by one continuous dynode with a conducting surface that acts as the array of continuous dynodes. The surface is constructed by glass doped with lead or coated with beryllium. A voltage is applied at either end of the horn and the ion beam strikes the surface near the entrance. Electrons are then ejected off the surface and reflected towards the opposite inner surface until they reach the other end of the horn. With each impact more electrons are released until a cascade effect like the SEM is produced, see Figure 10.13. The CEM design is much more compact than an SEM and can provide gains as high as 10^8 electrons for lower cost. Both designs can be prone to overload (particularly when high concentrations of analytes are used) and have a limited lifetime and will often need to be replaced if a large number of samples are put through the instrument.

10.2.5.2 Faraday Cup

A *Faraday cup detector* is a simple type of detector that is found is some mass spectrometers. It can detect ions by the direct measurement of charge with a conducting electrode. These electrodes are usually a conically shaped metal cup known as the collector electrode. The collector electrode is connected to a high-impedance amplifier and the incoming ion beam transfers its charge to the cup causing a voltage drop across a resistor. This discharge current is then amplified directly providing the measure of ion abundance. The response from this type of detector is quite slow and is unsuitable for the fast speeds of many mass analysers particularly those that scan through voltages such as the quadrupoles.

10.2.6 Anatomy of a Mass Spectrum

Put simply, a mass spectrum is a plot of ion intensity versus *m/z* and is the main output of a mass spectrometer. For lower molecular weight ions, the charge is likely to be singular, resulting in ion intensities that represent the mass of the detected ions. Larger molecules may exist in multiple charge states, however this is dependent on the number of ionisable

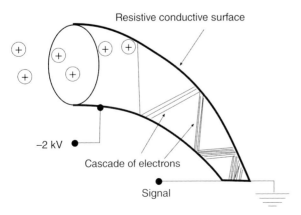

Figure 10.13 Schematic diagram of a continuous electron multiplier. When the ion beam strikes the specially coated dynode, several secondary electrons are emitted. These electrons go on to strike the surface of the dynode on the opposite side and several more electrodes are released. This whole process is repeated producing a cascade of electrons (excess of 10^8 electrons for each striking ion) that arrive at the anode.

sites and the ionisation technique. The signal intensity, depending on the type of detector used, could be recorded in a variety of ways, such as counts per second or electrical signal (volts or amps). As a relatively loose relationship between the recorded signal and the actual ion abundance exists, the axis is frequently labelled in arbitrary units that vary between instrument manufacturers. Alternatively, the abundance of all the ions can be displayed as a percentage of the base peak. The base peak is simply the most abundant peak in the spectrum and may or may not be the molecular ion. Given that most MS instruments in forensic laboratories are hyphenated to a chromatographic technique any mass spectrum is representative of a given time point in the chromatogram, i.e. the mass spectra of any analyte eluting from the column at that time. A chromatogram is a plot of signal, as the ion current at the detector against time. For a total ion chromatogram (TIC), the signal represents the sum of all the ions recorded by the detector at every time point in which the mass analyser separates ions. It is important to know that, unlike spectroscopic signal from an ultraviolet (UV) detector for example, this signal is not continuous. This concept is shown in Figure 10.14 and the distance between each sampling 'dot' is dependent on the speed of acquisition of the mass analyser, that is, how long it takes (usually in the millisecond range) to scan and separate the ions. The wider the mass range, the longer the scan will take.

10.2.6.1 The Molecular or Quasi-Molecular Ion

In a mass spectrum, it is both the relative intensities and the value of each m/z signal that is important in identifying the analyte. One of the most common features of a mass spectrum

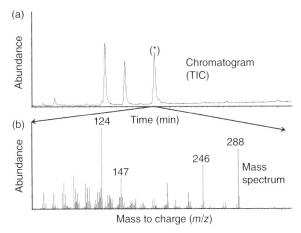

Figure 10.14 (a) Plot of ion abundance versus time. When coupled to a chromatographic system such as GC or LC, as the analytes elute from the column and enter the ion source this results in an increase in the production of ions and thus the associated signal, and the generation of peaks in the chromatogram. For a total ion chromatogram (TIC), the ion abundance signal represents the sum off all the ions that reach the detector at a given time point. (b) Plot of the abundance of any m/z signal at a given time point. The mass spectrum displayed is representative of the ions recorded in the chromatogram at the apex of peak marked (*). All the m/z signals at this time point and their relative abundances can be observed. The more abundant diagnostically important ions are highlighted, and these can be used to confirm the identity of the analyte (in this case the steroid testosterone from an EI spectrum).

is the molecular ion. This represents the intact analyte without any fragmentation, and will therefore likely be the largest m/z in the mass spectrum (if all ions visible are from the analyte). The situation may be more complex for larger analytes, which carry multiple charges. Depending on the distribution of charges within the ion, some fragments produced may contain a proportionally lower number of charges relative to the molecular ion and thus actually have a higher m/z. The abundance of the molecular ion will vary based on several factors including the structure of the analyte itself and the type of ionisation used. Hard ionisation techniques such as EI will induce more fragmentation thus lowering the molecular ion signal, conversely softer ionisation techniques such as ESI result in less fragmentation and so larger molecular ion signals. In addition to the molecular ion the mass spectrum will also include peaks with slightly higher m/z within the same region of the spectrum. These peaks are a consequence of heavier isotopes, particularly ^{13}C, ^{18}O and ^{15}N whose incorporation into the analyte increases its mass. These peaks are still molecular ions, and for this reason this part of the spectrum may be referred to as the molecular ion region. For small molecules, these isotopic molecular ions are likely to be of low abundance due to the low probability of the incorporation of these heavy isotopes into the analyte. Their abundance can be used to calculate the elemental composition of the ion (i.e. the number of carbon and oxygen atoms contained within the molecule). These spectra can be interpreted using some knowledge based around a set of rules since the fragmentation will follow a limited number of mechanisms involved in the breaking and rearrangement of molecular bonds.

It is also possible to calculate the number of carbon atoms present in the spectra of the unknown compound by observing the intensity of a peak 1 m/z value greater than the molecular ion, since this is likely to be the ^{13}C isotope. For every ^{12}C present there will be 1.1% of ^{13}C. The so-called *nitrogen rule* allows us to determine the number of nitrogen atoms that may be present and we can obtain further information on some elements by looking for isotopic patterns and clusters.

The detailed interpretation of spectra is beyond the scope of this chapter but the interested reader is encouraged to read Fred McLafferty's book 'Interpretation of Mass Spectra' (McLafferty and Turecek 1993). For larger molecules, the chance of the incorporation of these heavier isotopes rises due to the increasing number of atoms of any element they contain. For macromolecules, these isotopic peaks start to actually become more abundant than the nominal molecular ion. Depending on the resolving power of the mass spectrometer this will produce different outcomes. In low resolution and mass accuracy instrumentation, the inability to distinguish between all the isotopic molecular ion peaks may result in a broader peak, with an average mass representing the mean of all the ions it contains, and is significantly higher than the nominal or monoisotopic mass (sum of the most abundant stable naturally occurring isotopes for all the elements contained) due to the influence of the heavy isotopes. In higher resolution instruments all the isotopic molecular ion peaks may be resolved. In this case the monoisotopic mass and all the isotopologues may now be seen separately and offer a further method of analyte identification. High resolution, high mass accuracy instruments allow the determination of the measured accurate masses, i.e. the mass of the molecule to several decimal places that is sufficient to determine elemental composition.

10.2.6.2 The Fragment Region

The rest of the mass spectrum will contain ions produced by fragmentation of the analyte. Again, the abundance of these ions will depend on the analyte, type of ionisation and instrumental parameters. Fragment ions are very useful for identifying the analyte (particularly analytes which have the same mass), and are most common in GC-MS mass spectra due to the utilisation of hard ionisation techniques creating 'fingerprint' regions containing many fragments. These fragmented spectra can be compared with existing mass spectral libraries to aid the identification of unknowns, although formal identification may require the concurrent analysis of an appropriate standard. Should such approaches not be possible, mass spectra are full of information that can be used by the skilled mass spectrometrist to determine the structure of an unknown compound.

Fragmentation of the molecular ion by EI can occur by four types of mechanisms. Three of the mechanisms are bond cleavages. The fourth is by molecular rearrangements. Rearrangements typically involve the breaking of bonds and the subsequent formation of a new bond between two atoms that were not bonded previously. Simple bond cleavage is the dominating fragmentation process in the EI mass spectrum of organic compounds. This is where direct cleavage occurs of the bonds. The first type of cleavage is known as *sigma-bond cleavage*. Upon removal of a sigma-electron (an electron that resides in a sigma-bond molecular orbital) the C—C bond becomes elongated and much weaker leading to fragmentation at that site. This then results in a charged moiety that we observe in the spectra and a neutral fragment that we cannot see since one side of the cleavage will contain the ionisation site and the other will not. The next type of cleavage is known as *homolytic* or *radical-site-driven cleavage*. Here the cleavage results from the tendency of the odd electron associated with the heteroatom or pi-bond in the molecular ion to pair with another electron for better stability. The end result is the formation of a double bond or triple bond and the loss of a neutral radical species. The final type of cleavage is known as *heterolytic* or *charge site-driven cleavage*. This time it is the positive charge on the ion rather than the odd electron of an odd electron species that induces the fragmentation. Table 10.2 shows some common fragments that might be encountered in EI spectra.

A full discussion of these mechanisms is outside the scope of this chapter but it is possible to appreciate that fragmentation can help confirm an analyte's structure. In MS/MS instruments, CID spectra can be obtained over a selected mass range and the ion peaks present can be interpreted in a similar way to EI spectra. The fragments give an indication of the identity of the intact ion in the same way as the peaks in EI spectra. Likewise, a standard solution consisting of certified reference materials can also be injected or infused into the instrument and the corresponding spectra obtained compared with the unknown.

10.2.6.3 Full Scan Mass Spectra

If operated in full scan mode, the TIC signal will be the sum of all the ions within the operational range of the method employed (masses below mass 42 are rarely recorded as they are both abundant and diagnostically of little value). The mass spectrum recorded from a full scan acquisition will therefore represent all the ions produced at that time (for the mass range scanned). This mode of acquisition will record the maximum number of fragments, providing the highest degree of certainty for analyte identification. However, ions

Table 10.2 Common EI fragments.

m/z	Ion	m/z	Ion	m/z	Ion
14	CH_2	35	Cl	57	C_4H_9, C_2H_5C-O
15	CH_3	36	HC_L	58	H_2S
16	O	39	C_3H_3	59	C_3H_6OH, $CH_2OC_2H_5$
17	OH	41	C_3H_5	60	CH_2COOH
18	H_2O, NH_4	42	C_3H_6, C_2H_2O	61	CH_3COO
19	F	43	C_3H_7	65	C_5H_5
26	C_2H_2, CN	44	CH_2CHO	66	C_5H_6
27	C_2H_3	45	CH_3CHOH, CH_2CH_2OH, CH_2OCH_2	67	C_5H_7
28	C_2H_4, CO, N_2	46	NO_2	68	$CH_2CH_2CH_2CN$
29	C_2H_5, CHO	49	CH_2Cl	69	C_5H_9, CF_3
30	CH_2, $NH2$	51	CHF_2, C_3H_3	70	C_5H_{10}
31	CH_2OH	53	C_4H_5	71	C_5H_{11}, C_3H_7C-O
32	O_2	54	CH_2CH_2CN	76	C_6H_4
33	SH	55	C_4H_7	77	C_6H_5
34	H_2S	56	C_4H_8	78	CH_2, NH_2

Note, the ions here are shown without charge and may be present as neutral fragments.

from co-eluting compounds or background signals (i.e. column bleed) may also be recorded, complicating the spectrum. Such ions may be removed during the data processing stage by techniques such as spectral subtraction. All mass analysers are capable of producing full scan mass spectra and the corresponding TIC.

10.2.6.4 Product Ion Spectra

It should also be noted that full scan spectra can be obtained from MS/MS modes in instruments capable of undertaking tandem MS. Here the mass analyser is set across a full *m/z* range of the operator's choosing to capture all product ions produced following the CID process. The spectra obtained are known as *product ion spectra*. Once again, these spectra can be complicated if any co-eluting species happen to be present with the same nominal mass as these precursor ions as they will not be distinguished by the mass analyser if it is a low mass accuracy instrument. Fragments from those species with the same nominal mass will be present in the MS/MS spectra making it difficult to interpret. The forensic scientist should never assume that it is unlikely that a compound with the same nominal mass will elute the same and be present in the sample. Particularly with biological extracts, in the case of forensic toxicology work, there remains a reasonable probability of this occurring and introduces an element of uncertainty associated with the data.

10.2.6.5 Extracted Ion Chromatograms

While representing a true picture of what the MS is detecting, a TIC typically can be very noisy as it displays all analytes eluting from the MS, many of which may not be of interest.

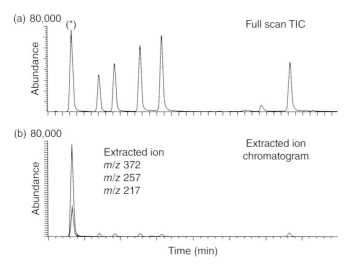

Figure 10.15 (a) Full scan TIC chromatogram. A full scan GC-MS run has been performed with the TIC displaying several peaks relating to multiple different analytes which are present in the preparation analysed. (b) The extracted ion chromatogram. Here three specific ions (m/z 372, 257, and 217) have been extracted post acquisition. A trace for the abundance of each ion is displayed. These ions are much more abundant in the peak of interest, denoted with a (*), than the other peaks and so the analyte of interest can be easily identified. Importantly, the specified ratio of the ions can also be observed which is an important identification criterion. Also, as this process is performed post acquisition, none of the original information is lost. Full scan mass spectra are still available and different ions can be extracted simply using software.

To increase the selectivity and simplify the chromatogram by removing noise, an extracted ion chromatogram (EIC) may be used. An EIC is a form of data analysis rather than a mode of acquisition, it is a way of simplifying the way the data is viewed, however it does not change what has been recorded. It can therefore be applied to both full scan and selected ion monitoring data. An EIC displays only the signal for a specified ion or ions selected, greatly simplifying the displayed data. This is achieved post acquisition by using the software to display the signal for the ions you wish to extract; an example of this is shown in Figure 10.15.

10.2.6.6 Selected Ion Chromatograms and Multiple Reaction Monitoring

A more targeted mode of operation is so-called *selected ion monitoring* (SIM), in which the TIC represents the sum of all the ions that are selected by the mass analyzer to reach the detector. By only detecting specified ions, the sensitivity of the method is increased relative to a full scan acquisition. This is because the mass analyzer will spend more time detecting the specified ions of interest as the electronics are no longer scanning through a set range of voltages. Further, only the selected ions are recorded, meaning those of a different m/z produced by co-eluting compounds are excluded provided the instrument has sufficient mass accuracy and resolution. This simplifies both the chromatogram and spectrum by reducing interfering matrix noise; see Figure 10.16 for an example of the spectra and Figure 10.17 for the corresponding chromatograms. Effectively there are no mass spectra to be obtained since the mass 'range' focuses on a single ion. This has the additional benefit

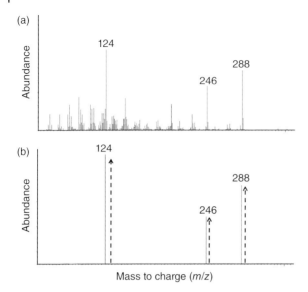

Figure 10.16 Full scan (a) versus selected ion monitoring (SIM) (b) spectra. The full scan mass spectrum represents all the ions recorded by the detector within a specified mass range. It therefore contains a large amount of information which can be used to identify the analyte. In GC-MS the large degree of fragmentation resulting from ionisation creates a highly complex mass spectrum which can be very useful for identifying compounds using library matching. The SIM spectra only record the specified ions, with all other ions being filtered out by the mass analyser and, therefore, do not reach the detector. In this example, three diagnostically important ions for testosterone have been chosen for SIM. By focusing on specified ions both the chromatogram and the mass spectrum are simplified. In addition, by focusing on a few specific ions the sensitivity of the method can be increased. However, much of the information from the mass spectra is lost.

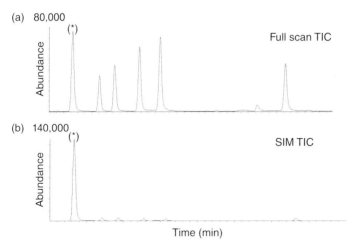

Figure 10.17 (a) Full scan TIC chromatogram and (b) SIM TIC chromatogram. A SIM GC-MS run has been performed and the mass analyser has been set to select only three specific ions (m/z 372, 357, and 217) which will reach the detector. The TIC therefore represents the sum of the abundance of only these three ions. The result is therefore a simplified chromatogram as none of the other analytes in this mixture contain a significant amount of these ions. The identification of the analyte of interest in the chromatogram has therefore been made easier. In addition, by focusing on only three specific ions the sensitivity of the method has improved with the abundance of the peak of interest (*) increasing.

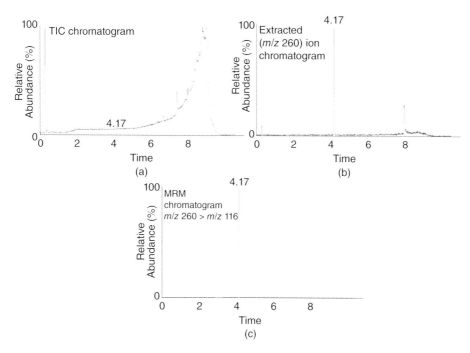

Figure 10.18 A full scan LC-MS analysis has been performed on a solution containing a low concentration of the drug propranolol and this is shown in (a). The TIC displays a very small peak for propranolol (4.17 minutes) and several other peaks relating to multiple different analytes that are present in the preparation analysed. In (b) the extracted ion chromatogram is displayed with the ion of m/z 260 extracted from the full scan data acquired post acquisition. By focusing on an ion specific for the analyte of interest the signal-to-noise ratio (S/N) is greatly improved and the analyte is easily identified in the chromatogram. As this process is performed post acquisition none of the original information is lost. The full scan mass spectrum is still available and different ions can be extracted simply using software.

In multiple reaction monitoring (MRM) mode the mass analyser has been set to select only an ion of m/z 260 in the first quadrupole (Q_1). Following fragmentation in the second quadrupole (Q_2) the third quadrupole (Q_3) has been set to monitor the fragment m/z 116. The chromatogram therefore represents the abundance of only this ion. The result is in a much- simplified chromatogram (c). The identification of the analyte of interest in the chromatogram has therefore been made easier and the sensitivity is also improved in terms of S/N.

of reducing noise in the resulting chromatograms. Scanning instruments such as ion traps and quadrupoles are capable of operating in SIM mode. In MS/MS mode, multiple precursor ions can be selected as individual precursor scans to undergo CID. The second stage of MS can be on a fixed m/z value if the product ions are known. This is known as multiple reaction monitoring or MRM, see Figure 10.18. (The terms MRM and selected reaction monitoring [SRM] are interchangeable. MRM will be used to avoid confusion with SRM, used elsewhere in this book for Standard Reference Materials). This has the potential to increase sensitivity in the resulting MRM chromatogram as all interfering noise from other ions has been eliminated. This process works to much greater gain on QQQ instruments as Q_1 and Q_3 can be fixed rather than scanning. This greatly decreases the duty cycle of the instrument allowing more of the target ions to reach the detector. An ion-trap instrument would still have to scan out all ions produced from the precursor's dissociation and this greatly increases the duty cycle of the mass analyser. Less time is spent filling the trap with the analyte and as such sensitivity in terms of signal-to-noise is lower in comparison with the equivalent mode of operation on a QQQ. Setting a QQQ to run in MRM mode is likely to be that instrument's main use. It can therefore facilitate high degrees of sensitivity and selectivity making it highly suitable for quantitative analysis down to very low concentrations. At the time of writing, many QQQ instruments are capable of detecting a few hundred pg/ml concentrations of small molecules from blood or urine extracts.

The downside of this targeted approach is that all other ions are lost, reducing the information associated with any peak of interest and contained throughout the chromatogram. From a forensic perspective, the data cannot be revisited to look for unknowns and this might not be desirable if there is only a limited amount of sample with which to investigate. Typically, to maintain the appropriate degree of certainty when identifying compounds, at least two (and if the fragmentation allows it three) ions per analyte are recorded. One of these is termed the quantifier (which is used for quantification if required) and the others are termed qualifiers. The presence, and more importantly the relative abundances of these ions to each other, are crucial for the correct identification of any compound.

10.2.6.7 Precursor Ion Detection and Neutral Loss Scanning

This mode of operation is only possible on QQQ instruments and provides a spectrum of all precursor ions that might fragment to a common diagnostic product ion. We can obtain this type of spectrum by adjusting the third quadrupole to transmit a chosen product ion and scanning the first quadrupole over a certain m/z range to transmit only those precursor ions that fragment to yield the chosen product ion. Q_1 scans the mass range and Q_3 filters a unique product ion and associates it back to the precursor ion that it originated from. Neutral loss scans are also only possible on QQQ instruments; all precursors that undergo a loss of a specific common neutral species are monitored. To obtain this type of data both Q_1 and Q_3 are scanned simultaneously but with an offset that correlates with the mass of the specified neutral. This type of scan is similar to precursor ion scanning in that classes of compounds can be determined from a mixture provided they have a common neutral loss.

10.3 Application to Analytes

The mass spectrometer when hyphenated to chromatographic techniques makes for a powerful analytical instrument to aid in identifying samples of forensic interest. Data acquired by mass spectrometers are admissible in court and are particularly useful for specifying the identity of seized controlled substances. The chromatographic component is particularly important as quite often drugs can be bulked out with cutting agents and identifying these cutting agents can help determine the source of origin. A further use here of the combination of chromatography and MS is for narcotics that originate from natural products such as heroin. The other opiate components that are extracted from the poppy resin can also be detected and the relative amounts of these substances can be used to trace the country or even region of origin.

Perhaps one of the most significant applications of chromatography–MS is for the detection of drugs, poisons and banned substances in biological samples such as blood and urine. This can provide evidence to help determine causes of death, determine whether a driving offence has taken place, or prevent cheating in sports. Collectively this type of analysis is known as *forensic toxicology* and the almost universal nature of MS detection allows for a wide range of compounds to be detected. These can be steroids in the case of anti-doping or fast acting sedatives such as benzodiazepines in cases of drug-facilitated sexual assaults. Urine has several advantages over blood that makes it a popular analytical matrix particularly in *ante*-mortem samples where it is easier to collect. Unlike blood it does not require an invasive procedure for collection, sample volumes will typically be much larger (up to several hundred millilitres); it typically requires simplified sample preparation and usually offers an increased detection window compared with blood containing a greater range of metabolites. Blood samples can, however, give information that urine samples cannot. Due to differences in excretion rates and the influence of urinary dilution, the concentration of drugs and their metabolites in urine may vary widely, meaning, unlike in blood samples, it can be difficult to determine the time or size of dose and thus the pharmacological species encountered.

Of significant concern, at the time of writing of this book, is the increasing use of so-called novel or new psychoactive substances (NPS). A wide variety of compounds have been developed to circumvent local drug laws and/or increase the potency of traditionally abused drugs such as cannabis and heroin. In the case of cannabis, a number of synthetic compounds have been made that are outlawed in many countries but are widely available on the internet or dark web. These synthetic cannabinoids are much more potent than their natural counterparts and are rapidly metabolised by the human body such that the parent compound is present only in small quantities. Here high-resolution accurate MS using Orbitrap or Q-TOF instruments becomes very useful as the structures of both the parent drug and the metabolite are unlikely to be known. The ability to undertake elemental analysis becomes highly advantageous. A number of groups have developed high mass accuracy methods for the targeting of NPS and the interested reader should look at the reviews and research articles by Pasin et al. and Cawley et al. (Cawley et al. 2016; Pasin et al. 2017).

The use of MS for the detection of forgeries both as works of art or as documents for criminal gain has been investigated. This presents a whole different range of chemical entities and sample matrices. MALDI-TOF-MS has been used to try to identify paints and dyes to assist in forgery detection of paintings (Soltzberg et al. 2007). Here small sub-µg amounts of sample can be taken to minimise damage. Laser desorption (without matrix) MS has been used on pigment-based pen inks to determine the chemical composition of colourants used in potential forgeries to help link them to the criminals involved (Papson et al. 2008). SIMS has also been used successfully to determine the chemical composition of paints (Gresham et al. 2000). The very small size of the primary ion beam, often in the tens of nanometres, makes this technique particularly useful for microscopic sized samples. The SIMS approach is able to determine distinctions in paint coatings but still preserve the bulk of the sample for further testing using techniques such as Fourier transform infrared spectroscopy.

MALDI MS has been successfully used for the identification of drugs deposited within fingermarks (Groeneveld et al. 2015). The ability to track or raster the laser across the fingermark allows the investigator to build up a colour-coded image based on the intensity of ions from the mass spectra. This 'imaging' approach can provide remarkable detail including ridge patterns from the latent fingermarks. Similar work by the same group has also shown that condom lubricants can be detected producing images of similar detail and identifying the brand of condom used (Bradshaw et al. 2013). MALDI has also been used to detect the presence of drugs in single hairs (Vogliardi et al. 2009). A small amount of matrix can be deposited onto the hair and the laser tracked along its length. This has provided a fast screening method for hair samples since very little sample preparation is necessary. However, it can be difficult to determine whether the drugs, cocaine in this instance, are present as external contamination or from genuine administration. SIMS has been used to help with this distinction as layers of the hair strand can be eroded by the ion gun (Flinders et al. 2015). This facilitates access to drugs actually embedded within the hair's structure with their presence more likely to have originated from incorporation by administration. DESI has been used to determine excipient contents of potentially counterfeited drugs (Cardoso-Palacios and Lanekoff 2016). The ability to mount the tablets into the DESI source and its ambient nature again reduces the need for any sample preparation and this cuts down on the analysis time. DART has been used in a similar way both with paints (Chen and Wu 2017) and intact drug tablets (Pavlovich et al. 2018). The consideration here is the degree of surface penetration that DESI and DART can achieve. In comparison with SIMS, it would be difficult to obtain information on layers of coating as the DART source cannot erode the surface.

Finally, another major area in which MS is a powerful tool for forensic analysis is in the area of explosives and combustible materials. Here MS has the potential to screen for potential explosive materials that may be used for terrorist activities in places of mass transportation such as airports or train stations (Zhao et al. 2017). Alternatively, MS can be used as a tool to help identify the origins of the explosives following a blast (Bridoux et al. 2016). Equally, MS can be used to help determine the cause of a suspicious fire by determining the source of ignition (Ferreiro-González et al. 2016). Ambient methods of ionisation have again proven useful in this area (Forbes and Sisco 2018). Since very little sample preparation is necessary, intact pieces of evidence such as pieces of clothing can be analysed directly.

10.4 Interpretation and Law

10.4.1 Chain of Custody

The term chain of custody (CoC) refers to chronological documentation that records the paper trail, custody, control, transfer, analysis, and disposition of physical or electronic evidence (Froede 1976; Chamberlain 1988; WADA 2019). Within forensic science, this could include a range of evidence including drugs, biological samples and various types of environmental samples associated with a crime or crime scene. The correct recording of the CoC is crucial in maintaining the integrity of the evidence in such a way that it remains beyond reasonable doubt linked to the person or location. Furthermore, it should demonstrate that the possibility of tampering or contamination of the evidence is minimised. In practise a specified person must be responsible for the evidence at all times and any movement of the evidence (or a portion of it) documented. For laboratories, the responsibility to maintain the CoC begins when the sample arrives and ends when the sample is either destroyed internally or transferred elsewhere. It is, though, important to note that while the laboratory may not have control of the CoC prior to its arrival, any break in the CoC up to this point can essentially undermine the analytical evidence subsequently produced. This may be particularly important if the evidence is collected by persons not normally involved in evidence collection such as physicians who may deal with victims of assault for example. Their primary priority will be the treatment of the individual rather than the collection of evidence in an appropriate manner. The effect of a break in the CoC prior to laboratory analysis is that a defendant may be able to argue that the analytical evidence cannot be linked to the case with sufficient certainty.

Laboratories must undertake measures to ensure the CoC and integrity of the sample during its reception. The transport container should be inspected and any irregularities/damage appropriately documented. The date and time of receipt of the delivery should be recorded along with the name and signature of the courier and the laboratory representative receiving the samples.

Once in the laboratory the internal CoC must include the sequence of persons in custody of the sample or any subsequent aliquot noting the individual involved, the action taken, and the location. While historically the CoC was documented and maintained by paper records more recently laboratory information management systems (LIMSs) are being used with records being maintained electronically. These systems are designed to maintain complete CoC records from sample receipt, including actions associated with laboratory process such as sample movement, splitting, analysis, and reporting automatically. Such systems are intended to reduce the possibility of errors or missed documentation. However, their successful implementation requires adequate staff training and usage of the system.

10.4.2 New Forensic Regulations

In February 2016 the Forensic Science Regulator of the UK released the Codes of Practise and Conduct (Issue 3), following the initial release in 2011 (Tully 2016). The stated aim was to ensure all the evidence produced by forensic science providers to the criminal justice system (CJS) is of the required standard, therefore, reducing miscarriages of justice.

To monitor compliance with these standards a system of accreditation is proposed, which will be administered by the United Kingdom Accreditation Service (UKAS), and include monitoring programmes such as inter-laboratory comparisons and proficiency testing. The Code of Practise and Conduct was wide ranging covering all areas of forensic science with most disciplines required to reach the appropriate level of accreditation (generally ISO 17025) by October 2017. This included the disciplines most reliant on MS such as firearm discharge residue analysis and drug analysis and toxicology. While laboratories will previously have had more independence in their procedures, the requirement for accreditation has created a significant workload for laboratories as they seek to align existing procedures with those outlined by the regulations.

10.4.3 ID Criteria – Screen and Confirmation

As an example of the requirements for mass spectral data we shall predominantly consider the criteria laid out by the World Anti-Doping Agency (WADA) for the field of anti-doping in the technical document TD2015-IDCR (WADA 2015). Despite the focus here on the mass spectral data, it must be remembered that the ability of a method to identify an analyte is a result of the entire procedure, which starts from sample collection and will include areas such as the adequate transportation of samples.

10.4.4 Chromatographic Criteria

The retention time (RT) of an analyte in a sample of interest within a chromatographic run should under identical conditions be closely matched to that of a standard. The standard used may be a spiked sample using the same matrix, an administration sample of the same drug, or a certifiable reference standard, depending on the availability of each of these resources for the compound in question. The variation in the RTs between the analyte and standard can be assessed either as a percentage or absolute difference, differing by no more than ±0.1 minutes or 1% whichever is greater. Alternatively relative retention times (RRTs) where the RT of the peaks of interest is given relative to a known reference compound may also be used.

10.4.5 Mass Spectrometric Identification Criteria

The identification of analytes by MS can be divided into two main approaches. Top down where the intact analyte is identified by direct analysis and the generation of diagnostic ions and bottom up where the analyte of interest is broken up chemically or enzymatically prior to analysis and the produced fragments identified. Regardless of the approach used, a summary of the criteria to be applied is presented below but for full details the reader is referred to the original document (WADA 2015).

- The measured mass of each ion used for identification shall be within ±0.5 Da of the corresponding ion acquired from a known standard analysed in the same analytical batch.
- For single-stage MS, at least three diagnostic ions shall be used.
- For tandem MS at least two precursor–product ion transitions shall be monitored. The isolation width of the precursor ion shall not be more than m/z 1.3 (unless required by its molecular mass and charge state).
- If the above criteria for the number of diagnostic ions or MRM transitions cannot be met a second analytical approach must be used. This may be an alternative derivative or ionisation approach but must be based on a different chemical or physical principle.
- The abundance of ions may be determined from the peak area or height of integrated selected ion chromatograms.
- The S/N of all diagnostic ions shall be greater than three to one (3:1).
- Relative abundances shall be calculated by dividing the area or height of the ion trace of each diagnostic ion by the area or height obtained from the base peak.

The relative abundances of any of the diagnostic ions shall not differ by more than the specified criteria that may vary dependent on the relative abundance of the ion being assessed. Between 100% and 50% the tolerance is ±10% in absolute terms (i.e. a relative abundance of 60% in the base peak gives a tolerance of 50–70% for the analyte of interest). Between 50% and 25% the tolerance is ±20% in relative terms (i.e. a relative abundance of 40% in the base peak gives a tolerance of 32–48%). Between 25% and 1% the tolerance is ±5% in absolute terms (i.e. a relative abundance of 10% in the base peak gives a tolerance of 5–15% for the analyte of interest).

10.5 Case Studies

10.5.1 Serial Killing by Poisoning

From 2004 to 2009, a Finnish nurse, Aino Nykopp-Koski, worked in a number of hospitals, care homes and in the homes of patients. During this period, she killed a number of elderly patients using opiates and sedatives. This case draws parallels to one of the UK's most prolific serial killers, Harold Shipman, who also used opiates to murder elderly patients. Nykopp-Koski was caught by the actions of a vigilant emergency room doctor who had saved the life of Nykopp-Koski's last victim and reported her to the police. This led to an extensive investigation of all the deaths of patients under Nykopp-Koski's care. This brought to the attention of the police five deaths that had occurred under suspicious circumstances and in two of the cases the body had been cremated. This meant that there were limitations in terms of sample availability apart from liver histology blocks. Histology blocks are paraffin embedded tissue used for microscopic analysis of cells and tissue. In the journal *Drug Testing and Analysis*, Erkki Vuori and co-workers report on the analysis of the paraffin embedded liver tissue by LC-TOF and GC-MS and the subsequent findings in relation to this case (Vuori et al. 2013). They were able to cut some of the embedded tissue from the blocks with a scalpel and then grate the tissue with a kitchen grater. The purpose of this was to produce a high degree of surface area from the tissue for extraction. About 200–300 mg of tissue was transferred to a test tube and was extracted using 94% ethanol. The ethanol was evaporated off and then the sample was reconstituted in a phosphate buffer.

Following solid-phase extraction from the buffer solution, qualitative accurate mass (LC-TOF-MS) drug screening was performed (m/z 50–800) hyphenated to reversed-phase liquid chromatographic separation. Substance identification was based on accurate mass, isotopic pattern and RT using full scan spectra. This data was compared with an in-house database of drugs, drugs of abuse and their metabolites. EI-GC-MS and negative ion chemical ionisation was also used to provide comparative data to the TOF analysis. This complementary approach would provide fragmentation information to help further elucidate the identity of the compounds from the sample that would not be possible from the TOF data alone. The liver sample was shown to contain oxazepam, temazepam and trimethoprim which were not on the patient's list of medications. The patient had only been prescribed intravenous diazepam for any epileptic seizures with the last time used being several weeks before death. A work's list indicated that Nykopp-Koski was present when the patient's condition worsened and would have had access to the other medications. Poisoning by benzodiazepines could also explain the patient's clinical condition and eventual death. In this case, MS proved useful for the high degree of sensitivity required to work with an unusual matrix. Furthermore, since it was unknown what was expected in the sample, library matching using high mass accuracy data was key in this investigation. The high mass accuracy work could be supported by EI fragmentation spectra for further confirmation.

10.5.2 Surreptitious Insulin Administration

The detection of insulin remains of great interest in the field of forensic toxicology as it is a compound that has been implemented in a number of murders (Marks 2009). It is also a compound that can be abused in sports for muscle building gain. Traditionally, insulin has been measured by immunoassay approaches together with a related compound known as c-peptide. Greater levels of insulin to c-peptide can indicate that endogenous insulin has been admitted. However, this approach lacks specificity and now that a number of different recombinant insulin analogues exist it has become challenging to use immunoassay as a means of insulin detection particularly in the forensic context. Thevis and co-workers have developed LC-MS/MS approaches to help selectively identify insulin analogues in anti-doping samples (Thevis et al. 2006) and in forensic cases (Green et al. 2010; Thevis et al. 2012). The use of MS is key here as the tandem MS used allows for the fragmentation of the insulin analogues to provide highly specific fingerprints of the insulin molecules involved. This can be undertaken to such a degree that the differentiation between endogenous insulin and an exogenous recombinant type that has two of its amino acids switched around is possible. Green and co-workers report on an eight-week-old infant presented to an accident and emergency department with lethargy, tachycardia and a low blood glucose concentration. Initial clinical tests did not report growth hormone deficiency, oral hypoglycaemic ingestion, or any inborn errors of metabolism. Furthermore, difficulties in restoring and maintaining glucose concentrations were observed and elevated insulin and low C-peptide concentrations seemed to suggest exogenous insulin administration. The use of tandem MS was applied for analysis of the infant's blood and the insulin analogue Humalog was detected. This data suggested surreptitious administration of insulin to the child. The child was later placed into foster care with no further episodes of hypoglycaemia.

10.6 Forensic Developments

10.6.1 Beyond Blood and Urine

In forensic toxicology when considering biological analysis it is usually blood and urine samples that are first thought of, but other matrices should be considered, and indeed are likely to become more popular depending on the application and circumstances (Parkin and Brailsford 2009). So-called *alternative matrices* may offer further advantages to the forensic toxicologist and one example is the use of hair. Drugs are incorporated into the hair at the time they are administered. As the hair grows (typically 1 cm a month), the drug which is incorporated into the shaft of the hair migrates away from the scalp as new hair is formed. The hair therefore can be considered a historical record of drug administration and frequently more hair will always be available should further analysis be required. This is particularly useful if knowledge of prolonged drug use (or abstinence) is required, and in cases where there is either delayed sample collection or the use of drugs with short detection windows, cases where traditional matrices will likely result in limited forensic evidence being available. Drawbacks with hair analysis must be considered though such as individuals who have short hair, the effect of hair treatment such as dyeing and washing on drug concentrations, environmental contamination and the differing incorporation rates of drugs into hair within the follicle. Other matrices may include sweat, which is non-invasive and very easy to collect using swabs or sweat patches. Sweat has been proposed as an alternative to urine samples, as it decreases the risk of sample manipulation. Drawbacks though remain, such as the need for highly sensitive instrumentation due to the limited sample volume and sample losses associated with some current collection devices. The use of dried blood spots (where small fingerprick volumes of blood are dried on filter paper) has widely been used clinically and may offer the advantages of blood sampling while providing an easier means of sample collection and transportation. As with sweat sampling though, the limited sampling volume places a high emphasis on the use of highly sensitive instrumentation. The analysis of oral fluid has also been proposed for applications such as roadside drug testing and monitoring of drug use. Again, it offers easier sample collection at the expense of sample volume. It was initially thought that oral fluid concentrations paralleled those in blood, making it an excellent alternative to blood. However, incorporation into oral fluid was later shown to be compound dependent and in addition may be affected by factors such as the pH of the fluid. These factors have so far limited the applications of oral fluid in toxicology. Finally, breath analysis is widely associated with the detection of ethanol, but the ease of use has attracted an increase in research in this area with the possibility to detect a greater variety of drugs in this matrix proving attractive.

10.6.2 High Mass Accuracy Mass Spectrometry

We are now in the second era of high mass accuracy, high resolution mass spectrometry (HRMS), this time coupled with degrees of sensitivity that allow us to probe complex samples with femtomolar sensitivity. In the use of MS for forensic analysis this high mass accuracy approach is highly useful both for the identification of unknowns and for the screening of drugs in toxicological and anti-doping analysis. The invention of the Orbitrap mass analyser by Alexander Makarov (Makarov 2000) competitively drove the development of the Q-TOF mass analyser (Morris et al. 1996) and together these instruments now make up the majority of high mass accuracy instruments used in forensic laboratories today.

LC is a popular approach to sample screening as it allows the analysis of a wide range of compounds, including larger or more polar analytes which may be problematic by GC, in a single analytical run. Initially LC-MS/MS was used for such purposes, but the commercial availability of LC-HRMS instruments such as LC-Q-TOF MS and LC-Orbitrap MS allowed their integration into forensic laboratories, and has facilitated a shift in approach from targeted to non-targeted screening (Polettini et al. 2008; Ojanpera et al. 2012; Li et al. 2013; Rosano et al. 2013; Mollerup et al. 2016). A targeted approach is commonly employed with single quadrupoles in SIM mode or triple quadrupoles in MRM mode. For compounds to be identified in a targeted approach they must not only elute during the known chromatographic time window but produce known ions which are preselected to monitor for their presence. Targeted approaches have been highly successful, with sensitive methods capable of screening for hundreds of compounds developed, but they have two principle drawbacks. First, the necessity to preselect the ions at the appropriate time in the chromatograph means this analytical knowledge must be acquired prior to sample analysis. This becomes a problem both in toxicology and anti-doping due to the rapid development of new drugs in the areas of NPS and performance enhancement. Secondly, the ever increasing number of transitions means that eventually you will reach the limit of MRM transitions that a QQQ may perform, preventing the addition of new analytes to the method. HRMS offer a solution to these two problems. First, due to their ability to determine the accurate mass of any fragment (rather than just the nominal mass), they are able to determine the elemental composition of any ion recorded. This reduces the reliance on spectral libraries and allows the confirmation of previously unknown novel compounds. Secondly, HRMS can be considered to represent total data capture, with all information filtering taking place post acquisition during data processing. There is, therefore, no limit on the number of compounds that can be detected. Furthermore, the total data capture allows the retrospective data mining of HRMS files. Should a new compound of interest come to light, data can be reprocessed using the mass of the new compound, negating the need for repeated sample extraction. This approach does though rely on the new compound being successfully extracted during the initial sample preparation.

While the advantages of both Q-TOFs and Orbitraps can be considered together as HRMS there are differences between the two instruments. Compared with an Orbitrap, Q-TOFs typically have faster mass spectral acquisition speeds and good isotopic abundance accuracy, making isotopic pattern fit useful for compound identification. However, many Q-TOFs currently struggle to perform rapid polarity switching unlike Orbitraps, meaning samples are often analysed twice, once in positive mode and again in negative mode increasing analysis times. Due to the advantages of LC-HRMS screening, the approach was adopted by the laboratory responsible for the anti-doping analysis prior to the London 2012 Olympic and Paralympic Games (Musenga and Cowan 2013). During the games, around 5000 urine samples were analysed with a required reporting time for results being just 24 hours following sample receipt by the laboratory. In anti-doping, testing usually begins with a screening procedure (also termed initial testing method), the purpose of which is to identify suspect samples (those which are suspected of containing a prohibited substance) for confirmation procedures. Screening methods must be able to detect the prohibited substances with great sensitivity (1 ng/ml under regulations in force in 2012) but also minimise the generation of false suspects that may result in unnecessary time-consuming confirmation procedures. To maximise efficiency, anti-doping laboratories have invested significant effort into combining different screening methodologies for multiple classes of compounds, therefore maximising the number of analytes that can be screened in the lowest number of analytical runs. The method developed for the Olympic and Paralympics Games in 2012 included a broad range of compounds such as anabolic agents, β-blockers, glucocorticoids, narcotics, stimulants, diuretics and β2-agonists. To cover this breadth of compounds, three types of analyses were performed within a single analytical run (negative mode ESI, positive mode ESI, and positive mode ESI with CID), taking advantage of the fast polarity switching offered by the Orbitrap. The use of CID was particularly important for the analysis of β-blockers, as their similarity to endogenous molecules meant that even using high mass accuracy, interferences of the same mass could still occur. However, using CID with non-target precursor ions resulted in the generation of full scan mass spectra, providing the required degree of selectivity.

10.6.3 Mobile Mass Spectrometers

When considering the use of MS in forensic science, the application of the technique in specialised laboratories by experienced and highly trained staff comes to mind. This situation has been a necessity for a variety of reasons, including equipment size, the requirement for gas flow and vacuum pumps, robustness, usability, and the need for sample preparation (Eckenrode 2001; Hoffmann and Jackson 2015). However, more recent developments in ambient ionisation techniques (such as DESI and DART discussed above), which require little or no sample preparation, and a physical reduction in the size of instrumentation (from large floor standing or benchtop devices to shoe box size) are leading to the development of mobile mass spectrometers. Fully mobile mass spectrometers offer several advantages over laboratory-based instrumentation. The reduced sample preparation, transport and analysis times will speed up analysis. This may be crucial in circumstances such as counter terrorism where rapid results can help dictate a response in real time, but also will reduce the total cost associated with analysis. In addition, the CoC associated with sample analysis will be simplified, decreasing the probability of issues arising regarding the integrity of the evidence. Several issues with current mobile MS instrumentation should still be considered. Due to their potential use in the CJS, the quality of traditional evidence from MS techniques must be maintained. Additionally, the appropriate infrastructure must be developed for their use (i.e. agreed standards required for their use and sufficient manufacturer support) and the evidence from these techniques must be established within the legal framework such that it is accepted with sufficient confidence. Further, a greater understanding of ambient ionisation techniques is required. Some analytes are known to perform more poorly in some instrumentation than in others, thus leading to potential false negatives or inappropriately high limit of detection (LOD). It is anticipated that the use of mobile mass spectrometers will continue to grow within the forensic community as instrumentation is developed further. Currently, the most widely used portable form of MS is ion mobility MS, which is used in airports to detect for traces of explosives. This approach has been successful as it is sensitive (detecting nanograms of material), portable and offers rapid analysis. However, specificity with this technique can be an issue as there are molecules that have similar mobility spectra to markers of explosives, which can therefore cause false positives.

Currently several ambient ionisation techniques are employed that may be used by mobile MS. The most popular being DESI, DART and extractive electrospray ionisation (EESI). DESI is the most widely used and employs a standard electrospray to desorb and ionise molecules from a surface. It has already been applied successfully to the analysis of explosives and counterfeit pharmaceuticals, and in document verification and biological analysis.

References

Aston, F.W. (1920). The constitution of atmospheric neon. *Philos. Mag.* 39 (6): 449–455.

Beynon, J.H. (1959). High resolution mass spectrometry of organic materials. In: *Advances in Mass Spectrometry* (ed. J.D. Waldron), 328–354. Pergamon.

Bradshaw, R., Wolstenholme, R., Ferguson, L.S. et al. (2013). Spectroscopic imaging based approach for condom identification in condom contaminated fingermarks. *Analyst* 138 (9): 2546–2557.

Bridoux, M.C., Schwarzenberg, A., Schramm, S., and Cole, R.B. (2016). Combined use of direct analysis in real-time/Orbitrap mass spectrometry and micro-Raman spectroscopy for the comprehensive characterization of real explosive samples. *Anal. Bioanal. Chem.* 408: 5677–5687.

Cappelle, D., Yegles, M., Neels, H. et al. (2015). Nail analysis for the detection of drugs of abuse and pharmaceuticals: a review. *Forensic Toxicol.* 33 (1): 12–36.

Cardoso-Palacios, C. and Lanekoff, I. (2016). Direct analysis of pharmaceutical drugs using Nano-DESI MS. *J. Anal. Methods Chem.* 2016: 3591908.

Cawley, A., Pasin, D., Ganbat, N. et al. (2016). The potential for complementary targeted/non-targeted screening of novel psychoactive substances in equine urine using liquid chromatography-high resolution accurate mass spectrometry. *Anal. Methods* 8 (8): 1789–1797.

Chamberlain, R.T. (1988). Legal issues related to drug testing in the clinical laboratory. *Clin. Chem.* 34 (3): 633–636.

Chen, T.H. and Wu, S.P. (2017). Forensic applications of direct analysis in real time (DART) coupled to Q-orbitrap tandem mass spectrometry for the in situ analysis of pigments from paint evidence. *Forensic Sci. Int.* 277: 179–187.

Cody, R.B., Laramee, J.A., and Durst, H.D. (2005). Versatile new ion source for the analysis of materials in open air under ambient conditions. *Anal. Chem.* 77 (8): 2297–2302.

Comisarow, M.B. and Marshall, A.G. (1974). Fourier transform ion cyclotron resonance spectroscopy. *Chem. Phys. Lett.* 26 (4): 489–490.

Dempster, A.J. (1918). A new method of positive ray analysis. *Phys. Rev.* 11 (4): 316–325.

Eckenrode, B.A. (2001). Environmental and forensic applications of field-portable GC-MS: an overview. *J. Am. Soc. Mass Spectrom.* 12 (6): 683–693.

Fenn, J.B., Mann, M., Meng, C.K. et al. (1989). Electrospray ionization for mass spectrometry of large biomolecules. *Science* 246 (4926): 64–71.

Ferguson, L.S., Wulfert, F., Wolstenholme, R. et al. (2012). Direct detection of peptides and small proteins in fingermarks and determination of sex by MALDI mass spectrometry profiling. *Analyst* 137 (20): 4686–4692.

Ferreiro-González, M., Barbero, G.F., Palma, M. et al. (2016). Determination of ignitable liquids in fire debris: direct analysis by electronic nose. *Sensors (Basel, Switzerland)* 16 (5): 695.

Flinders, B., Cuypers, E., Zeijlemaker, H. et al. (2015). Preparation of longitudinal sections of hair samples for the analysis of cocaine by MALDI-MS/MS and TOF-SIMS imaging. *Drug Test Anal.* 7 (10): 859–865.

Forbes, T.P. and Sisco, E. (2018). Recent advances in ambient mass spectrometry of trace explosives. *Analyst* 143 (9): 1948–1969.

Froede, R.C. (1976). The laboratory management of a medicolegal specimen. *Ann. Clin. Lab. Sci.* 6 (3): 251–255.

Green, R.P., Hollander, A.S., Thevis, M. et al. (2010). Detection of surreptitious administration of analog insulin to an 8-week-old infant. *Pediatrics* 125 (5): e1236–e1240.

Gresham, G.L., Groenewold, G.S., Bauer, W.F., and Ingram, J.C. (2000). Secondary ion mass spectrometric characterization of nail polishes and paint surfaces. *J. Forensic Sci.* 45 (2): 310–323.

Groeneveld, G., de Puit, M., Bleay, S. et al. (2015). Detection and mapping of illicit drugs and their metabolites in fingermarks by MALDI MS and compatibility with forensic techniques. *Sci. Rep.* 5: 11716.

Hardman, M. and Makarov, A.A. (2003). Interfacing the orbitrap mass analyzer to an electrospray ion source. *Anal. Chem.* 75 (7): 1699–1705.

Hemmersbach, P. (2008). History of mass spectrometry at the Olympic Games. *J. Mass Spectrom.* 43 (7): 839–853.

Hobbs, A.L. and Almirall, J.R. (2003). Trace elemental analysis of automotive paints by laser ablation-inductively coupled plasma-mass spectrometry (LA-ICP-MS). *Anal. Bioanal. Chem.* 376 (8): 1265–1271.

Hoffmann, W.D. and Jackson, G.P. (2015). Forensic mass spectrometry. *Annu. Rev. Anal. Chem. (Palo Alto, Calif)* 8: 419–440.

Huynh, C. and Halamek, J. (2016). Trends in fingerprint analysis. *Trac Trends Anal. Chem.* 82: 328–336.

Jennings, K.R. (1968). Collision-induced decompositions of aromatic molecular ions. *Int. J. Mass Spectrom. Ion Phys.* 1 (3): 227–235.

Karas, M. and Hillenkamp, F. (1988). Laser desorption ionization of proteins with molecular masses exceeding 10000 daltons. *Anal. Chem.* 60 (20): 2299–2301.

Karas, M., Gluckmann, M., and Schafer, J. (2000). Ionization in matrix-assisted laser desorption/ionization: singly charged molecular ions are the lucky survivors. *J. Mass Spectrom.* 35 (1): 1–12.

Kingdon, K.H. (1923). A method for the neutralization of electron space charge by positive ionization at very low gas pressures. *Phys. Rev.* 21: 408.

Lawrence, E.O. and Edlefsen, N.E. (1939). On the production of high speed protons. *Science* 72: 376–377.

Li, X., Shen, B., Jiang, Z. et al. (2013). Rapid screening of drugs of abuse in human urine by high-performance liquid chromatography coupled with high resolution and high mass accuracy hybrid linear ion trap-Orbitrap mass spectrometry. *J. Chromatogr. A* 1302: 95–104.

Makarov, A. (2000). Electrostatic axially harmonic orbital trapping: a high-performance technique of mass analysis. *Anal. Chem.* 72 (6): 1156–1162.

Marks, V. (2009). Murder by insulin: suspected, purported and proven-a review. *Drug Test Anal.* 1 (4): 162–176.

Matthews, B., Walker, G.S., Kobus, H. et al. (2011). The analysis of dyes in ball point pen inks on single paper fibres using laser desorption ionization time of flight mass spectrometry (LDI-TOFMS). *Forensic Sci. Int.* 209 (1–3): e26–e30.

McLafferty, F.W. and Turecek, F. (1993). *Interpretation of Mass Spectra*. University Science Books.

Mollerup, C.B., Dalsgaard, P.W., Mardal, M., and Linnet, K. (2016). Targeted and non-targeted drug screening in whole blood by UHPLC-TOF-MS with data-independent acquisition. *Drug Test Anal.* 9 (7): 1052–1061.

Moore, C. and Crouch, D. (2013). Oral fluid for the detection of drugs of abuse using immunoassay and LC-MS/MS. *Bioanalysis* 5 (12): 1555–1569.

Morris, H.R., Paxton, T., Dell, A. et al. (1996). High sensitivity collisionally-activated decomposition tandem mass spectrometry on a novel quadrupole/orthogonal-acceleration time-of-flight mass spectrometer. *Rapid Commun. Mass Spectrom.* 10 (8): 889–896.

Munson, M.S.B. and Field, F.H. (1966). Chemical ionization mass spectrometry I: general introduction. *J. Am. Chem. Soc.* 88: 2621–2630.

Musenga, A. and Cowan, D.A. (2013). Use of ultra-high pressure liquid chromatography coupled to high resolution mass spectrometry for fast screening in high throughput doping control. *J. Chromatogr. A* 1288: 82–95.

Nicol, J.D. (1959). Police science technical abstracts and notes. *J. Crim. Law Criminol.* 40: 109–112.

Ojanpera, I., Kolmonen, M., and Pelander, A. (2012). Current use of high-resolution mass spectrometry in drug screening relevant to clinical and forensic toxicology and doping control. *Anal. Bioanal. Chem.* 403 (5): 1203–1220.

Papson, K., Stachura, S., Boralsky, L., and Allison, J. (2008). Identification of colorants in pigmented pen inks by laser desorption mass spectrometry. *J. Forensic Sci.* 53 (1): 100–106.

Parkin, M.C. and Brailsford, A.D. (2009). Retrospective drug detection in cases of drug-facilitated sexual assault: challenges and perspectives for the forensic toxicologist. *Bioanalysis* 1 (5): 1001–1013.

Pasin, D., Cawley, A., Bidny, S., and Fu, S.L. (2017). Current applications of high-resolution mass spectrometry for the analysis of new psychoactive substances: a critical review. *Anal. Bioanal. Chem.* 409 (25): 5821–5836.

Paul, W. and Steinwedel, H. (1953). A new mass spectrometer without a magnetic field. *Z. Naturforsch.* 8a: 448–450.

Paul, W., Reinhard, H.P., and Zahn, O. (1958). The electric mass filter as mass spectrometer and isotope separator. *Z. Phys.* 152: 143–182.

Pavlovich, M.J., Musselman, B., and Hall, A.B. (2018). Direct analysis in real time-mass spectrometry (DART-MS) in forensic and security applications. *Mass Spectrom. Rev.* 37 (2): 171–187.

Polettini, A., Gottardo, R., Pascali, J.P., and Tagliaro, F. (2008). Implementation and performance evaluation of a database of chemical formulas for the screening of pharmaco/toxicologically relevant compounds in biological samples using electrospray ionization-time-of-flight mass spectrometry. *Anal. Chem.* 80 (8): 3050–3057.

Rosano, T.G., Wood, M., Ihenetu, K., and Swift, T.A. (2013). Drug screening in medical examiner casework by high-resolution mass spectrometry (UPLC-MSE-TOF). *J. Anal. Toxicol.* 37 (8): 580–593.

Soltzberg, L.J., Hagar, A., Kridaratikorn, S. et al. (2007). MALDI-TOF mass spectrometric identification of dyes and pigments. *J. Am. Soc. Mass Spectrom.* 18 (11): 2001–2006.

Sommer, H., Thomas, H.A., and Hipple, J.A. (1951). Measurement of e/m by cyclotron resonance. *Phys. Rev.* 82: 697–702.

Stephens, W.E. (1946). A pulsed mass spectrometer with time dispersion. *Phys. Rev.* 69: 11–12: 691.

Takats, Z., Wiseman, J.M., Gologan, B., and Cooks, R.G. (2004). Mass spectrometry sampling under ambient conditions with desorption electrospray ionization. *Science* 306 (5695): 471–473.

Tanaka, K., Waki, H., Ido, Y. et al. (1988). Protein and polymer analyses up to m/z 100 000 by laser ionization time-of-flight mass spectrometry. *Rapid Commun. Mass Spectrom.* 2 (8): 151–153.

Thevis, M., Thomas, A., Delahaut, P. et al. (2006). Doping control analysis of intact rapid-acting insulin analogues in human urine by liquid chromatography-tandem mass spectrometry. *Anal. Chem.* 78 (6): 1897–1903.

Thevis, M., Thomas, A., Schanzer, W. et al. (2012). Measuring insulin in human vitreous humour using LC-MS/MS. *Drug Test Anal.* 4 (1): 53–56.

Thomson, J.J. (1897). Cathode rays. *Philos. Mag.* 44: 293–316.

Thomson, J.J. (1911). Rays of positive electricity. *Philos. Mag.* 21 (122): 225–249.

Tully, G. (2016). Codes of Practise and Conduct. Issue 3. Forensic Science Regulator.

Vincenti, M., Salomone, A., Gerace, E., and Pirro, V. (2013). Application of mass spectrometry to hair analysis for forensic toxicological investigations. *Mass Spectrom. Rev.* 32 (4): 312–332.

Vogliardi, S., Favretto, D., Frison, G. et al. (2009). A fast screening MALDI method for the detection of cocaine and its metabolites in hair. *J. Mass Spectrom.* 44 (1): 18–24.

Vuori, E., Pelander, A., Rasanen, I. et al. (2013). A rare case of serial killing by poisoning. *Drug Test Anal.* 5 (9–10): 725–729.

WADA (2015). TD2015IDCR. https://www.wada-ama.org/sites/default/files/resources/files/td2015idcr_-_eng.pdf (Accessed 29 August 2020).

WADA (2019). International Standard for Laboratories version 10.0. https://www.wada-ama.org/sites/default/files/resources/files/isl_nov2019.pdf (Accessed 29 August 2020).

Waters Corporation (2020). DESI – MS Imaging for Biomedical Research. https://www.waters.com/waters/en_US/DESI%3A-MS-Imaging-for-Biomedical-Research-/nav.htm?cid=134988839&alias=Alias_desi&locale=en_US (accessed 29 May 2020).

Whitehouse, C.M., Dreyer, R.N., Yamashita, M., and Fenn, J.B. (1985). Electrospray interface for liquid chromatographs and mass spectrometers. *Anal. Chem.* 57 (3): 675–679.

Wiley, W.C. and McLaren, I.H. (1955). Time-of-flight mass spectrometer with improved resolution. *Rev. Sci. Instrum.* 12 (26): 1150.

Wilm, M. and Mann, M. (1996). Analytical properties of the nanoelectrospray ion source. *Anal. Chem.* 68 (1): 1–8.

Wolstenholme, R., Bradshaw, R., Clench, M.R., and Francese, S. (2009). Study of latent fingermarks by matrix-assisted laser desorption/ionization mass spectrometry imaging of endogenous lipids. *Rapid Commun. Mass Spectrom.* 23 (19): 3031–3039.

Yost, R.A. and Enke, C.G. (1978). Selected ion fragmentation with a tandem quadrupole mass spectrometer. *J. Am. Chem. Soc.* 100 (7): 2274–2275.

Yost, R.A. and Enke, C.G. (1979). Triple quadrupole mass spectrometry for direct mixture analysis and structure elucidation. *Anal. Chem.* 51 (12): 1251–1264.

Zhao, Q., Liu, J., Wang, B. et al. (2017). Rapid screening of explosives in ambient environment by aerodynamic assisted thermo desorption mass spectrometry. *J. Mass Spectrom.* 52 (1): 1–6.

11

Isotope Ratio Mass Spectrometry

Sarah Benson and Kylie Jones

11.1 Forensic Introduction

Forensic scientists face an enduring challenge to keep pace with changes in the criminal environment to ensure relevance and effectiveness. New criminal methodologies and a level of sophistication enabled by technology and globalisation need to be considered when assessing new opportunities. Also, to be considered is the effectiveness of traditional forensic processes and techniques in supporting evolving law enforcement challenges including the prevention and disruption of crime.

Scientists collaborate with partners, including industry, to identify science and technology solutions that enhance support to current and emerging challenges. Analytical techniques are an example where continuous evolution is required. The discrimination power and limits of detection that were once regarded as sufficient are constantly improving and new expectations are set. Where identification and comparison of materials to a certain level of confidence was once adequate, comparison or identification of source is now more commonly being asked of forensic scientists. The level of confidence in attributing a sample to its origin, either through direct comparison or using a database, continues to evolve and show benefit across a range of contemporary investigative challenges.

Isotope ratio mass spectrometry (IRMS) is one technique that has demonstrated value in supporting questions regarding *source* and *origin* with a level of confidence not previously seen. The value of using IRMS in an analytical sequence is the potential to differentiate two or more samples that have been identified as having the same chemical composition using traditional analytical techniques (e.g. samples of trinitrotoluene [TNT] recovered from an unexploded device compared with a sample recovered from a suspected bomb maker's premises). Whilst IRMS can be regarded as being in its infancy in the field of forensic science, it has solid foundations in other scientific disciplines including geology and ecology. This chapter provides an overview of the IRMS technique, including theory of operation; applications in forensic science; considerations for operational implementation; intelligence value and court readiness; case studies; and future developments.

Analytical Techniques in Forensic Science, First Edition.
Edited by Rosalind Wolstenholme, Sue Jickells and Shari Forbes.
© 2021 John Wiley & Sons Ltd. Published 2021 by John Wiley & Sons Ltd.

11.2 Basis of the Technique

11.2.1 Isotopes

The nuclei of atoms are made up of protons and neutrons. The number of protons in an atom's nucleus is a unique characteristic of a chemical element and corresponds to the element's position in the periodic table of elements.

Whilst atoms usually have the same number of neutrons and protons in their nuclei, some atoms of the same element vary by one or more neutrons compared with the common form of the element. These atoms are called isotopes and are observed in almost all elements within the periodic table. The increase or decrease in the number of neutrons results in a difference in the atomic mass between an element and its isotopes (number of protons plus neutrons). This mass variation extends to molecules and compounds made up of these elements.

There are two forms of isotopes: radioactive isotopes – those that have an unstable combination of protons and neutrons and undergo radioactive decay over time; and stable isotopes – which do not decay over time. The majority of isotopes are stable isotopes (Criss 1999). The International Union of Pure and Applied Chemistry (IUPAC) publish a valuable reference that contains all known isotopic species within the periodic table of elements (IUPAC 2013), with further technical information including abundance values available through the Commission on Isotopic Abundances and Atomic Weights (CIAAW 2015).

Elemental carbon is a good example to demonstrate stable isotope configurations. The common form of this element is carbon-12 with each atom containing six protons and six neutrons. However, the element can take two other isotopic forms: carbon-13 with one extra neutron, which is stable; and carbon-14 with two extra neutrons, which is radioactive (half-life of 5715 years). The isotopes are depicted by listing the number of protons plus neutrons (i.e. atomic mass) as a superscript to the left of the element's symbol. In the case of carbon, this is ^{12}C, ^{13}C or ^{14}C (Figure 11.1).

11.2.2 Isotopic Abundance and Delta Notation

For all elements that have stable isotopes, there are at least two forms: the light, common isotope, which has a relative abundance around 99%; and the heavy, rare isotope(s) (some

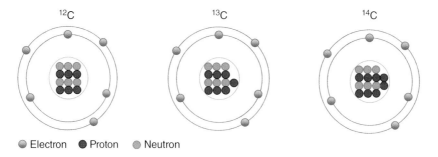

Figure 11.1 Elemental configurations of carbon, showing stable isotope (^{13}C) and radioactive isotope (^{14}C).

elements have more than one), which have relative abundances less than 1% of the total amount of the element. The most commonly measured light stable isotopes are the isotopes of hydrogen, carbon, nitrogen, oxygen and sulfur. The relative abundance of carbon isotopes for example are: 98.89% for the light isotope ^{12}C; 1.11% for the heavier ^{13}C; and less than one ten-billionth of 1% for ^{14}C (Criss 1999).

Due to the low relative abundance of most heavy isotopes, the percentage unit does not allow for a precise representation of the abundance when measured numerically. Thus, the natural abundance of stable isotopes is expressed as a ratio of the light to heavy isotopic forms (e.g. $^{13}C/^{12}C$). Given that this ratio is usually less than 0.01 for light elements, a delta notation is utilised (De Groot 2008).

Delta values are calculated using the following equation, which represents the relative difference between a standard material with known value and the measured ratio of an unknown sample.

$$\Delta^h E_{std} = (R_{sample}/R_{standard}) - 1 \quad (11.1)$$

where h refers to the heavy isotope, E refers to the element being measured, std refers to the international standard referenced, and R is the ratio of light to heavy isotope (Coplen 2011).

The values are expressed in parts per thousand or per mil (‰), which is a unitless number. Generally, delta values are quoted relative to a master or primary reference scale that is traceable through the use of international standards. In practical terms, a negative delta value represents depletion in the heavy isotope, whilst positive values are considered enriched in the heavy isotope relative to the standard. Using delta values as a means of expressing isotope ratio results enables more accurate interpretation with respect to the significance of very small differences.

Coplen (2011) outlines a range of description and calculation standards for working with isotopic results, recommended by the IUPAC Commission on the Isotopic Abundances and Atomic Weights. This paper is generally referred to as a key standard for all results published containing isotope ratio measurement results.

11.2.3 Standards and Reference Materials

The International Atomic Energy Agency (IAEA) and the National Institute of Standards and Technology (NIST) both supply a range of natural and non-natural isotopic standard materials. The international standards produced by these agencies have values that are traceable back to a master/primary scale or material (those that are set to 0‰). For comparability, all reported isotopic abundance values must be traceable back to these scales. Commonly referenced international standard scales and the stable isotopes they reference are shown in Table 11.1 (Werner and Brand 2001; De Groot 2004).

The international standards that directly scale against these primary reference materials are limited by the availability of the material in the environment and are cost prohibitive for routine use. Most laboratories choose to develop and use working standards or materials with consensus values determined through round robin trials (e.g. Forensic isotope ratio mass spectrometry [FIRMS] Network proficiency testing program samples), for day-to-day or research measurements. Where the same material is being measured in high frequency,

Table 11.1 International isotopic primary reference materials

Element	Standard name
Oxygen	Vienna Standard Mean Ocean Water (VSMOW)
Hydrogen	
Carbon	Vienna Pee Dee Belemnite (VPDB)
Nitrogen	Atmospheric nitrogen (Air)
Sulfur	Vienna Canyon Diablo Troilite (VCDT)

it is also an advantage for the precision of the results to matrix match both the reference materials and quality control standards measured within each analytical sequence.

Guidelines for selection and calibration of materials for use as laboratory standards are described by Kipphardt et al. (2000). Additionally, forensic laboratories that are accredited to international standards must ensure that any reference materials purchased or working standards calibrated meet international standard guidelines for the laboratory accreditation system in place (e.g. ISO/IEC 17025:2005 [2005]). Guidelines have been produced to assist in this process such as those set out in ISO 17034:2016.

11.2.4 Isotopic Variability – Fractionation and Mixing

Isotopic fractionation refers to any process that changes the relative abundances of the stable isotopes of an element. Fractionation can occur during chemical, physical or biological processes. Stable isotopes of various elements occur naturally in the atmosphere, earth, and organisms. The whole earth isotope ratios were determined at the time of the earth's formation. Although the overall abundance of light to heavy isotope in natural materials is fixed, the ratio of a material at any given time can vary, and in some cases (e.g. in the water cycle) is constantly shifting (Gat et al. 2001; Dawson and Siegwolf 2007).

Variability of the isotopic abundances of natural materials occurs due to fractionation and mixing effects resulting in the creation of specific isotope ratio values that are characteristic of the origin, purity and/or manufacturing processes of the products and their constituents. Although generally only the lighter elements are affected by isotopic fractionation, the increased precision of modern mass spectrometry (MS) instrumentation has enabled natural variation as a result of isotopic fractionation to be observed in a number of heavier elements (e.g. iron or strontium) (Vanhaecke and Heumann 2004; Tipple et al. 2013).

Fractionation occurs due to differences in the vibrational frequencies of the bonds between light and heavy isotope containing molecules. Heavier isotopes have a tighter bond and a lower vibrational frequency than lighter isotopes, creating a difference in the reaction rates between light and heavy isotopes, with heavier isotopes having less free energy available for reaction and hence requiring more energy to break a bond. In simpler terms, light isotopes react faster than heavy ones.

Selective enrichment and/or depletion of materials can occur through a variety of kinetic and equilibrium processes. The two main mechanisms by which isotopic fractionation occurs are: kinetic isotope effects – which are produced by differences in reaction rates; and thermodynamic isotope effects – which relate to the energy state of a system (Galimov 2012).

Kinetic isotope effects are those that take place in reactions that are irreversible. This causes permanent fractionation and a net depletion of the heavy isotope within the product of the reaction. Kinetic effects occur due to differences in the diffusion of molecules across a surface, diffusion of molecules through a substance and due to enzyme catalysed reactions. Other factors that can affect the levels of fractionation in these reactions include pressure, temperature, and the availability of reactants (Fry 2006).

Thermodynamic isotope effects are those that occur as a result of differences in the physicochemical properties of molecules such as infrared absorption, molar volume, and boiling and melting points. Thermodynamic effects are equilibrium exchanges – they are two-way reactions where the isotopic exchange between two different phases or product pools results in a net fractionation between the two pools. These effects are observed due to the mass differences of different isotopic species, with heavier isotopes being more tightly bonded and slower to react than lighter isotopes. For example, in the transformation of liquid water to vapour via evaporation, the reaction rate for $^1H^1H^{16}O$ would be faster than the deuteriated species $^1H^2H^{16}O$.

As discussed in greater detail in Sharp (2007), the fractionation factor of a particular reaction can be either theoretically or experimentally defined. Temperature can affect fractionation factors however partitioning of the heavier isotope into the side of the reaction with the highest bond stiffness generally occurs. Fry (2006) presents a series of calculations that can be performed to assess and quantify the fractionation factor for a particular reaction. Although not necessary for all materials, these types of calculations are useful in identifying and understanding the source and magnitude of fractionation in reactions, or for forensic work, in understanding the effects of manufacturing on the transformation of materials. An example of this would be an experiment and calculations to define whether fractionation exists along the manufacturing processes used to create polyethylene foils, to examine whether the starting materials (petroleum products) can be source attributed from the values of the manufactured output.

Conversely and also of relevance in forensic work is the mixing of isotopic sources to form different (and unique) bulk isotopic values. Mixed isotopic values, based on the proportion and individual isotopic values of the starting materials, are an important consideration for the measurement of bulk-manufactured materials that may be encountered in forensic casework. Work by the authors of this chapter into the manufacturing processes and mixing events in the production of office papers for example, has shown that common assumptions about bulk measured values (i.e. that document paper values are attributed to the cellulose used in paper) have been shown to be incorrect and that filler content (isotopic value and proportion) plays a significant role in the final bulk measured value (Jones et al. 2013b).

During interpretation of measured isotopic abundance values, fractionation of the starting materials and products during the following processes needs careful consideration when analysing and interpreting stable isotope ratio data. Examples from the literature that demonstrate sources of fractionation or imprecision of measurement include:

- Manufacturing, synthesis or growing conditions (Desage et al. 1991; Besacier and Chaudron-Thozet 1999; Meier-Augenstein 1999; Palhol et al. 2004; Collins et al. 2009).

- Sample collection, preparation, storage, handling (Dautraix et al. 1996; Meier-Augenstein 1999; Philp and Jardé 2007).
- Analysis (Brand 1996; Werner and Brand 2001; Schmitt et al. 2003).

11.2.5 Isotopic Variability of Natural Materials

Coplen and Shrestha (2016) present a valuable reference of the expected δ values for a range of materials. As an example, Figures 11.2 and 11.3 show the typical carbon and oxygen isotopic ranges measured for a series of naturally occurring materials. Also included in these figures are references to available standard materials and their published $\delta^{13}C_{VPDB}$ and $\delta^{18}O_{VSMOW}$ values.

The most fundamental natural cycle for isotopic distribution is the global hydrological cycle. This cycle encompasses how water is cycled between the oceans, air masses and ground or surface waters. The Global Meteoric Water Line is an equation defined by Craig (1961) that describes an average of the relationship between δ^2H and $\delta^{18}O$ in natural terrestrial waters. Figure 11.4 demonstrates some of the effects of a range of factors including altitude, latitude, and seasonality on the Meteoric Water Line (Professor John Gibson). It is this variability that provides the basis for the forensic comparison of water or materials where the isotopic abundance values have been contributed to from water such as hair, nails, or bone. References such as Gat et al. (2001), Dawson and Siegwolf (2007) and Sharp (2007) describe in greater detail the factors causing variation in the isotopic values of precipitation and land mass water.

Naturally occurring carbon dioxide cycling is primarily driven by the equilibrium exchange of dissolved inorganic carbon (bicarbonate, HCO_3^-) in oceans and CO_2 in the atmosphere. It is this cycling that provides the basis for carbon variability in broader systems including plant $\delta^{13}C$ values, which are derived from atmospheric uptake of CO_2 and fixation using photosynthesis. Typical inputs, fractionation and processes affecting the carbon cycle are shown in Figure 11.5 (Mackenson and Schmeidel 2019). Variability derived from these sources provides the basis for the forensic comparison of a number of plant-based materials including document papers, marijuana, and cocaine.

Active exchanges are also occurring between the atmosphere, terrestrial ecosystems and the ocean. In the last 50 years, anthropogenic inputs from terrestrial ecosystems have caused a net overall depletion in atmospheric carbon to −8‰ (Keeling et al. 1979; Keeling et al. 2001). The main cause of this depletion is the large increase in the concentration of CO_2 from burnt fossil fuels entering the atmosphere (Sharp 2007).

11.2.6 Instrumentation: Stable Isotope Ratio Mass Spectrometers

A number of analytical techniques can be used to measure stable isotopes, whereby selection is dependent on the desired application. Techniques continue to evolve and include nuclear magnetic resonance (NMR) with site-specific natural isotope fractionation (SNIF) and cavity ring down laser spectroscopy (CRDS). With the development of multi-collector inductively coupled plasma mass spectrometry (ICP-MS) instruments, heavy stable isotopic abundance measurements (including strontium and lead) are now beginning to emerge more frequently in forensic applications, particularly for human movement and

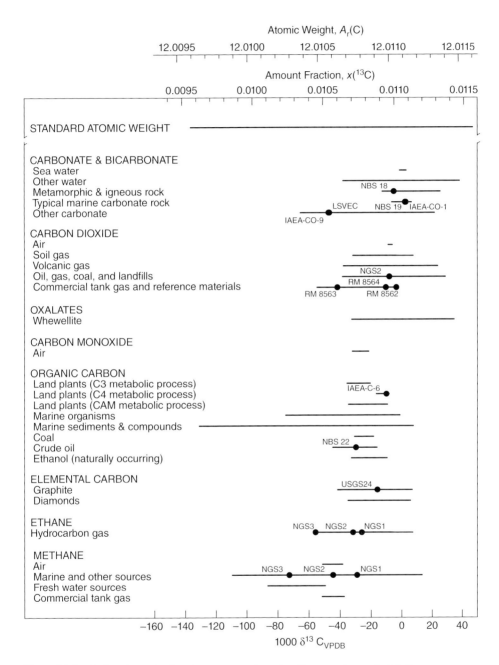

Figure 11.2 Carbon isotopic abundance ranges for naturally occurring materials Source: Reproduced with permission from Coplen and Shrestha (2016).

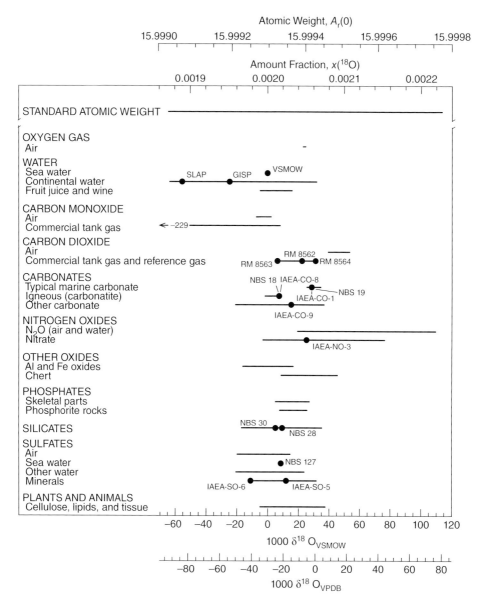

Figure 11.3 Oxygen isotopic abundance ranges for naturally occurring materials. Source: Reproduced with permission from Coplen and Shrestha (2016).

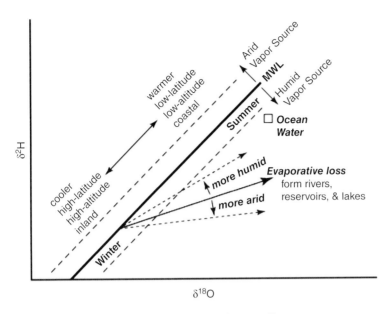

Figure 11.4 Effect of a range of factors on δ^2H and δ^{18}O values of water. Source: Reproduced with permission from Professor John Gibson.

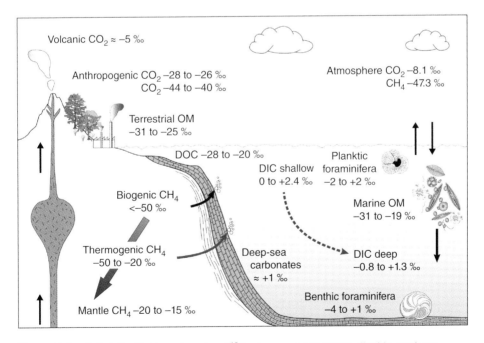

Figure 11.5 Global distribution of modern δ^{13}C value ranges in carbon dioxide, methane, dissolved inorganic carbon (DIC) and dissolved organic carbon (DOC), particulate organic matter (OM) and carbonates. Source: Reproduced with permission from Mackensen and Schmiedl (2019).

provenancing studies. Whilst each of these techniques adds value to the ultimate question, the focus of this chapter is on IRMS for the measurement of light stable isotopes.

11.3 Introduction to the Isotope Ratio Mass Spectrometer

The breakthrough in classical IRMS was the introduction of the dual inlet mass spectrometer by Urey in 1948. Since then, the instrument has been further developed and automated, and a range of peripherals developed that have streamlined sample preparation processes, leading to the systems commercially available today (Barrie et al. 1989).

Traditionally, isotopic studies were undertaken using a dual inlet IRMS, where the sample gas was loaded into a sample bellows and alternatively fed into the IRMS instrument with a standard (known) gas. High precision measurements are possible using dual inlet through continual calibration and fine-tuning of the signal by alternating between the sample and known gas. The main complication of these systems however was the need to combust solid samples and separate the combustion products into the molecules of interest offline, which requires a range of laboratory equipment including vacuum lines, cryogenic traps, and furnaces. This preparation work limits sample throughput and the accessibility of the instrument (De Groot 2004).

Continuous flow systems couple sample preparation instruments in line with the IRMS, providing fast and systematic measurement with minimal sample preparation required prior to introduction to the instrument. Depending on the sample or molecule to be measured, different combustion and separation techniques are used to convert a solid or liquid sample into a simple gas, which is then directly fed into the IRMS instrument.

The key difference between an IRMS instrument and a standard mass spectrometer is the presence of multiple detectors used to simultaneously collect, rather than scan, the mass weights of interest. Each of the detectors is set at a different position at the end of the MS flight tube, through which different path lengths are formed by the ionised molecules based on their molecular mass. Figure 11.6 is a schematic of an IRMS instrument.

11.3.1 IRMS – Detection and Measurement

To measure the isotopic ratios of a material, the gaseous samples first enter the ionisation chamber of the mass spectrometer. The sample impacts with a focused electron beam in a high vacuum environment, resulting in the loss of electrons from the molecules and producing positive ions. These ions are accelerated out of the chamber and through a flight tube between the poles of an electromagnet, where they are separated according to their mass-to-charge ratio (m/z). The ions are collected by an array generally consisting of three (sometimes up to seven) Faraday cup (FC) collectors.

The FCs are positioned so that the major ion currents simultaneously strike the middle of the entrance slit of the respective cups. Each incoming ion contributes one charge. No stray ions or electrons can enter the cup, and no secondary particles, formed from the impact with the inner walls of the cups, exit the cup. Set ion currents are continuously monitored, then amplified, digitised using a voltage-to-frequency converter (VFC), and finally transferred to a computer. The computer integrates the peak area for each isotopomer and calculates the

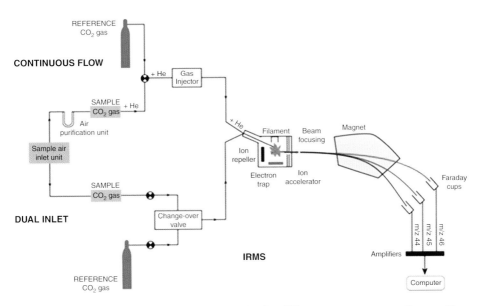

Figure 11.6 Schematic of an IRMS instrument showing different measurement options – either continuous flow or dual inlet. Source: Reproduced with permission from Voglar et al. (2019).

corresponding ratios. For example, when analysing CO_2, the data consists of three ion traces for the different isotopomers: $^{12}C^{16}O_2$, $^{13}C^{16}O_2$, and $^{12}C^{18}O^{16}O$, with their corresponding masses at m/z 44, 45, and 46, respectively (Brand 1996; Meier-Augenstein 2010).

The use of multiple detectors is key for IRMS work for two reasons – the first is that the precision of the detectors allows for the collection of very low abundance isotopes which, given the low relative abundance of the heavy isotopes, is vital in producing precise measurements. The second is the ability to collect all of the molecules from a given sample, to ensure that the ratio calculated accurately reflects the material being measured. This is important due to reaction rate differences between the light and heavy isotopes with the molecules containing heavy isotopes eluting later from the separation column than the lighter isotope.

11.3.2 Sample Preparation

Two types of measurements can be undertaken using IRMS: bulk measurements; and compound specific measurements. Each requires a different sample preparation instrument and interface to the IRMS.

11.3.3 Bulk Stable Isotope Analysis

Bulk stable isotopic measurements are achieved through elemental analysers that use either combustion or high temperature conversion (depending on the isotope to be measured) to convert the samples into the appropriate gas species measured by IRMS. Bulk samples are just that – the material is sampled, weighed and placed into a tin or silver capsule and measured as a bulk. This means that there is no sample transformation prior to measurement.

Table 11.2 Measured species of light stable isotope species and their molecular structure and mass

Element	Isotopes of interest	Gas species measured	Most common isotopologues of interest	Masses measured
Nitrogen	^{14}N, ^{15}N	N_2	$^{14}N^{14}N$, $^{14}N^{15}N$, $^{15}N^{15}N$	28, 29, 30
Carbon	^{12}C, ^{13}C	CO_2	$^{12}C^{16}O_2$, $^{13}C^{16}O_2$, $^{12}C^{18}O^{16}O$	44, 45, 46
Oxygen	^{16}O, $^{17}O^{a)}$, ^{18}O	CO	$^{12}C^{16}O$, $^{13}C^{16}O$, $^{12}C^{18}O$	28, 29, 30
Hydrogen	^{1}H, ^{2}H	H_2	$^{1}H^{1}H$, $^{1}H^{2}H$	2, 3

a) ^{17}O is of very low relative abundance (less than 0.0375) and is not the target isotope but may interfere with the measurement of both CO_2 and CO. This interference is, in general, automatically corrected by the instrument control software based on the methods proposed by Craig (1957).

There are some instances where sample cleaning or equilibration is required but in general, the material is still identifiable prior to being weighed. In some instances, sample weighing is not required. An example of this is using a 1 mm micropunch to cut accurate size pieces of polyethylene, which are placed into capsules for direct loading into the elemental analyser.

Table 11.2 details the actual gas species measured for each light stable isotopic element, including the isotopologues (isotopic species of the same element) measured and the molecular masses of interest for those species. While other isotopologues may be possible and contribute to measurements at these molecular masses, their abundances are very low (less than 0.1%) and hence have not been included here.

11.3.4 Bulk Measurements by Quantitative High Temperature Combustion

Elemental analysers were adapted for IRMS sample preparation in the early 1980s. The first systems were manual, where each sample was individually dropped into the sample combustion tube (Craig 1961). These prototype systems were very quickly followed by fully automated analysers which provided the basis for modern day instruments (Lloyd 1966).

High temperature combustion elemental analysers are used for the measurement of carbon, nitrogen and sulfur. For analysis, samples are weighed into tin capsules that are dropped into a combustion furnace heated between 900 °C and 1700 °C, though typically temperatures around 1000 °C are used. The combustion furnace is packed with oxidative materials such as chromium oxide on aluminium and silvered cobaltous oxide for carbon and nitrogen, and reduced copper or tungsten oxide for sulfur measurements.

The combustion tube is continually purged with an inert helium carrier gas and an additional pulse of oxygen is introduced with the sample to increase the temperature of the reaction. This causes combustion of the sample at over 1800 °C (termed 'flash combustion'). The resulting mixture of gases is swept by the helium carrier stream through a reducing section containing a reduced copper packing to remove excess oxygen and to reduce nitrogen oxides to elemental N_2. A heated molecular sieve is used to separate N_2 and CO_2 before they enter the IRMS instrument (Benson 2009).

Carbon and nitrogen isotopes can be measured from the same sample however sulfur must be measured separately due to the difference in internal packing material required for combustion. The internal packing materials and configurations used for the measurement of carbon and/or nitrogen isotopes varies for different instruments and materials and the published literature can guide instrument optimisation for specific applications.

11.3.5 Bulk Measurements by Quantitative High Temperature Conversion

High temperature conversion elemental analysers (TC/EA) are used for the conversion of bulk samples into the simple gases H_2 and CO for measurement of hydrogen and oxygen isotopes. The sample is weighed into a silver capsule and dropped into a hot glassy carbon furnace held at 1450 °C. A molecular sieve is used to separate the reaction gases before they enter the open split interface. As for flash combustion, the internal packing materials and conditions used in the measurement of hydrogen and oxygen isotopes can be varied for different materials; however, there are a lower number of configurations possible than for carbon and nitrogen. Liquid autosamplers are also able to be used to directly inject water samples into the TC/EA reactor. Published literature can assist and guide instrument optimisation (Gehre and Strauch 2003).

An example of a typical output for a bulk measurement is included in Figure 11.7. For each sample, a set of reference gas peaks (square peaks) are run to bracket the eluting sample peak. This provides a within sample scale and reference point to further increase the precision of the measurement.

11.3.6 Compound Specific Isotope Analysis

In compound specific isotope analysis (CSIA), the isotopic compositions of target molecules within a mixed sample are measured. The most common method of separating a sample online is by gas chromatography though other forms of chromatography such as liquid chromatography are becoming more prevalent. The chromatographic methods separate a bulk sample into its constituents and, during peak windows of interest, a switch diverts the

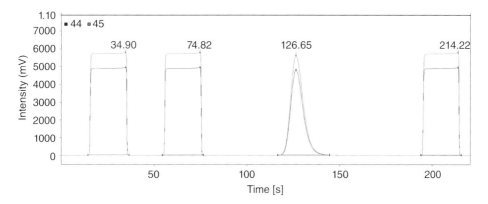

Figure 11.7 Typical output for a sample measured using IRMS. The square peaks are reference gas peaks (in this case, reference CO_2), which are used to scale the output of the sample peak (rounded peak). Intensity differences between the two detectors are observed with the lower trace for each peak for mass 44, and the higher for mass 45. A ratio is calculated for the peak area difference of the two traces.

separated sample into a combustion tube heated between 700 °C and 800 °C to convert the sample into a simple gas. The combusted sample is directed through an open split to the mass spectrometer. This method is most useful for volatile substances such as plant fatty acids, petroleum, food and perfume authentication and drug studies (Lichtfouse 2000).

11.4 Interpretation

The interpretation of IRMS results presents a number of challenges that are unique to this analytical technique.

There are fundamentally two questions that IRMS analysis seeks to support. These questions are:

1. Do the two (or more) samples share a common origin?
2. What is the specific origin of a sample(s)?

The primary question at hand will guide the user as to which resources are required to assist the interpretation process. Ideally, a forensic laboratory may hold a specific material background database, which is updated by the creation of a case database that matches the specific circumstances of the case more closely. These two types of databases can be defined as:

- Material databases – a population database of a specific material of interest (e.g. the explosive TNT) produced over time, by different manufacturers/sources, which can reasonably be expected to represent a particular market (e.g. TNT produced in Australia). This may also include studies of the inter- and intra-source variability (e.g. within brand homogeneity). Material databases are used to inform the variability of the material and can be used to infer source, if representative samples are collected.
- Case databases – a contemporaneous database of a material of interest created specifically for the purpose of the case at hand. This database contains representative samples of the material from case relevant sources (e.g. cling film collected from the Sydney area). Case databases are used to answer specific questions raised within an examination. They are typically single use and contemporaneous, but the values can be added to a material database to inform the broader picture of variability.

Due to the extensive resources required to establish and maintain material specific databases and the broad range of materials encountered in forensic science, case-based databases are often more feasible to construct than material databases. Using a case database however requires a sound understanding of the sampling considerations, knowledge of the manufacturing process and/or the origin of the materials to undertake the interpretation. Where this foundational piece is missing, this may limit the strength of the conclusions able to be given. Furthermore, this approach is only viable when the investigative question requires a comparison of two or more samples. Where the question relates to identification of the origin of a sample, a contemporaneous and well-maintained material database is required to not only identify but also to eliminate all potential sources.

Isoscapes, or isotopic landscapes, are visual representations of the isotopic variations observed within environmental cycles. The most predominant use of isoscapes to date has been in mapping the variations of δ^2H and δ^{18}O in the hydrological cycle (Bowen 2010, Bowen et al. 2014) introduce the concepts and statistical processing used to produce isoscapes. Forensically, isoscapes have been produced for the purposes of food authentication, to source drug crop growing locations, and to assist in human provenancing studies (Chesson et al. 2010; Hurley et al. 2010; Podlesak et al. 2012). Although very useful in comparing measured data, the isoscapes produced are still prediction models and some level of caution needs to be used when attempting an interpretation utilising these databases (Bowen 2010).

Central to interpretation are issues surrounding isotopic fractionation and its effect on the measured values of the samples of interest. When differences are observed between two samples, the analyst must consider whether these have occurred due to:

- Isotopic fractionation or mixing due to sample handling or contamination.
- Analysis effects, particularly if the samples being compared have been measured in different analytical sequences. Calculation of expanded measurement uncertainties is key in ensuring that these effects are taken into account during comparison.
- Differences in the constituents of the samples due to different growing or environmental conditions (fundamental source differences).
- Differences induced due to manufacturing or synthesis (batch differences).

A difference as a result of constituent, manufacturing or synthesis process may indicate that the two samples of interest have originated from different sources. If the isotopic compositions of two samples are found to be indistinguishable, and isotopic fractionation can be excluded from occurring during handling, analysis, etc. (usually achieved through laboratory quality assurance [QA] processes), then it may be concluded that the two samples may have originated from the same source. A material database can assist in strengthening this opinion, should the values be unique or of low abundance within the market from which it originated. Interpretation and opinion weighting can be assisted through the use of likelihood ratios, an example of which is included in Farmer et al. (2009) in relation to white architectural paints.

11.5 Case Studies

The use of stable isotope ratio analyses in forensic science can still be considered an emerging technology, particularly within traditional police forensic science laboratories. The principles and methods are generally accepted within the broader scientific community; however, as the technique is not routinely used, it may not yet be considered generally accepted amongst forensic scientists. The premature presentation of results as evidence in a court of law can challenge: the credibility of the technique; the scientist; the organisation the results have been produced for; and, more broadly, the field of forensic science. The use and presentation of results that include isotopic abundance measurements must be done within a QA system.

Isotope ratio results have only been presented in court on a handful of occasions, some of which are discussed by Ehleringer and Matheson (2010). This publication specifically reports on two cases (*Mejdrech vs. Lockfromer Co.* in the US in 2003 and *Ullises Shipping*

Corporation vs. Fal Shipping Co. in the UK in 2006) where the courts found that the isotopic evidence was not reliable due to shortcomings in the specific applications and interpretation of the science, rather than the reliability of the general principles and methods of stable isotope analysis. In addition to the general commentary around court admissibility, reliability and acceptance of forensic evidence, Ehleringer and Matheson (2010) provide an invaluable summary of the theory and application specifically related to the judicial admissibility of stable isotope abundance measurements. In Australian courts to date (2019), IRMS evidence has been reported in support of a number of cases and verbally presented in evidence by the authors in two cases. Though low, the number of cases reported and presented will continue to rise as the application of the technique continues to gain traction.

Ehleringer and Matheson (2010) discuss a number of other examples, detailing court judgements of IRMS and the varying applications of the technique including:

- The comparison and linkage of explosives seized from terrorist shoe bombers Richard Reid in 2001 and Saajid Badat in 2003. Mr Badat pleaded guilty before his trial began.
- The analysis of urine samples to determine whether natural or synthetic testosterone was detected in samples collected from world-class cyclist Floyd Landis during the 2006 Tour de France. The Court of Arbitration for Sport relied on stable isotope evidence to conclude that Mr Landis should be banned from cycling for two years.
- The comparison of questioned anthrax samples with known laboratory sources, mailed in 2001 to various targets after the 11 September 2001 attacks. The suspected perpetrator committed suicide before the case was presented to a court.
- The comparison of a number of hydrogen peroxide explosive samples relating to the 7 July 2005 London bombings. In the case of *R vs. Ibrahim (Muktar)*, an isotope expert was called to rebut the lead defendant's claim that he diluted the hydrogen peroxide with London tap water. The expert testifying said this was impossible based on his comparison of the isotopic profiles of the bomb residue and samples of London tap water. The Approved Judgement to the appeal by one of the defendants in Manfo Kwako Asiedu and The Queen (hearing dates: 9 and 10 February 2015) contains an overview of the application of IRMS in this matter and also the court's assessment and judgement of the related evidence (Calvert-Smith 2015). This case is a strong example of the care required when applying, reporting and presenting isotopic results for court as the expert in this case gave evidence beyond what the technique and results were reasonably able to provide.

In support of achieving acceptance of isotopic abundance comparisons in forensic science and the courtroom, the forensic and IRMS communities as well as the broader scientific community have roles to play. A range of challenges, which include practical as well as interpretative challenges, are faced by the forensic practitioners implementing these techniques. These challenges include: the development and maintenance of relevant, representative and contemporary databases; the availability of matrix matched and isotopically comparative reference materials, particularly for hydrogen measurements; the construction of interpretation and reporting frameworks; and science based knowledge gaps, e.g. consensus for a procedure to control hydrogen exchange within sample preparation methods.

The FIRMS Network is an international community of forensic practitioners who are involved in the measurement of stable isotopes to support forensic science enquiries. A number of the initiatives and goals of the Network are aimed at providing a solid foundation for IRMS results to be presented in court and include: inter-laboratory collaborative trials and proficiency tests to measure performance; establishing agreed principles and methodology; and providing guidance as to the use of isotopic abundance values for court purposes. The broader scientific community plays a key role in undertaking inter-disciplinary collaborative research to advance knowledge and applications in forensic science and participation in peer review and publications. Each of these aspects assists with demonstrating relevance, building expertise, and demonstrating reliability and thus general acceptance.

At the organisation and laboratory level, a fit-for-purpose validation should be completed for all methods, procedures and techniques utilised to produce forensic results (National Association of Testing Authorities 2009). In the absence of this, the reliability may be challenged in a judicial process, not only adversely affecting the outcomes of the trial at hand through a miscarriage of justice but also the future acceptance as expert evidence. This is somewhat easier, or more able to be achieved, in laboratories with existing accreditation schemes. For practitioners working outside of accredited facilities, the FIRMS Network conducts assessments of individual performance under an 'approved practitioner scheme' to assure the laboratories' customers that appropriate quality control procedures are in place (FIRMS Network 2017).

11.6 Applications in Forensic Science

The use of IRMS presents value in increasing the strength of the opinion forensic scientists can attach to conclusions relating to the common origin of materials. The technique has also demonstrated application to support contemporary challenges in determining the geographic origin and movement of people and materials. Like all applications within forensic science, the technique should be fit-for-purpose and applied at an appropriate stage of an analytical sequence to assist with determining the question at hand. Applications will continue to evolve with changing focus points for law enforcement and advances in technology and knowledge, thus ongoing research is vital.

The three key areas where IRMS has the potential to provide greatest effect in forensic science can be summarised as:

1. Determining whether two or more chemically identical samples have a common origin (common source).
2. Provision of information relating to the origin of a sample through database comparison (origin, authenticity, production method).
3. Determining a likely geographic origin of an unknown sample by geospatial mapping of predicted data (source attribution).

The number and variety of studies published utilising isotopic abundance comparisons for forensic applications continues to increase. In particular, significant bodies of research on the comparison of illicit drugs, explosives and human provenancing continue to be a focus. Comprehensive review articles by Benson et al. (2006), Carter et al. (2005), Daeid et al. (2010), Gentile et al. (2015), and Cerling et al. (2016) explore the work conducted in support of a broad range of forensic evidence types. A selection of publications is utilised in this section to provide an overview of the diverse applications in the field of forensic science.

11.6.1 Distinguishing between Naturally Occurring and Synthetic Materials in Doping, e.g. Endogenous and Exogenous (Synthetic) Testosterone

Piper et al. (2011) report that steroids are the most frequently detected class of doping agents and the detection of the misuse of endogenously occurring steroids (such as testosterone) is the most challenging but can be assisted through the use of isotopes. The authors present a comprehensive review of different methods for the analysis of a number of steroids, validation issues and sample preparation methods and parameters.

Brooker et al. (2014) present a study where carbon isotope ratios of seized illicit testosterone were compared against the known range for endogenous steroids to ensure doping control procedures could confirm misuse of testosterone in sport.

11.6.2 Determining Authenticity and Predicting Geographical Origin of Food, Pharmaceuticals and Other Materials, e.g. Counterfeiting

Bizjak Bat et al. (2012) present the application of IRMS in determining authenticity and geographical origin of food samples, specifically fruit juices. This application is significant due to the large global consumption and, thus, economic importance of correct labelling and maintenance of reputation and quality. The authors describe the stable isotope analysis and also multi-elemental analysis of freshly squeezed apple juices from Slovenia, providing a predictive ability through a database of values to nominate the region of origin.

Rees et al. (2016) demonstrate that stable isotopes can be utilised to indicate an animal's geographic origin. The results demonstrate that the global isotopic variability of stable isotopes in drinking water and feed are transferred into animal tissue; in this case, poultry. The application is used to verify origin labels for legislation and regulatory compliance and also to address public concern regarding rearing prior to slaughter. The authors demonstrate the value of combined stable isotope and elemental data to enhance the reliability to indicate the geographical origin.

Chiocchini et al. (2016) discuss the authentication and verification of the geographical origin of Italian extra virgin olive oils (EVOOs). The authors identified EVOOs from four distinct macro areas through the development of oxygen and carbon isoscapes to estimate the spatial distribution of stable isotope compositions.

11.6.3 Tracing the Geographic Origin and Movement of Wildlife, Persons and Materials

Lehn et al. (2015) present the use of light stable isotope analysis (H, C, N, S) of body tissue material including teeth, bone, hair and nail, to determine the provenance of unidentified persons. The authors highlight the value each tissue sample provides in getting different information spanning the entire life of a person. The application is important where traditional identification methods fail and information about the geographical location of a person at different times of their life is of assistance. The authors discuss the use of worldwide reference hair collections representing the dietary (food and water) heterogeneity among the regions reflected in the database. The authors also discuss the correction factors used in comparing different sample matrices, i.e. hair keratin and bone collagen. The study successfully demonstrates the application through case studies involving two adults and a newborn.

Font et al. (2015) demonstrate the successful application of isotope analysis of bone and teeth (Sr and Pb isotopes and ^{18}O, ^{13}C, ^{15}N) in providing regions of most probable origin for an unidentified male who had drowned in the Netherlands. A positive identification was later made by DNA with an individual from south-west Poland, which was within the area proposed by the study. The authors discuss the importance of the integration of all isotopic data in informing interpretation, including C and N values in bone collagen, which provided information about the dietary habits of an individual; O values in teeth, which reflected ingested water; and Sr and Pb isotopes, which reflect bioavailable Sr and Pb in the environment. Where increased spatial prediction is required, the authors outline that detailed studies of local drinking waters and geology would be required.

Ziegler et al. (2016) present a study on the analysis of light stable isotopes (^{13}C, ^{15}N, ^{18}O, ^{2}H, and ^{34}S) to predict the provenance of elephant ivory of unknown origin. The application is in support of law enforcement and wildlife forensics to prevent the illegal slaughter of elephants causing significant decline in some populations. Whilst some value is presented, the authors highlight that the within-site and within-individual variation in ivory is significant due to the diverse plant intake and movement patterns of elephants thus requiring an increase in sampling effort across different populations.

11.6.4 Identifying the Source of Environmental Contaminants

Philp and Jardé (2007) and Schmidt et al. (2004) provide reviews of the application of stable isotopes to environmental contamination challenges. Philp and Jardé (2007) discuss the application of isotopic measurements for the determination of the source of a contaminant in the environment and in evaluating the weathering pattern of the material to inform how long the contaminant may have been present. Schmidt et al. (2004) seek to answer the same problems, but focus on the use of compound specific IRMS to address the problem.

Li et al. (2009) demonstrate that stable carbon isotope profiles on n-alkanes can be useful tools for tracing the source of oil spills, particularly weathered oils. The significance of this application is increasing due to the increasing demand for and thus movement of crude oils.

Shin et al. (2013) report that the isotopic composition of methyl *tert*-butyl ether (MTBE), an additive in gasoline, in contaminated groundwater sites can be utilised to identify its production origin, even following weathering and loss by evaporation.

11.6.5 Determining the Geographical Origin of Plant Materials, e.g. Natural Illicit Drugs – Cannabis, Cocaine, and Heroin

Muccio and Jackson (2011) discuss the analysis of six cocaine samples using single quadrupole MS and IRMS for identification and classification. The authors demonstrate the value of carbon isotope ratios for the discrimination of samples which otherwise demonstrated identical chemical and physical properties. The authors highlight the value of combining more than one isotope ratio to provide additional information on the geographical location of crops being grown.

Tipple et al. (2016) studied variations in paraffin wax compositions in *Cannabis sp.*, grown indoors and outdoors across the US. In the US this is relevant, as increased regulation has led to additional incentives to certify the setting of the growth environment. The authors studied distributions, concentrations and carbon isotope ratios of n-alkanes isolated from *Cannabis sp. inflorescences* to assess if isotopic variations were related to known growth conditions. When used on unknown samples, the model identified the correct growth environment 90% of the time. The authors report that lipid markers can be used to trace cultivation methods, with further research required to understand the mechanisms.

Casale et al. (2006) report on the stable isotope analyses of heroin seized by Australian authorities from the merchant vessel *Pong Su*. The origin and the synthesis process of the cocaine were determined to be unknown utilising traditional techniques. The samples were identified to be isotopically distinct from known origin/process classifications of Southwest Asian, Southeast Asian, South American and Mexican samples, thus supporting the theory that the samples originated from a new region or new illicit production process.

11.6.6 Characterising Microorganisms

Kreuzer-Martin and Jarman (2007) present the results of research on the characterisation of the production environment of microorganisms using a range of analytical signatures that included stable isotopes. The objective of this application is to be able to associate recovered microorganisms with both the culture media and water used for growth. A database of carbon, nitrogen, oxygen and hydrogen stable isotope ratios of 247 cultures of *Bacillus subtilis* strain 6051 was established. The authors demonstrate variability between samples and cultures produced under different conditions, correlation between growth medium and spores, and examples of the ability to exclude growth substrates for specific cultures. The different factors in the process that require consideration in interpretation are also discussed.

11.6.7 Determining Synthetic Pathways Used to Manufacture Illicit Drugs, e.g. Ecstasy and MDMA, Methamphetamine, and Amphetamine

Collins and Salouros (2015) review the use of stable isotope ratios for profiling methylamphetamine to determine synthetic routes and the identity of precursor chemicals. This application ultimately informs law enforcement decision-making regarding the disruption of drug trafficking by establishing links between seizures based on the stable isotope profile. The authors demonstrate the value of this technique particularly where conventional analytical profiling techniques are of little use due to the highly refined nature of the methylamphetamine. The profiles can also inform policy makers on the diversion of legitimate industrial chemicals for illegitimate use.

Buchanan et al. (2011) report on research to assess the ability to link sources of methlenedioxymethamphetamine (MDMA) by measuring organic impurities utilising gas chromatography–mass spectrometry (GC-MS) and stable isotopes using IRMS. Samples of MDMA/HCl were synthesised under controlled conditions, varying the synthetic route, reaction conditions, and batch of starting materials to evaluate the effect. The authors determined that for the data set, discriminant analysis using a combination of IRMS and GC-MS data provided the most accurate sample discrimination.

Collins et al. (2016) present the findings of a study on the stable isotope analysis of cathinone derivatives to enable discrimination between different seizures and assist in linking samples from the same seizure. This application is growing in significance as new substances emerge within the illicit drug market, and as seen in this case, resulting in fatalities through recreational use.

11.6.8 Distinguishing between Two or More Samples of a Material to Infer Source or a Common Origin

Dietz et al. (2012) studied the variation in carbon isotope ratios in polyvinyl chloride (PVC) tape backings. This application is an example of enhanced certainty that can be achieved utilising stable isotopes in complement to traditional forensic comparisons of physical evidence. Negligible variation was observed within single rolls; however, large variations between tape brands and different product types were reported by the authors. The discrimination power was reported to be enhanced when combined with traditional chemical and physical analyses, thus highlighting the importance of ensuring that IRMS is utilised at an appropriate stage in an analytical sequence. The study also demonstrated that the carbon isotope values were mostly preserved in the tapes after an explosion.

Jones et al. (2013a, 2016) presented the results of the construction of a database of 125 document paper samples, collected from within Australia and New Zealand. To provide a basis for interpretation, the study also included an examination of inter- and intra-ream variability of document papers, identifying a risk of false negatives from the higher than expected natural variability within a single ream of paper. Using a pairwise comparison, the use of stable isotope abundances was shown to be effective in discriminating 68% of the samples using carbon and 82% of samples using oxygen.

Carter et al. (2014) discuss how likelihood ratios can be utilised to report the strength of results from a combination of physical, chemical and isotopic measurements. The authors presented work based on the analysis of duct tapes. The importance of an appropriate number of background samples was emphasised to increase the confidence of the output of the presented models.

11.6.9 Distinguishing Between Two or More Samples of Ignitable Liquids and Chemicals

Schwartz et al. (2013) present the results of a study analysing components of four flammable household chemicals that may be utilised in cases of arson. Discrimination was achieved utilising compound specific carbon isotope ratios between different sources of neat ignitable liquids; however, post-combustion comparison with the neat liquid was not possible due to fractionation during combustion. The authors emphasised that the potential false positive identification rate for two ignitable liquids is not known in the absence of a comprehensive database and thus should be utilised at present for exclusionary purposes only.

Kreuzer et al. (2012) demonstrated the value of stable C and N isotope ratios for associating common sources of commercially available sodium and potassium cyanide. The illegal use of these and other such chemicals as poisons demands enhanced analytical capabilities to determine their origin. The authors emphasise the importance of proper handling and storage of samples to ensure fractionation does not occur on account of these processes.

11.6.10 Determining Source Through Association of Starting Materials and End Products, e.g. Explosives

Stable isotope ratio analysis can provide value in determining common origin for homemade explosives, but also commercial and military explosives. Lock et al. (2012) present the results of a study whereby the homemade explosive hexamethyl triperoxide diamine (HMTD) was prepared using different sources of the precursor hexamethylenetetramine (hexamine). The HMTD was synthesised under both laboratory-controlled conditions and also less controlled conditions in the field. The authors discuss the impact of the reaction efficiency and variability in synthesis conditions on the fractionation of carbon and nitrogen isotope ratios, and thus the ability to correlate precursor with end product.

Howa et al. (2014) present a study comparing the reactants used in the commercial manufacture of hexogen (commonly referred to as RDX) and octogen (commonly referred to as HMX) explosives with the end products. Comparisons were also made between RDX and HMX and also between and within samples. Two RDX processes were compared. The different factors affecting the ability to correlate these samples are discussed, specifically relating to the application of associating a material between an event and a seized cache.

Benson (2009) and Benson et al. (2009a, 2009b, 2010) present a series of papers on the forensic analysis of explosives. Ammonium nitrate, pentaerythritol tetranitrate (PETN) and triacetone triperoxide (TATP) were analysed, with a focus on method validation and interpretation with respect to known manufacturing processes and source. These preliminary studies demonstrate the successful application of IRMS to the analysis of explosives of forensic interest to assist in discriminating samples from different sources and attribute samples to manufacturing source.

11.7 Future of IRMS and Stable Isotopic Comparisons

The application of IRMS to analytical challenges in forensic science continues to evolve. Targeted and collaborative research to enhance knowledge and understanding of materials and processes and to support the sound interpretation of results is essential. Collaboration to build, maintain and share databases is critical in ensuring effective casework implementation across the diverse set of applications in law enforcement and related fields. Partnerships between academia, industry, and law enforcement to inform the evolution of technology are key to ensure emerging requirements can be supported through new technology. Likewise, once developed, communication and education throughout the scientific and judicial communities is fundamental to achieving the full implementation of IRMS in forensic science.

References

Barrie, A., Davies, J., Park, A., and Workman, C. (1989). Continuous-flow stable isotope analysis for biologists. *Spectroscopy* 4 (7): 42–52.

Benson, S. (2009). Introduction of isotope ratio mass spectrometry (IRMS) for the forensic analysis of explosives. PhD thesis. University of Technology Sydney.

Benson, S., Lennard, C., Maynard, P., and Roux, C. (2006). Forensic applications of isotope ratio mass spectrometry – a review. *Forensic Science International* 157 (1): 1–22.

Benson, S., Lennard, C., Maynard, P. et al. (2009a). Forensic analysis of explosives using isotope ratio mass spectrometry (IRMS) – discrimination of ammonium nitrate sources. *Science & Justice* 49 (2): 73–80.

Benson, S., Lennard, C., Maynard, P. et al. (2009b). Forensic analysis of explosives using isotope ratio mass spectrometry (IRMS) – preliminary study on TATP and PETN. *Science & Justice* 49 (2): 81–86.

Benson, S.J., Lennard, C.J., Hill, D.M. et al. (2010). Forensic analysis of explosives using isotope ratio mass spectrometry (IRMS) – part 1: instrument validation of the DELTAplusXP IRMS for bulk nitrogen isotope ratio measurements. *Journal of Forensic Sciences* 55 (1): 193–204.

Besacier, F. and Chaudron-Thozet, H. (1999). Chemical profiling of illicit heroin samples. *Forensic Science Review* 11 (2): 105–120.

Bizjak Bat, K., Vidrih, R., Nečemer, M. et al. (2012). Characterization of Slovenian apples with respect to their botanical and geographical origin and agricultural production practice. *Food Technology and Biotechnology* 50 (1): 107–116.

Bowen, G.J. (2010). Isoscapes: spatial pattern in isotopic biogeochemistry. *Annual Review of Earth and Planetary Sciences* 38: 161–187.

Bowen, G.J., Liu, Z., Vander Zanden, H.B. et al. (2014). Geographic assignment with stable isotopes in IsoMAP. *Methods in Ecology and Evolution* 5 (3): 201–206.

Brand, W.A. (1996). High precision isotope ratio monitoring techniques in mass spectrometry. *Journal of Mass Spectrometry* 31 (3): 225–235.

Brooker, L., Cawley, A., Drury, J. et al. (2014). Stable carbon isotope ratio profiling of illicit testosterone preparations – domestic and international seizures. *Drug Testing and Analysis* 6 (10): 996–1001.

Buchanan, H.A., Kerr, W.J., Meier-Augenstein, W., and Daéid, N.N. (2011). Organic impurities, stable isotopes, or both: a comparison of instrumental and pattern recognition techniques for the profiling of 3,4-methylenedioxymethamphetamine. *Analytical Methods* 3 (10): 2279–2288.

Calvert-Smith, M. J. (2015). Approved Judgement: Manfo Kwako Asiedu and The Queen. www.judiciary.gov.uk/wp-content/uploads/2015/04/r-v-asiedu-judgment.pdf (accessed 29 May 2020).

Carter, J., Sleeman, R., Hill, J. et al. (2005). Isotope ratio mass spectrometry as a tool for forensic investigation (examples from recent studies). *Science & Justice* 45 (3): 141–149.

Carter, J.F., Doyle, S., Phasumane, B.-L., and NicDaeid, N. (2014). The role of isotope ratio mass spectrometry as a tool for the comparison of physical evidence. *Science & Justice* 54 (5): 327–334.

Casale, J., Casale, E., Collins, M. et al. (2006). Stable isotope analyses of heroin seized from the merchant vessel Pong Su. *Journal of Forensic Sciences* 51 (3): 603.

Cerling, T.E., Barnette, J.E., Bowen, G.J. et al. (2016). Forensic stable isotope biogeochemistry. *Annual Review of Earth and Planetary Sciences* 44: 175.

Chesson, L., Podlesak, D., Erkkila, B. et al. (2010). Isotopic consequences of consumer food choice: hydrogen and oxygen stable isotope ratios in foods from fast food restaurants versus supermarkets. *Food Chemistry* 119 (3): 1250–1256.

Chiocchini, F., Portarena, S., Ciolfi, M. et al. (2016). Isoscapes of carbon and oxygen stable isotope compositions in tracing authenticity and geographical origin of Italian extra-virgin olive oils. *Food Chemistry* 202: 291–301.

CIAAW (2015). Isotopic Abundances. http://www.ciaaw.org/isotopic-abundances.htm (accessed 27 October 2015).

Collins, M. and Salouros, H. (2015). A review of some recent studies on the stable isotope profiling of methylamphetamine: is it a useful adjunct to conventional chemical profiling? *Science & Justice* 55 (1): 2–9.

Collins, M., Cawley, A.T., Heagney, A.C. et al. (2009). δ13C, δ15N and δ2H isotope ratio mass spectrometry of ephedrine and pseudoephedrine: application to methylamphetamine profiling. *Rapid Communications in Mass Spectrometry* 23 (13): 2003–2010.

Collins, M., Doddridge, A., and Salouros, H. (2016). Cathinones: isotopic profiling as an aid to linking seizures. *Drug Testing and Analysis* 8 (9): 903–909.

Coplen, T.B. (2011). Guidelines and recommended terms for expression of stable-isotope ratio and gas-ratio measurement results. *Rapid Communications in Mass Spectrometry* 25 (17): 2538–2560.

Coplen, T.B. and Shrestha, Y. (2016). Isotope-abundance variations and atomic we flights of selected elements: 2016 (IUPAC technical report). *Pure and Applied Chemistry* 88 (12): 1203–1224.

Craig, H. (1957). Isotopic standards for carbon and oxygen and correction factors for mass-spectrometric analysis of carbon dioxide. *Geochimica et Cosmochimica Acta* 12 (1): 133–149.

Craig, H. (1961). Isotopic variations in meteoric waters. *Science* 133 (3465): 1702–1703.

Criss, R. (1999). *Principles of Stable Isotope Distribution*. New York: Oxford University Press.

Daeid, N.N., Buchanan, H.A.S., Savage, K.A. et al. (2010). Recent advances in the application of stable isotope ratio analysis in forensic chemistry. *Australian Journal of Chemistry* 63 (1): 3–7.

Dautraix, S., Guilluy, R., Chaudron-Thozet, H. et al. (1996). 13C isotopic analysis of an acetaminophen and diacetylmorphine mixture. *Journal of Chromatography A* 756 (1–2): 203–210.

Dawson, T. and Siegwolf, R. (2007). *Stable Isotopes as Indicators of Ecological Change*. Academic Press.

De Groot, P. (2004). *Handbook of Stable Isotope Analytical Techniques*, vol. 1. Elsevier Science.

De Groot, P. (2008). *Handbook of Stable Isotope Analytical Techniques*, vol. 2. Elsevier Science.

Desage, M., Guilluy, R., and Brazier, H. (1991). Gas chromatography with mass spectrometry or isotope-ratio mass spectrometry in studying the geographical origin of heroin. *Analytica Chimica Acta* 247 (2): 249–254.

Dietz, M.E., Stern, L.A., Mehltretter, A.H. et al. (2012). Forensic utility of carbon isotope ratio variations in PVC tape backings. *Science & Justice* 52 (1): 25–32.

Ehleringer, J.R. and Matheson, S.M. Jr., (2010). Stable isotopes and courts. *Utah Law Review* 2010 (2).

Farmer, N., Meier-Augenstein, W., and Lucy, D. (2009). Stable isotope analysis of white paints and likelihood ratios. *Science & Justice* 49 (2): 114–119.

FIRMS Network (2017). FIRMS Approved Forensic Practitioners. http://www.forensic-isotopes.org/fafp.html (accessed 5 May 2017).

Font, L., van der Peijl, G., van Leuwen, C. et al. (2015). Identification of the geographical place of origin of an unidentified individual by multi-isotope analysis. *Science & Justice* 55 (1): 34–42.

Fry, B. (2006). *Stable Isotope Ecology*. Springer Verlag.

Galimov, E. (2012). *The Biological Fractionation of Isotopes*. Elsevier.

Gat, J., Mook, W., and Meijer, H. (2001). *Environmental Isotopes in the Hydrological Cycle: Principles and Applications*, vol. II. UNESCO/IAEA.

Gehre, M. and Strauch, G. (2003). High-temperature elemental analysis and pyrolysis techniques for stable isotope analysis. *Rapid Communications in Mass Spectrometry* 17 (13): 1497–1503.

Gentile, N., Siegwolf, R.T., Esseiva, P. et al. (2015). Isotope ratio mass spectrometry as a tool for source inference in forensic science: a critical review. *Forensic Science International* 251: 139–158.

Howa, J.D., Lott, M.J., Chesson, L.A., and Ehleringer, J.R. (2014). Carbon and nitrogen isotope ratios of factory-produced RDX and HMX. *Forensic Science International* 240: 80–87.

Hurley, J., West, J., and Ehleringer, J. (2010). Stable isotope models to predict geographic origin and cultivation conditions of marijuana. *Science & Justice: Journal of the Forensic Science Society* 50 (2): 86.

ISO/IEC 17025:2005 (2005). General requirements for the competence of testing and claibration laboratories. http://www.iso.org/iso/iso_catalogue/catalogue_tc/caatalogue_detail.htm?csnumber=39883 (accessed 29 May 2020).

IUPAC (2013). IUPAC Periodic Table of the Isotopes. http://www.ciaaw.org/pubs/Periodic_Table_Isotopes.pdf (accessed 29 May 2020).

Jones, K., Benson, S., and Roux, C. (2013a). The forensic analysis of office paper using carbon isotope ratio mass spectrometry. Part 1: understanding the background population and homogeneity of paper for the comparison and discrimination of samples. *Forensic Science International* 231 (1): 354–363.

Jones, K., Benson, S., and Roux, C. (2013b). The forensic analysis of office paper using carbon isotope ratio mass spectrometry. Part 3: characterizing the source materials and the effect of production and usage on the $\delta 13C$ values of paper. *Forensic Science International* 233 (1–3): 355–364.

Jones, K., Benson, S., and Roux, C. (2016). The forensic analysis of office paper using oxygen isotope ratio mass spectrometry. Part 1: understanding the background population and homogeneity of paper for the comparison and discrimination of samples. *Forensic Science International* 262: 97–107.

Keeling, C.D., MOOK, W.G., and Tans, P.P. (1979). Recent trends in the 13C/12C ratio of atmospheric carbon dioxide. *Nature* 277 (5692): 121–123.

Keeling, C. D., S. C. Piper, R. B. Bacastow, et al. (2001). Exchanges of Atmospheric CO_2 and $^{13}CO_2$ with the Terrestrial Biosphere and Oceans from 1978 to 2000. I. Global Aspects. Scripps Institution of Oceanography.

Kipphardt, H., Valkiers, S., Taylor, P.D.P., and De Biévre, P. (2000). "Calibration" in isotopic measurements. *International Journal of Mass Spectrometry* 198 (1–2): 71–76.

Kreuzer, H.W., Horita, J., Moran, J.J. et al. (2012). Stable carbon and nitrogen isotope ratios of sodium and potassium cyanide as a forensic signature. *Journal of Forensic Sciences* 57 (1): 75–79.

Kreuzer-Martin, H.W. and Jarman, K.H. (2007). Stable isotope ratios and forensic analysis of microorganisms. *Applied and Environmental Microbiology* 73 (12): 3896.

Lehn, C., Rossmann, A., and Graw, M. (2015). Provenancing of unidentified corpses by stable isotope techniques–presentation of case studies. *Science & Justice* 55 (1): 72–88.

Li, Y., Xiong, Y., Yang, W. et al. (2009). Compound-specific stable carbon isotopic composition of petroleum hydrocarbons as a tool for tracing the source of oil spills. *Marine Pollution Bulletin* 58 (1): 114–117.

Lichtfouse, E. (2000). Compound-specific isotope analysis. Application to archaelogy, biomedical sciences, biosynthesis, environment, extraterrestrial chemistry, food science, forensic science, humic substances, microbiology, organic geochemistry, soil science and sport. *Rapid Communications in Mass Spectrometry* 14 (15): 1337–1344.

Lloyd, R. (1966). Oxygen isotope enrichment of sea water by evaporation. *Geochimica et Cosmochimica Acta* 30 (8): 801–814.

Lock, C.M., Brust, H., van Breukelen, M. et al. (2012). Investigation of isotopic linkages between precursor materials and the improvised high explosive product hexamethylene triperoxide diamine. *Analytical Chemistry* 84 (11): 4984–4992.

Mackensen, A. and Schmiedl, G. (2019). Stable carbon isotopes in paleoceanography: atmosphere, oceans, and sediments. *Earth-Science Reviews* 197: 1–23.

Meier-Augenstein, W. (1999). Applied gas chromatography coupled to isotope ratio mass spectrometry. *Journal of Chromatography A* 842 (1–2): 351–371.

Meier-Augenstein, W. (2010). *Stable Isotope Forensics: An Introduction to the Forensic Application of Stable Isotope Analysis*. Wiley.

Muccio, Z. and Jackson, G.P. (2011). Simultaneous identification and $\delta 13C$ classification of drugs using GC with concurrent single quadrupole and isotope ratio mass spectrometers. *Journal of Forensic Sciences* 56 (s1).

National Association of Testing Authorities (2009). Technical Note 17 – Guidelines for the Validation and Verification of Chemical Test Methods. http://www.nata.asn.au/phocadownload/publications/Guidance_information/tech-notes-information-papers/technical_note_17.pdf (accessed 29 March 2013).

Palhol, F., Lamoureux, C., Chabrillat, M., and Naulet, N. (2004). 15N/14N isotopic ratio and statistical analysis: an efficient way of linking seized ecstasy tablets. *Analytica Chimica Acta* 510 (1): 1–8.

Philp, R.P. and Jardé, E. (2007). Application of stable isotopes and radioisotopes in environmental forensics. In: *Introduction to Environmental Forensics* (eds. B.L. Murphy and R.D. Morrison), 455–512. OUP.

Piper, T., Emery, C., and Saugy, M. (2011). Recent developments in the use of isotope ratio mass spectrometry in sports drug testing. *Analytical and Bioanalytical Chemistry* 401 (2): 433–447.

Podlesak, D.W., Bowen, G., O'Grady, S.P. et al. (2012). $\delta 2H$ and $\delta 18O$ of human body water: a GIS model to distinguish residents from non-residents in the contiguous USA. *Isotopes in Environmental and Health Studies* 48 (2): 259–279.

Rees, G., Kelly, S.D., Cairns, P. et al. (2016). Verifying the geographical origin of poultry: the application of stable isotope and trace element (SITE) analysis. *Food Control* 67: 144–154.

Schmidt, T., Zwank, L., Elsner, M. et al. (2004). Compound-specific stable isotope analysis of organic contaminants in natural environments: a critical review of the state of the art, prospects, and future challenges. *Analytical and Bioanalytical Chemistry* 378 (2): 283–300.

Schmitt, J., Glaser, B., and Zech, W. (2003). Amount-dependent isotopic fractionation during compound-specific isotope analysis. *Rapid Communications in Mass Spectrometry* 17 (9): 970–977.

Schwartz, Z., An, Y., Konstantynova, K.I., and Jackson, G.P. (2013). Analysis of household ignitable liquids and their post-combustion weathered residues using compound-specific gas chromatography-combustion-isotope ratio mass spectrometry. *Forensic Science International* 233 (1–3): 365–373.

Sharp, Z. (2007). *Principles of Stable Isotope Geochemistry*. Upper Saddle River, NJ: Pearson Education.

Shin, W.-J., Lee, S.-W., Heo, S.-Y., and Lee, K.-S. (2013). Stable isotopic fingerprinting for identification of the methyl Tert-butyl ether (MTBE) manufacturer. *Environmental Forensics* 14 (1): 36–41.

Tipple, B.J., Chau, T., Chesson, L.A. et al. (2013). Isolation of strontium pools and isotope ratios in modern human hair. *Analytica Chimica Acta* 798: 64–73.

Tipple, B.J., Hambach, B., Barnette, J.E. et al. (2016). The influences of cultivation setting on inflorescence lipid distributions, concentrations, and carbon isotope ratios of cannabis sp. *Forensic Science International* 262: 233–241.

Vanhaecke, F. and Heumann, K. (2004). Precise isotope ratio measurements as a unique tool in modern analytical chemistry. *Analytical and Bioanalytical Chemistry* 378 (2): 227–228.

Voglar, G.E., Zavadlav, S., Levanič, T., and Ferlan, M. (2019). Measuring techniques for concentration and stable isotopologues of CO2 in a terrestrial ecosystem: a review. *Earth-Science Reviews* 199: 102978.

Werner, R. and Brand, W. (2001). Referencing strategies and techniques in stable isotope ratio analysis. *Rapid Communications in Mass Spectrometry* 15 (7): 501–519.

Ziegler, S., Merker, S., Streit, B. et al. (2016). Towards understanding isotope variability in elephant ivory to establish isotopic profiling and source-area determination. *Biological Conservation* 197: 154–163.

Part III

Chromatographic Techniques

12

Chromatographic Separation and Theory

Sue Jickells and Shari Forbes

12.1 Introduction

Chromatography is a powerful separation technique that has largely replaced the more classical separation methods of distillation and precipitation. The technique was first detailed in the early 1900s by Mikhail Tswett, who successfully separated the plant pigments in chlorophyll by passing solvent extracts of leaves through a glass column packed with finely divided particles of various sorbents including calcium carbonate, aluminium hydroxide, sugar and bone charcoal (Berezkin 1989). The resulting coloured bands led to the naming of the technique *chromatography*.

Chromatography as we recognise it today encompasses a series of techniques that can separate compounds in complex mixtures rapidly and, depending on the instrumentation used, simultaneously identify and quantify the separated compounds. In this section we will look at what causes separation to take place and how we can exploit this knowledge when using chromatographic techniques in chemical analysis.

Chromatographic separation is not a random process; various theories have been developed to identify the factors that influence separation. The most important theories will be presented here, with emphasis on how knowledge of these can be exploited to improve the efficiency and effectiveness of separations. We will concentrate our discussions on gas chromatography (GC) and high performance liquid chromatography (HPLC) and, to a lesser extent thin layer chromatography (TLC), because these are the chromatographic techniques most widely used in forensic science, although we recognise that other chromatographic techniques exist which differ in terms of how separation is achieved and the associated terminology applied.

The terminology associated with chromatography can seem overly extensive and complex to those new to the field. A scientist concerned with sample analysis using well established methods (standard operating procedures, SOPs) will use relatively few of these terms on a daily basis. Nevertheless, every analyst using chromatographic techniques should have a good understanding of what is happening to bring about separation such that they recognise when something unusual is happening in a particular separation, what is causing the problem, and how best to deal with it. It is even more critical for analysts involved in method development to have an in-depth understanding of the chromatographic process and to be able to calculate some of the fundamental parameters involved. Some simple calculations can often avoid hours or days of trial and error when trying to establish and optimise a new method.

12.2 Chromatography

Chromatography is defined by the International Union of Pure and Applied Chemistry (IUPAC) as 'A physical method of separation in which the components to be separated are distributed between two phases, one of which is stationary (the *stationary phase*) while the other (the *mobile phase*) moves in a definite direction'. The mobile phase 'may be a liquid (liquid chromatography) or a gas (gas chromatography) or a supercritical fluid (supercritical-fluid chromatography). In gas chromatography, the expression "carrier gas" may be used for the mobile phase. In elution chromatography, the expression "eluent" is used for the mobile phase'. The stationary phase 'may be a solid, a gel or a liquid. If a liquid, it may be distributed on a solid. This solid may or may not contribute to the separation process. The liquid may also be chemically bonded to the solid (bonded phase) or immobilized onto it (immobilized phase)' (IUPAC 2009).

In the types of chromatography that are discussed here, the mobile phase moves in one direction through the chromatographic system and the stationary phase is either coated onto a planar surface (*planar chromatography*), or held in a column (*column chromatography*). It is the interaction between the sample and the stationary and mobile phases which brings about separation of components in the sample. In GC, the mobile phase plays less of a role in terms of interaction with the sample components and it is largely the interaction between the sample and the stationary phase which influences separation.

Three types of chromatography are generally recognised: *frontal*, *displacement*, and *elution*. The analytical techniques used routinely in forensic science operate via elution chromatography which is defined as 'a procedure for chromatographic separation in which the mobile phase is passed through the chromatographic bed after the application of the sample' (IUPAC 2009). By far the most common format used for elution chromatography is column chromatography (e.g. GC and HPLC) but planar chromatography (e.g. TLC) is also used in forensic science.

Other classifications used to describe chromatographic separations are adsorption and partition chromatography. These terms indicate how the sample components interact with the stationary phase. In *adsorption* chromatography the 'separation is based mainly on differences between the adsorption affinities of the sample components for the surface of an active solid' (IUPAC 2009). In *partition* chromatography, the 'separation is based mainly on differences between the solubility of the sample components in the stationary phase (gas chromatography), or on differences between the solubilities of the components in the mobile and stationary phases (liquid chromatography)' (IUPAC 2009). The term partition is associated with the use of liquid stationary phases.

12.2.1 Planar Chromatography

In planar chromatography, 'the stationary phase is present as, or on, a plane. The plane can be a paper, serving as such or impregnated by a substrate as the stationary bed (paper chromatography) or a layer of solid particles spread on a support, e.g. a glass plate (thin layer chromatography, TLC)' (IUPAC 2009). TLC is the most commonly employed form of planar chromatography in forensic science and the most common format is for the planar surface to be a square or a rectangle of thin aluminium sheet or, less commonly, of glass, with the stationary phase coated on one surface. This format is commonly referred to as a TLC plate. The sample mixture, in solution, is applied to the stationary phase close to one edge of the plate as a spot or band (Figure 12.1a) and then the plate is stood vertically, with the bottom edge in a shallow layer of the mobile phase (Figure 12.1b). The mobile phase moves upwards and through the stationary phase by capillary action carrying with it components of the sample which are soluble in the mobile phase (Figure 12.1c). If the mobile and stationary phases are chosen correctly, the components in the sample mixture separate from one another and can then be detected on the plate.

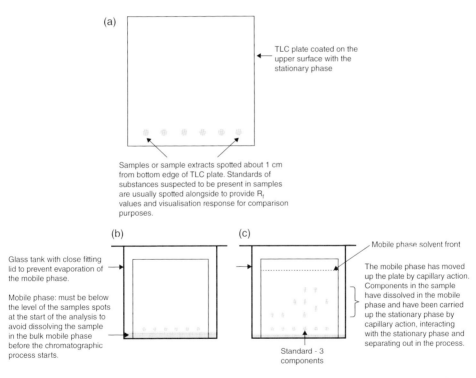

Figure 12.1 Thin layer chromatography (TLC). (a) TLC plate spotted with sample extracts and a standard ready for chromatographic analysis. (b) The TLC plate is placed in the mobile phase held in a tank with a close-fitting lid. (c) The developed TLC plate showing components which were present in the samples separated out. Note that some of the components present in the samples have the same R_f value (R_f = distance travelled by sample component from original position/distance travelled by the mobile phase front) as those in the standard analysed alongside, potentially indicating that the samples contain one or more of the components present in the standard.

12.2.2 Column Chromatography

HPLC and GC involve column chromatography which is defined by IUPAC (2009) as 'A separation technique in which the stationary bed is within a tube. The particles of the solid stationary phase or support coated with a liquid stationary phase may fill the whole inside volume of the tube (packed column) or be concentrated on or along the inside tube wall leaving an open, unrestricted path for the mobile phase in the middle part of the tube (open-tubular column)'.

In HPLC, the stationary phase is typically particulate, with particles packed in a short column made of a material which can withstand the high pressure required to force the mobile phase through the stationary phase. The particles fill the entire inside of the column, i.e. a *packed column* format. The sample mixture, in solution, is applied to one end of the column and is then carried though the column by the mobile phase which flows continuously from the inlet to the outlet end of the column. As the components of the sample travel through the column, they encounter the stationary phase where they interact through intermolecular forces. It is this interaction which brings about separation. In HPLC, monolithic stationary phases are also used (see Section 14.2.4.1). These phases are not particulate, but the same principles apply in terms of the mobile phase flowing constantly in one direction through the column and in intimate contact with the stationary phase.

In GC, the stationary phase is most commonly covalently bonded to the walls of the column such that the stationary phase forms a thin film on the inner walls of the tubular column and the centre of the column is open and unrestricted to the carrier gas (mobile phase) flow. This format is referred to as wall-coated open tubular (WCOT) (see Section 13.2.3) and the stationary phase is referred to as a liquid stationary phase because it behaves like a liquid even though it is covalently bonded to the column wall and cannot move. As is the case for HPLC, the sample is applied at one end of the column. It usually enters the column in the gas phase but, in some types of GC, may be introduced in solution. In GC, for separation to take place, the sample must transfer to the gas phase so that it travels along the column in the mobile phase (*carrier gas*), interacting with the stationary phase in the process. GC can be further subdivided according to the type of stationary phase employed, with the terms solid and liquid phase chromatography used to describe the use of solid and liquid stationary phases, respectively. In *packed column* GC, the stationary phase is particulate. Separations using packed column GC are relatively rare in forensic science, with open tubular column formats used for the majority of separations.

Various classifications used in chromatography are shown in Table 12.1.

12.3 The Separation Process

The separation of components in a sample during elution chromatography relies on the differential distribution of the components between the mobile and stationary phase, driven by physicochemical processes. The sample components, carried in the mobile phase, distribute from the mobile phase into the stationary phase and back again, this distribution process takes place many times along the length of the column or TLC plate. Where liquid stationary phases are used, components partition into the stationary phase. In the case of

Table 12.1 Terms associated with column chromatography.

Chromatographic technique	Mobile phase	Stationary phase	Classification	Type of interaction/ separation process
Gas chromatography	Gas	Liquid coated or covalently bonded to the inner walls of the tubular, usually narrow diameter, column	Gas–liquid chromatography	Partition (between gaseous MP and liquid SP)
	Gas	Solid. Usually in particulate form packed into a tubular column	Gas–solid chromatography	Adsorption (from gaseous MP onto active solid SP)
Liquid chromatography	Liquid	Liquid coated or covalently bonded to a solid support in particle form and packed into a tubular column	Liquid–liquid chromatography	Partitioning between (immiscible) liquid MP and liquid SP
	Liquid	Solid. Usually in particulate or monolith form inside a tubular column (HPLC) or on a planar surface (TLC)	Liquid–solid chromatography	Adsorption (from liquid MP onto active solid SP)
	Liquid	Liquid containing ionic groups coated or covalently bonded to a solid support in particulate or monolith form	Ion exchange chromatography	Ion exchange
	Liquid	Inert support, typically gel-like, polymeric, with pores	Size exclusion chromatography	Based on exclusion effects due to differences in molecular size and/or shape
	Liquid	Support with immobilised biological ligand	Affinity	
Supercritical fluid chromatography	Supercritical fluid	Organic compound adhering to solid particles		Partitioning between supercritical fluid and organic surface

MP, mobile phase; SP, stationary phase.

active, solid stationary phases, adsorption/desorption interactions take place between the sample components and active sites on the stationary phase. Differences in the distribution constants (or adsorption/desorption energies) of components between the mobile and stationary phases result in the separation of the components into discrete bands, also known as zones, as they pass through the chromatographic system. The higher the number of equilibrations taking place along a column, the greater the likelihood of separating substances with similar distribution constants.

Figure 12.2 illustrates a liquid chromatography (LC) separation of a sample mixture containing two substances (A and B) on a packed HPLC column containing a liquid stationary phase covalently bonded to a particulate, solid support. Sample component B has a larger distribution constant (K_c) than component A (i.e. B distributes more readily into the stationary phase than A). Whilst a component is distributed into the stationary phase, it is not

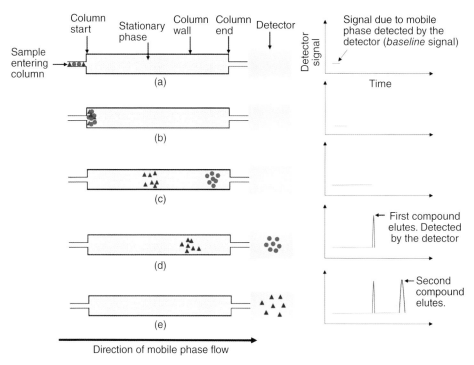

Figure 12.2 The process of chromatographic separation illustrated for a HPLC column packed with particulate stationary phase. (a–e) represent the time course of a separation of two substances. The sample is injected (a) with the mobile phase flowing in one direction. The detector is placed at the end of the column and detects components eluting from the column. When no compound is eluting, the signal output from the detector represents detection of the mobile phase only (a–c). This is referred to as a *baseline* signal. Whilst in the column, components in the sample interact with the stationary phase and, assuming they have different K_c values, separate out from one another as they pass along the column (b–e). As a compound elutes from the column, it enters the detector. Assuming the detector reacts to the presence of the compound, the detector signal increases as the number of sample molecules entering the detector increases and then decreases again to baseline level as the sample is swept out of the detector. This increase and decrease in signal produces a *peak* (d) on the *chromatogram* (plot of detector signal vs. time). As another substance elutes from the column and is detected, another peak is seen on the chromatogram (e).

carried along the column in the moving mobile phase. Because component B distributes more readily into the stationary phase, it spends more time in the stationary phase relative to component A and hence moves less distance along the column than A in a given time and elutes later than A.

In GC and HPLC, a detector is usually placed at the end of the column and, provided the detector reacts to the presence of the components of interest, it will detect them as they elute from the column. If the detector is operating all the time, as is usual for techniques such as GC and LC, a *chromatogram* will be obtained (Figure 12.2) which is a plot of detector signal (also referred to as detector response) (y-axis) versus time (x-axis). The flat portion of the chromatogram where no substances are eluting (or where substances may be eluting which do not react to the detector or are present at concentrations too low to be detected) is referred to as the *baseline*. When a compound elutes from the column and is detected by the detector, the chromatogram shows a rise from the baseline value to a maximum value (peak maximum) representing the point at which the maximum number of molecules of the compound are eluting and being detected. After this point, the number of molecules eluting decreases and hence the detector signal reduces and, unless another compound elutes, it will reduce back to the baseline value.

In the case of TLC, the separation process is similar, with components which distribute more readily into the stationary phase (larger K_c) travelling less distance along the TLC plate compared with components which distribute more readily into the mobile phase (smaller K_c). The detection process is somewhat different to GC and HPLC because a detector is not placed at the end of the plate and no chromatogram is obtained. Instead, detection is carried out after the chromatographic process is stopped (the TLC plate is removed from contact with the mobile phase) and then the position of compounds on the plate is detected in situ. This may be via visual observation for coloured compounds, by response to a light source (e.g. fluorescence) or by reaction with one or more reagents sprayed onto the plate to produce a coloured or fluorescent product in situ.

One of the primary mechanisms driving separation is molecular interaction between the sample components and the stationary phase and, in the case of LC, the mobile phase. Molecular interactions include van der Waals forces, dipole–dipole and dipole-induced dipole interactions, hydrogen bonding, and ionic interactions. Steric hindrance may also play a role in some separations and is particularly important in *size exclusion chromatography* (also referred to as gel permeation chromatography), where separation is mainly caused by the difference in hydrodynamic volume of the sample components, with intermolecular forces playing a lesser role than in partition and adsorption chromatography.

The mobile and stationary phases in chromatography are chosen to ensure that the sample components distribute themselves to varying degrees between the two phases such that the components of interest separate out from one another and from other components in the mixture. The skill of the analyst lies in being able to manipulate the chromatographic conditions to bring about this separation and to do this in a reasonable timeframe.

12.3.1 Distribution Constant

IUPAC defines the *distribution constant* (K_c) for chromatography as 'The concentration of a component in or on the stationary phase divided by the concentration of the component in the mobile phase' (IUPAC 2009). Hence K_c is denoted by:

$$K_c = \frac{c_S}{c_M} = \frac{n_S/V_S}{n_M/V_M} \tag{12.1}$$

where c_S is the molar concentration of the solute in the stationary phase and c_M is the molar concentration in the mobile phase. The values n_S and n_M refer to the number of moles of analyte in the two respective phases and V_S and V_M refer to the volumes of the two phases. Optimising the K_c value through the appropriate choice of mobile and stationary phases will enhance the distribution of components between phases and improve band separation.

12.3.2 Hold-Up Time (or Volume)

Compounds in a sample which do not distribute into the stationary phase at all and which spend all their time in the mobile phase will elute from the column rapidly (or in the case of TLC move with the solvent front). These compounds are said to be *unretained*. The time taken for an unretained compound to elute from the column is denoted by t_M, referred to as the *hold-up time* (Figure 12.3), and provides a measure of the migration rate of the mobile phase. Alternatively, this parameter can be expressed in volume terms denoted by

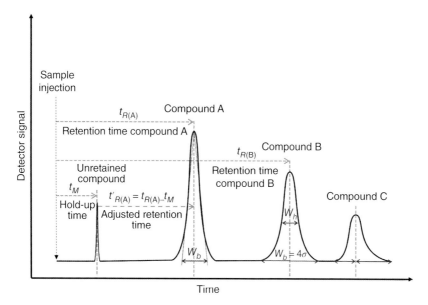

Figure 12.3 Sample chromatogram illustrating some important parameters in chromatography. The chromatogram shows three retained compounds detected (A, B, and C) and the corresponding retention times (t_R) for A and B. The first peak which elutes is non-retained, giving the hold-up time (t_M) for the system. The adjusted retention time (t'_R) is shown for compound A.
Compound A shows the peak width at baseline (w_b) calculated using the triangulation method. Compound B shows the width at half-height (w_h) and at baseline (w_b) calculated assuming the peak is Gaussian. Note that the peak for compound C is asymmetric in shape. Calculating N using the triangulation method or w_b or w_h for this peak would result in significant error. For a discussion of calculating N, see Section 12.4.2.

the *hold-up volume* or void volume (V_M), calculated as:

$$V_M = t_M F \tag{12.2}$$

where F is the flow rate of the mobile phase measured in ml/min.

12.3.3 Retention Time (or Volume)

A retained compound will spend a period of time associated with the stationary phase. When this compound elutes from the column it will do so at time t_R, referred to as the *retention time* (or total retention time), which is measured from the time of sample injection to the time of elution of the peak maximum for that compound (Figure 12.3). This retention parameter can also be expressed in terms of the volume of mobile phase required to elute a retained compound and is termed the compound's *retention volume* (V_R) (or total retention volume), calculated as:

$$V_R = t_R F \tag{12.3}$$

t_R (or V_R) is a constant value for a particular compound on a particular chromatographic system provided all parameters remain unchanged (i.e. same instrument, same column and stationary phase, same mobile phase composition and flow rate, same connecting tubing for injector and detector, same temperature conditions, etc.). This makes t_R (or V_R) exceedingly useful in terms of identifying a particular compound because if an analyst can purchase or otherwise obtain a *standard* of a compound suspected to be present in a sample, the standard can be dissolved in a suitable solvent, injected into the chromatographic system and its retention time determined under the specific set of chromatographic conditions. The term *standard* as used here refers to a substance of known identity and purity, usually obtained from a commercial source. If no compound elutes from a sample at the retention time of the standard substance, the compound corresponding to the standard is either not present in the sample, or is present but at a concentration below the limit of detection of the system. If a compound is detected in the sample with a similar, but not identical retention time to the standard compound, it cannot be considered to be the same substance. Even if the two compounds have identical mass spectra, unless the retention times match within very close agreement, the two substances cannot be considered to be identical if they have been analysed under identical conditions. Very high sample concentrations can sometimes affect retention times. If this is suspected to be the reason for a non-match in retention times, the sample should be diluted and reanalysed to see if this produces a match in retention time.

Because the values of t_R and V_R for an individual compound will vary between chromatographic systems, it can often be useful when comparing t_R or V_R to calculate the adjusted retention time, t'_R (Eq. (12.4)), or adjusted retention volume, V'_R (Eq. (12.5))

$$t'_R = t_R - t_M \tag{12.4}$$

$$V'_R = V_R - V_M \tag{12.5}$$

These values make it easier to compare separations carried out on different chromatographic systems because *extra column parameters* that influence t_R or V_R are excluded. Factors influencing t_R or V_R include the chemical structure of the sample components as well as the type of mobile and stationary phase used in the system, the flow rate of the mobile phase, the length of the chromatographic column, temperature, and extra column parameters. Extra column parameters include the internal volume of tubing between the point of sample introduction and the start of the column, the internal tubing volume between the end of the column and the detector, detector volume and any dead volume in tubing fittings.

Because t_R is dependent upon the particular combination of a chromatographic system, an analyst must not assume that the value for t_R will be the same if the analysis is carried out on a different instrument even if the same chromatographic column and mobile phase is used because extra-column factors can contribute to t_R. Standards of the components suspected to be present should always be analysed alongside samples to check for a match in t_R.

12.3.3.1 Retention Time and Sample Concentration

Equation (12.1) indicates that c_S should be directly proportional to c_M for a compound to ensure that the distribution constant (K_c) for that compound is constant over a broad range of solute concentrations. If this were not the case then t_R for a compound would differ depending on its solute concentration. This would not be acceptable for evidential purposes. In general, solute concentrations for most analyses in forensic science are low (trace analysis) and there is unlikely to be a problem. However, an analyst should always bear in mind the potential for problems to arise with t_R at high solute concentrations. Diluting the sample and reanalysing at a lower concentration should reveal if there is a problem: if sample concentration is influencing K_c the retention time, t_R, of the compound in the sample will change, whilst t_R for the standard will remain the same.

12.3.4 Retention Factor

The *retention factor* (k) of a compound is defined by IUPAC (2009) as 'a measure of the time the sample component resides in the stationary phase relative to the time it resides in the mobile phase; it expresses how much longer a sample component is retarded by the stationary phase than it would take to travel through the column with the velocity of the mobile phase. Mathematically, it is the ratio of the adjusted retention volume (V'_R) or time (t'_R) and the hold-up volume (V_M) or time (t_M)'.

$$k = \frac{t'_R}{t_M} \text{ or } \frac{V'_R}{V_M} \tag{12.6}$$

If a substance has $k < 1$, the substance will elute rapidly, close to the hold-up time for the column and, depending on the sample, with a high degree of probability that there will be co-elution of the substance with other poorly retained compounds. If k is >20, the elution time (and overall analysis time) will be long, leading to a delay in obtaining a result for the analysis and a higher than necessary cost. Ideally, k should be in the range of 1–10 for all sample components for isocratic HPLC separations to ensure efficient separation in a satisfactory time. For ultra high pressure/performance liquid chromatography (UHPLC) separations (see Section 14.2.4.3), some chromatographers recommend a minimum value for k of ≥ 3 for early eluting analytes. Calculating k is useful in method development because it can be used to contrast the migration rates of solutes in different columns and is independent of column geometry or mobile phase flow rates.

12.3.5 Separation Factor

The *separation factor* of a column (α) measures the degree of separation between two adjacent peaks. It represents the relative difference in retention of two solutes as they travel through the column and, by definition, is always greater than unity. The separation factor for two solutes A and B is defined as:

$$\alpha = \frac{K_B}{K_A} \quad \text{or} \quad = \frac{k_B}{k_A} \tag{12.7}$$

where K_A and k_A are the distribution constant and retention factor, respectively, for the more rapidly eluted sample component A; and K_B and k_B are the equivalent parameters for the more strongly retained sample component B. To calculate α from a chromatogram, the following equation is used.

$$\alpha = \frac{(t_R)_B - t_M}{(t_R)_A - t_M} = \frac{(t'_R)_B}{(t'_R)_A} \tag{12.8}$$

where $(t_R)_A$ and $(t_R)_B$ and $(t'_R)_A$ and $(t'_R)_B$ are the retention times and adjusted retention times, respectively, for components A and B (Figure 12.3).

12.4 Separation Theory

Since the advent of chromatography, researchers have tried to identify the processes which influence separation. The primary driver for this is to gain a better understanding of the important factors involved in the separation process so that separations and the associated equipment can be optimised in a more systematic way, rather than via experimental trial and error. To this end, various theories have been developed and modified. The primary theories which will be discussed are *Plate theory* and *Rate theory*. It should be recognised that other theories have been developed and other approaches used to investigate the chromatographic process, but space precludes a discussion of all of these. Plate theory is included because of its prominent place in the development of chromatography and because it was one of the earliest attempts to define a theory of chromatography. It also defined some important parameters which still have a prominent place in separation theory.

12.4.1 Plate Theory

Martin and Synge were pioneers in chromatography and developed a theory of chromatography, which is often referred to as *Plate theory*, although it was not called this in their pioneering 1941 paper describing partition chromatography. Their aim was to develop a theory that would 'give a picture of the concentration of solute at any time and place in the column, and of the way in which the resolution depends upon the length of the column' (Martin and Synge 1941). They based their theory on concepts developed to describe distillation separations, a field which had used the terminology *theoretical plates* to compare the efficiency of distillation columns. In their theory, Martin and Synge (1941) considered a chromatographic column to be composed of a series of theoretical plates, with a solute undergoing partition between two phases (what we now call mobile and stationary phases) within each theoretical plate. They introduced the term 'height equivalent to a theoretical plate' (HETP) and defined it as 'the thickness of the layer such that the solution issuing from it is in equilibrium with the mean concentration of solute in the non-mobile phase throughout the layer' i.e. a full equilibrium situation. In their theory, they considered that the HETP was constant throughout the column, i.e. each plate was the same height, with the exception of the column start and end. For the purposes of simplification, they made the assumption that the solute concentration and presence of other solutes did not impact on the distribution between the two phases. They showed that the larger the number of plates (N) a column has (low values of HETP), the more effectively it can separate compounds with similar partition coefficients. They also recognised that the flow rate of the mobile phase was important (HETP ∝ to flow rate), that a slow rate of flow led to diffusion of the solute band, and also that $HETP \propto \sqrt{d_p}$, where d_p is the diameter of the stationary phase particles. To ensure a low value for HETP, they recommended that the mobile phase flow rate should be fast and uniform and for the stationary phase to be composed of very small particles. They also recognised that the diffusion of the solute in the mobile phase was important and that lower values for HETP would be obtained with low molecular weight compounds.

12.4.2 Theory versus Practice: Band Broadening

In an ideal chromatographic system, separated bands would elute from the chromatographic column with a very narrow width and with all solutes having bands of equal width. However, as can be seen from Figure 12.4, this is not the case. This chromatogram shows the analysis of aerosols from electronic cigarettes analysed by thermal desorption gas chromatography–time of flight mass spectrometry (GC-TOF-MS) (Rawlinson et al. 2017). Peak widths increase with increasing retention time. This spreading out of solutes during the chromatographic separation process is called *band broadening*. This phenomenon is undesirable for several reasons, the main ones being that band broadening can result in incomplete separation of compounds, can lead to errors in quantification, potentially reduces the number of compounds that can be separated in a given time, and also affects the limit of detection of an analysis, i.e. the smallest measure of a substance that can be detected with reasonable certainty for a given analytical procedure (McNaught and Wilkinson 1997).

Theory predicts that peaks in chromatographic separation are Gaussian, i.e. there is a normal distribution of molecules around a central maximum (mean, μ) according to the following equation:

$$f(x) = \frac{1}{\sigma\sqrt{2\pi}} \exp\left(-\frac{(x-\mu)^2}{2\sigma^2}\right) \tag{12.9}$$

where σ is the standard deviation of the distribution around μ. For a normal distribution 100% of data lies within 4σ.

Because band broadening is undesirable, the propensity of a column to minimise broadening has been used as a way to express the *efficiency* of a column, with *plate number, N*, used to express column performance. The number of theoretical plates (N) for a column is calculated as:

$$N = (t_R/\sigma)^2 = (V_R/\sigma)^2 \tag{12.10}$$

and is determined from a chromatogram (Figure 12.3).

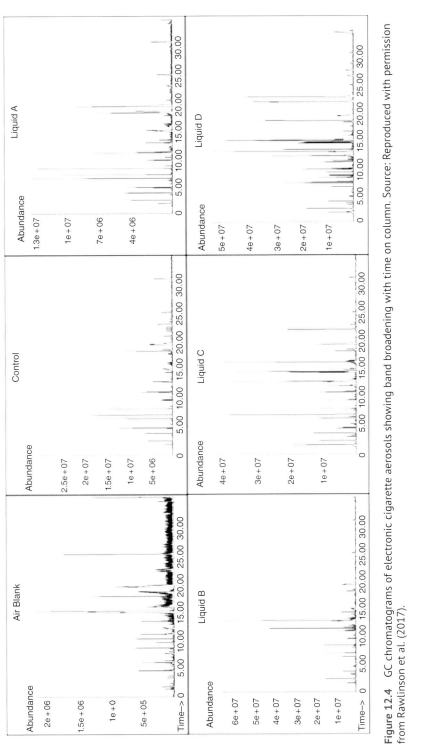

Figure 12.4 GC chromatograms of electronic cigarette aerosols showing band broadening with time on column. Source: Reproduced with permission from Rawlinson et al. (2017).

As can be seen from Eq. (12.10), N is directly related to the degree of the broadening of the chromatographic band as it passes through the system; reduced band broadening gives a greater value for N and vice versa.

There are several ways of measuring σ to determine N. One way is to measure the *peak width at base* (w_b) (Figure 12.3).

$$N = \left(\frac{4t_R}{w_b}\right)^2 = 16(t_R/w_b)^2 \qquad (12.11)$$

Caution must be observed to use the same units throughout, i.e. if t_R is expressed in minutes, peak width must be calculated in minutes.

In practice, measuring w_b is difficult and prone to considerable error because it is not always easy to determine at what point a peak 'starts' and 'ends'. This problem can be overcome by extending the sides of the peak to the point at which they intersect the baseline and measuring w_b between the intersection points (Figure 12.3). This procedure is sometimes referred to as the tangent or triangulation method.

An alternative approach is to measure the width of the peak at a point above the baseline to avoid the problems of identifying the peak start and end. The *peak width at half height*, w_h, is often used for this purpose. For a Gaussian peak, the width at 50% peak height = 2.3546σ. Hence:

$$N = 5.545\left(\frac{t_R}{w_h}\right)^2 \qquad (12.12)$$

The plate height (H), or HETP, is calculated as:

$$H = \frac{L}{N} = \frac{L}{16}\left(\frac{w_b}{V_R}\right)^2 = \frac{L}{16}\left(\frac{w_b}{Vt_R}\right)^2 \qquad (12.13)$$

where L is the column length.

A large value for N, and hence a small H, corresponds to good column efficiency, although a large value for N is not a guarantee that a particular column will be suitable for a particular analysis. If two compounds have exactly the same value for K_c (the distribution constant) on a column, they will not separate, irrespective of the number of theoretical plates, unless a parameter is changed which brings about a difference in K_c.

Peaks in chromatography are rarely exactly Gaussian and it is far more common for peaks to be asymmetric (Figure 12.3). Applying w_b, w_h or the tangent method to asymmetric peaks will give inaccurate values for N. Most chromatographers in their day-to-day work will not measure N or H but may wish to measure these parameters when first installing a new column because most columns are supplied with the value for N. Some laboratories determine the value when a column is first installed, measure it at regular intervals, and then replace the column when N drops below a predetermined value. Peak fit methods have been developed to determine efficiency and various software programs are commercially available. Gritti and Guiochon (2013) developed a numerical integration method which they claim offers improved accuracy for measuring efficiency.

It must be recognised that N and H are theoretical concepts and that a chromatography column cannot literally be divided into regions (plates) where a full equilibration takes place, as is assumed in the Plate theory of Martin and Synge (1941). In reality, full equilibration between the mobile and stationary phase never takes place in chromatographic separations because the thermodynamics of the distribution process mean that at least some molecules of a solute are always moving from the stationary phase into the mobile phase and vice versa. Nonetheless, the concept of N is useful in being able to evaluate and compare the efficiency of chromatographic columns and systems and an idealised situation of full equilibration is assumed.

12.4.3 Rate Theory

The Plate theory concept of Martin and Synge (1941) provides a useful measure of a column's efficiency and enables a comparison of the efficiency between columns. However, the theory is not particularly useful in terms of helping an analyst, or column and instrument manufacturers to identify the factors responsible for band broadening and how these parameters might be optimised to maximise the efficiency of separations. The concept of HETP was used by van Deemter et al. (1956) who proposed a Rate theory to explain the main parameters which influence band broadening. The Rate theory has proven exceedingly useful and has formed the basis for subsequent developments in chromatography theory and research to improve the efficiency of separations. The simplified form of the van Deemter equation is written as:

$$H = A + \frac{B}{u} + Cu \tag{12.14}$$

where H is the plate height and u is the linear velocity of the mobile phase across the average cross-section of the chromatographic column. A is a coefficient that describes *eddy diffusion* (also referred to as *eddy dispersion*) due to differences in flow within the column. According to the van Deemter equation, A is independent of the mobile phase velocity. B is the *longitudinal diffusion coefficient* and is inversely proportional to the mobile phase velocity (u). C is the *mass transfer coefficient* and is linearly proportional to the mobile phase velocity. The fundamental difference between the Plate and Rate theory is that the latter looks at 'the actual continuous column', whereas the former considers separation in terms of discrete plates.

There is considerable ongoing discussion in the chromatographic community about the applicability of the van Deemter equation to modern chromatographic separations and the parameters that it describes. Much of this discussion has been driven by the advent of new developments in stationary phases such as monolithic columns and superficially porous particles (also known as core shell™ or pellicular particles), and liquid chromatographic separations carried out at very high pressures. Although many authors consider the van Demeter equation as set out in Eq. (12.14) to be overly simplistic in terms of representing the factors influencing chromatographic separation, it still retains a central place in discussions of chromatography theory. The terms in the van Deemter equation are discussed in more detail below, including some more recently explored parameters which influence band broadening and shed more light on the parameters influencing separation. The van Deemter equation has been modified over the years to introduce additional terms but the simplified form is retained here. Interested readers are referred to the work of Giddings (1961a, 1961b, 1991), Golay (1958), Horvath and Lin (1976), Huber and Hulsman (1967),

Knox (1961, 1977), and Knox et al. (1978), amongst others, who were early pioneers in building on the work of van Deemter and colleagues.

12.4.3.1 Eddy Diffusion (*A*)

The *A* term is generally considered to apply to particulate and other solid stationary phases and hence is of greater importance in LC separations compared with GC carried out with capillary WCOT columns. A liquid flowing around particles travels different distances (Figure 12.5a) and at different velocities due to viscous drag reducing the mobile phase velocity close to particle surfaces (Giddings 2017). As the distance from a particle surface increases perpendicular to the direction of flow, velocity increases until it reaches a maximum (Figure 12.5a). Thus, solute molecules will travel along a column at different rates depending on whether they are carried close to the particle surface or in the bulk mobile phase flow, often referred to as slow or fast streams, respectively. Research has also shown that there is inhomogeneity in the packing of particulate stationary phase across the radius of a column. Some slurry packed columns have been shown to produce a lower density of packing in the centre. This results in higher mobile phase velocity at the centre compared with near the column wall (Gritti and Guiochon 2010a, 2010b). The

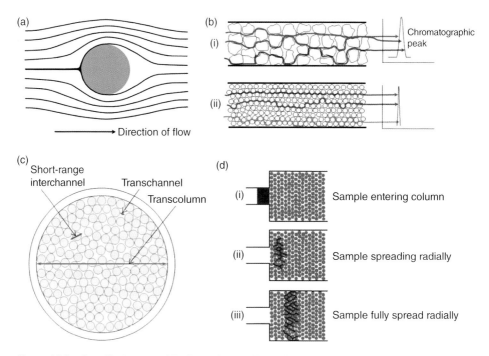

Figure 12.5 Contributions to eddy dispersion. (a) Flow of eluent around a particle. The flow rate is lowest at the particle surface. In addition, molecules in different streams travel different distances around particles. (b) Axial cross section of HPLC column and chromatogram showing different path lengths taken by solute molecules resulting in band broadening: (i) irregular shaped particulate phase; and (ii) smaller diameter, more regularly shaped phase producing less difference in path lengths and less band broadening. (c) Radial cross section of HPLC column indicating different distances over which dispersion arises. (d) Axial cross section of inlet end of HPLC column: (i)–(iii) show radial dispersion due to flow exchange as the sample travels along the column.

arrangement of particles at the column wall can also differ in comparison with other parts of the column (Gritti and Guiochon 2012a, 2012b); Gritti and Guiochon note that the average velocity between particles is about 10% higher over five particle diameters from the column wall compared with the centre. Added to this is the aspect that particles do not pack homogeneously in a column, even on a local scale plus, even for a given particle size, there will be some variation in size (particle size distribution). This means that some channels between particles are wider than others and also that the path length that some molecules take will be slightly longer or shorter than the path lengths taken by other molecules of the same compound (Figure 12.5b). Over the length of a column, these different path lengths and slight variations in mobile phase velocity lead to a dispersion of solute molecules from their original starting zone, resulting in band broadening by axial dispersion (Gritti and Guiochon 2012a, 2012b).

Giddings (1965) identified the following contributions to eddy dispersion: trans-channel (the distance between two particles), short-range inter-channel (across several particles), and trans-column (across the internal column radius) as illustrated in Figure 12.5c. Khirevich et al. (2010), in numerical simulation studies of packed beds, concluded that short-range inter-channel and trans-channel effects were the main contributors to eddy dispersion but their simulations did not take into account the influence of column walls.

Two major pathways have been identified as important in the dispersion of solutes across packed columns: (i) radial dispersion by flow exchange; and (ii) by diffusion driven by concentration gradients. At low flow velocities, diffusion dominates, whereas at high flow velocities flow exchange is the dominant dispersion mechanism (Gritti and Guiochon 2010b). In flow exchange, solute molecules are dispersed across the radius of the column by collisions with particles causing eluent streams carrying solutes to split and diverge (Figure 12.5d). As this process takes place, there is dispersion of the solute band.

Gritti and Guiochon (2012b) consider for small molecules, or large molecules excluded from the mesopore of porous particles at above optimum flow rate, that the contribution of the A term to H exceeds 75%, i.e. it is a major contributor to H in narrow diameter packed and monolithic columns.

Numerical simulations, made possible by increases in computer processing power and speed, are enabling investigations of packaging morphologies (particle shape, porosity, size and composition, including monolithic phases); packing geometries; column dimensions; mobile phase flow rates; pressure, temperature and other parameters, and their relationship to the eddy dispersion term. Such simulations are useful in helping chromatographers and column manufacturers identify which are the most critical aspects in terms of optimum column efficiency. The reader is referred to papers by Gritti and Guiochon (2010a, 2010b), Daneyko et al. (2012) and Khirevich et al. (2009) for more detailed discussions.

12.4.3.2 Longitudinal Diffusion (B)

Diffusion occurs as molecules disperse from a region of higher solute concentration to a region of lower concentration. In chromatography, this means that solute molecules migrate from the concentrated centre region of the band outwards, with axial diffusion (along the length of the column) contributing to band broadening (Figure 12.6). At low flow rates, a solute spends a longer period of time in the column allowing more time for molecules to disperse from the centre of the band, thereby causing the band to broaden. As the mobile

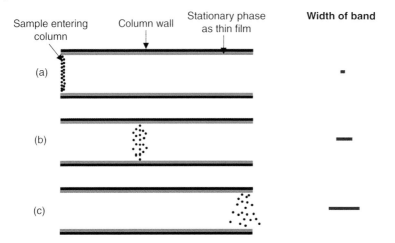

Figure 12.6 Band broadening due to longitudinal diffusion in the chromatographic column. (a)–(c) represent increasing time on column.

phase flow rate increases, dispersion decreases because there is less time for molecules to diffuse. This is indicated in the van Deemter equation by: $H \propto B/u$. In LC, longitudinal diffusion is of less concern compared with GC because molecules diffuse more slowly in the liquid, mobile phase.

12.4.3.3 Mass Transfer (C)

In the seminal paper by van Deemter et al. (1956), resistance to mass transfer was viewed as differences in the time it takes for the molecules of a compound to equilibrate between the stationary and mobile phase; these small differences in time give rise to band broadening because whilst partitioned into the stationary phase, a molecule does not move along the column. More recent research has shown that the situation giving rise to band broadening due to differences in mass transfer is considerably more complex. A series of papers by Gritti and Guiochon has explored the situation in more detail. They identify the following three contributions to the C term in LC using solid phases (particulate or monolithic): particle diffusivity, mesopore diffusion, and surface diffusion (Gritti and Guiochon 2012a). They view the stationary phase as composed of (i) a stagnant layer of eluent surrounding the particles of packed columns or skeleton of monolithic phases, (ii) eluent within the mesopore, and (iii) in reversed phase (RP) HPLC, the hydrophobic surface of the particles or skeleton.

A silica particle can be visualised as solid material containing a network of pores (mesopore), with mobile phase filling the mesopore. Molecules may diffuse into and across the mesoporous network before transferring to the stationary phase (covalently bonded to the silica gel surface). Diffusion through the mesopore may be hindered depending on the size of the analyte molecule and the size and tortuosity of the mesoporous network. Small molecules diffuse more readily through the network than large ones, particularly where the pores are narrow and tortuous. Whilst a molecule is in the mesoporous network of a particle, it is not carried along the column as rapidly as are molecules which are still in the bulk mobile phase. The heterogeneous nature of the mesopore means that some molecules will spend a longer period of time within the network than others and hence molecules will

Figure 12.7 Mass dispersion in porous, silica particles. The channels/pores within the particles represent the mesoporous network. In (a), the solute molecules have not long entered the column and are clustered in a fairly tight zone as they start to diffuse into the mesoporous network within particles. In (b), molecules which did not instantaneously diffuse into the mesopore are carried along the column in the bulk eluent flow with the result that the band starts to spread out.

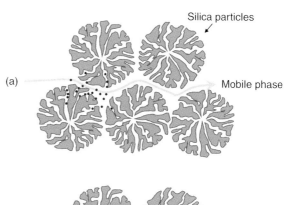

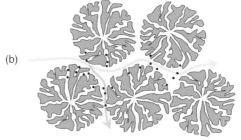

tend to disperse from their original, narrow solute zone as they travel down the column (Figure 12.7).

In RP-HPLC, it has been shown that molecules can diffuse along the surface of the hydrophobic stationary phase and that this can speed up diffusion (Miyabe and Guiochon 2000). Solute molecules also need to diffuse across the stagnant layer of mobile phase at the particle or monolith surface to reach the stationary phase.

The complex pathways and processes that solute molecules may experience when transferring from the mobile to the stationary phase and back again, plus the fact that partition is not an instantaneous process, all contribute to band broadening. Efficiency in terms of mass transfer decreases (band broadening increases) with increasing mobile phase flow rates because the flow of the eluent carries the solute more rapidly through the column and hence exacerbates the minute differences between the diffusion pathways of solute molecules.

The C term is less affected by flow rate in GC compared with LC because diffusion rates in the gas phase are more rapid than in the liquid phase. In addition, the stationary phase in WCOT columns acts like a thin film with no mesopore. Hence diffusion in and out of the stationary phase is likely to be less hindered than in HPLC.

Gritti and Guiochon (2012a) also propose that an additional HETP term should be derived to take account of the frictional heating which occurs in HPLC performed at ultra high pressure (UHPLC), also referred to as very high pressure HPLC (vHPLC), using columns packed with small diameter particles. Unless columns are well thermostatted, research has shown that temperature gradients can form both longitudinally and axially in columns. Axial temperature gradients give rise to higher temperatures at the column centre compared with the wall with a consequent decrease in viscosity and increased flow rate at the centre, plus higher diffusion coefficients at the centre compared with the walls. This can give rise to loss of efficiency and can also influence retention factors; sometimes adversely.

Although the Rate theory has helped to identify some of the major parameters which are involved in band broadening, it is clear from the literature that there are still many unanswered questions regarding the factors causing band broadening and what practical improvements can be made to produce even more efficient columns. The relatively recent adoption of UHPLC with the use of narrow bore, small particle size, shorter columns has re-energised the field of chromatography theory because these columns place considerable constraints on the chromatographic process. There is a need for a better understanding of the factors driving efficient separations so that more efficient and effective columns can be produced to match advancements in instrumentation.

12.4.3.4 Non-Column Parameters Contributing to Band Broadening

Other parts of the chromatographic system can contribute to band broadening such that:

$$\sigma^2_{Total} = \sigma^2_{Column} + \sigma^2_{Extra\ column} \tag{12.15}$$

Extra-column parameters contributing to band broadening include sample introduction (injection solvent choice and injection volume), the tubing connecting the injector to the column and the column to the detector (length and internal diameter), and the detector (volume, time constant and collection rate). Provided peak volumes are large, relative to the extra-column volume (ECV), the effects on efficiency will be minimal. This is not the case with UHPLC/vHPLC, where narrow diameter, short columns are often employed. For example, the maximum ECV for a 2.1 mm internal diameter × 50 mm long column is 5 µl (Taylor 2016), i.e. if the ECV contributed by the parameters listed above exceeds this volume there is likely to be a loss of efficiency; with losses estimated to exceed 30%. Instrument and column manufacturers offer considerable practical advice about how to reduce ECV.

12.5 Practical Applications of Chromatographic Theory

Many readers may have been wondering what the theory outlined above has to do with day-to-day practice in analytical chemistry. From a practical viewpoint, an analyst is concerned with being able to separate the analyte(s) of interest from one another and from other interferents which are present in a sample in order to detect, to identify and, if required, to quantify the analytes to a high degree of certainty and accuracy and to do this in the minimum time possible. Time is important because there is a cost to analysis; the shorter the total analysis time, the lower the cost. In addition, rapid analysis may help the police or other authorities; either allowing a suspect to be charged or released, or to confirm or guide lines of inquiry. An understanding of chromatographic theory can help an analyst in the following:

- Developing new methods of analysis.
- Troubleshooting current methods.
- Rapidly adjusting current methods if an interferent co-elutes with an expected analyte.
- Evaluating new chromatography columns which may offer significant advantages over current ones in use, with a view to adopting for use.
- Transferring current methods to alternative or new instrumentation.

- Recognising why current methods of analysis have been developed and what each part of the method and instrumental set-up contributes to the overall analysis and how small changes to a method may have a critical impact on the analysis.
- Responding to possible global shortages of or withdrawal of critical resources (e.g. particular solvents, carrier gases).
- Understanding manufacturers' advertising and chromatographic supply catalogues and technical brochures.
- Understanding journal articles reporting development of methods or applications in forensic science.
- Improving their performance in the workplace and gaining promotion.
- Training others.
- Presenting evidence in a judicial setting.

12.5.1 Optimising Chromatographic Separations

A commonly encountered problem in chromatographic separations is the incomplete resolution of two or more compounds. This may manifest as two incompletely resolved peaks or, in some situations, as a single peak which suitable detectors identify as two or more co-eluting compounds. In some instances, e.g. where a mass spectrometer is used as a detector, it may be possible to identify and quantify co-eluting compounds but, in many cases, it will be necessary to adjust the analytical method to fully separate, or resolve, all compounds.

12.5.1.1 Resolution

Resolution is a measure of the separation of two peaks in terms of their average peak width at baseline (Figure 12.3) and can be calculated according to Eq. (12.16).

$$R_s = \frac{2(t_{r2} - t_{r1})}{(w_{b1} + w_{b2})} \tag{12.16}$$

where w_{b1} and w_{b2} are the peak widths at baseline of the first and second closely eluting peaks, respectively. A value of $R_s = 1.5$ is considered to represent baseline separation of two Gaussian peaks, i.e. complete separation. For robust methods, a resolution of ≥ 1.7 may be desirable to ensure that separation is maintained if there are minor perturbations in analytical conditions.

Retention factor (k), separation factor (α), and plate number (N) can be mathematically related to the resolution R_s of a column (i.e. the ability of a column to separate two analytes) using the following equation:

$$R_s = \frac{\sqrt{N}}{4} \left(\frac{\alpha - 1}{\alpha}\right) \left(\frac{k}{1 + k}\right) \tag{12.17}$$

where k is the retention factor of the later eluting of the two peaks.

Equation (12.17) is valuable in guiding the choice of chromatographic conditions to yield an optimal separation. A graphical plot of this equation (Figure 12.8) indicates that adjusting the separation factor α will have the greatest effect on resolution. Increasing the efficiency (as measured by N) increases resolution but has less effect than increasing the separation factor. Adjusting parameters which influence the retention factor (k) has a significant

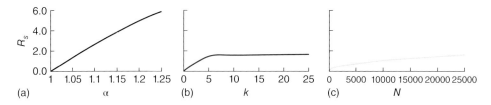

Figure 12.8 Graphical representation of Eq. (12.17) showing how changing (a) the separation factor (α), (b) the retention factor (k) and (c) the number of theoretical plates (N) influences resolution (R_s). The following fixed values were used: $\alpha = 1.05$, $k = 5$, and $N = 20\,000$.

effect on the resolution of early eluting peaks (in Figure 12.8, for peaks where $k = \geq 5$) but has minimal effect for later eluting peaks.

If two compounds are incompletely resolved, the implication is that they have very similar distribution constants (K_c) under the analytical conditions employed, i.e. both compounds have similar affinities for the stationary phase and mobile phase. Hence something needs to be done to change the parameters which influence the distribution constant. The most obvious parameters to change are the mobile phase composition and the stationary phase chemistry.

In HPLC, the simplest way to modify α is to adjust the mobile phase composition because this can be done without changing the column. A change in the relative proportions of aqueous solvent to organic modifier may be sufficient and is probably the first change that an analyst would make. If operating under a gradient elution (see Section 14.2.6), the change would be exceedingly simple and only requires a modification to the mobile phase gradient which can be done through computer control of the HPLC system. If making such a change does not resolve resolution problems, the next changes to try would be a change in the solvent used as the organic modifier, e.g. from acetonitrile to methanol (or vice versa) or, for compounds with ionisable groups, a change in pH or ionic strength of the mobile phase. If this fails to overcome the problem, the next thing to try is to change the chemistry of the stationary phase.

Most column manufacturers produce technical literature on column chemistry which can help analysts when selecting an alternative column to overcome challenges with resolution. The analyst should also consider the chemistry of the incompletely resolved compounds and the molecular interactions occurring between these compounds and the mobile and stationary phases. This should help them select a phase which can exploit other molecular interactions to enhance resolution. Some column manufacturers offer 'try before you buy' where the analyst can see if a column will solve their analytical problems before they purchase. Changing to a column with a different stationary phase chemistry requires a change of column. It takes time to change columns and, if a new column has to be purchased, there is the purchase cost and delivery time to be considered. There is also the possibility that, although the new column may overcome the problems of incomplete resolution, some analytes which were previously resolved in the original analysis might now co-elute. This is why changing the mobile phase composition to overcome problems with incomplete resolution is generally done before changing columns.

As noted in Section 12.4.1, the Plate theory of Martin and Synge indicated that efficiency is related to particle size, with smaller particles producing increased efficiency. Hence using a column with a smaller particle size should help improve resolution. However, reducing particle size will increase the pressure required to force mobile phase through the column. Equation (12.18) is the pressure equation for LC where η is the viscosity of the mobile phase, F is the mobile phase flow rate, L is the column length, r is the column radius, d_p is the particle size, and K^0 is the specific permeability.

$$\Delta P = \frac{\eta F L}{K^0 \pi r^2 d_p^2} \tag{12.18}$$

Reducing the particle size by half increases ΔP fourfold. As shown by Eq. (12.17), increasing N will also increase resolution. This change is most commonly made by using a longer column. Doubling L will double the *backpressure*: the pressure required to pump the mobile phase through the column. Hence care must be exercised in HPLC when using a column with smaller particle size or longer length to overcome resolution problems to ensure that this does not cause problems with backpressure. Increasing the backpressure puts additional strain on the HPLC pumps and may eventually damage them.

Temperature can also be used to overcome problems with incomplete resolution in HPLC but, in contrast to GC, the effect of increasing the temperature in terms of individual compounds and the effect on elution order is far less easy to predict. However, because changing temperature is relatively easy, it may be something to consider before changing to a different column.

12.5.1.2 GC

In GC, temperature is normally the first parameter to be adjusted to overcome problems with incomplete resolution. Where a temperature programmed analysis is employed, reducing the rate of temperature increase, including incorporating an isothermal period, may help separate poorly resolved compounds.

Unlike HPLC, changing the nature of the mobile phase in GC is not used as an option to overcome resolution problems. This is because the gases used as mobile phases in GC do not undergo significant molecular interactions with the compounds being analysed; the main interactions are those which take place between the compounds and the stationary phase. Hence changing the chemistry of the mobile phase in GC has much less effect on α than in HPLC.

If problems with incomplete resolution in GC cannot be overcome by adjusting temperature, the next parameter to change is the stationary phase chemistry. Stationary phases are available with considerably different chemistries and the same comments apply as for HPLC, i.e. the analyst should consider the chemistry of the incompletely resolved compounds and the molecular interactions they can undergo and choose a phase which can exploit different interactions.

Using a longer column can be considered (increase in N) but only improves R_s by a factor of the square root of the increase in length. Analysis time is directly linked to column length and although doubling column length will double N, it also doubles analysis time and will lead to increased band broadening. Hence using a longer column to overcome problems with resolution is not usually the first choice for GC, although, depending on the temperature conditions used, and the retention time of the compounds causing problems, it may be possible to make adjustments to the temperature programme to minimise the effect of a longer column on analysis time.

The k term in Eq. (12.17) is influenced by the thickness of the stationary phase, the internal diameter of the column, and temperature. In GC, open tubular columns can be purchased with different thickness of stationary phase (see Section 13.2.3) and hence can be another option to overcome resolution problems. However, using a thicker film increases t_R, and hence analysis time, unless other parameters are changed. If a new column needs to be purchased, the cost of the column and delivery time need to be taken into account.

When using SOPs and accredited methods, as is often the case in forensic science, changing any part of a method must be considered carefully and no changes should be made without suitable authorisation. Any changes made, and the reasons for changes, should be fully documented.

12.5.1.3 Mobile Phase

Flow rate

The A, B and C terms in the van Deemter equation can be measured experimentally. These can be plotted against H in a so-called van Deemter plot. A typical van Deemter plot for GC is shown in Figure 12.9.

When all values are combined, a plot of H versus the average mobile phase velocity, u (solid line in Figure 12.9) shows that H is large at low mobile phase velocities, reduces to an optimum value, u_{opt}, as the mobile phase velocity increases and then increases again as velocity increases above u_{opt}. The reason for this profile is that at very low values of u, analytes pass through the chromatographic system slowly giving time for longitudinal diffusion (the B term in the van Deemter equation) to take place resulting in considerable band broadening and high values for H (dashed line in Figure 12.9). As the mobile phase velocity increases, longitudinal diffusion is constrained and band broadening reduces (i.e. H reduces), as can be seen for the B/u term in Figure 12.9. The C term (resistance to mass transfer) is directly proportional to u, with H increasing as flow rate increases (grey line in Figure 12.9) because the processes involved in mass transfer become disrupted.

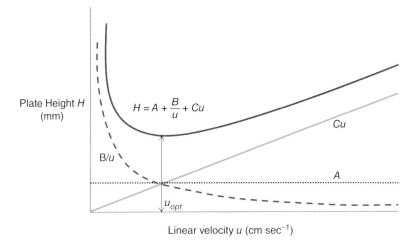

Figure 12.9 van Deemter plot for a GC separation showing the change in plate height, H, versus the linear velocity of the mobile phase, u, for the van Deemter equation ($A + B/u + Cu$) and the individual contributions A, B/u, and Cu.

For GC separations carried out on open tubular columns, H is independent of A because eddy diffusion is negligible. Golay (1958) proposed a simplified form of the van Deemter equation for open tubular columns ($H = B/u + Cu$) and a Golay plot against linear velocity looks very similar to Figure 12.9 minus the A term.

The important aspect to note from Figure 12.9 is that there is an optimum mobile phase velocity at which H is a minimum (u_{opt}). Operating at a mobile phase velocity $\ll u_{opt}$, will result in poor efficiency and give a high degree of band broadening. However, operating at flow rates $\gg u_{opt}$ also gives rise to band broadening due to disruption of solute mobile phase$\leftrightarrow$stationary phase equilibration processes. Thus, the implication is that analyses should be carried out at u_{opt}. However, several additional factors need to be considered. The elution of substances from a column is directly proportional to mobile phase flow rate. Hence, the faster the flow rate, the faster the analysis. Given that analysis should be carried out in the minimum time possible so as to reduce costs and provide an analytical result for the customer as quickly as possible, the mobile phase should be operated at as fast a flow rate as possible whilst providing acceptable efficiency and avoiding possible damage to the instrumentation and, in the case of HPLC, avoiding damage to tubing and column packings. Manufacturers strive to produce columns which can be operated at flow rates $\gg u_{opt}$ with minimal loss of efficiency, i.e. with a flat curve for H above u_{opt} and provide information about optimal flow rates for a particular column.

GC carrier gas choice

A van Deemter plot for the three carrier gases commonly used in GC, namely nitrogen (N_2), helium (He), and hydrogen (H_2) (Figure 12.10), shows that N_2 gives the lowest value for H, with the implication that it is the optimum choice as a carrier gas for GC. However, efficiency decreases markedly for N_2 above u_{opt}. This is an important consideration because analysis time is proportional to u. N_2 is not generally used for GC analyses because it can only be operated at relatively low linear velocities compared with He or H_2 without giving rise to considerable band broadening.

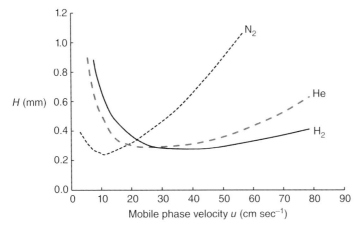

Figure 12.10 van Deemter plot (plate height, H, vs. mobile phase velocity, u) for the common carrier gases used in GC: nitrogen (N_2), helium (He), and hydrogen (H_2).

Since He and H_2 can be operated at significantly higher linear velocities (i.e. faster analysis times) without causing significant band broadening, they are the carrier gases of choice for most GC separations. H_2 can be operated at considerably higher flow rates than He whilst still offering good efficiency (flatter curve above u_{opt}) and hence is the choice for fast GC separations whilst minimising band broadening. In practice, He is more commonly used as a carrier gas, partly due to safety concerns with using H_2 at the high temperatures used in GC and because GC-MS manufacturers traditionally recommend using He for compatibility with some parts of their MS systems.

Difficulties with the worldwide supply and cost of He in 2012–2013 led to the re-exploration of N_2 as a carrier gas for GC. The general consensus is that N_2 can be used as an alternative to He or H_2 for less demanding GC analyses but changes to column diameter and length may be required to maintain similar analysis times. Most GC-MS manufacturers do not recommend N_2 as a carrier gas because it gives rise to reduced sensitivity through MS analysis.

Another reason for operating $>u_{opt}$ in GC is that the carrier gas flow rate may fall below u_{opt} during temperature programmed analysis because the viscosity of a gas increases with increasing temperature. For GC systems where carrier gas flow rate is regulated by column head pressure, if the flow rate is set at u_{opt} at the start of the temperature programme, the likelihood is that the flow rate will fall below optimum as the temperature is raised, resulting in considerable band broadening. Most modern GC systems are equipped with electronic pneumatic control systems which keep the carrier gas flow rate constant through temperature programmed analyses. Nevertheless, because WCOT columns offer high efficiency at flow rates $>u_{opt}$, the time advantage offered by operating at high linear velocity means that most GC analyses are carried out at flow rates $\gg u_{opt}$ irrespective of whether the flow rate is controlled by column head pressure or flow control.

HPLC: particle size

For HPLC where particulate stationary phases are employed, chromatographers have modified the van Deemter equation to take account of particle size. One well recognised modification is that of Knox:

$$h = Av^{1/3} + \frac{B}{v} + Cv \tag{12.19}$$

where $h = H/d_p$. h is the reduced plate height and d_p is the particle diameter. v, the reduced velocity, is defined as $v = ud_p/D_m$ (where D_m is the diffusion coefficient).

A plot of the Knox equation for varying particle sizes is shown in Figure 12.11. This shows that smaller size particles give considerably greater efficiency. They also exhibit flatter profiles above u_{opt} indicating that they can be operated at flow rates significantly above optimum without significantly affecting efficiency, although caution must be exercised about the effect of backpressure on the LC system and column. Use of very small particle sizes (e.g. <2 μm has led to the development of UHPLC/vHPLC, with systems designed to handle the significantly higher pressures required to achieve the high flow rates at which these columns can operate whilst maintaining high efficiency offering rapid separations and narrow peak widths. Although small particle sizes can offer advantages in terms of minimising band broadening, there is considerable debate about optimum particle composition,

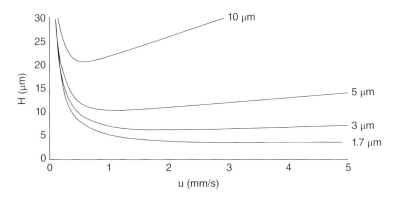

Figure 12.11 Theoretical Knox curves plotted for HPLC with particle sizes 1.7, 3, 5, and 10 μm. Source: Reproduced with permission from Nguyen et al. (2006).

size, shape and size variation, column length and diameter, plus indications that column walls and packing method can significantly influence efficiency. Analysts working with UHPLC/vHPLC are advised to pay particular attention to the scientific literature to keep abreast of developments in this area so that they can select columns and instrument parameters which maximise chromatographic efficiency, i.e. sufficient resolution of analytes to give acceptable accuracy and precision and other parameters required from a particular analytical method in the minimum time possible.

For UHPLC/vHPLC using wide bore columns with small particles, efficiency may reduce at high mobile phase flow rates if columns are not well thermostated due to temperature gradients forming within columns via frictional heating.

12.6 Conclusion

The theory underlying chromatographic separations is still the subject of considerable debate and research, driven by the desire of column manufacturers to gain a competitive edge by producing more efficient columns and for analysts to gain a better understanding of how they can improve separations.

Having read the above chapter, you should have a better understanding of how chromatographic separation works and the important parameters which influence separation and that can be manipulated by an analyst in their day-to-day work. We encourage you to take advantage of the excellent literature and advice provided by instrument and column manufacturers to support working analysts.

References

Berezkin, V. (1989). Biography of Mikhail Semenovich Tswett and translation of Tswett's preliminary communication on a new category of adsorption phenomena. *Chemical Reviews* 89 (2): 279–285.

Daneyko, A., Hlushkou, D., Khirevich, S., and Tallarek, U. (2012). From random sphere packings to regular pillar arrays: analysis of transverse dispersion. *Journal of Chromatography A* 1257: 98–115.

van Deemter, J., Zuiderweg, F., and Klinkenberg, A.V. (1956). Longitudinal diffusion and resistance to mass transfer as causes of nonideality in chromatography. *Chemical Engineering Science* 5 (6): 271–289.

Giddings, J.C. (1961a). Lateral diffusion and local nonequilibrium in gas chromatography. *Journal of Chromatography A* 5: 61–67.

Giddings, J.C. (1961b). The role of lateral diffusion as a rate-controlling mechanism in chromatography. *Journal of Chromatography A* 5: 46–60.

Giddings, J.C. (1965). Comparison of theoretical limit of separating speed in gas and liquid chromatography. *Analytical Chemistry* 37 (1): 60–63.

Giddings, J.C. (1991). *Unified Separation Science*. New York: Wiley.

Giddings, J.C. (2017). *Dynamics of Chromatography: Principles and Theory*. CRC Press.

Golay, M.J. (1958). Theory of chromatography in open and coated tubular columns with round and rectangular cross-sections. *Gas Chromatography* 2: 36–55.

Gritti, F. and Guiochon, G. (2010a). Mass transfer mechanism in liquid chromatography columns packed with shell particles: would there be an optimum shell structure? *Journal of Chromatography A* 1217 (52): 8167–8180.

Gritti, F. and Guiochon, G. (2010b). Relationship between trans-column eddy diffusion and retention in liquid chromatography: theory and experimental evidence. *Journal of Chromatography A* 1217 (41): 6350–6365.

Gritti, F. and Guiochon, G. (2012a). Theoretical and experimental impact of the bed aspect ratio on the axial dispersion coefficient of columns packed with 2.5 μm particles. *Journal of Chromatography A* 1262: 107–121.

Gritti, F. and Guiochon, G. (2012b). Mass transfer kinetics, band broadening and column efficiency. *Journal of Chromatography A* 1221: 2–40.

Gritti, F. and Guiochon, G. (2013). The van Deemter equation: assumptions, limits, and adjustment to modern high performance liquid chromatography. *Journal of Chromatography A* 1302: 1–13.

Horvath, C. and Lin, H.-J. (1976). Movement and band spreading of unsorbed solutes in liquid chromatography. *Journal of Chromatography A* 126: 401–420.

Huber, J. and Hulsman, J. (1967). A study of liquid chromatography in columns, the time of separation. *Analytica Chimica Acta* 38: 305–313.

IUPAC (2009). *IUPAC Compendium of Chemical Terminology: Gold Book*. Research Triangle Park, NC: IUPAC.

Khirevich, S., Höltzel, A., Seidel-Morgenstern, A., and Tallarek, U. (2009). Time and length scales of eddy dispersion in chromatographic beds. *Analytical Chemistry* 81 (16): 7057–7066.

Khirevich, S., Daneyko, A., Höltzel, A. et al. (2010). Statistical analysis of packed beds, the origin of short-range disorder, and its impact on eddy dispersion. *Journal of Chromatography A* 1217 (28): 4713–4722.

Knox, J.H. (1961). The speed of analysis by gas chromatography. *Journal of the Chemical Society (Resumed)*: 433–441.

Knox, J.H. (1977). Practical aspects of LC theory. *Journal of Chromatographic Science* 15 (9): 352–364.

Knox, J.H., Done, J.N., Fell, A.F. et al. (1978). *High-Performance Liquid Chromatography*. Edinburgh University Press.

Martin, A. and Synge, R.M. (1941). A new form of chromatogram employing two liquid phases: a theory of chromatography. 2. Application to the micro-determination of the higher monoamino-acids in proteins. *Biochemical Journal* 35 (12): 1358.

McNaught, A.D. and Wilkinson, A. (1997). *IUPAC Compendium of Chemical Terminology*. Blackwell Science.

Miyabe, K. and Guiochon, G. (2000). Fundamental interpretation of the peak profiles in linear reversed-phase liquid chromatography. *Advances in Chromatography* 40: 1–113.

Nguyen, D.T.T., Guillarme, D., Rudaz, S., and Veuthey, J.-L. (2006). Fast analysis in liquid chromatography using small particle size and high pressure. *Journal of Separation Science* 29 (12): 1836–1848.

Rawlinson, C., Martin, S., Frosina, J., and Wright, C. (2017). Chemical characterisation of aerosols emitted by electronic cigarettes using thermal desorption–gas chromatography–time of flight mass spectrometry. *Journal of Chromatography A* 1497: 144–154.

Taylor, T. (2016). System and column volumes in HPLC - we still haven't got the message! https://www.crawfordscientific.com/technical/chromatography-blog/hplc-chromatography-tips/hplc-troubleshooting/system-column-volumes-in-hplc (accessed 29 May 2020).

13

Gas Chromatography

Shari Forbes

13.1 Introduction

The general principles and mathematical equations described in Chapter 12 can be applied to gas chromatography (GC) with some minor modifications. In GC, sample components are separated based on the partitioning between a gaseous mobile phase and a stationary phase held in a long, tubular column. Sample injection generally requires the sample to be vaporised and transferred to the start of the column. The sample components move through the column with the assistance of the gaseous mobile phase. Unlike other chromatographic techniques, the mobile phase only acts as a carrier and minimal interaction between the sample and mobile phase occurs.

The most widely used GC columns are referred to as capillary columns. If the stationary phase of the column is a liquid, the technique is traditionally referred to as gas–liquid chromatography (GLC) whereas if the stationary phase is a solid, it is referred to as gas–solid chromatography (GSC). In forensic science, GLC is the only type of GC widely employed and will therefore be the only technique discussed in this chapter (herein simply referred to as GC). GC is used extensively in forensic laboratories for the analysis of volatile and semi-volatile organic compounds including alcohol, drugs and poisons, fire debris (including ignitable liquids), explosives, pesticides, and inks. The list of applications is constantly expanding and some of the newer applications will be discussed later in this chapter.

13.2 Gas Chromatography Components

Figure 13.1 is a schematic of the basic components found within a GC instrument. An inert carrier gas (mobile phase) flows consistently through the injection port, column and detector. In the most common form of GC analysis, the sample is injected and vaporised in the heated injection port and carried into the column in the mobile phase. The column is coated with a thin film of high boiling liquid stationary phase. The analytes in the sample partition between the mobile and stationary phase, thus separating into individual components. This separation is based on the relative solubility in the liquid phase and

---

Analytical Techniques in Forensic Science, First Edition.
Edited by Rosalind Wolstenholme, Sue Jickells and Shari Forbes.
© 2021 John Wiley & Sons Ltd. Published 2021 by John Wiley & Sons Ltd.

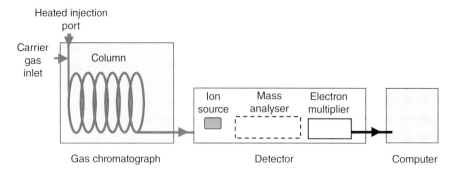

Figure 13.1 Schematic of the components of a GC-MS instrument.

the relative vapour pressure of the analytes. The analytes then pass from the column to a detector which, assuming the detector responds to the analyte, generates an electrical signal which is recorded. A data system integrates the signal into a chromatogram and typically includes a report with peak area, retention time, and quantitative results. Each of these components of the system will be discussed individually with reference made to the variable configurations available.

13.2.1 Mobile Phase System

The mobile phase in GC is a chemically inert carrier gas. These include helium, nitrogen and hydrogen, with helium most commonly used. Gases are typically purchased in pressurised tanks with associated regulators, gauges and flow meters to control the flow of gas into the system. If hydrogen is used as a carrier gas, many laboratories prefer to use a hydrogen generator. The carrier gas system also incorporates a molecular sieve upstream of the GC instrument to remove impurities and water which can degrade the liquid phase in the column and be detrimental to the functioning of the instrument. Purifier traps are also available which remove traces of oxygen which can degrade the stationary phase. Electronic flow meters in the GC instrument regulate the flow of gas to the desired rate. The flow rates for a capillary column should be determined experimentally but optimal values are typically around 1 ml/min for a 0.25 μm internal diameter column. Inlet pressures are typically in the range of 10–50 psi (0.68–3.45 bar) and, assuming they remain constant, will ensure a consistent flow rate within the system. Electronic flow controllers enable the flow to remain constant during temperature programmed analysis; an important consideration because the viscosity of a gas increases with increasing temperature and hence, with older GC systems where there is no facility to increase gas flow during an analytical run, there is the potential for the flow rate to decrease as the column temperature is increased.

The relationship described by the van Deemter equation is applicable to GC and gaseous mobile phases (see Section 12.4.3). The longitudinal diffusion term (B/u) (where B is the longitudinal diffusion coefficient and u is the mobile phase velocity) becomes more important in GC because of the larger diffusion rates in gases. Diffusion rates in gases can be 10^4–10^5 times greater than in liquids. As a result, the lowest point of the van Deemter plot which relates plate height to the mobile phase velocity will be considerably broader in GC than in liquid chromatography (LC) and a wider range of flow rates can be used to produce

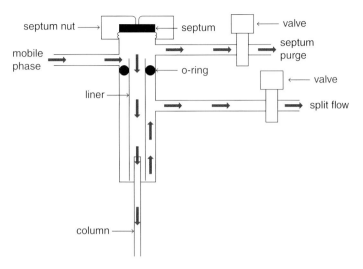

Figure 13.2 Vaporising injector for gas chromatography. Source: Reproduced with permission from Darling et al. (2013).

efficient separations. With capillary columns containing thin films, the efficiency as measured by plate number, N, is already sufficient and the focus is instead on speed. Hence, these columns can be run at flow rates above optimal with minimal impact on separation.

13.2.2 Sample Injection System

The sample injection system should permit a rapid injection of gases or liquids into the carrier gas stream. The most commonly used injector is referred to as a vaporising injector (Figure 13.2). Other injector systems are available including on-column and pyrolysis injectors. In the vaporising injector, samples are injected into the GC system using a calibrated microsyringe. The syringe needle is pushed through a rubber or silicone septum to allow the sample to be directly injected into the vaporisation chamber which typically contains a quartz glass tube referred to as a 'liner'.

The liner forms a heated vessel into which the liquid sample is injected. This enables rapid vaporisation of the sample and transfer onto the head of the GC column as a tight band. Differences in the physical and chemical properties of samples, along with variation in the injection volume and technique used (i.e. split vs. splitless), gas flow rate, and inlet temperatures necessitates the availability of a range of liners. It is important to choose the appropriate liner to ensure efficient transfer of the sample onto the column as inefficient transfers can lead to peak tailing, band broadening or splitting.

For most analyses, as part of sample preparation, the final sample to be analysed is dissolved in a solvent. The volume of the vaporised solvent must be considered when determining the appropriate liner. The liner volume should be sufficient to accommodate the solvent volume as it increases during heating and vaporisation. If the diameter is too small, the volume will expand beyond the capacity of the liner, causing sample loss, peak tailing and poor reproducibility of the peak area. If the liner diameter is too large, there will be a

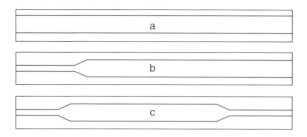

Figure 13.3 Gas chromatography inlet liners (a) untapered split liner; (b) tapered splitless liner; (c) double-tapered liner.

large dead volume which increases the time taken to transfer the sample onto the column and leads to band broadening and peak tailing.

A second consideration is the type of injection used. Some manufacturers recommend different liners for split and splitless injections (discussed further below), although this is not always necessary. Liners for split injections are typically open at the bottom (Figure 13.3a) while liners for splitless injections will be tapered (Figure 13.3b) at the bottom and the column inserted into the taper. Single-tapered liners can assist in minimising contact of the sample with reactive metal components in the injector system. Double-tapered liners (Figure 13.3c) help to contain the vapour, preventing loss of sample through the top of the liner and into the septum purge flow (Figure 13.2). The final consideration is whether to use a packed or unpacked liner. Packed liners are available containing deactivated quartz wool onto which the liquid sample is deposited as an aerosol. This increases the surface area available for heating and vaporising the aerosol into a gaseous sample, ensuring the sample is fully vaporised before introduction onto the column. The outcome is enhanced injection reproducibility and symmetrical peak shapes with minimal boiling point discrimination.

To achieve high efficiency and reduce the likelihood of band broadening, poor resolution, and discrimination, the sample should be introduced as a 'plug' of vapour into the inlet, regardless of the technique used. This ensures that a narrow band of sample flows onto the column. Rapid injections are necessary for high column efficiency and are best achieved using autoinjectors. These are often referred to as autosamplers and are a standard feature on modern GC instruments. Autosampler injections provide significantly better precision than manual syringe injections as the volumes injected are more reproducible and human error becomes negligible. The loss of sample through evaporation at the needle tip is also reduced due to the rapid injection process. Autosamplers have the added advantage of freeing the analyst to carry out other work and also continue working through the night when most analysts are at home asleep!

13.2.2.1 Liquid Samples

Since liquids expand when vaporised, small volumes of liquid samples are desirable when using capillary columns, typically in the range of 0.01–3 μl. The standard volume injected is generally 1 μl, although microsyringes come in a range of volumes including 1, 5, and 10 μl. A volume of 1 μl of sample generates approximately 200–1000 μl of vapour (Grob and Barry 2004), depending on the solvent used for sample dissolution.

Most samples will have been subjected to some sort of extraction procedure prior to GC analysis such that the sample injected consists of a low concentration of the compounds of interest dissolved in solvent. However, samples such as arson accelerants can be analysed directly because they are a mixture of co-soluble organic compounds. Injecting 1 μl of

Figure 13.4 (a) Sharp, Gaussian peak shape expected and (b) shark-fin shaped peak indicative of column overloading.

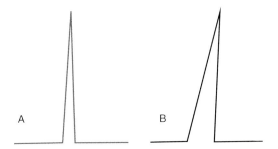

an accelerant, or a solution of one or two compounds present at very high concentration (mg/ml), can be problematic because 1 μl (~1 mg) of injected sample will have a similar mass to the total mass of the liquid stationary phase in the column. This results in 'overloading' of the column and poor chromatographic separation such as that shown in Figure 13.4. Sample splitting can avoid this issue by introducing a smaller volume of sample using a split/splitless injector, the more common name given to the vaporising injector.

Split injections allow a fraction of the sample to be passed onto the column head while the remainder is discarded as waste (Figure 13.5). The process involves injecting sample into the heated injection port containing a liner. The sample is vaporised and a percentage of the sample of the vapour enters the head of the column while the rest of the vaporised sample passes out through a split outlet or valve. Typical split ratios range from 1:10 up to 1:100 (i.e. for a 1:100 ratio, only one part of the vaporised sample passes to the column and the rest, 99 parts, is discarded through the split outlet flow, see Figure 13.5).

Split injection is recommended for relatively concentrated (i.e. ppm and higher) samples. Such injections offer several advantages including high resolution separations, introduction of undiluted samples, and analysis of 'dirty' samples by inserting a plug of deactivated quartz wool into the inlet liner (i.e. a packed liner) to trap particulate or non-volatile compounds. However, split injections are not recommended for trace analysis since only a fraction of the sample enters the column.

The split/splitless injector also allows for *splitless injections*. A splitless injection may be necessary to improve the sensitivity of analysis of a sample, i.e. where there is such a small concentration of analyte present in a sample that a high proportion is required to be transferred to the column for the analyte to be detected. The same set-up is used as for split injection, except the split flow outlet remains closed during sample introduction and for approximately 30–45 seconds thereafter to ensure the majority of sample vapour is passed onto the column (Figure 13.6). About 95% of the injected sample will reach the column before the split valve is opened and any remaining vapour is swept out of the valve. This is the preferred method for trace analysis in forensic science where low ppm to ppb or ppt concentrations are expected. However, splitless injections have several disadvantages. First, optimisation of the column temperature and purge time is required to ensure that the sample passes to the head of the column as a narrow band. Secondly, splitless injection is not well suited for highly volatile compounds. Finally, for good separation, the early eluents must have boiling points at least 20–30 °C higher than the solvent used. Caution needs to be exercised with splitless injection in terms of potential interferents which might be present in the sample to be analysed. If non-volatile compounds are present at high concentrations,

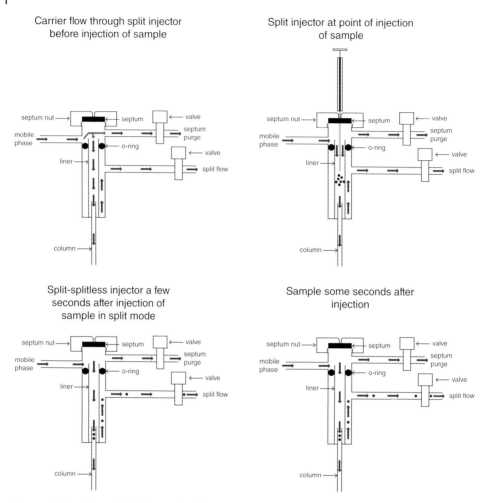

Figure 13.5 Schematic of the carrier flow of sample through a split injector. Source: Reproduced with permission from Darling et al. (2013).

they will not be volatilised and will remain on the column. If many samples are being analysed, these compounds can build up on the column to such an extent that they start behaving like a stationary phase and may cause unwanted interactions with the analytes of interest.

Liquid samples can also be injected by direct injection, on-column injection, and cold on-column injection. Direct injection allows the sample to be introduced through a glass liner so that the vapours are carried directly to the column, whereas on-column injection involves inserting the syringe needle directly into the capillary column without the use of a liner. Both techniques require wide-diameter columns with thick film coatings and flow rates around 5-10 ml/min (which is considerably faster than the standard flow rate of 1 ml/min). These techniques can provide better trace analysis and quantitation but the resolution is typically not as good as with split or splitless injections.

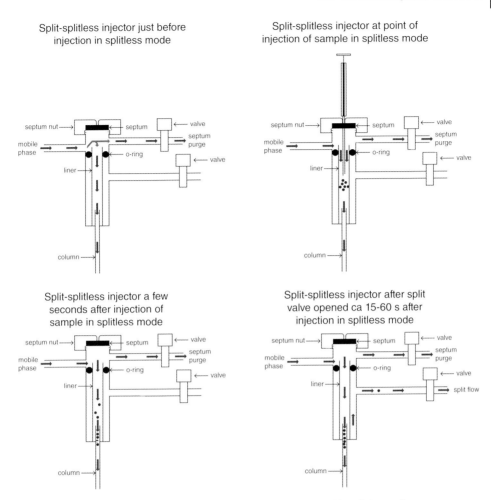

Figure 13.6 Schematic of the carrier flow of sample through a splitless injector. Source: Reproduced with permission from Darling et al. (2013).

Cold on-column injection allows the entire sample to be placed on the column without a separate vaporisation chamber. The region around the syringe is cooled to allow the sample to be introduced directly onto the head of the column as a liquid. The liquid is subsequently vaporised using the temperature programming function for the column or inlet. This ensures that, essentially, 100% of the sample is available for detection and measurement. The method is particularly valuable for thermally labile or wide-boiling range samples. However, this method may not be practical for all situations, especially with 'dirty' samples, i.e. those containing particulate material or a high proportion of non-volatile compounds, because all of the sample material will reach the column. As a result, any compounds which are not volatilised at the temperature at which the system is operated remain on the column. If many samples are analysed this leads to a build-up of non-volatile material which can give rise to unwanted interactions, potentially causing band broadening or, in some circumstances, chemical changes to the analytes.

A programmed temperature vaporisation (PTV) injector is a hybrid of the above techniques and is, essentially, a split/splitless injector which has been modified to allow sample injection while the inlet is cold. The sample is injected into the cold chamber and rapidly heated to desorb the sample onto the column. The advantage of this type of inlet is its versatility and capability to inject hundreds of microliters of sample; a distinct advantage for samples containing analytes at very low concentrations. Although similar to the on-column inlet, it reduces the peak broadening that is often seen with on-column injections. Thermal discrimination is also eliminated.

Samples which are considered 'dirty' or are clearly not suitable for injection into the column are often pretreated prior to GC analysis. Liquids such as water, blood, and other biological fluids can be treated using liquid–liquid extraction (LLE), solid-phase extraction (SPE), or solid-phase microextraction (SPME) as discussed in Chapter 4. LLE and SPE are useful for transferring analytes from water into a volatile solvent or to separate the analytes from other non-volatile substances. The methods can also be used for preconcentration of the analyte to assist with detection of trace quantities by evaporating the solvent to reduce the final volume of the extract. SPME can be used to remove analytes from water or other non-volatile substances. This is often referred to as *direct immersion SPME*. The use of SPME has the advantage of avoiding column overload effects as minimal liquid is injected with the analyte.

13.2.2.2 Gases and Volatile Compounds

Thus far, emphasis has been on the analysis of compounds which are not volatile at ambient temperature, with such compounds typically being analysed in the liquid phase. However, a major advantage of GC, particularly in comparison with high performance liquid chromatography (HPLC), is that it can be readily applied to the analysis of samples which are volatile at ambient temperature, or which can be readily volatilised at temperatures not far above ambient temperature. This encompasses the analysis of gases and of volatile compounds. For the discussion which follows, 'gas' is taken to mean a compound which is entirely in the gaseous phase at ambient temperature and is not absorbed on, or into, any type of sample matrix. A volatile compound is taken to indicate a compound which has some volatility at ambient temperature but which would not be entirely in the gas phase at ambient temperature, or which has sufficient volatility to enter the gaseous phase at temperatures of the order of 20–80 °C.

13.2.2.3 Gas Samples

Gaseous samples are ideal candidates for GC analysis because of the requirement that the analyte is in the vapour phase for separation. At moderate to high concentrations, gas samples can be directly injected onto the column by passing the samples through a gas-tight valve (also known as a gas-sampling valve). The valve is initially set to the sampling position (Figure 13.7a) and the gas is allowed to flow through a loop of known volume. Once the valve is switched to the analysing position (Figure 13.7b), the gas in the loop is carried into the GC column by the carrier gas. Gas-tight valves can be used for both qualitative and quantitative analyses. Alternatively, a gas-tight syringe can be used if this type of valve is not available. Syringes can be purchased in a range of volumes from 1 µl up to 100 ml depending on the sample size.

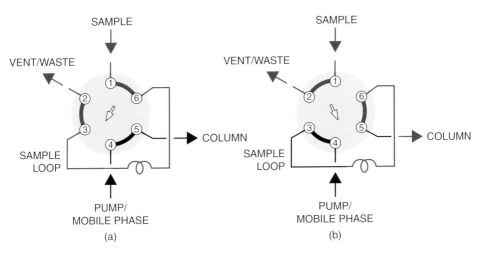

Figure 13.7 Direction of flow through a gas sampling valve. (a) At the point of sampling, mobile phase flows to the column, sample is directed through the external loop; and (b) mobile phase is switched through the external loop and flushes the sample onto the column. Source: Reproduced with permission from Vitha (2016).

For gases that are present in trace quantities, such as volatile organic compounds (VOCs) in air, the sample must be collected and concentrated prior to injection into the GC system. Commonly used methods include trapping on solid absorbents or in liquids, or cold trapping.

Where a *cold trap* is used, sometimes referred to as a cryogenic trap, the gas sample is passed over a hollow coil which is maintained at very low temperatures so that the analytes condense and are collected on the trap. The trap is then rapidly heated to release the compounds and the analytes are transferred directly into the GC injection port.

13.2.2.4 Volatile Compounds: Headspace Analysis

Although gas samples could be analysed by headspace analysis, this technique is more commonly applied to the analysis of volatile compounds.

Headspace analysis relies on the principle that the volatile components present in a liquid or solid sample will also be present (or can be forced into) the vapour phase above the sample. By collecting the headspace, the volatile components can be analysed by GC without interference from non-volatile compounds which may be present in the sample. This technique can be used for the analysis of volatile components in water or biological fluids and is commonly employed in forensic toxicology to measure the blood alcohol concentration (BAC) of an individual where a confirmatory test is required. In environmental forensics, this method is often used to measure pollutant VOCs in water or soil. Another common use of headspace analysis is the analysis of suspected arson ignitable liquid residues (ILRs) in fire debris.

Headspace analysis can be separated into static (also referred to as passive) and dynamic methods.

13.2.2.5 Static Headspace Analysis

Static headspace extraction (SHE) methods involve placing the sample in a sealed vial or other sealed container so that the volatile components distribute between the sample and vapour phase (i.e. the headspace) (Figure 13.8).

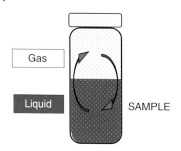

Figure 13.8 Distribution of volatile compounds from the liquid sample into the gas phase, also referred to as the headspace.

The system is allowed to reach equilibrium and then a portion of the vapour phase is collected using a gas-tight syringe and injected into the GC system. The analyte is not exhaustively removed in SHE; rather a portion is collected based on its distribution constant (also referred to as partition coefficient).

Many of the issues encountered with this method relate to poor reproducibility as a result of the system not reaching equilibrium. Method development therefore requires optimisation of the equilibration time to ensure reproducible results are achieved. Care needs to be taken to avoid the volatile components recondensing in the gas-tight syringe, if used. Automated SHE systems have heated syringes to prevent such condensation and are recommended over the use of manual gas-tight syringes.

Many SHE methods incorporate a sorbent to trap analytes for transfer to the GC system. This is particularly useful for trace analysis where being able to concentrate volatile compounds from a sample before injecting onto the GC system will improve sensitivity of analysis. Those most commonly used sorbents in forensic applications include activated charcoal strips (ACS) and headspace solid-phase microextraction (HS-SPME) fibres (see Section 4.2.4). The sorbent is placed directly into the sealed container and exposed to the vapour phase for an optimised period of time (Figure 4.6). For ACS, the analytes are extracted from the sorbent using solvent extraction and injected into the GC system as a liquid. In the case of HS-SPME, the fibre can be injected directly into the heated injection port of the GC and the analytes desorbed at high temperatures.

Both methods require optimisation of the equilibrium and exposure times, as well as the vial temperature (e.g. room temperature or heated) since the distribution constant will increase with increased temperature. Determining the volume of sample is also important so as not to overload the headspace and subsequently the GC column. Selecting the optimal extraction solvent for the ACS method is an important step in the method development. In contrast, HS-SPME requires optimisation of the injector temperature to ensure all of the analytes are desorbed from the fibre. The benefit of HS-SPME is that the fibre can be reused up to a maximum number of desorptions (typically ~100 but recommended by the manufacturer).

13.2.2.6 Dynamic Headspace Analysis

Dynamic headspace analysis is also referred to as 'purge-and-trap' because an inert gas is used to 'purge' the analytes from the sample and 'trap' the analytes on a cold trap or solid adsorbent material (Figure 13.9). The analytes are subsequently desorbed into the inlet of the GC system for separation and identification.

Figure 13.9 Schematic of dynamic headspace sampling using 'purge-and-trap'.

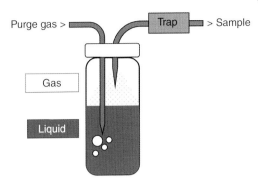

Purge-and-trap analyses are typically designed to exhaustively extract all, or a high proportion of, volatile components from the sample. Hence there is no requirement for compounds to reach an equilibrium between the sample matrix and the headspace, as is typically the case for static headspace methods.

Much work has focused on developing sorbents which allow a broad range of analytes to be extracted from the vapour phase. Sorbents can be either a single layer, such as Tenax, or multilayer where a series of weak, moderate or strong materials are used (e.g. Tenax, silica gel, and charcoal). Purge-and-trap analyses are generally more reproducible than static headspace methods and are better suited to low volatile analytes and to samples containing volatile analytes at low concentrations.

In both static and headspace analysis, and in gas analysis involving trapping on sorbents, the trapped volatiles need to be transferred to the GC instrument for separation and detection. This typically involves thermal desorption (TD) where the trapped samples are heated rapidly and transferred to the GC column. When coupled directly to a GC system, TD has the power and versatility to combine sample collection, sample preparation, selective concentration, and efficient GC injection, into a completely automated procedure. TD can be configured in such a way as to enable repeat extractions and desorptions of analytes into smaller volumes of gas, thus concentrating the analytes and increasing the sensitivity and detection limits of the system. A further advantage of TD is the reduction in background interference because analytes can be retained during the trapping process while interferents are selectively purged.

Besides those already described, sample collection options for TD can include sorbent tubes, canisters, and bags (Figure 13.10).

These techniques can be used to collect sample vapour from gas, liquid and solid samples. Sample tubes can be used as either passive (diffusive) or dynamic sample collection techniques to trap volatile components. This is particularly useful for collecting samples in the field such as for environmental air sampling. The tubes are then sealed and returned to the laboratory for analysis. Automated TD units allow the analytes to be thermally desorbed from the sorbent tube onto a cryogenic focusing trap followed by desorption directly into the GC inlet. Key applications of TD include analysis of ILRs in fire debris, fugitive air/gas monitoring in environmental forensics and scent/odour analysis for canine detection training (described further in Section 13.6).

(a) (b) (c)

Figure 13.10 Sample collection options including (a) sorbent tube containing Tenax and Carbograph 5TD; (b) gas sampling canister; and (c) Tedlar® bags.

13.2.2.7 Pyrolysis GC

Pyrolysis occurs when a substance undergoes thermal decomposition in an inert environment. It occurs at high temperatures and causes the solid material to decompose producing lower molecular weight, gaseous compounds. A GC instrument can operate at temperatures up to 350 °C (and in some cases 450 °C) but there are many substances which do not vaporise at these temperatures and, as a result, cannot be analysed by GC. Pyrolysis GC can, however, operate at much higher temperatures (~700–1000 °C), allowing substances such as hairs, fibres, paints, plastics, and other polymers to be analysed. At high temperatures and in the absence of oxygen, the sample will decompose into a stable fragment, called a pyrolysate. If repeated under the same conditions, the number and size of pyrolysates should be the same for a specific sample type.

A small portion of the sample is placed into apparatus which can be inserted directly into the injector port of the GC instrument. An inert mobile phase ensures there is no oxygen present in the injector and the pyrolyser is heated rapidly to a high temperature. The sample will decompose and the fragments will be separated in the same way as in a standard GC instrument (Figure 13.11).

The resulting data series is referred to as a pyrogram, rather than a chromatogram. Pyrolysis GC is still used in fibre analysis to isolate monomers, dimers, and other polymer fragments and degradation products. However, concerns about the lack of reproducibility of the method, combined with the fact that infrared spectroscopy and microspectrophotometry can provide the same information, means that pyrolysis GC is becoming less frequently used in forensic laboratories.

13.2.3 Columns and Chromatographic Separation

Although capillary columns were introduced in 1959, it was only in the 1980s that their popularity increased and today more than 90% of all GC applications involve capillary columns (McNair and Miller 2008). In forensic laboratories, capillary columns are predominantly, if not exclusively, used and this section will therefore only focus on capillary columns,

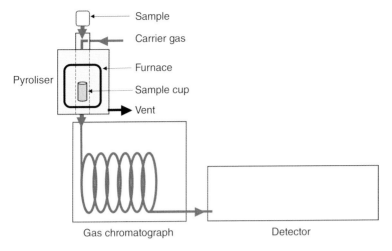

Figure 13.11 Configuration of a pyrolysis GC-MS system.

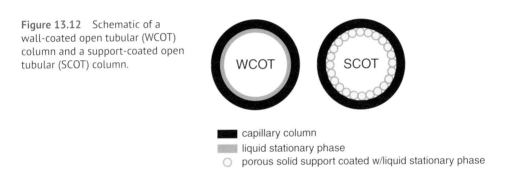

Figure 13.12 Schematic of a wall-coated open tubular (WCOT) column and a support-coated open tubular (SCOT) column.

as opposed to the original packed columns used in GC. The most commonly used capillary columns are open tubes that contain a thin film of liquid phase coated to the inner wall. They are more correctly referred to as wall-coated open tubular or WCOT columns, as opposed to support-coated open tubular (SCOT) columns where a porous solid support coated with liquid stationary phase is attached to the capillary's inner wall (Figure 13.12). Most commercial columns are now made of fused silica although earlier capillary columns included glass or stainless steel. The thin, uniform liquid phase on the inert fused silica surface generates a much higher efficiency (typically 3000–5000 theoretical plates per metre) than packed columns, explaining their popularity for most GC applications.

WCOT columns provide the highest resolution of all GC columns commercially available. Numerous manufacturers produce a diverse range of column sizes and types with internal diameters of 0.10, 0.20, 0.25, 0.32 and 0.53 mm and lengths varying from 10 to 50 m. Film coating thickness can range from 0.1 to 5.0 µm and is selected based on the analytes of interest. A thin film coating offers high resolution and fast analysis times but limited sample capacity. In contrast, a thick film coating will increase the sample capacity but reduce the resolution; hence this type of film is typically only used for heavy volatile compounds.

13.2.3.1 Column Selection

When selecting a column, five parameters must be considered: (i) internal diameter; (ii) column length; (iii) film thickness; (iv) stationary phase; and (v) flow rate. Each of these will be discussed briefly below with the exception of flow rate which has been detailed previously.

Internal diameters of commercially available capillary columns range from 0.10 to 0.53 mm (100–530 µm). The smallest diameter column, i.e. 0.10 mm, offers only a limited sample capacity and is not well suited for trace analysis, hence is not frequently used in forensic laboratories. The more commonly used internal diameter columns include 0.25 and 0.32 mm as they represent the best compromise between resolution, sample capacity, speed of analysis and ease of operation. Since busy forensic laboratories do not have the time to regularly change columns between analyses, these internal diameters are most commonly seen in GC systems for forensic analysis. The 'widebore' column with an internal diameter of 0.53 mm is also infrequently used in forensic laboratories due to the loss in resolution. Such columns are typically employed for the on-column syringe injections discussed previously.

Column length, L, is an important consideration because the longer the column, the more theoretical plates, N, and thus the better the separation. However, *resolution* is only proportional to the square root of the column length so doubling the column length, while also doubling the plate number, will have a smaller increase in resolution. Given that retention time, t_R, is also proportional to column length, a longer column will also increase the analysis time. However, if high resolution is imperative, then a longer column should be employed. Column lengths of 30 m are recommended for most applications, particularly in forensic laboratories as they provide a good compromise between resolution and analysis time. Short columns can be used for fast analysis of non-complex samples (i.e. with a limited number of compounds present) if warranted. Columns of 60 m or longer are typically only recommended for natural product analysis and are avoided in forensic laboratories because of their lack of efficiency due to lengthy analysis times.

In forensic chemistry applications, the standard film thickness for most capillary columns is 0.25 µm as this represents a suitable middle-ground between the high resolution of thin films and the high sample capacity of thick films. Such films offer the use of practical operating temperatures with reduced column bleed since the amount of bleed will be proportional to the amount of liquid phase in the column. Column bleed is typically visualised as a temperature-dependent increase in the baseline and is generally caused by degradation of the stationary phase, such that it is 'lost' from the column and elutes from the column where it is detected by the detector. Factors causing column bleed include heating at temperatures higher than recommended for the particular stationary phase, heating columns in the absence of carrier gas flow and in the presence of oxygen or exposure to high or low pH. Most liquid phases do not start to bleed until oven temperatures exceed ~250 °C. Thick films can be employed when greater retention of sample compounds is required, particularly volatile compounds. However, due to the higher temperatures required to elute compounds from thick films, a higher degree of column bleed also typically results.

Ideally, liquid stationary phases should be highly selective for the analytes of interest and able to operate at high column temperatures with minimal column bleed. This is particularly important for trace analysis which requires the use of sensitive detection systems such as flame ionisation detectors (FIDs) (Section 13.2.4) and mass spectrometers discussed in Chapter 10. The most stable stationary phases are covalently bonded to the fused silica and crosslinked, although phases can also be coated onto the column walls and then cross linked. Liquid phases are crosslinked by heating the capillary column to a high temperature which causes the methyl groups to form free radicals and subsequently crosslink, forming a stable, high-molecular-weight stationary phase. Chemical bonding with silanol groups on the fused silica surface can also occur, further strengthening the phase. Nowadays, most capillary columns are crosslinked providing a more temperature stable and longer lasting column.

13.2.3.2 Column Temperature and Programming

Column temperature is a very important variable in GC and, together with the stationary phase, is the most important factor in effecting a successful separation. As temperature increases, retention times and retention factors will decrease because the distribution constants are temperature-dependent. The temperature at which a given GC analysis is carried out depends on the boiling point of the sample components and the degree of resolution needed for each component. On non-polar stationary phases, the order of elution is closely linked to the volatility of the analyte, with a mixture of analytes eluting in order of decreasing volatility. Where the sample to be analysed consists of a single or a few analytes with similar chemical composition and similar, but different volatilities, it may be possible to carry out the analysis at a single temperature, i.e. isothermally. A complex mixture containing a broad range of boiling points will require the use of *temperature programming* to increase the temperature step-wise or continuously throughout the separation process. Figure 13.13 demonstrates the improvement in separation of sample components by using temperature programming rather than an isothermal method.

The method commences with sample injection where the column is typically held at a low isothermal temperature to allow highly volatile components time to interact with the column and for the sample solvent to vaporise and pass through the system ahead of the analytes. A temperature gradient is then incorporated into the method which is conducive to separating the analytes with intermediate to high boiling points by allowing their retention on the stationary phase for longer. A linear temperature gradient is often used to elute compounds over a large range of volatilities. Certain analytes may, however, require a non-linear temperature programme or a series of step-wise linear increases. For example, if a sample contains analytes with a range of volatilities, but two or more of the analytes have very similar volatilities, an isothermal period within the temperature programme may aid separation of these analytes.

As noted in Section 4.2, it is difficult to engineer an extraction procedure that extracts the analyte only. Usually, some other compounds in the sample matrix are co-extracted. Depending on their volatility, and the injection method, they may be transferred to the column. Provided these compounds have similar volatilities to the compounds of interest, they will be eluted through the temperature programme used to elute the analytes. However, if the compounds are less volatile than the analytes, the column needs to be heated to a temperature sufficient to elute these compounds but not higher than the maximum recommended temperature for the stationary phase. To avoid adding considerably to analysis

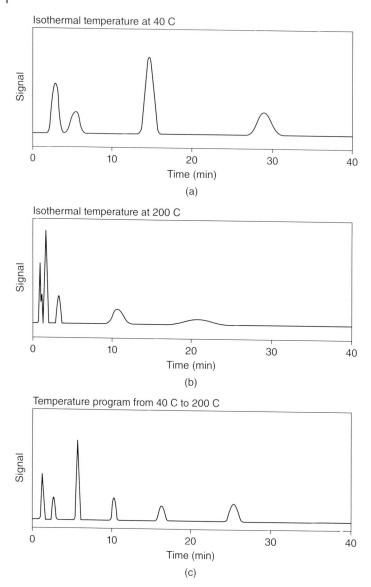

Figure 13.13 Comparison of a low and high isothermal analysis compared with a temperature programmed analysis: (a) isothermal temperature at 40 °C; (b) isothermal temperature at 200 °C; and (c) temperature programme from 40 to 200 °C. Source: Reproduced with permission from Dunnivant and Ginsbach (2011).

time, a rapid heating step is added to the temperature programme after elution of the analytes. It is important to elute co-extracted compounds from the system because otherwise they remain on the column. If many samples are injected, these compounds can build up on the column and cause problems in the analysis either by unwanted interactions with analytes or by eluting in subsequent injections of other samples producing 'ghost peaks'.

The aim is to separate (resolve) all analytes of interest from one another, and from any interferents, in the minimum time. Some experimentation may be required to achieve this aim. Software is available to model and optimise chromatographic separations. If the temperature programme cannot be adjusted to resolve analytes, a different stationary phase should be tried. Knowledge of the chemical nature of the analytes can be used to select the most appropriate stationary phase. Column manufacturers can also offer advice. If the analysis has been carried out on a non-polar stationary phase (e.g. 100% polydimethylsiloxane), and one of the analytes which cannot be resolved is polar and the other non-polar, using a slightly more polar stationary phase should bring about an improved separation.

Most GC systems allow the temperature of the analytical column to be heated in a controlled manner between 50 °C and 350 °C. This upper temperature limit is governed by the maximum temperature at which fused silica capillary GC columns can be operated. Temperature has a critical effect on separation in GC, so the temperature of the GC oven must be controlled for good separation. Different stationary phases have different maximum operating temperatures and the stationary phase will degrade rapidly if the column is heated above the maximum temperature. Non-polar stationary phases can be heated to higher maximum operating temperatures than polar phases. Column manufacturers specify the maximum operating temperatures for columns, usually specifying a slightly higher (circa 25 °C) temperature for isothermal analysis compared with temperature-programmed use.

When using a temperature programme for analysis, it must also be borne in mind that the column needs to be brought back to the starting temperature between analyses. If the temperature range of the programme is large, temperature re-equilibration can add considerable time to the analysis. Thus, the temperature should only be taken as high as is necessary to elute the analytes of interest and unwanted components which may be present.

13.2.4 Detectors and Detection Systems

Gas chromatographs can be coupled to numerous different detectors and the choice of detector will often depend on the field of application. Traditionally, GC has been coupled to FIDs, electron capture detectors (ECDs), nitrogen–phosphorous detectors (NPDs) and flame photometric detectors (FPDs) to name a few. It is now common for GC systems to be coupled with mass spectrometric detectors to enable both detection and identification of the components as they elute from the column. MS detectors commonly employed in forensic applications are discussed in detail in Section 10.2.4 and only a brief overview will be given here.

Ideally, a GC detector should demonstrate the following characteristics (Skoog et al. 2017):

- Good sensitivity.
- Good reproducibility and reliability.
- Limited fluctuation or variation over time.
- A linear response over several orders of magnitude.
- A rapid response time.
- A consistent response to all solutes or a selective response to one or more classes of solutes.

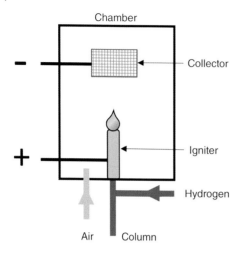

Figure 13.14 Schematic of a flame ionisation detector.

- Non-destructive.
- Easy to use.
- Not readily contaminated by samples and/or easy and quick to clean.
- The unequivocal identification of analytes.

It is not realistic to think that one detector would portray all of these characteristics and as a result, compromises are made based on the application. Some detectors are well-suited to specific applications such as the FPD for the analysis of air and water pollutants. Other detectors, such as mass spectrometers, can be used for a broad range of applications. This section will provide an overview of the FID, ECD, NPD and mass spectrometric detectors since they are the most commonly used detectors in forensic laboratories.

13.2.4.1 Flame Ionisation Detectors

Other than mass spectrometers, the FID is the most widely used detector for GC, particularly in forensic science. The flame within the FID is a mixture of air and hydrogen. As the solutes elute from the column, they are directed into the flame where they produce ions at high temperature (Figure 13.14). A voltage applied between the anode and cathode causes the ions to move towards the collector. The signal response or output is a plot of the current produced by the detector as a function of time. When no sample is being burned, there should be minimal ionisation occurring, and any small signal observed is a result of impurities in the air and hydrogen gases.

An FID is highly sensitive to compounds containing C—H bonds. The more bonds present, the stronger the signal response. It is classed as a mass-sensitive detector rather than a concentration-sensitive detector because it responds to the number of carbon atoms entering the detector over time. As a result, changes in flow rate will have minimal impact on the detector response which is advantageous for some analyses.

An FID is insensitive to carbonyl, alcohol, and amine functional groups, as well as halogens, as they produce few or no electrons in a flame. Non-combustible gases such as noble gases, SO_2, NO_x, CO_2, CO, and H_2O will not be detected by an FID either. The detector is therefore ideal for the analysis of most organic samples, especially those contaminated with water or the oxides of nitrogen and sulfur.

The advantages of an FID include a high sensitivity to organic compounds, large linear response range, minimal background noise, and ease of use. The major disadvantage of an FID is the destruction of the sample, the additional gases required and, for forensic science, the lack of information provided about the chemical identity of detected compounds. GC-FID is largely being replaced by gas chromatography–mass spectrometry (GC-MS) in most forensic laboratories because of the power of mass spectrometry to provide simultaneous compound identity.

13.2.4.2 Electron Capture Detectors

The ECD functions by passing the eluted solutes over a radioactive β emitter such as nickel-63 (^{63}Ni). The ^{63}Ni emits beta particles in the absence of analytes which collide with the nitrogen carrier gas to produce many more electrons. These electrons produce a constant current when collected by a positive electrode. The current will decrease when an electronegative analyte elutes from the column and enters the detector because it will capture some of the free electrons. The negative ions that are formed have slower mobilities than free electrons and will not be collected by the positive electrode. The degree of capture is subsequently proportional to the concentration of the analyte.

The ECD produces a selective and highly sensitive response, particularly to compounds such as halogens, peroxides, and nitro groups. The ECD has found widespread use in environmental forensics due to its selective response to halogen-containing compounds present in pesticides (particularly chlorinated insecticides) and polychlorinated biphenyls. In forensic laboratories, it is often employed for the analysis of explosives and explosive residues and halogen-containing drugs. In addition to being highly sensitive and selective to certain functional groups, the ECD does not destroy the sample to the same extent as an FID although, in practice, post-detector trapping of compounds is rare and virtually unknown in forensic science. The major disadvantages of ECDs is the necessity of a radioactive source and their proneness to contamination requiring ultrapure gases and clean samples for analysis.

13.2.4.3 Nitrogen–Phosphorous Detectors

As the name suggests, an NPD is selective towards compounds containing phosphorous and/or nitrogen. The eluting compounds are mixed with hydrogen and introduced into an ignited flame, similar to an FID. The hot gases flow around an electrically heated bead of an alkali salt forming a plasma with temperatures in the range of 600–800 °C. Any phosphorous- and nitrogen-containing compounds within the plasma produce a large number of ions and subsequent current signals, allowing the detector to determine when these compounds are present.

The NPD is approximately 500 times more sensitive to phosphorous-containing compounds and 50 times more sensitive to nitrogen-containing compounds than the FID. These characteristics make the NPD ideal for detecting phosphorous-containing pesticides in environmental samples and nitrogen-containing drugs and metabolites in toxicological samples.

13.2.4.4 Mass Spectrometric Detection Systems

Mass spectrometers represent a powerful detection system with the capabilities to both identify (through structure, elemental composition, and molecular weight) and quantify separated components of a mixture. In forensic science, it is the preferred detector for GC analysis of evidence. The coupling of GC with low resolution mass spectrometry is more commonly referred to as GC-MS. There are many configurations for mass spectrometers as detectors but the following discussion focuses on the type of benchtop GC-MS system that is used by virtually all forensic science laboratories. Other laboratories may have more complex, expensive mass spectrometers with more extensive capabilities for use in specialised analyses but they will probably have several of the following GC-MS systems and the bulk of analyses will be carried out on such instruments.

The mass spectrometer consists of an ion source, a mass analyser, a detector and a data system for reporting the output (see Figure 10.3 for a schematic). Due to the low flow rate from the column (~1 ml/min), the eluting components from GC can be directly transferred under vacuum to the ionisation chamber of the mass spectrometer. The most commonly used technique is electron ionisation (EI) where ionisation of the molecules is accomplished by impact with a highly energetic electron beam. The electrons excite the analyte molecule causing ionisation and fragmentation. The ions are then attracted or repelled by charged lenses into the mass analyser.

An alternative means of ionisation that is occasionally used in explosives analysis and in some other applications is chemical ionisation (CI). This technique employs a reagent gas (such as methane) in the ion chamber which produces a cation that will undergo further reactions with the production of secondary ions. These secondary ions impart less energy than EI. The result is a far simpler mass spectrum than EI with less, and sometimes no fragmentation. This is valuable for determining the molecular mass of an analyte as the major peak is a readily identifiable protonated molecular ion peak, $[M+1]^+$.

A mass analyser can be thought of as a mass filtering device which separates ions by magnetic or electric fields based on their mass-to-charge ratio (m/z). Common mass analysers include quadrupole and ion trap analyzers although the time-of-flight mass analyzer is also becoming more popular in forensic applications. Triple quadrupole mass spectrometric systems can also be coupled to GC, offering possibilities for increased sensitivity of analysis, but with an increased cost compared with single quadrupole systems. Further details about each type of mass analyser are provided in Section 10.2.4.

Mass spectrometers can typically be operated to monitor a wide range of ions or to monitor only a limited range. Both ways of operating have advantages and disadvantages. In scan mode, the mass analyser is set to analyse a wide m/z range. In so-called *selected ion monitoring* (SIM) mode the mass analyser is set to monitor ions of m/z specific for the analytes of interest. The two different modes of operation have consequences in terms of sensitivity and specificity of analysis and data output.

Taking a single quadrupole mass analyser as an example, the mass analyser scans across the m/z range continuously and repeatedly. The full scan range is of the order of 1–1000 amu for modern fast scanning GC-MS systems, typically taking 1 second to scan across this mass range. The scan time must be balanced against the mass range being scanned. A large mass range will reduce the amount of time spent scanning each m/z, thereby reducing the overall sensitivity of the analysis because, effectively, less ions of each m/z will be allowed through the quadrupole to reach the electron multiplier during the elution of a substance compared with scanning a narrow mass range. Hence, to maximise sensitivity, the mass range selected for scanning should be as small as possible but encompassing the anticipated mass range of analytes. The scan time is also important in terms of the time it takes for a GC peak to transit the mass spectrometer. The generally recognised recommendation is that about 6–10 scans should be collected across the width of a peak. This helps ensure that the area under the peak calculated by the mass spectrometer software is a precise and accurate value.

In scan mode with EI there is generally extensive fragmentation of the parent compound. An electron multiplier detector detects the ions transiting the quadrupole and their abundance (the number detected). This enables a mass spectrum to be generated – a plot of m/z of the ions detected versus their abundance, plotted relative to the m/z of the most abundant ion detected. The mass spectrum can be searched against mass spectral databases or compared with the mass spectrum of a standard of the suspected compound analysed under similar conditions.

SIM analysis offers higher sensitivity and specificity compared with scan analysis. Higher sensitivity results from SIM analysis because, effectively, the system is monitoring just a few ions and hence, in a given time period, more ions of that m/z will be allowed to transit the quadrupole and reach the detector compared with a scan analysis. On account of the higher sensitivity offered, SIM analysis is recommended for quantitative analysis of targeted compounds at trace levels (down to ppb).

There are some potential downsides to SIM analysis which the analyst must consider. Since the system is set to monitor specific ions, only compounds containing those ions are detected. Whilst this has the benefit of decreasing the background noise and hence contributing to increased sensitivity, it also means that a compound whose presence in a sample might be of interest will not be detected if it does not fragment to produce the ions monitored, i.e. useful evidence might be missed. It is also possible that a compound may be present in a sample which fragments to produce one or more of the ions being monitored. If this compound elutes at the same, or very similar, retention time as an expected analyte, without additional information, the analyst will assume that the expected analyte is present. Hence it is recommended that at least two, or possibly three ions, should be monitored for each analyte. Peak area ratios (ion ratios) can be measured for each compound and compared with the ratios for standards. There should be reasonably good agreement between ion ratios for samples and standards. If there is not, the analyst should reconsider whether the compound detected is what they think it is. SIM analysis also means that it is not possible to generate a full scan mass spectrum if any unexpected peaks are seen in the SIM chromatograms. However, the sensitivity advantages offered by SIM outweigh the possible disadvantages, plus a full scan analysis can be carried out to check if any unexpected compounds are present.

GC-MS is used extensively for the analysis of natural and biological samples. In a forensic laboratory it is regularly employed for the analysis of fire debris, explosives and explosives residues, drugs and toxicological analyses, trace evidence such as inks and dyes, and environmental pollution. In the broader scientific arena, it is used by the food and beverage, petrochemical, environmental, and medical industries to characterise samples of interest.

13.3 Application to Analyte

When GC is employed to analyse a complex mixture, it has two primary functions. The first is to separate the sample components of the mixture. GC is the method of choice for separating complex organic, organo-metallic and biochemical mixtures, assuming the species are volatile at the temperatures operated during GC or can be derivatised (chemically reacted) to produce volatile species. The second function is both qualitative and quantitative. The retention times of target compounds can be compared with those of known standards to help confirm the identity of the compounds. Peak areas can be measured and compared with the peak area of a standard of the compound to quantify the amount of compound present in a sample.

For samples to be analysed by GC, they must first demonstrate several characteristics. Initially, the sample must be transferred to the vapour phase in order that chromatographic separation can take place. This requires that the sample analytes be sufficiently volatile within the temperature range of the instrument (up to 350 °C depending on the stationary phase of the column). The volatility of a compound is dictated by its vapour pressure and boiling point. Highly volatile compounds have high vapour pressures and low boiling points which make them ideal for separation using GC.

The size of the analyte can assist in determining its volatility and potential suitability for analysis by GC. Generally, a compound with a molar mass greater than 600 g/mol will have a level of volatility below that considered satisfactory for GC. Additionally, smaller mass compounds which have polar functional groups are sometimes difficult to analyse using GC because they have very high boiling points and low vapour pressures. Hence, the boiling point (or estimated boiling points) of the analytes should be used to indicate whether GC is a suitable analysis technique. If the compound has a boiling point below 500 °C at 1 atm, it should be sufficiently volatile for analysis by GC. If in doubt, a sample can be injected onto the column to determine whether the compound elutes in a reasonable time.

In addition to volatility, the analyte must also be thermally stable and not prone to thermal degradation. An unstable analyte may degrade under the higher temperatures used in the injection port or column, introducing additional unknown peaks into the chromatogram and making it difficult to identify the analyte(s). Thermal degradation can be minimised by varying the sample injection technique or by derivatising the analytes as discussed below. Cold on-column injection can overcome problems of thermal degradation in vaporising injectors.

13.3.1 Sample Derivatisation

While numerous analytes are sufficiently volatile for injection into a gas chromatograph, there are many that are not well suited to GC analysis due to their low volatility or thermal instability. This problem can often be solved by changing the chemical structure of the compound through *derivatisation* (a chemical reaction) to make it more volatile or thermally stable. Derivatives are formed by replacing the more polar groups of an analyte (e.g. hydroxyl or amine groups) with non-polar groups. Derivatising agents usually target the hydrogen atoms that may form hydrogen bonds with the stationary phase or with uncoated silanol groups on the fused silica column. Hydrogen bonding interactions are relatively

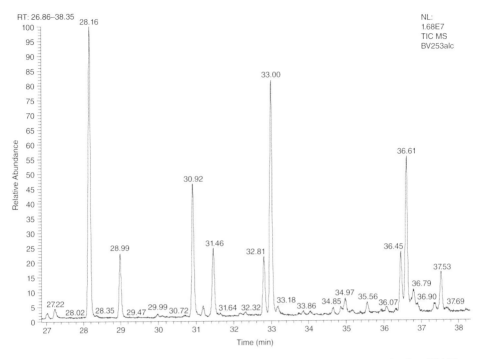

Figure 13.15 Total ion chromatogram showing TMS derivatised sterols analysed using GC-MS. Source: Reproduced with permission from von der Lühe et al. (2013).

strong and such interactions generally produce long retention times or, in extreme cases, complete retention of compounds on the column. The derivatised analyte will have lower volatility, be more thermally stable and undergo less hydrogen bonding interactions with the column.

The main chemical reactions involved in derivatisation processes include silylation, acylation, or alkylation. A common example of derivatisation is the silylation of sterols. Large sterols have low volatility due to their relatively high molecular mass and the presence of a hydroxyl group which can form hydrogen bonds as described above. Sterols injected directly into a gas chromatograph typically demonstrate broad peaks and unreasonably long retention times. To overcome these issues, the alcohol group is replaced with a trimethylsilyl (TMS) group to form a *TMS derivative* of the sterol. The addition of the less polar TMS group will increase the thermal stability of the sterol. Although the mass has increased, the volatility of the derivative increases because of reduced intermolecular hydrogen bonding and minimised hydrogen bonding interactions with so-called 'active sites' (e.g. silanol groups) on the GC column. The chromatogram of the TMS derivatives will therefore demonstrate well defined, narrow peaks in a shorter retention time, improving the overall analysis of the sterol compounds (see Figure 13.15 for example chromatogram).

In addition to increasing volatility and thermal stability, derivatisation can also improve the chromatographic behaviour of a compound, for example by reducing peak tailing and improving peak shape. It can also be used to alter the separation process by changing the positions of overlapping or poorly resolved peaks. Derivatisation can be used to enhance

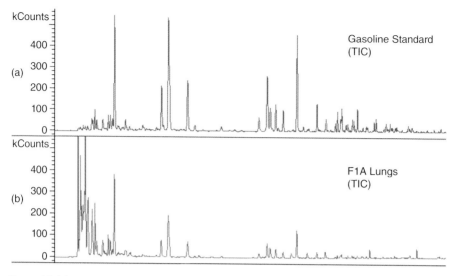

Figure 13.16 (a) The chromatogram of a gasoline standard analysed using GC-MS. (b) The chromatogram of a toxicological analysis of a lung sample taken from a deceased fire victim analysed under the same analytical conditions. Identification of peaks with the same chemical structure at the same retention times confirmed the presence of gasoline residue in the victim's lungs. Source: Image courtesy of K. Pahor.

detection of sample components by specific detectors. For example, halogen groups may be added to an analyte to enhance their detection by an ECD.

13.3.2 Qualitative Analysis

GC is a valuable technique for confirming the presence or absence of compounds in a mixture. Target analysis of these compounds can be achieved by purchasing pure standards and determining their retention times under the same conditions under which the sample is analysed. A match in retention time for a peak in a sample and standard is indicative that the compound is present in the sample (Figure 13.16).

However, unless a detector is used which provides information on the chemical identity of the substance, a match in retention time only may not be sufficient to confirm the peak identity unequivocally. Ideally, the result in terms of retention time match should be able to be replicated under different temperatures and on different columns. You would not expect the same retention time for a particular compound under different temperature programmes on different columns and that is the strength of the approach of carrying out analysis under different conditions to confirm identity. The chance of two different compounds having the same retention time under different analytical conditions reduces relative to the number of different conditions tested. Changing the nature of the stationary phase can be particularly discriminating because two compounds that have similar vapour pressures but differ in terms of functional groups are likely to have different retention times on markedly different stationary phases. However, the approach of repeating analysis on different columns is very time consuming and the power of mass spectrometry to provide information about the identity of a substance can usually obviate the need to analyse on different systems to confirm identity.

GC-MS is the preferred methodology for confirmation of chemical species within a mixture. While other GC detectors may be available in a forensic laboratory for screening purposes, GC-MS is considered the only confirmatory GC technique for the qualitative and quantitative analysis of forensic evidence. GC is also useful for determining the purity of compounds or the effectiveness of a purification process. The gas chromatogram will show additional peaks that have resulted from contamination, assuming these substances are volatile and are detected by the detector. The lack of a peak at the retention time expected for a standard of the compound analysed under the same conditions demonstrates the absence of that compound in the mixture or that the compound is present at a concentration below the limit of detection of the instrument. Hence, just because a peak is not present at the expected retention time, the analyst should not report that the compound is not present in a sample. Instead, it is usual to determine the limit of detection for the analyte for the method employed and then report the analyte as not detected, quoting the limit of detection of the method.

13.3.3 Quantitative Analysis

Quantitative data analysis can be carried out by: (i) measuring the peak height; or (ii) integrating the area under the peak. Integration of peak area is the preferred method, as it accommodates for variation in the chromatographic analysis such as column temperature, flow rate, or sample injection reproducibility.

Quantitative analysis generally requires a good separation of sample components in order to satisfactorily measure the peak areas of analytes, i.e. other sample components in the mixture should not co-elute with the analytes to be measured (see Figure 13.17 for an example of co-elution). This aspect is particularly important when using an FID which responds to the presence of virtually all organic compounds. If two or more compounds co-elute, the detector cannot differentiate between these compounds and the area under the peak becomes the sum of all the co-eluting compounds. ECDs are relatively insensitive to many compounds and, in effect, do not detect many compounds unless they are present at relatively high concentrations. Nevertheless, a co-eluting compound may influence the response to the detector and it is good practice to ensure full separation of compounds when using an FID or ECD for quantitative analysis. The same holds true when using an NPD.

Ensuring full resolution of compounds is not as critical, although still good practice, when using GC-MS because ions can be mass selected for identification and quantification. It is possible to use the mass analyser to permit only ions from the analyte to pass to the electron multiplier for detection and to filter out ions from co-eluting compounds (SIM analysis). However, this is only useful for targeted analysis where the expected compounds are known prior to analysis.

Ion suppression does not occur in EI GC-MS and so co-elution of compounds does not cause the problems that can often occur in LC-MS. The general recommendation for quantification in GC-MS is to optimise the GC separation to resolve all analytes of interest from each other and from other compounds present in the sample. As noted above, for SIM analysis, two or three ions can be monitored for each analyte and ion ratios calculated for samples.

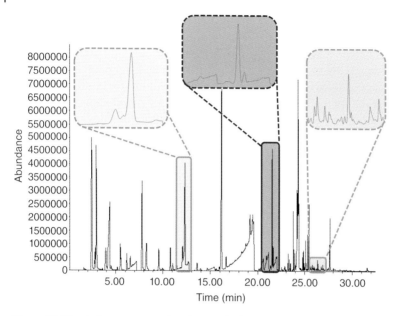

Figure 13.17 A chromatogram showing co-elution of peaks (in highlighted sections) and poor resolution leading to inaccurate quantification of peak areas. Source: Image courtesy of K. Perrault.

When making quantitative measurements, due attention should be paid to sampling, sample preparation, method validation, and quality assurance. Errors, in the sense of analytical errors, are an unavoidable part of these processes but can be accounted for as long as they are identified. As the volume of sample analysed decreases, especially in trace analysis of forensic samples, additional care must be taken to reduce or avoid errors as their magnitude can easily invalidate the data. Here we will focus on highlighting the types of errors that can be avoided in GC analysis.

Errors can occur at any step in the total procedure including during sampling, sample workup, the chromatographic separation process, detection, data analysis including peak area determination, and data calculations and interpretation. Given the major improvements in GC systems over the past few decades, particularly relating to their software capabilities, the major sources of error are usually in sampling and sample preparation. The objective of sampling is to collect a small sample that is still representative of the whole. Sample preparation can include a range of procedures such as dissolving, filtering, extracting, concentrating, and chemical derivatisation. Each of these procedures, if not conducted correctly, is prone to loss of sample (and hence analyte[s]) or to contamination. The addition of an internal standard (described in Section 13.3.3.1) is valuable for identifying or taking into account such losses as long as it is added before sample processing commences.

In forensic science, the collection of the original sample and handling of the sample before it is received in the laboratory and stored may not have been done by the scientist who carries out the analysis. Hence the analyst needs to bear this in mind and, if necessary, get in contact with the person who collected the sample should there be suspicions at any point during the analysis and data interpretation that there may have been contamination or some other procedure, as part of sample collection and transport to the laboratory, which may have adversely affected the analytical findings.

13.3.3.1 Methods of Quantitative Analysis

Area normalisation is the simplest and least accurate method of quantitative analysis and is essentially a calculation of the area per cent, which is assumed to be equal to the weight per cent (see also Chapter 2). The area per cent of an unknown analyte, X, is calculated as follows:

$$\text{Area}\%X = \left[A_x / \sum(A_i)\right] \times 100 \tag{13.1}$$

where A_x is the area of X and the denominator is the sum of all other areas. The accuracy of this method assumes that all compounds have eluted and been detected adequately and that they have the same sensitivities (i.e. response/mass). This assumption is rarely accurate, but the method provides an easy and fast semi-quantitative analysis, which may be suitable for certain applications (such as when pure standards of an analyte are not available). If standards are available, they can be analysed to obtain the relative response factor, f, for the analyte of interest which improves the accuracy of this method. The f value is calculated by assigning one of the analytes an arbitrary f value of 1.00. A mixture of this analyte (deemed the standard) and other analytes is made and analysed using GC. The peak areas for the standard, A_s, and the unknown analyte of interest, A_x, are measured and the relative response factor for the unknown, f_x, is calculated using the following formula:

$$f_x = f_s \times [A_s / A_x] \times [W_x / W_s] \tag{13.2}$$

where w_x / w_s is the weight ratio of the unknown to the standard.

Using external standards is another method of quantitative analysis which is particularly valuable in GC because the results can be included and interpreted by the software of the data system. A calibration curve is constructed by analysing known amounts of an analyte across a range of concentrations and plotting the peak areas (y-axis) against the concentration (x-axis). The curve can then be used to determine the concentration of an unknown amount of the same analyte in a mixture. The phrase 'calibration curve' is commonly used but, in reality, most analysts work within the concentration range where the detector response versus concentration gives a linear relationship. The concentration range of the standards should encompass the concentrations found in samples. If the concentration in a sample falls outside the concentration range of the calibration curve, further standards should be prepared at a more suitable concentration and analysed alongside the sample. Alternatively, if the concentration of the sample is higher than that of the highest standard, the sample can be diluted and reanalysed alongside selected standards. Reproducibility and accuracy of the injections are important for constructing calibration curves and manual injections are therefore not recommended for this method. Nowadays, most forensic laboratories only use autosamplers as these provide superior results.

Where manual injections are necessary or the method is not particularly reproducible, internal standards (ISs) are a better option for quantitative analysis. The compound selected for use as an IS should not be present in the mixture being analysed but, ideally, should be chemically similar to the analytes of interest. It should be well-resolved from all analytes in the mixture but should also elute close to the peaks of interest so as to not unnecessarily increase the analysis time. The IS is added to each sample at a concentration which approximates the concentration of the analytes and should be included prior to any sample preparation (such as derivatisation). In this way, if there is loss of the analyte through the sample preparation procedures, or if there is variability in the exact volume of sample injected, there should be the same variation for both analyte and IS. The IS is used to determine the concentration of analytes from measured ratios. The ratio of the peak area of the unknown analyte concentration in the sample, (A_x), to the peak area of the IS (A_{IS}) is obtained (A_x/A_{IS}). The ratio is converted to a concentration by comparing it with the area ratios of known analyte concentrations which include the same amount of IS. A graph can be generated with the y-axis representing the ratio of the peak area of the analyte (A_n) to the peak area of the IS (A_{IS}) and the x-axis representing the ratio of the weight of the analyte (W_n) to the weight of the IS (W_{IS}). From this graph, the concentrations of the analyte(s) (A_x) can be determined.

Where GC-MS is used for analysis, isotopically labelled standards of the analytes of interest are preferred, if available. Those most commonly employed are deuterated analogues. They are advantageous because, assuming they are appropriately labelled, they should behave in the same way as the unlabelled analogue in samples and standards through all stages of sample preparation and analysis. Caution must be observed to ensure that no label can exchange between the isotopically labelled IS and the non-labelled analogue in samples or standards (e.g. via H exchange) as this would invalidate quantification.

13.4 Interpretation and Law

GC is extensively used in forensic laboratories and is almost universally employed in the analysis of fire debris as well as drug and toxicological evidence. Until recently, GC-FID was considered an acceptable technique for the analysis of fire debris and even today, some laboratories still use this detector for analysis. Hydrocarbon accelerants/ignitable liquids, such as gasoline, kerosene, diesel, and fuel oils, are complex samples which contain hundreds of components. The resulting chromatogram will be a pattern of characteristic peaks for a particular ignitable liquid type (Figure 13.16). The accelerant can be identified based on pattern recognition because the sample components typically occur reproducibly in specific ratios and a recognisable pattern. However, this method does not allow for specific components to be identified. Such information is particularly useful in determining the manufacturer of an ignitable liquid or for comparing with a known sample. For this reason, GC-MS is now the preferred technique for fire debris analysis as it allows individual components to be identified, even among background contaminants or within mixtures of ignitable liquids. SIM is commonly used for identifying these components by selecting ions that are characteristic of a particular ignitable liquid.

In the forensic laboratory, GC-MS has been considered the 'gold standard' in terms of sample analysis and data presentation for court reporting purposes. However, it is not sufficient to analyse samples only. Blanks, controls, references, and standards must also be analysed along with the sample to maintain quality control and quality assurance of the method, instrument, analyst/scientist, and the laboratory.

Blank and/or control samples are used to determine the background levels of organic compounds of interest, or compounds present in the control which might be detected and which could interfere with quantification of the compounds of interest. In a drug analysis, a blank sample might simply be the solvent used in the sample preparation. In a toxicological analysis, a blank sample might be urine or blood known not to contain the compound of interest.

A blank sample should not contain the compounds of interest otherwise it is not possible to confirm the presence of the analytes in the sample because they could be present due to contamination during the analysis. In fire debris analysis, the sample being analysed might be the carpet on which a suspected ignitable liquid was poured. A part of the carpet considered to be free of the suspected ignitable liquid residue must also be sampled and analysed to check whether it contains any compounds corresponding to those present in the liquid. A false positive can result if this information is not known, i.e. the presence of an ILR will be reported when, in fact, none was present, the compounds detected having been derived from the carpet itself.

Standards are generally considered necessary for all quantitative chromatographic analyses in forensic science. For example, if a cocaine sample was submitted as evidence and the forensic scientist wanted to determine the per cent composition of cocaine in the sample, an IS would be added prior to analysis. A known amount of a reference sample of cocaine including the IS is analysed by GC-MS and compared with the evidence sample containing the same concentration of IS. A calculation of peak area proportions will assist in determining the per cent of cocaine in the evidence sample.

The per cent composition may be required by the investigating agency to determine where in the distribution chain a seized drug is represented. A higher concentration of cocaine in a sample generally represents a higher position in the distribution chain. For example, a sample containing 80% cocaine would more likely be seized from a distributor compared with a sample containing 40% cocaine which would more likely be seized from a user; the sample having been diluted as it passes through the distribution chain to maximise profit. Depending on the legal system, this may have implications in terms of the penalty imposed for possession and/or distribution.

Quantitative analysis is also imperative when GC-MS is used to confirm BACs. The level of alcohol detected in an individual's blood will dictate whether that individual was driving under the influence of alcohol and the subsequent penalty imposed.

In fire debris analysis, a comprehensive library of reference samples and chromatograms is necessary for the effective interpretation of data. Given the significant number of ignitable liquids commercially available including variations in manufacturers and suppliers, a forensic laboratory should attempt to analyse a large proportion of available materials on the same GC-MS with the same parameters as those used for sample analysis. Importantly, samples of neat, weathered, partially burned, and almost completely evaporated ignitable liquids should be analysed to cover the breadth of samples that may be submitted to a laboratory. Chromatograms of these references are stored in the library for subsequent comparison with fire debris samples.

Difficulties in compound identification also arise in the analysis of ILR extracted from fire debris. Often the GC chromatogram of the extracted sample looks distinctly different to reference standards since it has been subjected to high temperatures during the fire and some of the more volatile components may have evaporated. Contaminants from the fire scene may also be present in the extracted sample and can co-elute or mask compounds of interest. In order to conclude that a class or type of ignitable liquid is present, the scientist must be able to identify a majority of characteristic peaks for that type of ignitable liquid with few unidentifiable or unexplainable peaks. If a majority of the characteristic peaks are missing or are not identifiable due to a poor-quality chromatogram, then an inconclusive or negative finding is warranted. Importantly, a negative finding does not exclude the use of an ignitable liquid in the fire. ILR may not have been detected by GC-MS because it was completely consumed in the fire; its concentration was below the detection limits of the instrument; or the fire debris containing the residue was not collected near the point of origin. All of these issues must be considered when reporting the findings of the analysis.

When identifying a compound using a mass spectral library, a probability or quality index will be provided to indicate the level of similarity between the analyte mass spectrum and the closest library mass spectrum. This value will rarely be 100% since the library mass spectrum was produced under different conditions and on a different instrument to the analyte spectrum. However, a high similarity is required between the spectra to confirm the identity of the unknown compound. Identification of common drug compounds can be relatively straightforward because mass spectra are likely to be present in commercial reference libraries, or in-house libraries can be made from drug samples of known origin and identity. However, the continual influx of new designer drugs on the market makes some identifications difficult until reference samples are available.

In forensic laboratories, it is good practice (and often required) that evidence analysis follows ASTM standard guidelines, or standards from other nationally or internationally recognised organisations. ASTM International is a standards organisation that develops and regularly updates standard practices, methods, and tests in the laboratory. Use of ASTM guidelines provides the laboratory with a level of validity and quality assurance which can be demonstrated to accrediting organisations and the criminal justice system if required. ASTM methods commonly employed for the sampling and analysis of seized drugs and clandestine drug laboratory evidence include ASTM E2548 *Standard Guide for Sampling Seized Drugs for Qualitative and Quantitative Analysis,* ASTM E2329 *Standard Practice for Identification of Seized Drugs*, and ASTM E2882 *Standard Guide for Analysis of Clandestine Drug Laboratory Evidence*. Methods commonly employed for the analysis of ILRs include ASTM E1387 *Standard Test Method for Ignitable Liquid Residues in Extracts from Fire Debris Samples by Gas Chromatography* and ASTM E1618 *Standard Test Method for Ignitable Liquid Residues in Extracts from Fire Debris Samples by Gas Chromatography–Mass Spectrometry*. The United Nations Office on Drugs and Crime (UNODC) have also produced a series of recommended methods for the analysis of a number of controlled drugs.

13.5 Case Studies

13.5.1 Case Study 1

GC-MS is used in forensic toxicological laboratories for the analysis of unknown exogenous compounds (e.g. illicit drugs, prescription drugs, alcohol, etc.) in biological tissue.

However, the analysis and interpretation of toxicological results in post-mortem cases can be a challenging task due to the rapid degradation of soft tissues following death, even following embalming of the remains. A buried body may be exhumed in an attempt to establish concentrations of drugs and their metabolites in the body at the time of death to establish cause of death. This presents an even greater challenge due to the time that has elapsed between death and exhumation of the body, which is often weeks, months, or years. Typically, samples of blood or urine are used for drug quantification but in longer post-mortem cases liver, kidney or other soft tissue will be analysed. An interesting case study by Cippitelli et al. (2018) reports on the toxicological analysis of opiates collected from a range of matrices in an exhumed body that showed a satisfactory degree of preservation after one year of burial and without embalming.

The case involved a 45-year-old man who had died from cardiac arrest with pulmonary oedema. New evidence which came to light after the decedent was buried suggested a history of drug abuse and the body was exhumed one-year post-mortem. As a result of the moderate preservation of the body, a venipuncture mark was still visible on the decedent's arm with underlying blood infiltrate. Although blood, urine and gastric contents were no longer available, samples of hair, liver, kidney, and teeth were collected (Cippitelli et al. 2018). Liver and kidney samples were subjected to a two-step pre-injection process prior to analysis by GC-MS. Two aliquots of the same samples were subjected to LLE in a basic and acidic environment. The basic extracts were analysed by GC-MS before and after derivatisation, while the acidic extracts were analysed by GC-MS before and after methylation on-column. All results suggested the presence of opiates.

In an attempt to quantify the amount of opioids in the different matrices, a further analysis was performed involving pretreatment of the samples with acid hydrolysis followed by SPE. Hair samples were finely cut, and teeth samples were pulverised for analysis. All samples were subjected to an SPE extraction, followed by derivatisation and analysis by GC-MS. Morphine and 6-monoacetylmorphine (a metabolite of heroin) was detected in all matrices analysed. Their presence in hair indicated repeated use over a long timeframe, in this case at least 5–6 months. It was hypothesised that the metal-shell coffin in which the body was buried contributed towards a slower rate of decomposition, and thus preservation of useable tissues for the detection of opioids one-year post-mortem (Cippitelli et al. 2018).

13.5.2 Case Study 2

GC-MS is a commonly employed technique for the analysis of volatile and semi-volatile compounds. In fire investigations, GC-MS is used to detect the presence of ILRs in fire debris. Detection of ILRs can be challenging due to the presence of pyrolysis products and combustion-derived volatile substances (CVSs) which result from the burning of buildings and the thermal decomposition of furniture. However, GC-MS can also be employed to detect the presence of CVSs in a victim's blood post-mortem for the diagnosis of burning death. This diagnosis is typically based on anatomical autopsy findings such as inhalation injury and soot in the respiratory tract, as well as toxicological findings such as blood carboxyhaemoglobin (COHb) levels >10% and blood cyanide concentrations >0.25 µg/ml. In some cases, however, findings can be incomplete or contradictory. A case study by Suzuki et al. (2017) was able to demonstrate the value of GC-MS for the analysis of blood CVSs in fire victims. Such analysis enabled the determination of the type of fire-related death, a diagnosis not possible with traditional autopsy findings.

The study analysed and compared blood collected from the heart of 61 cadavers during autopsy – 36 building fire victims, 17 victims of self-immolation using kerosene, 6 victims of self-immolation using gasoline, and 2 outdoor fire victims, to examine a cross-section of fires involving buildings and furniture. The combustion products were pre-concentrated using a specialised extraction needle packed with a sorbent phase to trap the CVS. Following collection of the CVS, the needle was attached to a gas-tight syringe and the volatile compounds directly desorbed into a GC-MS system.

A range of 39 types of aliphatic and aromatic hydrocarbons were detected in the post-mortem blood sampled. Detection of α-methylstyrene and large concentrations of styrene was indicative of a building fire. Detection of C9–C12 aliphatic hydrocarbons and C3 alkylbenzenes was indicative of a kerosene-based fire, while detection of large concentrations of toluene with C3 alkylbenzenes was indicative of a gasoline-based fire. This case study demonstrated that it was possible to differentiate the type of fire based on the CVS detected in the victim's blood using GC-MS, even in victims who showed no signs of soot in their respiratory tract and who returned COHb and cyanide levels below those typically associated with a burning death (Suzuki et al. 2017).

13.6 Forensic Developments

GC provides one of the highest levels of resolution for the separation of sample components in complex mixtures. Most GC systems can separate hundreds of peaks in a single chromatogram. However, there are still many complex samples, such as petroleum-based fuels, which cannot be adequately separated using GC as they comprise thousands of individual substances. GC analysis of these complex samples results in significant co-elution and peak overlap due to the inability of GC columns to separate the sheer number of compounds present (Figure 13.17). Attempts to resolve these issues have led to the introduction of selective detectors (e.g. ECD or mass spectrometer) but this will typically decrease the general use of the instrument and increase the cost. Using an alternative stationary phase may improve the resolving power for the co-eluting peaks but may also lead to new co-eluting peaks. Hence, new multidimensional GC techniques have been developed which can satisfactorily resolve thousands of compound peaks as outlined in the following.

13.6.1 Multidimensional GC

Introducing an additional chromatographic separation stage is the most effective approach for improving the resolving power for very complex samples. Multidimensional chromatography utilises more than one column to separate the components of a single sample. Depending on the method employed, an aliquot of the sample or the entire sample will be collected after the first column and transferred onto a second column. The two columns contain stationary phases with different polarity or other physical properties that provides a second mechanism for separation.

Multidimensional GC can employ two GC columns in series (Figure 13.18) or can combine an HPLC column with a GC column. The major advantage of these methods is an increased peak capacity (i.e. the number of peaks that will fit in the space of a

13.6 Forensic Developments

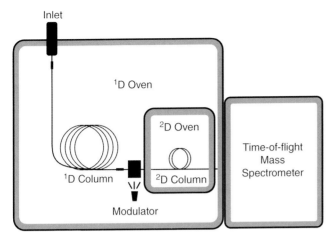

Figure 13.18 Schematic of the components of a GCxGC-TOFMS system including two columns in series coupled by a modulator. Source: Image courtesy of K. Perrault.

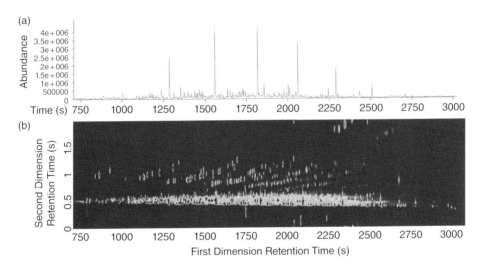

Figure 13.19 Example first dimension (a) and second dimension (b) chromatogram of unweathered kerosene showing increased peak capacity by an order of magnitude: GC = 231 peaks detected; and GCxGC = 2384 peaks detected. Source: Reproduced with permission from Nizio et al. (2016).

chromatogram). In two dimensions, peak capacities can be considered separately or combined. Comprehensive two-dimensional separations are superior because the peak capacities of the two columns are multiplied (rather than added), generating potentially thousands of peaks in a single chromatogram (Figure 13.19). Ideally, the two columns should be orthogonal which is best achieved by using two columns with different stationary phases.

Selectivity and sensitivity are also significantly increased by using two columns with different stationary phases. A non-polar first-dimension column combined with a polar second-dimension column will allow for the separation of, for example, polar alcohols from non-polar hydrocarbons, a separation that could not be achieved using one-dimensional GC. Retention of the polar compounds will be much greater on the second column than the first, assisting with their separation from the hydrocarbons. Sensitivity is enhanced by having a second column with smaller dimensions. The eluent from the first column is focused onto the second column which sharpens the peaks by reducing their width (peaks are typically 100 ms in width) and increasing their peak heights. Narrower peaks are readily detected by the fast scanning detectors commonly employed with multidimensional GC such as time of flight mass spectrometry (TOF-MS).

Multidimensional GC can be described as either heart cutting or comprehensive. Heart cutting refers to the collection of a single fraction from the first column and its re-injection onto the second column. In LCxGC this can be as simple as collecting the fraction from the LC system in a vial and injecting an aliquot onto the GC system. In GCxGC this is accomplished by a fast switching valve or cold spot between the two columns.

Comprehensive two-dimensional GC differs from heart cutting because the entire sample, rather than just a fraction, is transferred from the first column onto the second column. This means that the second-dimension chromatogram is sampled continuously throughout the separation. In comprehensive LCxGC, this is achieved by collecting the entire eluted sample from the LC system and injecting it onto the GC system, following appropriate sample preparation if required. In comprehensive GCxGC, this is achieved by connecting the two columns in series by a press-fit connector and introducing a modulator between the columns (Figure 13.18). The modulator continuously transfers the primary column effluent to the secondary column, at the same time focusing it into narrow bands for separation on the second column before it is passed to the detector.

While not commonly employed in forensic laboratories, use of multidimensional GC (and, in particular, comprehensive two-dimensional GC) has the potential to significantly advance the analysis of forensic evidence. One of the current areas of interest is the analysis of petroleum-based products often encountered in fire investigations (Frysinger and Gaines 2002; Nizio et al. 2016). The use of comprehensive two-dimensional GC coupled with TOF-MS not only improves the peak capacity of the sample, allowing a larger number of compounds to be accurately identified, but, with the use of the accompanying software, can demonstrate peak clusters that represent specific compound classes. This can assist in separating and identifying the hundreds or even thousands of compounds that may be present in an extract of fire debris, allowing rapid identification and fingerprinting of ignitable liquids.

Additional areas of interest include the forensic profiling of drugs, particularly methylenedioxymethamphetamine (MDMA) (Mitrevski et al. 2011; Schäffer et al. 2012, 2013); screening of drugs and their metabolites in biological fluids (Gröger et al. 2008; Guthery et al. 2010); the chemical profiling of decomposition odour with application to cadaver-detection dogs (Dekeirsschieter et al. 2012; Stadler et al. 2012); and its use in environmental forensic investigations involving the illegal dumping of waste (De Vos et al. 2011).

13.6.2 Portable GC

Portable and transportable GC systems have been considered since as early as 1963, although the original models were essentially gas analysers designed for specific compounds of interest. Although these portable GC systems can be coupled to a range of detectors including photoionisation detectors, ECDs, FIDs, and NPDs, current mobile GC technologies are combined with mass spectrometry. The addition of the mass spectrometer increases the weight, cost, and support requirements but has the unparalleled advantage of increased power with respect to the detection and identification of compounds in the field.

Samples of liquid or headspace gas can be injected using a field syringe. Purge and trap samplers are also available with portable GC systems for the analysis of liquid or solid samples. Current technologies incorporate a range of accessories to ensure portable and transportable GC systems are capable of collecting and analysing a large range of samples. The Inficon HAPSITE® ER portable GC-MS instrument offers headspace, SituProbe, SPME,

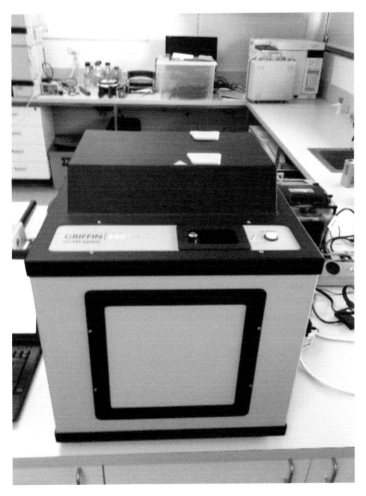

Figure 13.20 Example of an older model Griffin 450 portable GC-MS instrument.

and Thermal Desorber sampling systems in addition to their standard Air Probe. The Torion Technologies TRIDION-9 portable GC-TMS instrument (toroidal ion trap mass spectrometer) incorporates a SPME fibre or needle trap for sample collection and introduction. The FLIR Griffin™ 450 transportable GC-MS instrument (Figure 13.20) also offers multi-modal sample introduction. Although it is not field portable, the instrument is the first to offer the capability of analysing liquid, solid and vapour samples and incorporates a MS/MS detection system. Choice of sample introduction accessories will be dependent on the type of sample being analysed or the instrument availability.

Forensic applications for portable GC systems are still relatively new and include the analysis of chemical warfare agents (CWAs) (Sekiguchi et al. 2006; Smith et al. 2004), and fire debris (Visotin and Lennard 2016). An investigation of their capabilities for emergency responders, the military, and law enforcement is reported by Leary et al. (2016). The major limitations of portable GC systems are the electrical requirements and the carrier gases employed. As these limitations are resolved and the miniaturisation and portability of these instruments improves, the application of portable GC systems in forensic investigations is likely to increase.

References

Cippitelli, M., Mirtella, D., Ottaviani, G. et al. (2018). Toxicological analysis of opiates from alternative matrices collected from an exhumed body. *Journal of Forensic Sciences* 63 (2): 640–643.

Darling, S., Jickells, S., and Negrusz, A. (2013). Gas chromatography. In: *Clarke's Analytical Forensic Toxicology* (eds. A. Negrusz and G. Cooper). London: Pharmaceutical Press.

De Vos, J., Dixon, R., Vermeulen, G. et al. (2011). Comprehensive two-dimensional gas chromatography time of flight mass spectrometry (GC× GC-TOFMS) for environmental forensic investigations in developing countries. *Chemosphere* 82 (9): 1230–1239.

Dekeirsschieter, J., Stefanuto, P.-H., Brasseur, C. et al. (2012). Enhanced characterization of the smell of death by comprehensive two-dimensional gas chromatography-time-of-flight mass spectrometry (GCxGC-TOFMS). *PLoS One* 7 (6): e39005.

Dunnivant, F.M. and J.W. Ginsbach (2011). Gas Chromatography, Liquid Chromatography, Capillary Electrophoresis - Mass Spectrometry: A Basic Introduction. https://sites.google.com/whitman.edu/gclccems/home (accessed 29 July 2020).

Frysinger, G.S. and Gaines, R.B. (2002). Forensic analysis of ignitable liquids in fire debris by comprehensive two-dimensional gas chromatography. *Journal of Forensic Science* 47 (3): 471–482.

Grob, R.L. and Barry, E.F. (2004). *Modern Practice of Gas Chromatography*. Wiley.

Gröger, T., Schäffer, M., Pütz, M. et al. (2008). Application of two-dimensional gas chromatography combined with pixel-based chemometric processing for the chemical profiling of illicit drug samples. *Journal of Chromatography A* 1200 (1): 8–16.

Guthery, B., Bassindale, T., Bassindale, A. et al. (2010). Qualitative drug analysis of hair extracts by comprehensive two-dimensional gas chromatography/time-of-flight mass spectrometry. *Journal of Chromatography A* 1217 (26): 4402–4410.

Leary, P.E., Dobson, G.S., and Reffner, J.A. (2016). Development and applications of portable gas chromatography–mass spectrometry for emergency responders, the military, and law-enforcement organizations. *Applied Spectroscopy* 70 (5): 888–896.

McNair, H.M. and Miller, J.M. (2008). Capillary columns and inlets. In *Basic Gas Chromatography*, 2 edn Edition, Wiley, pp. 84 – 103.

Mitrevski, B., Wynne, P., and Marriott, P.J. (2011). Comprehensive two-dimensional gas chromatography applied to illicit drug analysis. *Analytical and Bioanalytical Chemistry* 401 (8): 2361–2371.

Nizio, K., Cochran, J., and Forbes, S. (2016). Achieving a near-theoretical maximum in peak capacity gain for the forensic analysis of ignitable liquids using GC× GC-TOFMS. *Separations* 3 (3): 26.

Schäffer, M., Gröger, T., Pütz, M. et al. (2012). Comparative analysis of the chemical profiles of 3, 4-methylenedioxymethamphetamine based on comprehensive two-dimensional gas chromatography–time-of-flight mass spectrometry (GC× GC-TOFMS). *Journal of Forensic Sciences* 57 (5): 1181–1189.

Schäffer, M., Dieckmann, S., Pütz, M. et al. (2013). Impact of reaction parameters on the chemical profile of 3, 4-methylenedioxymethamphetamine synthesized via reductive amination: target analysis based on GC-qMS compared to non-targeted analysis based on GC×GC–TOF-MS. *Forensic Science International* 233 (1–3): 201–211.

Sekiguchi, H., Matsushita, K., Yamashiro, S. et al. (2006). On-site determination of nerve and mustard gases using a field-portable gas chromatograph-mass spectrometer. *Forensic Toxicology* 24 (1): 17–22.

Skoog, D.A., Holler, F.J., and Crouch, S.R. (2017). *Principles of Instrumental Analysis*. Cengage Learning.

Smith, P.A., Koch, D., Hook, G.L. et al. (2004). Detection of gas-phase chemical warfare agents using field-portable gas chromatography–mass spectrometry systems: instrument and sampling strategy considerations. *TrAC Trends in Analytical Chemistry* 23 (4): 296–306.

Stadler, S., Stefanuto, P.-H., Brokl, M. et al. (2012). Characterization of volatile organic compounds from human analogue decomposition using thermal desorption coupled to comprehensive two-dimensional gas chromatography–time-of-flight mass spectrometry. *Analytical Chemistry* 85 (2): 998–1005.

Suzuki, Y., Ishizawa, F., and Honda, K. (2017). Semiquantitative screening of trace combustion-derived volatile substances in the blood of fire victims using NeedlEx® headspace gas chromatography/mass spectrometry. *Forensic Science International* 278: 228–239.

Visotin, A. and Lennard, C. (2016). Preliminary evaluation of a next-generation portable gas chromatograph mass spectrometer (GC-MS) for the on-site analysis of ignitable liquid residues. *Australian Journal of Forensic Sciences* 48 (2): 203–221.

Vitha, M.F. (2016). *Chromatography: Principles and Instrumentation*. Wiley.

von der Lühe, B., Dawson, L.A., Mayes, R.W. et al. (2013). Investigation of sterols as potential biomarkers for the detection of pig (*S. s. domesticus*) decomposition fluid in soils. *Forensic Science International* 230 (1–3): 68–73.

14

High Performance Liquid Chromatography and Ultra-High Performance Liquid Chromatography Including Liquid Chromatography–Mass Spectrometry

Sophie Turfus and Luke N. Rodda

14.1   Introduction

High performance liquid chromatography (HPLC) allows for the identity of unknown substances in a range of matrices including physical and biological matter. HPLC is commonly employed in pharmaceutical science to aid drug discovery and pharmacokinetic studies; in anti-doping; in the food and packaging industry; in the environmental sciences; as well as in forensic science. It is used as an alternative, and as a complement, to gas chromatography (GC) and is particularly useful when compounds are not amenable to GC. Although gas chromatography–mass spectrometry (GC-MS) has been described as the 'golden standard' (Maurer 1998) for detection of low molecular weight and volatile drugs and poisons in forensic and clinical toxicology, non-volatile and polar compounds are not amenable to this technique and require analysis by HPLC-based techniques such as liquid chromatography–tandem mass spectrometry (LC-MS/MS).

The principles of HPLC are similar to those for GC, although the mobile phase in HPLC is a liquid, usually a mixture of water, an organic solvent, and other additives depending on the type of detection. The stationary phase can either be a solid (liquid–solid chromatography) or a layer of thin liquid, immiscible with the solvents used as the mobile phase, covalently bound to a solid support (liquid–liquid chromatography). Both techniques are variants of partition chromatography but in HPLC, unlike GC, interactions also occur between the mobile phase and stationary phase, and between the analyte and the mobile phase, as represented in Figure 14.1. As with GC, temperature settings, as well as different columns, are exploited to tailor separation between compounds, but there are many more options for stationary phases and more variability in column dimensions in HPLC. Mobile phase pH, composition, and gradient elution employing multiple solvents can also be exploited in HPLC, which also offer more options for optimising separations compared with GC.

Reversed-phase chromatography is most often employed in HPLC, where the mobile phase is more polar than the stationary phase. However, some applications employ polar stationary phases such as silica, a common stationary phase that was popularised by column and thin layer chromatography. The stationary phase in reversed-phase chromatography is most commonly silica modified with octadecyl (C_{18}) or octyl (C_8) functional groups.

Analytical Techniques in Forensic Science, First Edition.
Edited by Rosalind Wolstenholme, Sue Jickells and Shari Forbes.
© 2021 John Wiley & Sons Ltd. Published 2021 by John Wiley & Sons Ltd.

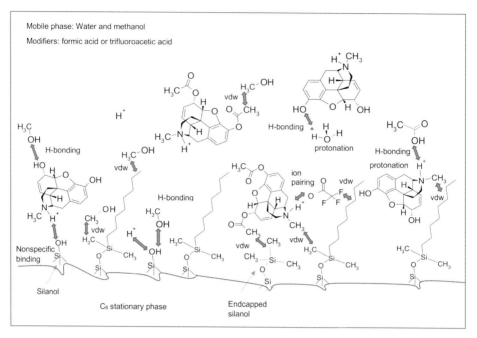

Figure 14.1 Interactions between stationary phase (C_8), drug analytes (morphine and diacetylmorphine) and mobile phase (methanol, water and trifluoroacetic acid) typical in reversed-phase liquid chromatography. A number of interactions are notable such as van der Waals' (vdw), hydrogen bonding (H-bonding), protonation, ion pairing and non-specific binding (which in this case includes a specific type of hydrogen bonding with silanol groups). Some stationary phases contain residual non-endcapped silanol (R_3Si-OH) groups that have not been functionalised in the manufacturing process. These can cause non-specific polar binding with analytes and mobile phase. In instances where the silanol groups have been endcapped, trimethylsilyl functionalities have been added which prevent non-specific binding and favour hydrophobic van der Waals interactions. The scenario illustrated may be typical in chromatographic separations which are followed by UV detection, or mass spectrometry where the protonation of analytes and charging of bases is a prerequisite for mass spectral detection while at the same time ensuring that there is favourable retention on a reversed-phase stationary phase.

The technique developed from column chromatography performed under gravity, where mobile phase moved through a column holding a stationary phase. Later, flash chromatography was developed which employed pressurised gas (10–15 psi) to force the mobile phase through the column, usually containing a stationary phase of silica gel particles.

The early liquid chromatographs were equipped with variable ultraviolet (UV), refractive index and fluorimetric detectors (Curry 1974). The superior power of HPLC combined with mass spectrometry (MS) was noticed a long time ago with one of the earliest technologies described by Ray Scott who dried HPLC eluent on wires which were subsequently inserted through pressure-reducing chambers into a quadrupole mass spectrometer (Curry 1974). Since that time, the state-of-the art instrumentation has become HPLC coupled to MS. Nevertheless, the detector used will depend on the application. For example, with bulk drug analysis, a UV detector or a photodiode array detector (see Chapter 6) is likely to provide sufficient sensitivity. However, these detectors may lack sensitivity for trace analysis of extracts from complex biological matrices.

One very useful application of HPLC involves the analysis of polar and acidic drugs, as well as polar metabolites of basic drugs in metabolite-rich matrices such as urine and liver. Often, detection of very polar metabolites is necessary. For example, by monitoring metabolites, greater diagnostic value is achieved compared with monitoring only the drug itself, particularly for drugs which may be unstable (e.g. antipsychotics such as olanzapine), or rapidly metabolised (e.g. heroin). Furthermore, determining metabolite ratios can aid in interpreting chronic versus occasional use of a particular drug, or time elapsed since administration. Thus, HPLC-based techniques are ideal for the analysis of polar metabolites, typically requiring no derivatisation or hydrolysis; modifications being made to the mobile phase to effect separation and improve sensitivity.

HPLC, as well as being amenable to analysis of polar compounds, has other benefits. For example, certain compounds, such as oxazepam, are subject to pH and temperature-dependent instability when analysed by GC and may degrade to a benzophenone. Since GC analysis of polar drugs and metabolites may require hydrolysis and/or derivatisation, modifications which may not always go to completion, HPLC has the potential to be a more sensitive technique when compared with GC, exhibiting lower limits of detection for some analytes. However, in the instances where GC is performed with multiple mass spectrometers (GC-MS/MS), comparable sensitivities can be attained with LC-MS/MS.

Presently, LC-MS/MS forms the basis of many multi-analyte methods. Targeted screening methods using selected reaction monitoring (SRM) (see also Figure 10.18) have become more commonplace in recent years and allow the detection of over 100 compounds in a single analysis (Figure 14.2). (In this chapter and in Chapter 10 SRM refers to selected reaction monitoring, elsewhere in this book it refers to Standard Reference Material. SRM and Multiple Reaction Monitoring [MRM] can be used interchangeably, therefore, MRM will be used to avoid confusion).

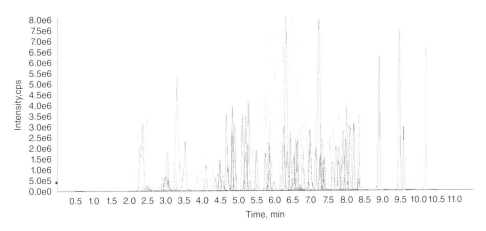

Figure 14.2 A typical chromatogram resulting from the analysis of spiked mobile phase used as a control sample in a targeted screen employing LC-MS/MS. Such samples are used in LC-MS/MS to establish the retention times at the start of a chromatographic run to assist in the verification of a positive result in multi-analyte drug screens. Although not all compounds are fully resolved from one another, incomplete resolution in LC-MS/MS analysis is often deemed acceptable if the compounds are of distinct masses and possess sufficient differences in their fragmentation properties. The method allows the simultaneous detection of over one 100 drugs in blood and urine, which is common among state-of-the-art methods.

14.2 Components of an HPLC instrument and their Optimisation

The components of a HPLC system are shown in Figure 14.3. The mobile phase, which is forced through the column using a pump, is delivered either at a constant composition, or is mixed prior to (quaternary pump), or after (binary pump), pumping. The pump can also be programmed to deliver the mobile phase at different speeds at pressures well above atmospheric pressure. The sample is introduced into the column via a sample loop attached to a 6-port valve before it travels through the column and is detected by one of many types of available detectors (such as UV or MS). The signal is converted to an output and data is processed using a computer. The main components are described in more detail in the following sections.

14.2.1 Pump and Mixer

Typical pumps used in HPLC consist of binary or quaternary pumps (Figure 14.4). Binary pumps employ high-pressure mixing of two solvents which are then pumped through the column. Quaternary pumps employ any combination and composition of four solvents which are mixed under low pressure before being pumped under high pressure through the column. In the binary pump, there is some ability to mix more than two types of solvents using a solvent selection valve. This enables the operator to connect two bottles to each pump instead of one to allow changing between solvents in between chromatographic runs. The solvents are combined in a designated mixer which is on the high-pressure side of the pumps in the binary pump format, and on the low-pressure side of the quaternary pump format. There must be thorough mixing of solvents to ensure that the composition of the mobile phase is fully homogeneous before it reaches the column.

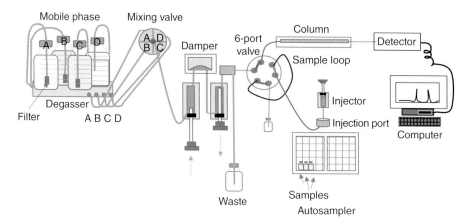

Figure 14.3 Typical features of an HPLC instrument: mobile phase, typically stored in screw-top glass bottles and protected from light, is pumped through a mixing valve (on the low-pressure side of the pump), 6-port valve, through the column to the detector. The instrumentation features a mixing valve where up to four mobile phases can be combined, as is typical of quaternary pumps traditionally used in HPLC that involve low-pressure mixing. The sample is introduced from the autosampler via the injection port into the mobile phase flow by means of the 6-port valve.

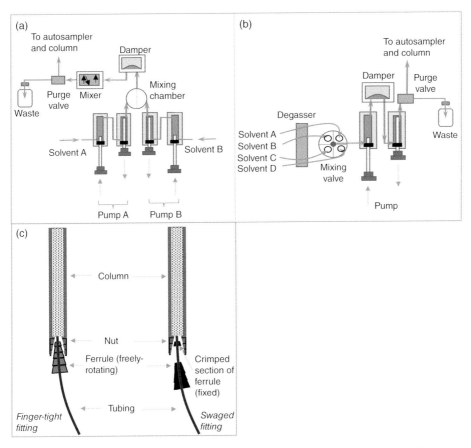

Figure 14.4 Typical pumps and connections used in HPLC. (a) Binary pumps. Two mobile phases are used, A and B. Each solvent is drawn into the pump separately by means of reciprocating pistons. As the piston pulls back from the pump, solvent enters the first chamber. As the piston pushes in, it pushes the solvent into a second chamber where the piston in that chamber simultaneously moves outward. The piston from that second chamber then pushes inward forcing the mobile phase into the mixing chamber. Pump B acts the same way for mobile phase B. The pumping rates for each pump can be varied over the course of the analysis to produce a gradient elution. The advantages of this type of pump are that mixing occurs on the high-pressure side of the pump enabling a lower dwell volume (total volume pre-column) which allows the composition of the mobile phase to be changed much more rapidly allowing faster equilibration and more effective separation using gradient elution. (b) Quaternary pumps possess only one dual piston pump which pumps mobile phase in a similar manner as described in (a), but after it has been mixed on the low pressure side of the pump. Another difference is that the damper can be located between the pistons as there is no contribution to the 'flow pulse' from a second dual piston pump as in (a). More combinations of solvents can be mixed although the process is generally slower and requires a larger volume for mixing to take place. (c) Typical connections used in HPLC to withstand high pressures. The diagram on the left depicts a 'finger tight' ferrule whereas a typical swaged fitting is shown on the right, where part of a two-piece ferrule is crimped onto the tubing to maintain an adequate length between the tip of the ferrule and the beginning of the stationary phase. This minimises extra-column volume and minimises band broadening and leakage of mobile phase from the column.

14.2.2 Autosampler and Inlet

In older systems, manual injection of samples was typically performed through a 6-port valve, with the sample, in liquid form, entering the mobile phase flow just ahead of the column inlet. However, in modern instruments, the sample injection is an automated process. Most HPLC instruments are equipped with an autosampler containing plates with a common format consisting of 96 positions per plate. Usually, at least two plates can be loaded into the autosampler which allows chromatographic runs to be performed unattended, including overnight analysis. The sample is injected into the inlet via a valve as shown in Figure 14.5 and the inlet is washed in between injections.

14.2.3 Injector

The injector consists of a sample introduction valve that allows flow-switching, an example of which is shown in Figure 14.5 for a 6-port valve. This valve alternates between a sample loading position, where the sample is loaded into a sample loop while the mobile phase flows through the column, and the sample injection mode where the sample is emptied from the sample loop onto the column while the injection port gets washed of excess sample to avoid contamination and carry-over into the injection of the next sample.

14.2.4 Column

In HPLC, columns are short in comparison with GC columns. They need to withstand high pressure and therefore are usually made of steel. The stationary phase is either present on

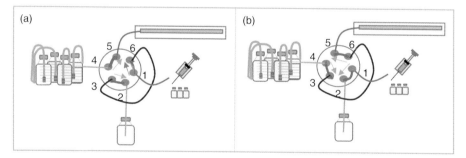

Figure 14.5 The 6-port valve used to inject samples. There are two settings in the 6-port valve; sample loading and sample injection. (a) Sample loading. The mobile phase flows from the solvent reservoir through ports 4 and 5 and to the chromatographic column. The sample is injected into a sample loop connected to the autosampler via ports 3 and 6. The sample loop is removable and can be replaced by one with a different volume. If full loop injection is used, the sample volume injected into the loop is typically set to exceed the loop volume, with excess sample directed to waste via port 2. The sample is held in the loop and is not connected to the column during sample loading. (b) Sample injection. The connections through the sampling valve are changed so that the mobile phase is now directed through port 4 to port 3 and through the sample loop. The mobile phase carries the sample from the loop, out through port 5 and onto the column and the sample loop is washed with mobile phase. After sample injection, the port is switched back to the sample loading position ready for the next loading and injection sequence. Similar valves with 8 or 10 ports can also be used to enable more complicated analyses with additional features, such as multiple columns being connected to the same pumps.

particles packed into the column or is coated on the inside wall of the column. Traditional stationary phases are based on silica, in particulate form, closely packed into the column with the silica being chemically modified to tailor separation of compounds with different chemical properties, a popular phase being octadecyl silica (C_{18}). The efficiency and resolution of a chromatographic separation can be modified by altering the length, diameter, particle size and stationary phase of the column. Generally, use of longer columns and those with smaller particle size is a suitable approach for more efficient separations. However, the challenge exists in maintaining a high efficiency (see Section 12.1), short analysis time, and acceptable inlet pressure as well as a low detection limit (see Section 12.4.2) required for many applications in forensic science (Guiochon 2006). The HPLC column is normally placed in a temperature-controlled oven to avoid fluctuations in retention time and pressure.

14.2.4.1 Stationary Phase

The stationary phases pertaining to the sorbents used in solid-phase extraction have been described in Section 4.2.3.1. In HPLC, similar stationary phases are employed. Silica-based phases are widely used. Many silica-based stationary phases contain residual silanol groups (Si—OH). During the manufacture of silica-based reversed-phases, silica is reacted with alkylchlorosilane compounds, typically C_4, C_8, or C_{18}, which provide the surface functionality of the stationary phase. The bulkiness of the alkyl moiety typically results in some unreacted silanol groups remaining on the silica surface. These residual silanol groups can cause unwanted, and uncontrolled, molecular interactions with analytes, so manufacturers carry out an additional 'endcapping' reaction with, for example, trimethylchlorosilane, which results in trimethyl groups on the surface of the stationary phase (Figure 14.1). However, the process is not 100% efficient and usually results in some residual silanol groups. These 'non-endcapped' silanol groups can interact with analytes. In this instance, non-specific binding can occur between these ionised silanol groups and analytes so that the predominant interactions are no longer hydrophobic in nature and peak tailing can occur which impacts on peak shape and resolution. For this reason, method development usually commences in the pH range of 2–3 where the ionisation of acids and silanol groups are suppressed, and basic compounds are ionised.

Due to issues such as thermal instability, columns based on monolithic packing material and alumina have also been adopted, particularly for HPLC methods employing extremes of pH. Alumina-based packing material simplifies retention mechanisms due to the absence of silanol groups, removes the requirement for amine modifiers, and reduces irreversible binding. Alumina can be coated with polybutadiene to give reversed-phase characteristics. The column used is dependent on the application such as the pH needed to effect separation. For example, alumina has a pH range of use from pH 2 to 12 compared with a traditional working range of pH 2–8 for silica. Columns with an extended pH range have been developed which use modified silica stationary phases, for example a range of pH 1–12 for bridged ethylsiloxane silica which is a stationary phase with fewer surface silanol groups. Immobilised metals and metal oxides are used as stationary phases in proteomics in the analysis of phosphopeptides.

Monolithic columns, in contrast to packed columns, are porous rod structures that are produced inside a column. They are highly porous, with many channels and a high surface area for intermolecular interaction of analytes with the stationary phase. Monoliths are produced by polymerisation and can be chemically modified, for example by using a technique called 'molecular imprinting' to enhance specificity for a particular compound or groups of compounds. Monoliths can be inorganic (produced from silica or metal oxides) or organic. Polymeric monoliths are often based on polymethacrylates or polydimethylsiloxane modified with functional groups such as C_{18}, zwitterions, cations or anions. Monoliths give comparable sensitivities with packed columns employing sub-2 μm particles. In addition, immobilisation of proteins such as membrane proteins or human serum albumin on the monolith can be used for screening for drugs with specific binding properties. Monoliths reduce plate heights (see Section 12.4.2), analysis times and the pressures required for delivery of mobile phases, due to their high permeability. However, they are expensive and suffer from non-uniformities and have not yet found widespread use outside the research environment (Guiochon 2007).

LC column manufacturers are continually trying to produce improved stationary phases. 'Core shell technology' which involves the use of 1.3–5 μm solid particles possessing thin coatings of stationary phase, reportedly minimises band broadening contributions of all three parameters (eddy diffusion, resistance to mass transfer, and longitudinal diffusion described in Section 12.4.3). One example of columns that use this technology are the Poroshell columns available from Agilent Technologies that contain particles with a thin porous layer (0.5 μm) of stationary phase. The improved particle morphology and limited solute penetration, dispersion and a reduced variance in particle size, along with reduced pressure requirements and improved thermal conductivity/heat dissipation of particles, are notable advantages of this technology (Chester 2013). The approach reportedly saves up to two thirds of the analysis time compared with analysis for which totally porous packing material is used, as has been demonstrated for shellfish toxins (Chester 2013).

Mixed-mode columns, where the stationary phase is engineered to provide more than one mode of interaction to help resolve analytes, have provided an added dimension of selectivity to a chromatographic separation by increasing hydrophilicity and promoting electrostatic interactions. Modified cationic or anionic functional groups improve peak shapes of acids and bases, respectively, and may occur on a stationary phase which is otherwise hydrophobic, providing two types of interaction, reversed-phase and ion-exchange interactions. For example, thioethers and amides provide weak anion exchange mechanisms and carboxylate and urea groups provide cation exchange mechanisms. The approach requires subtle modifications of pH and exploitation of the pK_as and pK_bs of compounds under analysis and that of the stationary phase, or the adjustment of the ionic strength of the mobile phase to tailor the retention of charged species. Non-charged species can still be retained and separated by controlling the organic strength of the mobile phase. As with reversed-phase separations, gradients (see Section 12.4.3) can be employed to modify pH, ionic strength and organic strength throughout the separation to exploit the full capacity of mixed-mode columns.

Another development in stationary phases is that of hydrophilic interaction liquid chromatography (HILIC) which uses stationary phases of silica, or surface-modified silica with, for example, peptide, polysaccharide, cyclofructan moieties, and similar solvents to reversed-phase chromatography. An excess of water provides the surface of the stationary phase with its binding characteristics. To provide zwitterionic characteristics, diphenylphosphoniumpropylsulfonate or imidazoline modifications have been employed. Interactions are typically ionic, and analytes elute in order of increasing polarity. HILIC was used to aid detection of the very polar endogenous substance creatinine in fingerprints to aid in the study of inter-individual variability of drug deposition in fingerprints (Goucher et al. 2009). Such phases are also typically used in supercritical fluid chromatography which typically exploits polar stationary phases and acetonitrile or alcohol/aqueous solvents and carbon dioxide mixtures. These conditions, along with high temperatures and pressure conditions, enhance fluidity of the mobile phase improving separation.

14.2.4.2 Column Dimensions

The dimensions of an HPLC column are limited by the pressure required to force the mobile phase through the stationary phase. Lengths of columns typically range from 35 to 250 mm. Typical internal diameters are 1.0, 2.1, 3.0, and 4.6 mm. Many companies also offer larger columns with matching phases to enable preparative chromatography and purification of compounds from injection of larger sample volumes, with advice on how to scale up mobile phase conditions such as flow rate and mobile phase composition. Some of the more common stationary phases are also available in narrow capillary (0.3, 0.5, and 0.8 mm) and nano (0.075 and 0.1 mm) internal diameter formats to enable analysis of very low concentrations of analyte, as might be expected for certain cases, for example in the analysis of hair samples or in cases of drug-facilitated sexual assault where much time has elapsed since drug administration.

14.2.4.3 Particle Size

Traditional HPLC columns employ particle sizes of between 2.5 μm and 5 μm. For particles based on silica gel, the pores of the silica range from 80 to 120 Å for the separation of small molecules with molecular weights of less than 3000 amu such as drugs, with larger pore sizes of 300 Å available for larger molecules such as proteins.

Small particle sizes offer greater chromatographic efficiency. The lower practical limit of particle size of 2.5 μm was reduced with the introduction of ultra high pressure/performance liquid chromatography (UHPLC) and completely porous particles. UHPLC is defined by the use of sub-2 μm particles as packing material and is characterised by the use of pumps capable of pumping mobile phase at a higher pressure through the column to exploit the minimum theoretical plate height in the van Deemter equation (see Eq. 12.14). Because the eluting peaks in UHPLC are narrower, detectors with fast detection capabilities are required to obtain as many data points as possible across a peak to enable accurate quantification. UHPLC offers significant improvements to efficiency, resolution and sensitivity compared with conventional HPLC. Table 14.1 outlines the differences between UHPLC and HPLC which are also summarised elsewhere (Kele et al. 2004; Plumb et al. 2004; Castro-Perez et al. 2005; Churchwell et al. 2005).

14.2.4.4 Pre-Column/Guard Column

Often, an additional column, referred to as guard column, is connected before the main analytical column to preserve the life of the latter. The guard column is shorter than the column itself (e.g. 12.5 or 17 mm), but has the same packing material, stationary phase and width as the column. Importantly, because the guard column contributes to separation, it

Table 14.1 Reported differences between HPLC and UHPLC.

Property	HPLC	UHPLC
Typical particle size (μm)	>2	≤2
Maximum operational pressure (psi)	5 000	15 000
Number of theoretical plates[a] in unit time at $k'=4$ (N/t)	800	9 000
Specific permeability (m^2)	4×10^{-15}–8×10^{-14}	2–3.5×10^{-15}
Sensitivity^a	Typically 3- to 5-fold less than that of UHPLC	Sensitivity increase up to 10-fold (typically 3–5).
Speed of analysis	Slower: typically 7–30 min	Faster: typically 2–6 min
Resolution and number of peaks	Co-elution of peaks results in lower signal intensities of analytes	Reduction of spectral overlap
	Typically half as many peaks as UHPLC.	Number of resolved peaks doubled
Peak widths	Wider peaks – several seconds to minutes	Typically 6 s
Detector used	Larger choice of compatible detectors including UV-Visible, diode array, fluorescence and MS including quadrupoles/time of flight/ion trap etc.	MS, diode array and other detectors available but fast detectors required (e.g. Waters Quattro Premier allows 5 ms scan times or quadrupole time of flight allows fast duty cycles). More difficult to couple to detectors from a different manufacturer
Phases available	Large variety-C_{18} most commonly used.	Phenyl, C_8, polar-embedded reverse phase – fewer due to novelty and difficulty in manufacturing small particles

a) See Section 12.4.2.

should be factored into method development and, thereafter, not be substituted with a guard column of another type. One manufacturer recommends that the guard column should be changed if either the resolution, pressure or the number of plates shift by at least 10% (Agilent 2012).

14.2.5 Fittings

The individual components of an HPLC system need to be linked together with tubing to enable a continuous connection between the HPLC pump, any mixing devices, the injector, the column and the detector. In modern HPLC systems, the pump, mixers and injector connections are generally fixed and the only connection/disconnection which needs to be made routinely is to replace the column (Figure 14.4c). Plastic (PEEK™, Delrin®, and Teflon®) tubing and metal tubing (stainless steel) are used to connect the components of the chromatographic system. The length and diameter of the tubing are critical and as far as possible both are kept to a minimum, particularly for connections post-column, to reduce band broadening. Post-column, tubing of internal diameter of 0.17 mm is common. This is compatible with the use of columns with internal diameters of 2.1 or 4.6 mm in standard use. Tubing of wider diameter (e.g. 0.25 mm) can reduce clogging and is typically used between the pumps and column. To enable use of the narrower tubing, and also to prolong the life of the column, filtering of mobile phases and samples prior to injection is recommended. Typically, connections which are not frequently changed, or areas of the instrument subjected to high pressure, employ metal fittings which are tightened using spanners.

In UHPLC, high pressures occur particularly between the pump and the column inlet and therefore metal fittings are often used. On the other hand, plastic tubing is often used in the lower pressure zone between the outlet of the column and the detector. However, universal 'finger tight' plastic tubing that fits components from different manufacturers and can withstand pressures of up to 10 000 psi is currently favoured. These finger tight fittings remove the necessity of 'swaging' i.e. where the ferrules are crimped to the tubing.

Fittings are generally composed of male nuts and ferrules secured into female fittings. The length of the tubing can be extended using a union piece, which can also aid in connecting multiple columns. In addition, a tee-piece can be used to combine flows from two sources and has a variety of uses such as the post-column infusion of an organic solvent into the eluent to improve ionisation in MS. The fittings and tubing have an influence on the extra-column volume, which is defined as the mobile phase volume. In which is defined as the mobile phase volume. In which analytes spend time in outside the column (previously called 'dead volume'). The extra-column volume should be kept to a minimum to reduce post-chromatographic band broadening. Connections which are susceptible to over-tightening or under-tightening lead to damaged tubing, unwanted additional mixing chambers, and, ultimately, leakage of the mobile phase. Plastic tubing may become loose at high temperatures and so the chromatographer must pay attention to the fittings.

14.2.6 Mobile Phase

Mobile phases in HPLC typically consist of an organic solvent mixed with water. The water may be buffered or additives such as formic acid or ammonia may be added to the aqueous component to modify the pH. Mobile phases are prepared using HPLC-grade solvents, additives and distilled deionised water. Buffers, which are sometimes used as the aqueous component, are prepared by weighing salts and diluting with water, pH adjustment or titration and filtration through 0.5 µm filter paper. The mobile phase lines are also fitted with filters made of glass or stainless steel. The mobile phase can have a consistent composition, as in isocratic elution. Alternatively, the instrument can be programmed to incorporate a change in the mobile phase during the analysis of a sample by using gradient elution. In reversed-phase chromatography, this usually involves the gradual increase of the organic solvent component and is particularly useful in aiding the elution of the most strongly retained substances. This results in more acceptable chromatography of substances that bind strongly to the column and reduces analysis time.

The mobile phase can be prepared as a solution in a single bottle (for isocratic elution) or be automatically mixed from two bottles (or more, for more complex mobile phases) after it is pumped into the mixer. Most commonly, the aqueous component is referred to as 'mobile phase A' whereas the organic solvent, or predominantly organic solvent mixture, is referred to as 'mobile phase B'. These are outlined in the following sections.

14.2.6.1 Mobile Phase A

Mobile phase selection is important in HPLC because many substances contain ionisable functional groups. Drugs usually contain either a basic or an acidic group and the charge can be manipulated using either a buffer (often in the 5–10 mM range) or an acid or base. It is important to buffer the sample as it is injected to avoid precipitation of salts, and for this reason, samples are often prepared in the matrix of the mobile phase. Alternatively, an acid often at 0.1% v/v is added to mobile phase A and sometimes B to achieve a low pH, or a base to achieve a high pH. As a rule of thumb, basic compounds are considered unionised (>99%) 2 pH units above their pK_a and are considered ionised 2 pH units below their pK_a. The reverse is true for acidic compounds. The unionised form is exploited in reverse-phase chromatography where organic compounds will be more strongly retained through van der Waals forces. If a mobile phase pH is employed around the pK_a of a substance, small fluctuations of 0.1 of a pH unit can cause retention time instability and difficulties in validating methods (Dolan n.d.).

The various modifiers which can be added to mobile phase A need to be selected carefully so that they are compatible with detectors and mode of operation. For example, while nonylamine has been previously used effectively as the amine modifier in isocratic chromatography employing photodiode array (PDA) detection, it was incompatible with chromatography employing gradient programming which revealed the presence of several extraneous peaks in the baseline. This was observed with different batches, instruments and operators; triethylammonium phosphate was suggested as a suitable alternative (Elliott and Hale 1997). Other modifiers such as those used to adjust pH can also produce an unacceptable elution ramp where the signal of the baseline increases in gradient elution producing unacceptable peak shapes and signal-to-noise ratios, particularly at later retention times. When coupled to MS, care must be taken to ensure that buffers are volatile. Mobile phase modifiers that are compatible with MS-based techniques are outlined in Table 14.2. Modifiers not only serve to effect separation but they can also be used to modify detector responses. For example the addition of formic acid to a system employing MS serves in the generation of $[M + H]^+$ ions required for detection of protonated basic drugs.

14.2.6.2 Mobile Phase B

Traditionally mobile phase B contains an organic solvent either as the sole solvent, or mixed with a lower proportion of aqueous solution. The organic component of the mobile phase is often methanol or acetonitrile. These solvents are the most favoured organic solvents for HPLC because they are miscible with water and offer low viscosity, making it practical to use these solvents at the pressures which can be generated readily by conventional HPLC pumping systems. Both solvents have low UV cut-offs (see footnote to Table 14.2); acetonitrile absorbs at <190 nm and methanol at <205 nm.

Table 14.2 Mobile phase modifiers typically used in HPLC.

	Operating pH range	Detectors	Notable advantages (Adv.) or disadvantages (Disadv.)
Buffers			
Phosphate buffer	2<pH<3.1	UV	Adv.: UV <220 nm[a]
	6.2<pH<8.2		Disadv.: pH range 11.3–13.3 cannot be exploited with some columns due to dissolution of silica
Ammonium formate	2.7<pH<4.7	UV/MS	
Acetate buffer	3.8<pH<5.8	UV/MS	Adv.: UV <220 nm
Citrate	2.1<pH<6.4	UV	Disadv.: >220 nm only, problems with check valves
Ammonium bicarbonate	6.6–8.6	UV/MS	
Borate	8.3–10.3	UV/MS	
Pyrrolidine	10.3–12.3		Adv.: extends lifetime of column at high pH
Acids			
0.1% Trifluoroacetic acid	2	UV/MS	Adv.: ion pairing reagent to minimise peak tailing, useful for proteins/peptides
			Disadv.: can cause ion suppression in MS
0.1% Phosphoric acid	2	UV	
0.1% Formic acid	2.7	UV/MS	Adv.: method of choice as can easily ionise a range of compounds in MS with minimal detector interferences

a) Many compounds only absorb in the low UV region. To maximise signal-to-noise ratio and hence permit detection, the absorption region of the mobile phase should not coincide with that of the analyte of interest. Hence modifiers which absorb at <220 nm, the lower end of the UV spectrum used in most UV-visible detectors, will not interfere with detection for analytes with absorptions >220 nm.

Acetonitrile is a poorer solvent for buffers/salts than methanol, being 'less-polar' in nature. For example, 15 mM phosphate buffer (pH 7) can be solubilised in 80% methanol but when using acetonitrile, the upper limit is 5 mM. Ammonium acetate and ammonium phosphate salts are more soluble in acetonitrile than potassium salts. If isocratic elution is used, particularly with a high percentage of organic phase, there is a risk of precipitation of buffer salts. This can also occur during mixing of mobile phases and at the end of a mobile phase gradient in reversed-phase chromatography where there is a high component of organic mobile phase present. Precipitation can be avoided by ensuring a low enough concentration of salts in the mobile phase B while ensuring the same concentration of salts in mobile phase A to prevent precipitation of salts upon mixing of the two mobile phases.

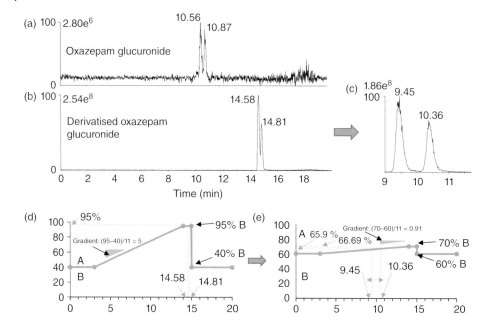

Figure 14.6 An example of method development for the analysis of oxazepam glucuronide, a major metabolite of oxazepam. Oxazepam glucuronide is an acidic molecule which is poorly analysed by LC-MS/MS methods tailored for the vast majority of drugs which are basic.
a) Chromatogram showing the peaks for the two diastereoisomers of oxazepam glucuronide. Not only is the signal height split between two diastereoisomers, but oxazepam glucuronide is poorly ionised in the acidic mobile phase. b) Derivatisation with an agent containing a charged quaternary phosphonium group results in much improved signal-to-noise ratios but the diastereoisomers cannot be resolved with the original mobile phase gradient. c) Changing the mobile phase gradient results in both a more favourable (baseline) separation and analysis time – from 14.58 and 14.81 minutes in (b) to 9.45 and 10.36 minutes in c). d) The original gradient employed for the separation in (a) and (b). e) The modified separation for (c) employing a shallower gradient but starting from a higher composition of organic solvent. During method development, it is helpful to plot the mobile phase gradient on a graph and establish the mobile phase composition at a particular retention time, as shown in (d) and (e). For example, 9.45 minutes in the chromatogram shown in (c), the elution time of the first diastereomer to elute, corresponds to 65.9% B. At 10.36 minutes, the elution time of the second diastereomer, is 66.69% B. Knowing the elution times and the composition of the mobile phase at these times enables changes to the gradient to be made to optimise resolution.

An example of optimisation of a chromatographic run employing two solvents A and B is displayed in Figure 14.6. In this example, the gradient of the chromatographic run has been optimised following the derivatisation of the predominant metabolites of oxazepam, oxazepam glucuronide which exists as two diastereoisomers.

Although organic solvents other than acetonitrile and methanol, can, theoretically, be used for mobile phase B, many other water miscible solvents either have higher UV cut-offs or higher viscosity, making them less suitable. For example, the viscosity of ethanol at 1.074 mPa.s is twice that of methanol (0.544 mPa.s) and over three times that of acetonitrile (0.306 mPa.s), and the UV cut-off (the wavelength at which the absorption in the UV region is effectively 0) is slightly higher at 210 nm (cf. acetonitrile <190 nm and methanol <205 nm).

The function of the organic modifier is to alter the partition of analytes. In reversed-phase separations, when the aqueous phase concentration is high, organic analytes in their neutral form partition readily into hydrocarbon liquid stationary phases. The more apolar/hydrophobic the analyte is, the more readily it will partition into the stationary phase. As the percentage of organic modifier increases, partition from the stationary phase to the mobile phase increases, thereby eluting more apolar compounds. The relative difference in hydrophobicity of analytes can be exploited in gradient elution, where the percentage of organic modifier is increased over the course of the analysis. The more polar compounds elute first and, as the organic solvent composition is increased, the less polar compounds elute. An increase in the organic content of the mobile phase by 10% will decrease the retention value/factor, k (see Section 12.3.4) of an analyte by a factor of two to three.

Organic solvents can also be classified in a variety of ways which can help determine the most useful solvent for a particular analysis. One classification is that of elutropic strength, $\varepsilon°$, a measure of the relative elution strength of a solvent, $\varepsilon°$ for methanol on a C_{18} phase is 1.0 and for acetonitrile is 3, i.e. acetonitrile has 3 times the elution strength relative to methanol. Thus, an analysis using acetonitrile as an organic modifier can be carried out in a considerably shorter time than if methanol is used to analyse the same sample under the same elution gradient. On silica phases, methanol is the more effective solvent, with an $\varepsilon°$ value of 0.7 compared with that of 0.52 for acetonitrile.

Another measure used to characterise solvents is solvent selectivity. This involves plotting values for the acidity (α), basicity (β) and dipolar nature (π) of solvents in a triangle known as the solvent selectivity triangle, where α is a measure of a solvent's ability to act as a hydrogen bond donor and β is its ability to act as a hydrogen bond acceptor. π relates to a solvent's ability to interact with a compound via apolar and polarisation forces. Methanol is considered to be slightly more 'acidic' than acetonitrile, i.e. to act as a better hydrogen bond donor. Both solvents are less 'acidic' than water. Acetonitrile is considerably more dipolar in nature than methanol and water. Software is available commercially which can optimise separations based on a few exploratory chromatographic analyses.

14.2.7 Effect of Temperature/Flow Rate

In chromatographic separations, there is a direct relationship between flow rate and elution – the higher the flow rate, the faster the analysis (see Section 12.4.3.2). The optimal flow rate is a trade-off between reduced longitudinal diffusion of analytes which occurs at higher flow rates and reduced resistance to mass transfer which occurs at lower flow rates. Either extreme of flow rate can lead to a loss of efficiency and band broadening according to the van Deemter equation but the overall decrease in efficiency is more marked at lower flow rates with loss of efficiency at higher flow rates being more gradual, as explained by the shape of the curve discussed in Section 12.5.1.2. In HPLC, problems arise if the flow rate is too high due to the back pressure generated across a column. UHPLC pumps have enabled instruments to cope with higher back pressures, but there are still limits to the back pressure that can be tolerated (Table 14.1). The back pressure of the system can be reduced by changing the column to a shorter column, a wider column, or a column with increased particle size.

Temperature has an effect on the retention characteristics of compounds since the temperature of a liquid has an effect on its viscosity and molecular diffusivity. The relationship is: diffusivity = ψ × (temperature/viscosity), where ψ is a parameter dependent on the physicochemical properties of the analyte and mobile phase. The product of diffusivity × viscosity is roughly proportional to temperature but increases gradually with temperature, e.g. with a 100 °C increase the product increases by about 30%. At higher temperatures, lower pressures are required to achieve the same mobile phase velocity and an increase in velocity is required to achieve maximum efficiency. In practical terms, if the temperature is raised, the optimum velocity increases significantly. The overall effect is that the efficiency remains the same as when lower column temperatures are employed but the retention times decrease exponentially and thus analysis time is faster (Guiochon 2006). On account of the sensitivity of retention with temperature changes, columns are usually held at a constant temperature during a chromatographic analysis. This is often well above ambient temperature to speed up analysis, typically 30–60 °C. Using the example of beer markers, Figure 14.7 shows the effect of modifying the slope of the gradient as well as altering the temperature of the column and flow rate. In general, if the column temperature is modified, this usually is accompanied by the need to increase flow rate to maintain the required resolution between peaks. Furthermore, the effect of temperature on ionisable compounds cannot be ignored because not only are the pK_a^s of compounds and the pH of buffers influenced by temperature, but the hydrophilicity of water decreases with increasing temperatures; all parameters which will influence retention, and possibly elution order. Stationary phases that rely on multiple mechanisms of retention of compounds often show decreased selectivity at higher temperatures. There may also be a heterogeneous temperature distribution inside the column as a result of compression in the pump, although a heat exchanger is usually employed between the pump and injector to ensure temperature homogeneity of the mobile phase with that of the column wall.

14.2.8 Detector

The most common detectors used are mass spectrometers and UV-based detectors such as PDA detectors (Figure 14.8). Along with MS, PDA detection is a useful technique for 'systematic toxicological analysis' and 'general unknown screening', terms used to describe a rational approach for screening samples for large numbers of substances (Lambert et al. 1997).

The most common types of detectors are described in the following sections.

14.2.8.1 Mass Spectrometer

Single or multiple mass spectrometers are often popular detectors used in conjunction with HPLC. The commonly employed method of electrospray ionisation (ESI), which is often used in conjunction with quadrupole mass spectrometers, is a soft ionisation technique that requires two quadrupole mass spectrometers connected in tandem to obtain useful information. This is because the ESI typically produces the analyte ion with no fragments. Therefore, fragmentation of the first ion in a collision cell is followed by the detection of diagnostic fragments in another quadrupole mass spectrometer. Fast atom bombardment mass spectrometry (FAB-MS) can give a spectrum containing the quasi-molecular ion (protonated molecule) as well as diagnostic fragments, as demonstrated for phenothiazines

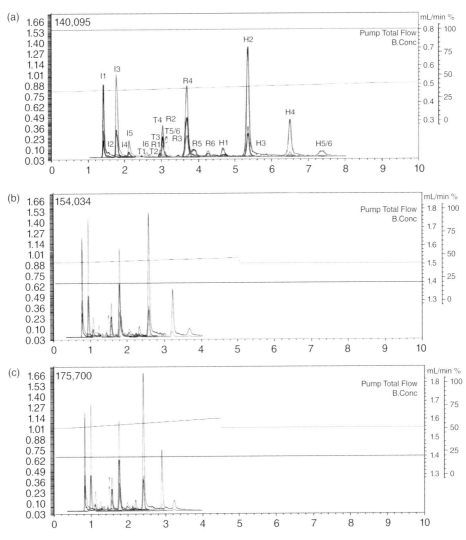

Figure 14.7 An example of the optimisation of chromatography for isohumulones (also known as iso-α-acids), potential markers of beer consumption. A C_{18} column (3 × 150 mm, 2.6 µm) along with a C_{18} guard column (3 × 10 mm, 2.6 µm) was employed. Mobile phases were 50 mmol/l ammonium formate in 90:10 water/acetonitrile, pH 2.8 (eluent A) and acetonitrile containing 0.1% formic acid (eluent B). (a) Chromatogram obtained employing a gradient of 42.5–50% mobile phase B over 10 minutes at 0.8 ml/min. (b) Chromatogram obtained following an increase in mobile flow rate to 1.4 ml/min and temperature of 60 °C. (c) Chromatogram obtained following a higher initial mobile phase B composition and gradient elution from 50 to 60% and a sharper gradient with the top of the gradient at 4.5 minutes, flow rate of 1.4 ml/min and temperature of 60 °C. Using an increase in flow rate, temperature and gradient, the analysis time has been more than halved compared with (a). I, iso-α-acids; R, rho-iso-α-acids; T, tetrahydro-iso-α-acids; and H, hexahydro-iso-α-acids.

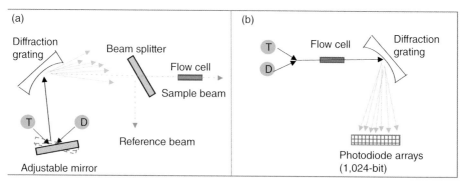

Figure 14.8 Typical detectors used traditionally in HPLC analyses. (a) UV detector. A tungsten (T) lamp provides wavelengths of 380–900 nm (i.e. visible region of the spectrum) whereas a deuterium (D) lamp enables coverage of 190–380 nm (i.e. the UV part of the spectrum). A mirror is used to select the source and the beam is passed through a diffraction grating that splits the beam into different wavelengths. The required wavelength is selected for absorption measurements by adjusting the angle of the mirror and diffraction grating. The beam is split to enable two identical beams, one which acts as the 'reference' beam and the other which passes through the flow cell containing the analyte(s) as they elute from the HPLC column. The difference in intensities of the beams indicates absorption of a particular wavelength by components in the sample. (b) Photodiode array detector. Unlike the UV detector, all the wavelengths are passed through the flow cell after which the beam is split by the diffraction grating and detected by photodiode arrays (1024 in number). The light intensity at each wavelength is converted into an electrical signal. Data is typically covered at a rate of 1 spectrum at each wavelength/second. Unlike UV detection, with photodiode arrays there is no splitting of the beam into multiple components prior to light entering the flow cell and this precludes the use of a reference beam.

(Mizuno et al. 2002). Sometimes 'in-source' fragmentation can be effected through manipulation of the cone voltage to produce fragments that can be detected in the first mass spectrometer. Following the optimisation of mass spectrometric parameters performed by direct infusion into the mass spectrometer, when coupling MS to HPLC, larger flow rates are used, and mass spectral parameters need to be re-optimised specifically for this purpose. For a more detailed discussion of mass spectrometric detectors, see Section 10.2.5.

14.2.8.2 UV Detector

UV was one of the first detectors employed with HPLC. Many HPLC instruments come equipped with UV or PDA detectors and are therefore readily available, and less costly than MS detectors. In addition, confirmation of findings in casework often requires two independent means of confirming the presence of a substance in a sample and sometimes the mobile phase flow is diverted through a UV spectrophotometer, a non-destructive detection method, after which the flow enters the mass spectrometer. Although this adds to extra-column volume and may increase the likelihood of band broadening, the tubing used occupies a very small volume with respect to the total column volume and the advantage of employing a method with additional confirmation is attractive to most practitioners. The detecting UV wavelength is chosen with regard to any mobile phase modifiers that interfere with the absorbance spectra of the compounds, and it is important that matrix effects are studied during the validation of methods. Thus, in the detection of protein biomarkers, the addition of urea, n-octylglucoside, and various ampholytes as mobile phase modifiers, required detection to be performed at 280 nm, whereas a mobile phase containing trifluoroacetic acid (TFA) and no other modifiers enabled visualisation of chromatograms using a wavelength of 214 nm (Legg et al. 2014).

14.2.8.3 PDA Detector

PDA detection employs UV detection, although in this set-up (Figure 14.8) the light is split into many different wavelengths after it has passed through the flow cell, enabling spectra to be collected at many different wavelengths. The rays hit semiconductors made of positive-type silicon (containing electron holes) and negative-type silicon (containing an excess of electrons) and this results in electron–hole pairs being generated which form the basis of the one-way electrical current produced. This form of detection is very common for HPLC systems and has been employed in conjunction with UHPLC. The advantages are that peak purity can be ascertained, as well as the structure of the compound.

Single wavelength detection, as exemplified by conventional UV detectors, and identification based on retention time is based on the assumption that the eluting peak is pure. If two or more compounds elute simultaneously, and if both compounds absorb at the wavelength used for detection, the analyst will not realise that more than one compound is present. In contrast, because PDA detection monitors multiple wavelengths, it allows the analyst to verify consistency of spectra at multiple wavelengths across a peak to check for peak homogeneity. In addition, the detector enables a compound to be quantified post-analysis based on the most ideal wavelength for that analyte and the best trade-off between a low background noise from impurities and high absorption of the analyte. This post-analysis optimisation is not possible with conventional UV detectors.

PDA detection is used to identify a wide range of compounds such as benzodiazepines, antidepressants, azides, antimalarials and drugs of abuse (Lambert et al. 1997). A method developed for the detection of benzodiazepines (He et al. 1998) had a detection limit of 10–30 ng and a limit of quantification of 50 ng/ml, typical of many drugs using this detector. The authors noted that for the benzodiazepines bromazepam, flunitrazepam, and flurazepam that had the worst sensitivities, detection could be improved by using optimised monitoring wavelengths of 242, 224, and 225 nm, respectively. They also discussed the relative compatibility of extraction solvents used for blood samples with the PDA detector, with ethyl acetate/n-hexane mixtures resulting in best recoveries and fewer interfering peaks compared with other solvents investigated. More recently a method used to detect tetrahydrocannabinol (THC), THC-acid A, cannabinol (CBN) and cannabidiol (CBD) in cannabis was described, which was used to provide expert evidence in court (Ambach et al. 2014). The use of this detector in a robust and high-throughput method for the identification of compounds in seized cannabis material also illustrates its usefulness in applications where a high concentration of analyte is expected. In this situation the sensitivity of the detector was sufficient to distinguish those plants containing THC below the legal limit of 1% from illegal marijuana plants. Similarly, the detector has sufficient sensitivity to determine cause of death in cases of drug overdose and poisoning and has applications in clinical toxicology. Disadvantages, nevertheless, are a lack of sensitivity for trace analysis in complicated matrices. In addition, spectra of compounds, especially ionisable ones, are dependent on pH. Hence any fluctuations in the preparation and storage of mobile phases may hinder the detection and use of library matching for positive identification.

14.3 Related Techniques

Other techniques are based on similar principals to those used in traditional HPLC employing a stationary phase supported in a column, a liquid mobile phase, and partitioning of analytes between the two phases. They include ion chromatography, affinity chromatography, and chiral chromatography which have all been previously applied in the field of forensic science. In each case, a modification of the stationary phase, and accompanying mobile phase, has allowed the separation of analytes possessing more specific properties – ionic compounds, those that show a high affinity to particular target molecules, and enantiomers that possess too similar physicochemical characteristics to be resolved by conventional HPLC.

14.3.1 Ion Chromatography

Some analytes such as inorganic explosives are more amenable to ion chromatography where ions are separated based on charge and size. Charged molecules bind to the charged stationary phase by electrostatic interactions. A mobile phase of increasing ionic strength containing ions which replace tightly bound analytes is often used. Conductivity, electrochemical detection, molecular spectroscopy and atomic spectroscopy have been used for detection. However, most recently this form of chromatography has been used in a forensic setting in conjunction with ESI and with inductively coupled plasma mass spectrometry (ICP-MS) (Barron and Gilchrist 2014) as well as high resolution mass spectrometry (HRMS) (Gilchrist et al. 2015). Commonly employed stationary phases are those containing tertiary/quaternary ammonium functional groups to retain anions and carboxylic/sulfonic acid to retain cations.

The approach has been used in the identification of explosives in sweat and fingerprints using a miniaturised capillary method (Gilchrist et al. 2012). Chelation ion chromatography uses ligands on the stationary phase, or ligands in the mobile phase, to disrupt binding of analyte with the stationary phase, and is used to separate metal ions. Ion-interaction, or ion-pair, chromatography involves the addition of substances to the mobile phase such as didodecyldimethylammonium for the retention of anions, and dioctylsulfosuccinate for cations. The hydrophobic portion of these agents binds to a hydrophobic stationary phase whereas the ionic moiety aids retention of ions.

14.3.2 Affinity Chromatography

In affinity chromatography, compounds are separated according to their specificity with an immobilised target, for example an enzyme, antibody, or receptor, or, alternatively, the substrate, antigen, or ligand can be immobilised. Recently, this principle has been used to detect ribosome-inactivating protein toxins from plants, such as ricin (from the castor plant *Ricinus communis*) and abrin (from *Abrus precatorius*), where the binding specificity of the B-chain, a lectin, to galactose (the ligand) was exploited in designing the stationary phase. Fractions were prepared from these chromatographic separations which were further analysed by LC-MS/MS following trypsin digestion to generate peptide fragments. The method was used successfully in forensic casework where illegal production of these toxins was investigated (Fredriksson et al. 2015). Affinity chromatography has also been harnessed into a commercially available point-of-care test used to detect the level of glycated haemoglobin (HbA1c), a more stable biomarker than glucose for determining metabolic state in the weeks leading up to death. Glycosylated haemoglobins bind with a higher affinity to immobilised boronic acid than non-glycosylated haemoglobins. Results of the method can help support an interpretation of diabetic ketoacidosis, starvation, alcoholic ketoacidosis, or acetone or isopropanol intoxication when used in conjunction with the detection of ketone bodies (Keltanen et al. 2013).

14.3.3 Chiral Chromatography

Separation of enantiomers can be performed in HPLC by derivatising an enantiomeric compound with chiral derivatisation agents to form diastereoisomers. Each diastereoisomer interacts slightly differently with the stationary phase, thereby enabling resolution. The disadvantage of this approach is the need for a derivatisation step. The analysis of enantiomers has been simplified with the introduction of chiral chromatography, where separation occurs using a chiral stationary phase. These phases contain a single enantiomer of a chiral compound, and are typically based on cellulose, amylose or cyclodextrin substituted with modifying groups. One forensic application, as exemplified by amphetamine (Hädener et al. 2017), involves the differentiation between the abuse of illicit (generally a 1:1 racemic mixture) and pharmaceutical preparations (containing the S-enantiomer or a defined proportion of S-/ R-enantiomers), with the S-enantiomer possessing stronger stimulant properties. This would be a suitable approach for the determination of the enantiomeric composition of powders and tablets, and biological samples for well-studied compounds such as methamphetamine. Specialised columns are available for the separation of members of specific drug families, such as the New Lux AMP chiral column (available from Phenomenex) that mediates separation of enantiomers of amphetamine and its analogues. Methods also have been developed for enantiomeric opioids such as methadone and tramadol. For example, separating R- and S-methadone can determine the concentration of the active R-enantiomer in methadone-related deaths (Jantos and Skopp 2013).

14.4 Chromatography Theory

As already discussed in Chapter 12, the van Deemter equation, Eq. (12.14), is used to describe the efficiency of packed chromatographic systems, typically GC (Knox et al. 1976). There are three parameters, namely A, B, and C, which describe, respectively, the effects of eddy diffusion (diffusion around particles of the packing), longitudinal diffusion, and resistance to mass transfer (both in the solid and stationary phases). The plate height, H, is a measure of efficiency related to the spreading of the chromatographic band as it travels through a chromatographic column (band broadening). Subsequent to the van Deemter equation, the Giddings, Huber, Hovath and Knox equations were developed (see Eq. 12.19 for the Knox equation). Giddings and his co-workers were dissatisfied with the van Deemter equation and the finite contribution of the A term (longitudinal diffusion) which is independent of solvent diffusion and they incorporated the interstices of the packing in the path length. The Huber (Huber and Hulsman 1967) and the Hovath

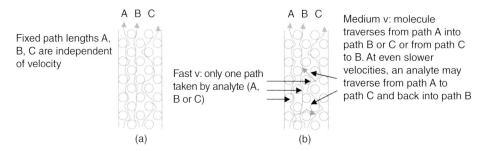

Figure 14.9 The differences between the *A* term in (a) the van Deemter theory ($H = A + B/v + Cv$) and (b) the Knox theory. A, B, and C are different paths taken by a molecule through a packed stationary phase. The path lengths of different analyte molecules are different. The Knox equation considers that the longer the molecule stays in the column, the more likely it is to diffuse into another path which may be of a longer length (as in path C) or shorter (paths A and B).

equations also modified the *A* term and incorporated an additional term to describe the resistance to mass transfer resulting from "turbulent mixing" which takes place in the eluent flow between closely packed particles. The Knox equation describes the effect of particle size on the reduced plate height (*h*) and reduced velocity (*v*) and also couples the *A* term to velocity. It is a better approximation of efficiency for HPLC than the van Deemter equation. The argument for this is that the path an analyte molecule takes around the stationary phase packing can vary and the faster the analyte travels the fewer paths it may traverse as shown in Figure 14.9, which would make the 'A' term dependent on flow rate (Knox et al. 1976). The parameters 'reduced plate height' and 'reduced velocity' have also been introduced into the equation to allow comparison of different columns packed with different particle sizes, highly relevant for comparison between UHPLC and HPLC. Although other approximations exist for HPLC systems, they are beyond the scope of this chapter.

14.5 Detection

The output of a HPLC separation is called a chromatogram regardless of the type of detector that has been used. Rather than absolute retention times, the use of relative retention time (RRT), or retention index (RI), is sometimes advocated in order to account for retention time drifts and to standardise the results within a laboratory, between laboratories, and when screening for substances against databases. RRT is generally based on the ratio of the retention time of a substance compared with a reference compound. In the calculation of RI, the retention behaviour of the analyte is compared against a number of reference compounds, each with assigned indices. Multiple RI scales have been recommended for basic drugs and acidic/neutral drugs, which have been applied to screen for more than 250 drugs relevant to clinical and forensic toxicology (Elliott and Hale 1997, 1998).

Chromatograms can be visualised in a number of ways depending on the information that has been detected. For example, if PDA detection has been used, chromatograms can be visualised at multiple wavelengths. It is important to check for peak purity by observing the homogeneity of spectra at different positions across the peak. This can be ascertained with modern software. It is also possible to use software to perform deconvolution of multiple spectra obtained from co-eluting substances.

If MS has been employed where multiple fragments of molecules are monitored in MRM, an 'extracted ion chromatogram' (Figure 14.10) is used to display the presence of a particular fragment ion across the whole chromatographic run (see Section 14.6 on coupling

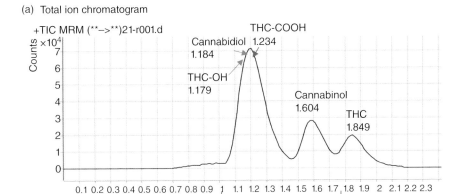

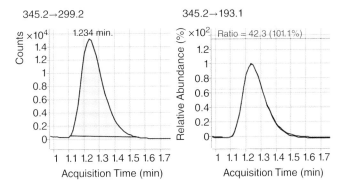

Figure 14.10 Distinction between a total ion chromatogram a) and extracted ion chromatograms (b) for mass spectrometric analysis using cannabinoids as an example. MRM was employed with a total of 20 transitions; two each for THC, THC-OH, cannabinol, cannabidiol, THC-COOH and their respective trideuterated internal standards. Here, the total ion chromatogram (a) depicts peaks generated from all the eluting compounds, i.e. the total ion chromatogram is a summation of all ions monitored. Peak integration of the total ion chromatogram for quantification would be impossible in this scenario especially for the co-eluting THC-OH, cannabidiol and THC-COOH (first peak in a)). b) Extracted ion chromatograms are shown for the two MRM transitions for THC-COOH, the main metabolite of THC, with the quantifier ion (ion used for quantification) on the left and the second ion (qualifier) on the right. The second ion must fall within the limits of the expected ratio with respect to the intensity of the quantifier ion, as represented by blue lines. The expected ratio is obtained from a standard analysed alongside the sample. The extracted ion chromatogram approach allows chromatograms to be viewed for one ion only, eliminating interference from other ions, and making determination of peak area possible for all compounds of interest, even if they overlap with other compounds in their elution from the column.

with MS). This reduces the noise compared with when a 'total ion chromatogram' is visualised and allows the analyst to 'integrate' a peak. This describes the processing by computer algorithms that ascertain the peak retention time, area and height, necessary for both identification and quantification. This can also be accompanied by use of other algorithms that allow functions such as peak smoothing. When employing such algorithms, it is essential that the analyst understands the reasoning behind their use and maintains consistency throughout the processing of results from all samples in an analysis. Importantly, the use of such algorithms should be included in the validation procedures for that method.

14.6 Coupling of Liquid Chromatography to Mass Spectrometry

Mass spectrometric detection requires compounds to be in the gaseous phase and to be in ionic form. Hence, for HPLC systems coupled to MS, interfaces are required to aid transfer of molecules from the liquid phase into the gaseous, high vacuum conditions in the mass spectrometer and to assist in the ionisation of molecules during the process. Two main ionisation techniques used in combination with HPLC are ESI and atmospheric pressure chemical ionisation (APCI).

Both APCI and ESI are considered 'soft' ionisation techniques because they do not produce extensive fragmentation of the compound. Although this can be exceedingly advantageous in terms of confirming the identity of an analyte through its molecular mass, if a high resolution MS instrument capable of accurately determining mass is not used, there is the possibility that other compounds with a similar mass could be present. Being able to identify the presence and relative abundance of several fragment ions for a compound is diagnostically much more useful. Hence, some types of mass spectrometer incorporate ways to fragment precursor ions following their formation during ESI or APCI, into product ions.

Once the ions are formed, they need to be separated according to their mass-to-charge ratio (m/z) and then detected. Mass separation in MS can be carried out using ion traps, quadrupoles or time-of-flight instruments. Quadrupole mass analysers can be coupled in sequence and the mode of detection referred to as tandem MS (MS/MS). The most common format is where three quadrupoles have been coupled in sequence (triple quadrupole) where the middle quadrupole is modified to fragment ions transiting from the first quadrupole to produce more characteristic ions, thereby enhancing the specificity of identification and sensitivity of the analysis.

ESI is commonly used for thermally labile and non-volatile compounds because evaporation is not required prior to ionisation. Rather, ionisation begins in the liquid phase and proceeds in solvent droplets of decreasing size. The sample is forced through a hypodermic needle which has a high potential difference relative to the surrounding counter electrode. This results in a spray of charged droplets with the same polarity as the charge on the needle and repulsion from the needle towards the counter electrode and the sampling cone. As the droplets reach the cone, solvent evaporation occurs, and the droplet diminishes until it is followed by 'Coulombic explosion' which causes the droplet to be ripped apart to form smaller droplets. These events are then repeated by the smaller droplets, eventually resulting in solvent-free charged, analyte molecules.

In APCI, the sample is heated in a quartz tube and desolvated by a pneumatic nebuliser. This is followed by interaction with a plasma known as the corona discharge where both solvent and nebulising gas (e.g. nitrogen) contribute to ion formation. Ions can be generated from low- to medium-polarity analytes. APCI is less suitable than ESI for lower analyte concentrations, due to the higher gas and mobile phase flow rates typically used, resulting in dilution of analyte through the MS system. For the more non-polar analytes, APCI overcomes some problems of ion suppression associated with ESI that are often caused by biological matrices.

In tandem MS using triple quadrupole instruments, the ability to use different modes allows the representation of data in a number of different formats. For example, MS/MS scanning collects information about all fragments produced from precursor ions, allowing the visualisation of either a total ion chromatogram or an extracted ion chromatogram (see Sections 10.2.4.8–10.2.4.11), where the elution profile and intensity of a single fragment derived from an analyte is viewed. Alternatively, SRM, otherwise known as multiple reaction monitoring (MRM) by some instrument manufacturers, can be employed to select only particular product ions derived from a precursor ion, most often the most abundant diagnostic two or three ions. In this scenario, a total ion chromatogram can be viewed as a combination of all the selected ions generated as well as an extracted ion chromatogram. Although similar information is generated as in MS/MS scanning, the MRM approach is more sensitive and specific and often provides a superior signal-to-noise ratio and limit of detection by eliminating other masses and reducing background noise. In some scenarios, a further mode, neutral loss (NL) scanning may be useful which detects only those compounds that have a common feature such as a sulfate or glucuronide moiety and exploits a common loss of a particular part of the molecule upon fragmentation. In the latter case, glucuronides monitored in positive mode MS lose a mass of 176 to form a fragment with a m/z 176 units lower than the m/z of the precursor ion. Like MRM, NL scanning provides greater sensitivity and specificity than MS scanning, and sometimes MS/MS scanning, but has more limited use. In forensic toxicological analyses, NL scanning is typically used for detecting types of metabolites and related molecules, with the important distinction that the absolute mass of the compound does not need to be known. One useful approach is to detect metabolite types that have longer detection windows, or are present at higher concentrations following drug administration, than parent compounds.

Ion trap instruments allow the provision of selected ion monitoring by fragmentation of a precursor ion, although the detection relies on ejection of all other product ions from the trap before the selected ion can itself be ejected and detected. Although ion traps are sensitive instruments because they are able to accumulate a greater number of eluting molecules, this detector is slower than a triple quadrupole mass spectrometer when operated in the equivalent mode. However, the advantage of the ion trap in being able to accumulate a greater number of ions prior to detection can be combined with the speed and specificity of the triple quadrupole instruments in linear ion trap instruments and other quadrupole–ion trap hybrid instruments.

HRMS instruments are also highly favourable detectors used in conjunction with HPLC. Recently, a popular combination of mass spectrometers involves the use of the hybrid Quadrupole time-of-flight (QTOF) instrument, due to its high mass accuracy, superior resolution and fast acquisition times. The instrument has found use in untargeted analysis due to its ability to differentiate similar masses and be used in a number of modes. 'MS all' approaches have been recommended which include obtaining data that do not rely on a priori knowledge of the compounds that are (or could be) in the samples (such as the precursor ions), and employ many collision energies. In data-independent (or information-independent) acquisition (DIA), all ions within a selected mass range (typically 20-35 m/z) are fragmented and this is repeated in step-wise fashion until the whole range of masses of interest are fragmented, as exemplified by 'sequential window acquisition of all theoretical fragment-ion spectra' (SWATH) (Boxler et al, 2019). This is in contrast to information-dependent acquisiton (IDA), where the precursor ions and collision energies are automatically selected in real time based on pre-defined criteria, for example to include fragmentation of ions above an intensity threshold, the most intense ion in a given retention time window, or to exclude interfering masses. In this approach, survey scans are used to trigger dependent scans, allowing both data acquisition and optimisation of settings to occur within the same run.

14.7 Types of Analytes

14.7.1 Basic Analytes

As the vast majority of drugs possess a basic functional group, HPLC coupled to PDA or MS forms the basis behind systematic toxicological analysis in the screening for hundreds of basic drugs simultaneously (Elliott and Hale 1998; Gerostamoulos et al. 2010). Particularly where MS is used, formic acid is often used in the mobile phase, aiding the protonation of basic drugs during MS ionisation. Mobile phases used commonly for basic drugs are applicable for detecting plant toxins and were used in a method detecting 34 active principles in toxic plants. Commonly used mobile phases of acetonitrile containing 0.1% formic acid (mobile phase A) and a 2 mM ammonium formate buffer adjusted to pH 3 (mobile phase B) were employed (Carlier et al. 2015).

14.7.2 Acidic Analytes

Acidic compounds can be typically analysed using the same hardware as that of basic or neutral analytes but the pH is increased to favour the ionised form when used in conjunction with MS. Mobile phase modifiers such as ammonia may be used to deprotonate and therefore ionise acidic drugs such as barbiturates and non-steroidal anti-inflammatory drugs, which are difficult to analyse using conventional mobile phases employed for basic drugs (Di Rago et al. 2014). A liquid chromatography–mass spectrometry (LC-MS) screen targeting 132 common acidic and neutral drugs used mobile phases of 25 mM ammonium acetate buffer adjusted to pH 7.5 (mobile phase A) and acetonitrile (mobile phase B) (Di Rago et al. 2014). An endcapped stationary phase was used to avoid non-specific binding to uncovered silanol groups and, in addition, it had butyl side chains to improve peak shape and separation of compounds that also possess basic functional groups. Butyl side chains are shorter than the typically used C_8 or C_{18} phases and thus are more 'polar' as they bring the analyte into closer contact with the oxygen-containing silica backbone. In addition, the system required optimisation of MS to assign detection in the negative or positive electrospray mode, after which the parameters were combined into a method requiring two injections, one operated in positive mode and another for the negative mode.

14.7.3 Proteins

LC-MS-based techniques can be particularly useful for the initial identification of protein biomarkers and can be used to direct the development of other, more traditional approaches such as enzyme-linked immunosorbent assays (ELISAs). Liquid chromatography offers the advantages of sensitivity, reproducibility and resolution over the more conventional technique of electrophoresis, particularly for lipophilic proteins, those present at very small quantities, and those larger than 150 kDa or smaller than 5 kDa in size (Legg et al. 2014). Thus, a greater number of compounds can be recovered, fractionated, and visualised providing a more complete profile. The technique also reduces problems of solubility encountered with gels and with band distortion encountered with higher amounts of total protein.

Proteins are typically digested with trypsin before analysis and the products are analysed in their denatured form as polypeptides. Specialty HPLC columns are available that have wide pores (e.g. 300 Å) to enable complete access of proteins to the stationary phase. The columns also offer compatibility with low mobile phase pH, for example due to TFA, an ion-pair reagent used to assist with retention. They are also compatible with high temperatures (e.g. 80 °C) to enable simultaneous denaturation of proteins. From a forensic application standpoint, three peptide biomarkers of vaginal fluid were identified using liquid chromatography-electrospray ionisation time of flight mass spectrometry (LC-ESI-TOF-MS). These biomarkers were not present in other body fluids such as nasal secretions, saliva, urine, semen and sweat (Igoh et al. 2015).

14.7.4 DNA

The first forensic applications of LC-MS in characterising polymerase chain reaction (PCR) amplified DNA products were published in 2001 and are reviewed elsewhere (Oberacher and Parson 2007). The stationary phases typically employed are C_8 on silica particles or polystyrene/divinylbenzene particles, or monoliths. Amines such as butyldimethylamine and cyclohexyldimethylamine have been suggested as suitable mobile phase additives for their role as ion-pairing reagents, at 10–25 mM which provides a good trade-off between resolving power and a lack of ion suppression in ESI. The technique has been used for the typing of short tandem repeats (STRs) and single nucleotide polymorphisms (SNPs) in nuclear DNA and the hypervariable sequences HV1 and HV2 in mitochondrial DNA (Kristinsson et al. 2009). Another method has recently been published employing ion pair high performance liquid chromatography–time of flight mass spectrometry (HPLC-TOF-MS) to profile PCR amplicons of mitochondrial DNA (Oberacher et al. 2006). The chromatographic system comprised a capillary column with dimensions of 50×0.2 mm (internal diameter) with a monolithic packing, flow rates of 2 μl/min and a column temperature of 68 °C to denature the DNA. Sometimes, an ion pairing reagent, triethylammonium bicarbonate, is used to improve detection by LC-MS (Premstaller et al. 2000).

14.7.5 Chiral Compounds

As noted previously, it is possible to separate enantiomers of compounds on chiral columns. In addition, derivatisation has been used in LC-MS. For example $R(-)$-3,4-methylenedioxymethamphetamine (R-MDMA) and S-MDMA and eight pairs of enantiomeric metabolites have been separated on traditional reversed-phase columns following derivatisation with the reagent N-(2,4-dinitro-5-fluorophenyl)-L-valinamide (DNPV). Stereo-selective metabolism for the S-enantiomer has been determined and the method applied successfully to human plasma (Steuer et al. 2015). This has led to the ability to determine differences in effect, toxicity, metabolism and distribution of enantiomers. As a result, interest in such studies has increased in the context of forensic science with investigations addressing the use of differential metabolism and enantiomeric ratios to establish time since drug administration.

14.7.6 Bulk Drugs and High-Concentration Analytes

As previously discussed, the analysis of bulk drugs and other samples possessing high concentrations of substances are amenable to HPLC coupled with UV or PDA detection. This instrumentation is advantageous in these situations as it is less prone to detector saturation and is less expensive than mass spectrometric techniques. Nevertheless, mass spectrometric analyses can be employed where large concentrations of analyte are expected. For example, methamphetamine and ecstasy (MDMA) form part of the roadside testing programme in Victoria, Australia, where drivers can be randomly stopped to provide oral fluid specimens. In such instances, very high concentrations of these drugs can be encountered in the oral cavity and an issue of over-sensitivity of the technique arises where the detector response is not linear to the concentration and detector saturation results. The Victorian Institute of Forensic Medicine (Australia) has developed a method of monitoring the +1 isotope of these drugs (Di Rago et al. 2016). Ecstasy, for example, has an elemental composition of $C_{10}H_{15}N$. Based on the major contributor of the +1 isotope being ^{13}C in any one of the positions and ^{13}C having a natural abundance of 1.109%, the +1 isotope would be expected to have a relative abundance of approximately 1% of the most abundant isotope. Targeting this isotope in MRM experiments therefore reduces detector saturation, and if a high signal persists, the sample is diluted to enable analysis.

14.7.7 Low-Concentration Analytes

HPLC in conjunction with MS is an effective tool in the analysis of low concentrations of analytes. For example, lorazepam and its polar glucuronide has been detected in fingerprints following the administration of 2 mg doses to volunteers, where a method with a limit of detection of 1.5 and 8 pg/fingerprint, respectively, was developed (Goucher et al. 2009). HPLC-MS is also useful for the detection of drugs in cases of drug-facilitated crime where low-dose drugs (for example low-dose benzodiazepines such as flunitrazepam, nitrazepam, triazolam and alprazolam), short half-lives, anterograde amnesia, and involvement of vulnerable groups may result in samples which contain low concentration of analytes, and which are collected at late time points following drug administration. The advantage of HPLC-MS is that it allows the simultaneous detection of drugs of differing polarities, particularly when coupled with some of the newer developments in chromatography, such as the use of Poroshell columns which possess high efficiencies enabling fast analysis.

14.8 Accreditation and Method Validation

In order to become accredited (see Chapter 2), a laboratory must adhere to international guidelines such as ISO/IEC 17025:2005: 'General requirements for the competence of testing and calibration laboratories'. A number of guidelines exist for the validation of HPLC methods including those from the United Nations Office on Drugs and Crime (UNODC) (UNODC 2009), the Society of Forensic Toxicologists (SOFT) (SOFT/American Academy of Forensic Sciences [AAFS] 2006) and the Scientific Working Group for Forensic Toxicology (SWGTOX) (SWGTOX 2013). For example, SWGTOX specifies as a bare minimum for qualitative analyses; the identification of the limit of detection, dilution integrity, carryover, stability, ionisation suppression/enhancement (when MS is used), and interference studies. The latter two are particularly important in HPLC methods because a number of mobile phases and reconstitution solvents can be used in analysis, which, along with sample components (both endogenous and exogenous), may have a different impact on the detection of any given analyte. More recently, the Organization of Scientific Area Committees (OSAC) for Forensic Science developed an AAFS Standards Board (ASB) and Approved American National Standard Board (ANSI) standard prescribing validation procedures that build on, and supersede, the SWGTOX guidelines (ASB 2019). For this reason, matrix enhancement and matrix suppression and specificity are important tests to be performed, and a minimum of 10 sources of blank sample (for example sources of blood or urine) should be used in the validation. Another difference between requirements for GC-based and HPLC-based methods are that requirements regarding deviations in retention time and ion ratios in MS with respect to calibrators and controls are slightly less stringent for HPLC (for example >2% and ± 30%) than GC (for example 1–2% and ± 20%) (SOFT/AAFS 2006). Some of the more important issues pertaining to HPLC are discussed in the following.

14.8.1 Use of Internal Standards

Some chromatography-specific issues pertain to the use of internal standards. For example, if MS is used as the detection method, a deuterated internal standard with an overlapping retention time, and a distinct molecular mass, is preferred. One of the earliest examples of a stable isotope labelled compound being used as the internal standard was the use of 'deuterated tetrahydrocannabinol' in the quantification of THC in blood (Curry 1974). In contrast, this would be detrimental for an analysis which employs another detector such as UV detection where the absorbance of the co-eluting internal standard would interfere with the detection of the analyte. In this case, an internal standard with a distinct retention time must be used.

14.8.2 Effect of Sample Matrix

Often, the sample matrix can influence whether a substance is detected and may alter the chromatography – for example by the presence of an interfering peak, a shift in retention time of the compound under study, a signal enhancement or suppression. This is often assessed by preparing a dilution of the drug standard in the mobile phase, and in an extract of blank sample matrix (e.g. of blood or urine containing no drug) and comparing the response. This is a vital part of method development and can reveal whether a method developed for one matrix is transferable across other sample types. In MS-based methods, a sample containing an isobaric molecule (same nominal mass) to the compound under study is an instance where a distinct retention time is crucial in distinguishing between the analyte and interferent and successfully making a positive identification.

14.8.3 Ion Ratios

The determination of ion ratios following fragmentation of a precursor ion can aid in the detection of a compound and is often used as a tool in forensic laboratories to aid in a positive identification. This can be illustrated for lysergic acid diethylamide (LSD) where despite three MRM transitions being detected consistent with LSD and acceptable ion ratios among two of the three ion combinations, the relative intensities of all three ions were not within the limits of those expected for LSD and there was an absence of iso-LSD which is in natural equilibrium with LSD (Sauvage et al. 2008). The signal was attributable to a zolpidem metabolite and this led to the search for other zolpidem metabolites, five of which were detected. Similarly, benzoylecgonine was mistaken for atropine which had a similar retention time but poor compliance in ion ratio between two transitions. This led to a second chromatographic separation to ascertain the identity of the compound. Longer chromatographic runs were also required to elucidate the identity of clomipramine because many of the fragment ions from this compound were also common to phenothiazines such as trimeprazine which eluted at similar retention times (Sauvage et al. 2008). This highlights the need for stringency in detecting specific ion ratios for routine methods as well as the need to tailor the analytical approach in certain scenarios.

14.9 Interpretation of Results in the Forensic and Legal Context

An important issue in the legal context relates to the diagnostic value of a method. In the case of MS or PDA as the detection method, a high diagnostic value may be easily achieved. Some authors have commented that different chromatographic methods should be awarded a different number of 'identification points' (IPs). An IP is given for each ion monitored for each drug, for example three MRM transitions combined with the retention time would give a method a total number of four IPs. Two IPs would be awarded for an ion if HRMS is used (Peters 2011). This is based on the assumption that the retention time and ion ratios are within the required ranges specified by repeat quality control analyses. Three IPs are required for a positive identification in the food industry. The World Anti-Doping

Agency, Society of Forensic Toxicologists, Gesellschaft für Toxikologische und Forensische Chemie, and the Food and Drug Administration have all published guidelines, and many refer to the superior diagnostic power of LC-MS/MS-based techniques (Sauvage et al. 2008).

Isobaric compounds, those possessing the same masses, may be detected rather than the analyte if they have similar retention times and mass spectral fragmentation patterns. Promazine/promethazine and ephedrine/pseudoephedrone are examples of pairs of isobaric compounds. In such scenarios, chromatographic resolution is paramount to the detection of these compounds and interpretation of the results, unless HRMS is used. When low resolution mass spectrometry (LRMS) is used, even compounds with a different elemental composition, for example 6-acetylmorphine and naloxone, may appear as isobars. While 6-acetylmorphine is a metabolite of heroin, an illegal drug, naloxone is used as an antidote to opiate overdose and to reverse the effect of opiates given during surgery. This illustrates a difficult scenario where the interpretation may have distinct legal outcomes. Sometimes, differentiating between the compounds present may be achieved through tandem MS where even if the same fragments are produced, they may be present at different ratios. If HRMS is available, this can resolve compounds with different elemental composition such as naloxone and 6-acetylmorphine.

In addition to the detection of particular drugs, many other biomarkers should be targeted by liquid chromatography in order to interpret toxicological findings. This is particularly true in cases where a drug has a short half-life, or $t_{1/2}$ (e.g. heroin or cocaine). Another challenge relates to metabolites originating from multiple sources (e.g. morphine). In these scenarios, better diagnoses may be obtained by targeting other metabolites of the same drug or by-products of illegal production processes. For example, morphine is a drug in its own right, a metabolite of codeine (a legitimate drug available over the counter) as well as a metabolite of heroin (a Class A drug in the UK but also available in some hospital settings in palliative care under the generic name diamorphine). In light of the very short half-lives of heroin and its diagnostic metabolite 6-acetylmorphine, morphine is often the only metabolite that is commonly detected following heroin consumption, along with its glucuronide conjugates. Therefore the detection of ATM4-glucuronide, a glucuronide metabolite derived from acetylated thebaine present in impure heroin, is recommended to counter the poppy seed defence (used to dispute an illegitimate source of morphine) (Chen et al. 2014).

A similar concept applies to compounds that are discovered to be unstable. For example, antipsychotics can produce diagnostic break-down products as exemplified for olanzapine (Saar et al. 2012). Ingestion of thermodegradable compounds such as benzodiazepines may be detected by targeting their benzophenones instead if a particularly high temperature is required in the method.

Some biomarkers are only present in unique scenarios. For example, cocaethylene has a longer half-life than that of cocaine and can provide the added dimension of co-administration of cocaine and ethanol. LC-MS/MS-based methods have also been used to address the hip-flask defence whereby alcohol is consumed intentionally by a driver between the time of a suspected offence and a breathalyser test, to mask impairment at the time of driving. Markers of beer consumption such as iso-α-acid congeners derived from the hop plant have been proposed to help with the interpretation of the true source of alcohol and can be used to corroborate case circumstances which make mention of particular alcoholic beverages (Rodda et al. 2013).

A number of reference books and compendia are used to interpret the toxicological findings obtained from liquid chromatographic methods. Baselt's 'Disposition of Toxic Drugs and Chemicals in Man' (Baselt 2014) and 'Clarke's Analysis of Drugs and Poisons' (Moffatt et al. 2011) are commonly used by many forensic toxicologists to interpret the concentrations obtained from analyses of various biological matrices. Both sources provide information on concentrations of drugs found in various scenarios such as cases, clinical studies and information on therapeutic and toxic concentrations. In addition, they can also be used to guide method development as they provide information on the pharmacokinetic and analytical properties of drugs as well as liquid chromatographic methods, coupled with various detectors, that can be adapted.

Importantly, it should be noted that the results of toxicological investigations should never be used in isolation; case circumstances and medical records must be consulted when making interpretations. For example, if cocaine is detected in a hospital patient, medical records and personnel should be consulted to determine whether this drug was employed as an anaesthetic in oral, laryngeal or nasal surgery. In certain scenarios, ad hoc tests may be necessary to narrow down the search for the drug that was administered but these tests should always be performed under the same stringent laboratory conditions with minimum validation protocols fulfilled, appropriate literature cited and justification of why the test was performed.

14.10 Case Studies

HPLC, particularly LC-MS and UHPLC-MS, is highly valuable in forensic casework. In this section, three scenarios are highlighted by presenting background information, the use of HPLC in the analysis of the case, and the resulting data and interpretation.

14.10.1 Case Study 1: Post-Mortem Death Investigation – Poly-Drug Overdose

After no contact for several days, the family of a 60-year-old male requested a well-being check to be performed. The subject was found unresponsive in his bed with a bag over his head in an apparent suicide. A beverage and multiple prescription medication containers were discovered on a bedside table, several of which were empty and showed signs of abuse based on prescription fill date and pill count. Medications were mirtazapine, pantoprazole, lamotrigine, escitalopram, zolpidem, oxymorphone, lorazepam and oxycodone. No signs of forced entry or foul play were discovered in the residence. The decedent was last contacted by his family five days earlier, with whom he typically stayed in close communication. A search of the decedent's computer revealed emails from four days earlier being opened, with all thereafter left unopened. No signs of tobacco, alcohol or illicit drug use were noted throughout the residence. Early signs of decomposition were present. A yellow/brown discharge was noted around the decedent's mouth and lips.

Table 14.3 Substances and their concentrations detected in a case involving poly-drug overdose.

Specimen	Compound	Result	Units	Analytical technique
Blood (peripheral)	Ethanol	0.02	% (w/v)	HS-GC-FID
Vitreous humour	Ethanol	Not detected		HS-GC-FID
Blood (peripheral)	Oxycodone	455	ng/ml	LC-MS/MS
Blood (peripheral)	Oxymorphone free	530	ng/ml	LC-MS/MS
Blood (peripheral)	Zolpidem	232	ng/ml	LC-MS/MS
Blood (peripheral)	Lorazepam	25	ng/ml	LC-MS/MS
Blood (peripheral)	Citalopram	1050	ng/ml	LC-MS/MS
Urine	Oxycodone	Confirmed present		LC-MS/MS
Urine	Oxymorphone total	Confirmed present		LC-MS/MS
Urine	Zolpidem	Confirmed present		LC-MS/MS
Urine	Lorazepam	Confirmed present		LC-MS/MS
Urine	Citalopram	Confirmed present		LC-MS/MS
Gastric contents	Oxycodone	185	mg	LC-MS/MS
Gastric contents	Oxymorphone	81	mg	LC-MS/MS
Gastric contents	Zolpidem	56	mg	LC-MS/MS
Gastric contents	Lorazepam	<5	mg	LC-MS/MS
Gastric contents	Citalopram	25	mg	LC-MS/MS

HS-GC-FID, headspace gas chromatography flame ionisation detector.

According to information obtained from the family, the decedent had a known history of depression, and suicidal ideations but no known suicide attempts. Death was determined as suicide caused by polysubstance abuse leading to fatal central nervous system depression (Table 14.3).

The versatility of LC-MS/MS in analysis of a wide range of different drug classes, in this example, opioids, benzodiazepines and antidepressants, makes it a useful tool for forensic toxicologists to comprehensively analyse many biological specimens in death investigations.

14.10.2 Case Study 2: Post-Mortem Death Investigation – No Derivatisation Needed for LC-MS

The subject, a 47-year-old male, was pronounced deceased in his residence after being found on the floor in front of the door to his bathroom. The subject was nude, except for a pair of shorts. There was a small pool of what appeared to be blood near the subject's face. The subject was cool to the touch. Rigor mortis was not apparent. According to the subject's friend, the subject had not been seen since two days earlier at which time he did not voice

Table 14.4 Substances and their concentrations detected by LC-MS/MS in a case involving a post-mortem death investigation.

Specimen	Compound	Result	Units	Analytical technique
Blood (peripheral)	Volatiles	None detected		HS-GC-FID
Vitreous humour	Volatiles	None detected		HS-GC-FID
Blood (central)	Benzoylecgonine	50	ng/ml	LC-MS/MS
Blood (peripheral)	Alprazolam	15	ng/ml	LC-MS/MS
Blood (peripheral)	Quietapine	<10	ng/ml	LC-MS/MS
Urine	Benzoylecgonine	Confirmed present		LC-MS/MS
Urine	Alprazolam	Confirmed present		LC-MS/MS
Urine	α-OH alprazolam	Confirmed present		LC-MS/MS

any complaints or issues and seemed to be in his normal state of health. The subject was reported to work security jobs and keep an inconsistent schedule. There was no evidence of foul play or another's involvement in the death. No medical records or notes of intent were discovered.

There was evidence of illicit drug use including a crack pipe and a white powdered substance on the floor. Additionally, the prescription medications alprazolam and Seroquel were also present but did not show signs of having been abused. Following the toxicological analysis, a friend of the deceased informed investigators of the subject's battle with cocaine for several years. Following histopathology and the toxicology results, the cause of death was given as likely myocardial infarction. Scarring of the cardiac tissue in conjunction with cocaine history and the presence of the cocaine metabolite benzoylecgonine in trace amounts in the blood and urine provided pivotal information in this case (Table 14.4).

In this analysis, the detection of benzoylecgonine was performed by LC-MS/MS without the need for derivatisation. Unlike GC, HPLC does not require the compound to be volatile prior to introduction into the instrument, making sample preparation less labour-intensive while reducing the number of resources required.

14.10.3 Case Study 3: Driving Under the Influence of Drugs – Increased Sensitivity with LC-MS

A female was driving home after a long day out with friends partying at a dance music festival. She had a history of recreational drug use and had been taking ecstasy and cocaine during the day. In an attempt to counterbalance the effects of the stimulants, she smoked a joint quickly as she wanted to get home. On the way home she encountered a random 'driving under the influence of drugs' (DUID) checkpoint where a police officer also saw the female noticeably weaving in and out of her lane as she was pulled over. The officer questioned the female's activities during the day, about which she lied. Following a negative breath alcohol reading, the officer then asked her to step out of the car to perform a Field

Sobriety Test. The female refused which resulted in the non-invasive collection of oral fluid for toxicological testing.

The chromatograms depicting the substances detected are shown in Figure 14.11. The driver was found guilty of driving under the influence of illicit drugs (Table 14.5).

In addition to the lack of derivatisation required for LC-MS/MS analysis, as previously described, the high throughput capabilities of these techniques are significant. The utilisation of improvements in MS capabilities allows for UHPLC peak widths below 10 seconds with sufficient mass spectral analysis across the peak for quantitation and identification purposes. The run time allowed for the quantitation of over 30 drugs of abuse was within five minutes per sample. This was prudent as the turn-around time of days rather than weeks was required for the reporting of blood specimens from suspected DUID-drivers. Typical LC-MS/MS and GC-MS analysis generally performs separation of compounds within 15–30 minutes, meaning a potential increase of 3–6 times the number of samples was feasible using UHPLC-MS/MS.

14.11 Forensic Developments

Much emphasis in the development of HPLC-based techniques is placed on coupling to instruments with ever-increasing sensitivity of detection, or resolution in the case of detection by MS (for example use of time of flight mass spectrometers). This is covered in Chapter 10. However, many developments relating to the column, particularly miniaturisation, multidimensional chromatography and automation, may also have future applications in forensic science. A few of these developments are summarised in the following sections.

14.11.1 Column Switching and Two-Dimensional HPLC

Some modern HPLC methods employ column switching. In this method, columns are connected in tandem and the set-up consists of a separation column followed by a concentration column (Mizuno et al. 2002). As well as employing different stationary phases in the second column, the system has the advantage of using different mobile phase systems and flow rates in the second column as another solvent can be supplemented via a T-junction. If appropriate flow rates are used, the second mobile phase can deliver the entire effluent of the first separation to the detector interface without the requirement of splitting the flow or the issue of detector saturation. Column switching has been used to analyse THC and 11-nor-9-carboxy-THC (THC-COOH) in cannabis users to minimise sample preparation and total run time by improved sample clean-up and concentration in the chromatographic stage of analysis. A large volume could be injected removing the need for pre-concentration or evaporation prior to loading onto the column (Hädener et al. 2016).

Another interesting application involves two-dimensional chromatography which is used to resolve substances in very complicated matrices, particularly where the concentration range of two co-eluting substances may span 10 orders of magnitude. The first column is used to separate components, visualised with a non-destructive detector such as UV, which are divided into fractions upon elution. This can be achieved by collecting multiple fractions

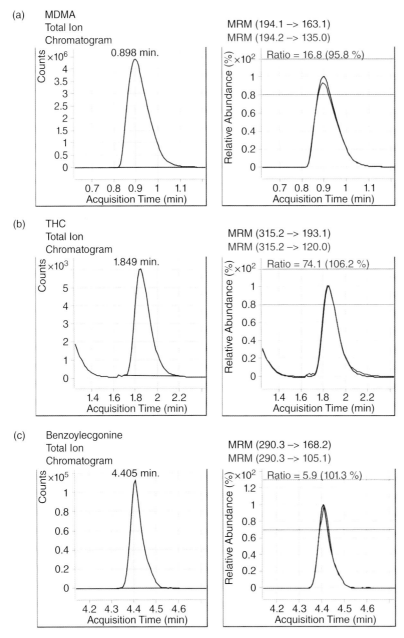

Figure 14.11 Three pairs of total ion chromatograms and MRM chromatograms for (a) MDMA (ecstasy), (b) THC (cannabis) and (c) benzoylecgonine (cocaine metabolite) depicting characteristic product ions at retention times of 0.9, 1.85, and 4.4 minutes as detected in Case Study 3. The expected ion ratio is calculated based on analysis of positive quality control specimens, together with the permitted tolerance in the ratio (typically ±20%). In each case, the second ion falls in the range (blue lines) of the expected intensity and thus fulfils ion ratio criteria.

Table 14.5 Substances and their concentrations detected by UHPLC-MS/MS in a case involving driving under the influence of drugs.

Specimen	Compound	Result	Units	Analytical technique
Oral fluid	3,4-Methylenedioxymethamphetamine (MDMA)	140	ng/ml	UHPLC-MS/MS
Oral fluid	Δ^9-Tetrahydrocannabinol (THC)	350	ng/ml	UHPLC-MS/MS
Oral fluid	Benzoylecgonine	220	ng/ml	UHPLC-MS/MS

in a 96-well plate (Legg et al. 2014). Subsequently the fractions are injected into a second column containing a different stationary and mobile phase system and the chromatograms can be visualised to determine characteristic combinations of retention times for particular compounds. The fractions from the second separation can further be analysed by LC-MS, as has been achieved for protein biomarkers important in forensic science (Legg et al. 2014).

14.11.2 Capillary Liquid Chromatography

There is a trend towards miniaturisation in chromatographic systems, as well as in the interfaces, used to couple chromatographic systems with detectors, such as nanoelectrospray ionisation sources. As ESI is a concentration-dependent process, the reduction of the mobile phase flow rates in combination with miniaturisation using columns of 25–150 μm internal diameters results in increased sensitivity. The droplets formed from the nanoelectrospray source are much smaller, improving ionisation and desolvation, and the entire solvent spray can enter the inlet of the mass spectrometer. This allows the ionisation process to be much more efficient. For example, this has been used to detect ketamine and metabolites present at low concentrations in hair samples (Parkin et al. 2013). The approach has applications in cases such as drug-facilitated sexual assault to elucidate drug administration after much time has elapsed.

DNA has been analysed using capillaries packed with monolithic material as well as particles with a C_{18} stationary phase. The monolithic columns resulted in a column efficiency of more than 190 000 plates/m and an improved performance of approximately 40% for an 18-mer oligodeoxynucleotide compared with the capillary packed with polystyrene-based particles with a diameter of 2.1 ± 0.12 μm (Premstaller et al. 2000).

14.11.3 Column-on-a-Chip Technologies

One aspect of miniaturisation involves the development of chips etched with grooves that form the basis of the chromatographic system. The advantages include the ability to analyse minute volumes of sample and improved sensitivity of detection. One area where this was used was in the discovery of 29 candidate protein biomarkers of saliva, urine, seminal fluid, vaginal fluid, peripheral blood and menstrual fluid. First, two-dimensional chromatography was used where samples were fractionated into 96-well plates and the fractions were further separated on another column and detected using UV absorbance.

After fraction collection from the second column, the fractions were injected onto a column contained on a chip (dimensions 150 or 43 mm × 300 Å, C_{18}) with a 40 nl or 60 nl trapping column and were separated using gradient elution (3–45% acetonitrile) (Legg et al. 2014). The use of multidimensional chromatography in the initial separation combined with detection using the column-on-a-chip provides highly specific and sensitive detection.

14.12 Conclusion

HPLC is a very powerful tool which can be used to separate mixtures of substances in complicated matrices. Its use is well substantiated, particularly as a result of its compatibility with a wide range of analytes that can be detected in many sample types within the field of forensic sciences. There is wide scope of coupling this technique with other developments in the field which has been enabled by parallel advances in detection methods and miniaturisation, ultimately leading to improved sensitivity, accuracy and precision in detection. In closing, it is difficult to know whether the scientific community has reached the zenith in HPLC-related methods. The ability to detect an ever-increasing number of analytes comes at the cost of specificity and sensitivity and vice versa. Likewise, the ability to detect ever-decreasing concentrations of analytes with highly sensitive methods comes at the cost of detector saturation, detecting contaminants, and in some cases, insignificant concentrations of analytes, interferents and endogenous compounds which add to the complexity of the techniques and interpretation of results. There is also a fine balance between miniaturised methods with fast analysis times, and fully automated, robust processes which do not require frequent trouble shooting and are compatible with problematic and 'messy' samples such as those routinely encountered in post-mortem toxicology.

References

Agilent (2012). LC and LC/MS. Your Essential Resource for Columns & Supplies. Canada.

Ambach, L., Penitschka, F., Broillet, A. et al. (2014). Simultaneous quantification of delta-9-THC, THC-acid A, CBN and CBD in seized drugs using HPLC-DAD. *Forensic Sci. Int.* 243: 107–111.

ASB (2019). Standard Practices for Method Validation in Forensic Toxicology. http://www.asbstandardsboard.org/wp-content/uploads/2019/11/036_Std_e1.pdf Accessed 4 8 20.

Barron, L. and Gilchrist, E. (2014). Ion chromatography-mass spectrometry: a review of recent technologies and applications in forensic and environmental explosives analysis. *Anal. Chim. Acta* 806: 27–54.

Baselt, R. (2014). *Disposition of Toxic Drugs and Chemicals in Man*. Foster City, CA: Biomedical Publications.

Boxler, M.I., Schneider, T.D., Kraemer, T. and Steuer, A.E. Analytical considerations for (un)-targeted metabolomic studies with special focus on forensic applications. *Drug Testing and Analysis* 11, 5: 678–696.

Carlier, J., Guitton, J., Romeuf, L. et al. (2015). Screening approach by ultra-high performance liquid chromatography-tandem mass spectrometry for the blood quantification of thirty-four toxic principles of plant origin. Application to forensic toxicology. *J. Chromatogr. B Anal. Technol. Biomed. Life Sci.* 975: 65–76.

Castro-Perez, J., Plumb, R., Granger, J.H. et al. (2005). Increasing throughput and information content for in vitro drug metabolism experiments using ultra-performance liquid chromatography coupled to a quadrupole time-of-flight mass spectrometer. *Rapid Commun. Mass Spectrom.* 19 (6): 843–848.

Chen, P., Braithwaite, R.A., George, C. et al. (2014). The poppy seed defense: a novel solution. *Drug Test. Anal.* 6 (3): 194–201.

Chester, T.L. (2013). Recent developments in high-performance liquid chromatography stationary phases. *Anal. Chem.* 85 (2): 579–589.

Churchwell, M.I., Twaddle, N.C., Meeker, L.R., and Doerge, D.R. (2005). Improving LC-MS sensitivity through increases in chromatographic performance: comparisons of UPLC-ES/MS/MS to HPLC-ES/MS/MS. *J. Chromatogr. B Anal. Technol. Biomed. Life Sci.* 825 (2): 134–143.

Curry, A.S. (1974). Chromatography and forensic chemistry. *J. Chromatogr. Sci.* 12 (10): 529–534.

Di Rago, M., Saar, E., Rodda, L.N. et al. (2014). Fast targeted analysis of 132 acidic and neutral drugs and poisons in whole blood using LC-MS/MS. *Forensic Sci. Int.* 243: 35–43.

Di Rago, M., Chu, M., Rodda, L.N. et al. (2016). Ultra-rapid targeted analysis of 40 drugs of abuse in oral fluid by LC-MS/MS using carbon-13 isotopes of methamphetamine and MDMA to reduce detector saturation. *Anal. Bioanal. Chem.* 408 (14): 3737–3749.

Dolan, J. (n.d.). A Guide to HPLC and LC-MS Buffer Selection. https://www.hplc.eu/Downloads/ACE_Guide_BufferSelection.pdf (accessed 29 May 2020).

Elliott, S.P. and Hale, K.A. (1997). Development of a high-performance liquid chromatography retention index scale for toxicological drug screening. *J. Chromatogr. B Biomed. Sci. Appl.* 694 (1): 99–114.

Elliott, S.P. and Hale, K.A. (1998). Applications of an HPLC-DAD drug-screening system based on retention indices and UV spectra. *J. Anal. Toxicol.* 22 (4): 279–289.

Fredriksson, S.A., Artursson, E., Bergstrom, T. et al. (2015). Identification of RIP-II toxins by affinity enrichment, enzymatic digestion and LC-MS. *Anal. Chem.* 87 (2): 967–974.

Gerostamoulos, D., Beyer, J., Woods, J.L. et al. (2010). Overnight toxicology – fast targeted screening of drugs and poisons in post-mortem blood. *Pathology* 42 (S1): PS28.

Gilchrist, E., Smith, N., and Barron, L. (2012). Probing gunshot residue, sweat and latent human fingerprints with capillary-scale ion chromatography and suppressed conductivity detection. *Analyst* 137 (7): 1576–1583.

Gilchrist, E.S., Nesterenko, P.N., Smith, N.W., and Barron, L.P. (2015). Organic solvent and temperature-enhanced ion chromatography-high resolution mass spectrometry for the determination of low molecular weight organic and inorganic anions. *Anal. Chim. Acta* 865: 83–91.

Goucher, E., Kicman, A., Smith, N., and Jickells, S. (2009). The detection and quantification of lorazepam and its 3-O-glucuronide in fingerprint deposits by LC-MS/MS. *J. Sep. Sci.* 32 (13): 2266–2272.

Guiochon, G. (2006). The limits of the separation power of unidimensional column liquid chromatography. *J. Chromatogr. A* 1126 (1–2): 6–49.

Guiochon, G. (2007). Monolithic columns in high-performance liquid chromatography. *J. Chromatogr. A* 1168 (1–2): 101–168; discussion 100.

Hädener, M., Weinmann, W., Schurch, S., and König, S. (2016). Development of a rapid column-switching LC-MS/MS method for the quantification of THCCOOH and THCCOOH-glucuronide in whole blood for assessing cannabis consumption frequency. *Anal. Bioanal. Chem.* 408 (7): 1953–1962.

Hädener, M., Bruni, P.S., Weinmann, W. et al. (2017). Accelerated quantification of amphetamine enantiomers in human urine using chiral liquid chromatography and on-line column-switching coupled with tandem mass spectrometry. *Anal. Bioanal. Chem.* 409 (5): 1291–1300.

He, W., Parissis, N., and Kiratzidis, T. (1998). Determination of benzodiazepines in forensic samples by HPLC with photo-diode array detection. *J. Forensic Sci.* 43 (5): 1061–1067.

Huber, J. and Hulsman, J. (1967). A study of liquid chromatography in columns. The time of separation. *Anal. Chim. Acta* 38: 305–313.

Igoh, A., Doi, Y., and Sakurada, K. (2015). Identification and evaluation of potential forensic marker proteins in vaginal fluid by liquid chromatography/mass spectrometry. *Anal. Bioanal. Chem.* 407 (23): 7135–7144.

Jantos, R. and Skopp, G. (2013). Postmortem blood and tissue concentrations of R- and S-enantiomers of methadone and its metabolite EDDP. *Forensic Sci. Int.* 226 (1–3): 254–260.

Kele, M., U. Neue, K. Wyndham, et al. (2004). Practical limits of high performance liquid chromatography – current status. Waters: poster WA 40161.

Keltanen, T., Sajantila, A., Valonen, T. et al. (2013). Measuring postmortem glycated hemoglobin – a comparison of three methods. *Leg. Med. (Tokyo)* 15 (2): 72–78.

Knox, J., Laird, G., and Raven, P. (1976). Interaction of radial and axial dispersion in liquid chromatography in relation to the infinite diameter effect. *J. Chromatogr. A* 122: 129–145.

Kristinsson, R., Lewis, S.E., and Danielson, P.B. (2009). Comparative analysis of the HV1 and HV2 regions of human mitochondrial DNA by denaturing high-performance liquid chromatography. *J. Forensic Sci.* 54 (1): 28–36.

Lambert, W.E., Van Bocxlaer, J.F., and De Leenheer, A.P. (1997). Potential of high-performance liquid chromatography with photodiode array detection in forensic toxicology. *J. Chromatogr. B Biomed. Sci. Appl.* 689 (1): 45–53.

Legg, K.M., Powell, R., Reisdorph, N. et al. (2014). Discovery of highly specific protein markers for the identification of biological stains. *Electrophoresis* 35 (21–22): 3069–3078.

Maurer, H.H. (1998). Liquid chromatography-mass spectrometry in forensic and clinical toxicology. *J. Chromatogr. B Biomed. Sci. Appl.* 713 (1): 3–25.

Mizuno, Y., Sato, K., Sano, T. et al. (2002). Identification and characterization of 17 phenothiazine compounds by capillary high-performance liquid chromatography/fast atom bombardment mass spectrometry. *Leg. Med. (Tokyo)* 4 (4): 207–216.

Moffatt, A.C., Osselton, M.D., Widdop, B., and Watts, J. (2011). *Clarke's Analysis of Drugs and Poisons*. London: Pharmaceutical Press.

Oberacher, H. and Parson, W. (2007). Forensic DNA fingerprinting by liquid chromatography-electrospray ionization mass spectrometry. *Biotechniques* 43 (4): vii–xiii.

Oberacher, H., Niederstatter, H., Pitterl, F., and Parson, W. (2006). Profiling 627 mitochondrial nucleotides via the analysis of a 23-plex polymerase chain reaction by liquid chromatography-electrospray ionization time-of-flight mass spectrometry. *Anal. Chem.* 78 (22): 7816–7827.

Parkin, M.C., Longmoore, A.M., Turfus, S.C. et al. (2013). Detection of ketamine and its metabolites in human hair using an integrated nanoflow liquid chromatography column and electrospray emitter fritted with a single porous 10 mum bead. *J. Chromatogr. A* 1277: 1–6.

Peters, F.T. (2011). Recent advances of liquid chromatography-(tandem) mass spectrometry in clinical and forensic toxicology. *Clin. Biochem.* 44 (1): 54–65.

Plumb, R., Castro-Perez, J., Granger, J. et al. (2004). Ultra-performance liquid chromatography coupled to quadrupole-orthogonal time-of-flight mass spectrometry. *Rapid Commun. Mass Spectrom.* 18 (19): 2331–2337.

Premstaller, A., Oberacher, H., and Huber, C.G. (2000). High-performance liquid chromatography-electrospray ionization mass spectrometry of single- and double-stranded nucleic acids using monolithic capillary columns. *Anal. Chem.* 72 (18): 4386–4393.

Rodda, L.N., Gerostamoulos, D., and Drummer, O.H. (2013). The rapid identification and quantification of iso-alpha-acids and reduced iso-alpha-acids in blood using UHPLC-MS/MS: validation of a novel marker for beer consumption. *Anal. Bioanal. Chem.* 405 (30): 9755–9767.

Saar, E., Gerostamoulos, D., Drummer, O.H., and Beyer, J. (2012). Identification of 2 hydroxymethyl-olanzapine as a novel degradation product of olanzapine. *Forensic Sci. Int.* 220 (1–3): 74–79.

Sauvage, F.L., Gaulier, J.M., Lachatre, G., and Marquet, P. (2008). Pitfalls and prevention strategies for liquid chromatography-tandem mass spectrometry in the selected reaction-monitoring mode for drug analysis. *Clin. Chem.* 54 (9): 1519–1527.

SOFT/AAFS (2006). Forensic Toxicology Laboratory Guidelines.

Steuer, A.E., Schmidhauser, C., Liechti, M.E., and Kraemer, T. (2015). Development and validation of an LC-MS/MS method after chiral derivatization for the simultaneous stereoselective determination of methylenedioxy-methamphetamine (MDMA) and its phase I and II metabolites in human blood plasma. *Drug Test. Anal.* 7 (7): 592–602.

SWGTOX (2013). Standard Practices for Method Validation in Forensic Toxicology.

UNODC (2009). Guidance for the Validation of Analytical Methodology and Calibration of Equipment Used for Testing of Illicit Drugs in Seized Materials and Biological Specimens.

15

Capillary and Microchip Electrophoresis

Lucas Blanes, Ellen Flávia Moreira Gabriel, Renata Mayumi Saito, Wendell Karlos Tomazelli Coltro, Nerida Cole, Philip Doble, Claude Roux and Robson Oliveira dos Santos

15.1 Capillary Electrophoresis: Introduction

Capillary electrophoresis (CE) is a separation technique employed for the analysis of charged and uncharged species, ranging from small inorganic ions to complex biomolecules. The versatility of CE arises from the many modes of operation such as capillary zone electrophoresis (CZE), micellar electrokinetic capillary chromatography (MEKC), isoelectric focusing (IEF), isotachophoresis (ITP), capillary electrochromatography (CEC) and capillary gel electrophoresis (CGE) (Baker and Foret 2015; Landers 2007).

A typical CE instrument uses a 50–100 cm long fused silica capillary with an internal diameter ranging from 25 to 100 μm immersed in a background electrolyte (BGE) containing electrodes connected to a high voltage power supply (Felhofer et al. 2010). The majority of commercially available systems employ ultraviolet (UV) detectors. However, laser-induced fluorescence (LIF) (Ban and Song 2013), capacitively coupled contactless conductivity detection (C^4D) (Brito-Neto et al. 2005a, 2005b; Kubáň and Hauser 2013) and mass spectrometric detectors (Klepárník 2015; Pejchinovski et al. 2015) are also widely available. The detector and operation mode is chosen for the targeted substance. For example, CGE with LIF detection is particularly appropriate for DNA analysis (Klepárník 2015), CZE with C^4D is widely used for the analysis of small ions, and MEKC is used for analysis of explosives (Casamento et al. 2003). A schematic diagram of a CE system is shown in Figure 15.1.

In contrast to chromatographic techniques, where the separation is based on differential partitioning between the mobile and stationary phases, in conventional CE or microchip-capillary electrophoresis (ME) the separation process occurs due to differences in electrophoretic mobility (μ_E) of the ionic species and is dependent on the charge and the size of the compound (Baker and Foret 2015). The μ_E is determined by the equation:

$$\mu_E = q/6\pi\eta r \tag{15.1}$$

where q, η, and r represent the ion charge, the viscosity of the buffer solution, and the hydrated radius of the ion, respectively.

Analytical Techniques in Forensic Science, First Edition.
Edited by Rosalind Wolstenholme, Sue Jickells and Shari Forbes.
© 2021 John Wiley & Sons Ltd. Published 2021 by John Wiley & Sons Ltd.

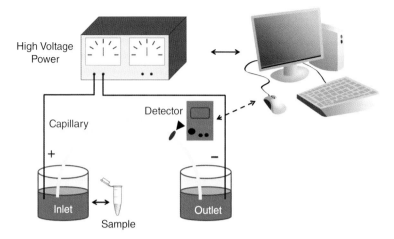

Figure 15.1 Schematic diagram of a capillary electrophoresis system. A few nanoliters of sample is injected into a fused silica capillary previously filled with the running electrolyte. A high-power supply imposes an electric field, which promotes the separation of the species towards the detector.

The potential difference applied across the capillary causes a phenomenon known as electro-osmotic flow (EOF). The EOF is the bulk flow of liquid along the capillary as a result of the surface charge of the silanol groups on the capillary wall and the differences in potential applied to the capillary extremities. At low pH (<2.5) the silanol groups present in the inner surface of the tube are normally protonated and neutrally charged making the EOF negligible. Above pH 3, the inner wall of the capillary becomes negatively charged due to deprotonation of these groups forming an electrical double layer at the interface between the capillary wall and the bulk solution. When an electric field is applied the EOF moves through the capillary in the cathodic direction transporting all species towards the detection point (Baker and Foret 2015). Depending on the magnitude of the EOF, cationic, anionic and neutral compounds can be simultaneously separated in a single electrophoretic run (Baker and Foret 2015; Kitagishi 1996).

The apparent mobility of each compound may be calculated as the vector sum of the electrophoretic and electro-osmotic mobilities, as demonstrated in Eq. (15.2):

$$\mu_{ap} = \mu_E + \mu_{EOF} \tag{15.2}$$

where μ_{ap} and μ_{EOF} are the apparent and electro-osmotic mobilities, respectively. The μ_{EOF} may also be calculated by Eq. (15.3):

$$\mu_{EOF} = \varepsilon \zeta / 4\pi \eta \tag{15.3}$$

where ε is the dielectric constant of the solution and ζ is the zeta potential, which is the potential drop across the mobile part of the double layer that is responsible for electrokinetic phenomena (McNaught and Wilkinson 1997).

The magnitude of the EOF can be estimated by a number of methods, for example, measuring the migration time of neutral markers or by generation of thermal markers (Saito et al. 2007). This allows calculation of an estimate of the electro-osmotic mobility for a given

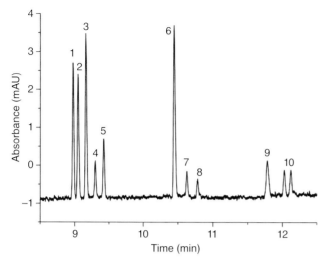

Figure 15.2 Electropherogram of amphetamine, methamphetamine, 3,4-methylenedioxymethamphetamine (1.25 ppm) and seven amphetamine analogues (2.5 ppm) using UV detection. CE conditions: BGE as 75 mM acetic acid+25 mM sodium acetate pH 4.55 with 30 mM hydroxypropyl-β-cyclodextrin; 50 μm internal diameter 90 cm (81.5 cm to detector) fused-silica capillary; 25 °C; voltage 30 kV; detection at UV-visible 200 nm; and 30 mbar sample injection for 5 seconds: (1) 2-bromo-*N*-methylbenzenemethanamine; (2) 2-methoxyphenethylamine; (3) 3-methyl-*N*-methylbenzylamine; (4) 2-phenethylamine hydrochloride; (5) 5-aminomethyl-7-chloro-1,3-benzodioxole hydrochloride; (6) 2-(4-methoxyphenyl)ethylamine; (7) amphetamine; (8) methamphetamine; (9) 2-methoxy-*N*-methylaniline; and (10) 3,4-methylenedioxymethamphetamine. Source: Reproduced with permission from Epple et al. (2010).

system from the following relationship:

$$\mu_{EOF} = \frac{l}{t_m \cdot E} \qquad (15.4)$$

where l is the effective length of the capillary (from its inlet to the detector), t_m is the migration time of a neutral marker, and E is the electrical field. A neutral marker is a neutral specimen that has the same mobility as the EOF.

In CE the sample is introduced into the capillary by temporarily replacing the buffer with a sample vial (in general 5–30 seconds) and applying an injection pressure (hydrodynamic injection or HD injection) or a potential difference (electrokinetic injection or EK injection). HD injection is widely used in conventional CE because the injected sample plug composition remains identical to the original sample. In EK injection the ions migrate into the capillary as a function of their electrophoretic mobility and consequently, the injected sample may differ in composition from the original sample. Following separation, compounds appear as peaks in an electropherogram with differing migration times (Baker and Foret 2015; Kitagishi 1996; Landers 2007). Figure 15.2 shows a typical CE electropherogram for the analysis of amphetamine-type substances (Epple et al. 2010).

Due to the importance of a reliable EOF, in the majority of separations a pretreatment (flushing the capillary with a solution containing 0.1–1 M NaOH for ~15 minutes) is necessary. Flushing the capillary between runs and periodically replacing the buffer is recommended to avoid changes in migration times due to electrolysis of the buffer components.

15.2 Microchip-Capillary Electrophoresis

Microchip-Capillary Electrophoresis (ME) is a miniaturised version of CE where the capillary is replaced by a chip capillary device, which was initially proposed in the early 1990s by Manz and co-workers (Manz et al. 1990a, 1990b).

ME is also called lab-on-a-chip (LOC), reflecting its potential for integration and miniaturisation of chemical and biological reactions. This design is able to increase sample throughput, improve performance, and reduce costs. ME offers other potential advantages, particularly for compounds of forensic significance. These include shorter analysis times and lower sample volumes, as well as the potential for creation of disposable devices and automation.

LOC devices in general are made of glass or polymers such as polydimethylsiloxane (PDMS) (Duffy et al. 1998; McDonald et al. 2000; McDonald and Whitesides 2002) or poly(methyl methacrylate) (PMMA) (Fiorini and Chiu 2005). Paper-based and polyester-toner microchips have also been proposed by some research groups (Lucio do Lago et al. 2003; Martinez et al. 2007, 2008; Coltro et al. 2010).

15.2.1 Sample Injection Modes in ME

EK injection is the preferred method of sample introduction in ME devices due to its simplicity (Saito et al. 2012). EK injection is performed in different modes known as floating, pinched and gated modes (Fu et al. 2002). An example of a simple ME device is shown in Figure 15.3. Sample injection occurs when a potential difference is applied between the sample and waste reservoirs for a few seconds, creating a sample flow towards the waste reservoir (i.e. a flow across the horizontal channel in Figure 15.3a). The electric field is then switched off and a second voltage is applied along the separation channel (Fu et al. 2002) causing part of the sample to be deflected into the separation channel (the vertical channel in Figure 15.3b). A short, well-defined sample plug is crucial to effective separation in ME devices. Pinching voltages may be applied at the buffer inlet and outlet to induce flow to the sample waste and reverse voltages may then be applied to prevent diffusion of sample into the separation channel (Figure 15.3b). However, in the pinched injection schemes, transportation times longer than 10 seconds in general, are required to migrate the analyte from the sample reservoir to the injection intersection (Zhang and Manz 2001).

Short injection times can be achieved using a gated injection mode. Here, analytes are continuously moving through the cross of the channels and are diverted at 90° by the buffer being electro-osmotically pumped towards the sample waste outlet, preventing sample leakage into the separation channel. For injection, the analyte flow is deflected into the separation channel for a short time, typically 0.1–1 seconds allowing a sample plug to enter the separation channel (Figure 15.4) (Karlinsey 2012).

Figure 15.3 Electrokinetic injection procedures in ME: (a) floating and (b) pinched modes. Inset charge-coupled device (CCD) images of the injection cross section during the sample loading. Left-hand figures show the electrical field operating during sample injection: the sample flows from the sample inlet (+ potential) towards the waste reservoir (−). The right-hand figures show the electrical field shortly after injection. Source: Reproduced with permission from Karlinsey (2012).

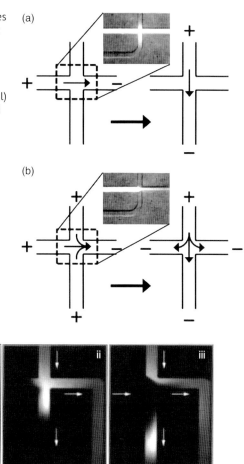

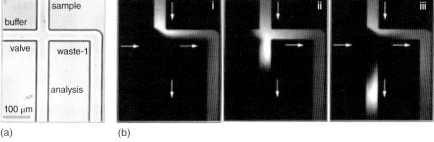

Figure 15.4 Gated injection scheme using a cross injection design. (a) Image of microchip channels. (b) CCD images of (i) the sample flowing to waste-1 channel, (ii) sample is injected into the analysis channel, and (iii) gate returns to its original state and thus starting analysis mode. Arrows depict direction of flow. Source: Reproduced with permission from Karlinsey (2012).

15.3 Detection Systems

Electrochemical detectors (amperometry and conductometry) are commonly used in ME devices due to their high sensitivity and relatively easy implementation (Guijt et al. 2001; Kubáň and Hauser 2011; Coltro et al. 2012; da Silva et al. 2013). LIF and light-emitting-diode-induced fluorescence (LED-IF) have also been widely used in ME for their high sensitivity and selectivity, especially for analysis of biomolecules in clinical and forensic applications (Felhofer et al. 2010; Kim et al. 2011; Shang et al. 2012). There have been various instrument arrangements reported for coupling CE and ME with mass spectrometry (MS) and their applications include: the analysis of proteins, including monoclonal antibodies, peptides, biomarkers, and glycans; metabolomics and forensic and pharmaceutical applications; as well as food and single cell analysis (Ramautar et al. 2009; Gottardo et al. 2012; He et al. 2014; Klepárník 2015).

15.4 CE and ME in Forensic Analysis

CE systems have been applied to forensic investigations requiring analysis of gunshot residue, explosives, toxins, inks, dyes, DNA and other complex samples, including an expanding number of drugs. The large number of applications using different modes of CE analysis demonstrates the versatility of this powerful analytical technique. Some examples of applications of CE and ME in forensic analysis are listed in Table 15.1.

15.5 Case Study: Lab-on-a-Chip Screening of Methamphetamine and Pseudoephedrine in Clandestine Laboratory Samples

The illicit manufacture of drugs leaves a variety of materials at a crime scene that may be a rich source of information for forensic investigations. In this case study, the suitability of a LOC instrument for the screening of methamphetamine, pseudoephedrine and ephedrine in a variety of sample types found in typical clandestine laboratory scenarios is presented.

Experiments were performed using a commercially available portable LOC instrument, the Agilent 2100 Bioanalyzer, with Agilent 2100 Expert software (Agilent Technologies, Waldbronn, Germany). Detection was by LED-Induced Fluorescence (LED-IF) (λ_{ex} 450 nm, λ_{em} 525 nm). Injection was carried out at 1.5 kV for 2 seconds with a separation voltage of 1.5 kV. Standard DNA 1000 microchips obtained from Agilent Technologies (Forest Hill, Australia) were used. These chips have a separation channel length of 15 mm, micro-channel depth of 10 μm and width of 50 μm. They are fabricated from soda lime glass and are etched with an array of microchannels that interconnect 12 sample wells to the separation channel. The 2100 Bioanalyzer contains individual electrodes that control the sequential movement of each sample into the separation channel through the application of a high voltage.

Different sample types were subjected to differing preparation procedures prior to derivatisation with fluorescein isothiocyanate isomer I (FITC; 3 minutes, 90 °C) to enable detection of the analytes. Selection of the preparation method for liquid samples was determined by the miscibility of the sample with sodium tetraborate buffer (50 mM) in combination with estimation of the pH using indicator strips at the collection site. The decision tree for determining the method is outlined in Figure 15.5. Solid samples for analysis were prepared using a single method (Figure 15.5 and Table 15.2), where two-layered liquids refer to an aqueous and an organic layer. The preparation procedure for each sample type is described in Table 15.2.

Electrophoretic separations were performed by MEKC using a buffer comprised of sodium tetraborate (50 mM; pH 9.66) containing sodium dodecyl sulfate (SDS; 50 mM) as the BGE. Nile Blue chloride dye (10 μg/ml) dissolved in the electrolyte buffer was primed through the microchannels prior to analysis for instrument focusing.

Table 15.1 Applications of CE and ME in forensic analysis.

Application	Mode of analysis	Detection	References
Toxicological screening of biological samples	CZE	MS	Polettini et al. (2008)
Enantiomeric separation of ephedrines, phenylacetone and amphetamine derivatives	MEEKC	UV	Wongwan and Scriba (2010)
Enantiomeric separation of amphetamine derivatives (ecstasy) in real clandestine samples	CZE	C^4D, UV	Epple et al. (2010)
Screening and quantitation of multiple drugs of abuse in urine including opiates, methamphetamine and derivatives, and cocaine and its metabolites	CZE	MS, UV	Alnajjar et al. (2007) and Kohler et al. (2013)
Separation of cocaine and its metabolites in human urine	CZE, MEKC	MS, UV	Su et al. (2008) and Hezinová et al. (2012)
Determination of *Ephedra* alkaloids in tablets and urine	ME-MEKC	LIF	Belder et al. (2011)
Determination of lysergic acid diethylamide	CZE, MEKC	LIF, FD	Fang et al. (2003) and Airado-Rodríguez et al. (2015)
Determination of γ-hydroxybutyric acid	CZE, MEKC	UV	Dahlén and Vriesman (2002) and Baldacci et al. (2003)
Determination of psychotropic tryptamine derivatives (tryptamine, serotonin, *N,N*-dimethyltryptamine, bufotenine, 5-methyltryptamine, *N,N*-diethyltryptamine, 5-methoxy-tryptamine, and other analogues) in synthetic drugs and pharmaceutical formulations	CZE	UV-LIF	Huhn et al. (2005)
Determination of creatinine in urine	CZE	UV	Liotta et al. (2009)
Determination of growth hormone	CZE	UV/MS	Catai et al. (2007)
Determination of nicotine alkaloids	NACE	UV/MS	Chiu et al. (2007)
Simultaneous determination of bromide, nitrate, nitrite, iodide and thiocyanate in saliva	ITP	UV	Xu et al. (2008)
Screening of toxic anions in biological samples: thiosulfate, bromide, chromate, iodide, chloride, sulfate, sulfite, pyrosulfate, sulfide, nitrite, nitrate, oxalate, perchlorate, azide, thiocyanate, tetrathionate, chlorate, citrate, tartrate, fluoride, bromate, formate, monohydrogen phosphate, chlorite, phthalate, bicarbonate, fluoroacetate, acetate, and benzoate	CZE	UV	Gillette et al. (2006)
Determination of natural toxic oligopeptides ama- and phallotoxins in extracts of dried mushroom samples	CZE	MS	Rittgen et al. (2008)

Table 15.1 (Continued)

Application	Mode of analysis	Detection	References
Determination of metabolites of ethanol: ethyl glucuronide and ethyl sulfate	CZE	UV	Křivánková et al. (2005) and Esteve-Turrillas et al. (2006)
Determination of biological marker of alcohol abuse – carbohydrate-deficient transferrin	CZE	UV	Pascali et al. (2011)
Screening of products of chemical warfare agents in water and soil	CZE	FPD	Hooijschuur et al. (2001)
Determination of inorganic ions present in explosives and its residues	CZE	C^4D, UV	Morales and Vázquez (2004) and Hutchinson et al. (2008)
Determination of organic components of gunshot residues	MEKC	UV	Morales and Vázquez (2004)
Determination of nitrate ester explosives	ME–MEKC	AD	Piccin et al. (2009)
Determination of azo and methine basic dyes	CZE	UV, MS	Stefan et al. (2009)
Determination of inkjet printing inks	CZE, MEKC	MS, UV	Szafarska et al. (2011) and Kula et al. (2014)
DNA analysis	ME-STR	LIF	Liu et al. (2011) and Le Roux et al. (2014)

AD, amperometric detection; C^4D, capacitively coupled contactless conductivity detection; CZE, capillary zone electrophoresis; FD, fluorescence detection; FPD, flame photometric detector; ITP, isotachophoresis; LIF, laser-induced fluorescence; ME, microchip-capillary electrophoresis; MEEKC, microemulsion electrokinetic chromatography; MEKC, micellar electrokinetic capillary chromatography; MS, mass spectrometry; NACE, non-aqueous capillary electrophoresis; STR, short tandem repeat.

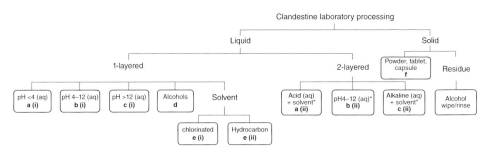

Figure 15.5 Decision tree for processing of samples from a clandestine laboratory. Source: Reproduced with permission from Lloyd et al. (2013).

The electropherogram presented in Figure 15.6 illustrates the separation of a mixed standard containing ephedrine, pseudoephedrine and methamphetamine. Relative standard deviations (RSDs) of 3–5% and 5–10% were calculated for intra- and inter-chip variability, respectively. For each unknown, a positive screening result was inferred when the migration time was within 4% RSD of the corresponding standard from the same chip. This method is

Table 15.2 Preparation procedures used for sample types determined using the decision tree in Figure 15.5.

	Sample type	Sample amount	Step 1: mix with toluene	Step 2	Step 3: mix with buffer	Step 4: remove layer for analysis	Analyse
a (i)	Acid (aq) pH <4	0.5 ml[a]	0.5 ml	Remove toluene	0.5 ml	Remove buffer (bottom) layer	Buffer layer[b]
a (ii)	Acid (aq)+solvent	0.5 ml of acid (aq) layer[a]	0.5 ml	Remove toluene	0.5 ml	Remove buffer (bottom) layer	Buffer layer[b]
b (i)	Neutral (aq) pH 4–12	0.5 ml	N/A	N/A	0.5 ml	N/A	Buffer mixture
b (ii)	Neutral (aq)+solvent	0.5 ml aq layer	N/A	N/A	0.5 ml	N/A	Buffer mixture
		0.5 ml solvent layer	N/A	N/A	0.5 ml	Remove buffer layer (refer to e)	Buffer layer
c (i)	Alkaline (aq) pH >12	0.5 ml	0.5 ml	Remove toluene	0.5 ml	Remove buffer (bottom) layer	Buffer layer[b]
c (ii)	Alkaline (aq) +solvent	0.5 ml of solvent layer	N/A	N/A	0.5 ml	Remove buffer layer (refer to e)	Buffer layer[b]
d	Alcohol/alcohol rinse	0.5 ml	N/A	N/A	0.5 ml	N/A	Alcohol/buffer mixture[b]
e (i)	Chlorinated solvent (e.g. chloroform)	0.5 ml	N/A	N/A	0.5 ml	Remove buffer (top) layer	Buffer layer
e (ii)	Hydrocarbon solvent (e.g. toluene, heptane)	0.5 ml	N/A	N/A	0.5 ml	Remove buffer (bottom) layer	Buffer layer
f	Solid	5–10 mg	N/A	Add 1 ml methanol	0.5 ml	N/A	Methanol/buffer solution

N/A, not applicable.
a) Basify to pH 14.
b) Adjust to pH 8–10, if required, prior to derivatisation.
Source: Reproduced with permission from Lloyd et al. (2013).

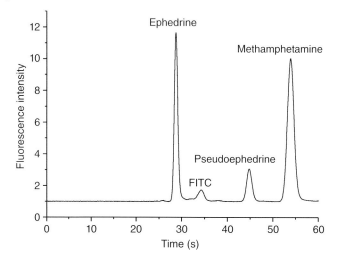

Figure 15.6 Electropherogram of a standard mixture of ephedrine, pseudoephedrine and methamphetamine (circa 50 µg/ml), derivatised with FITC in excess. Source: Reproduced with permission from Lloyd et al. (2013).

suitable for various types of pharmaceutical preparations containing pseudoephedrine, e.g. cold and flu liquids, gel capsules, tablets, and capsules.

15.5.1 Screening of Methamphetamine and Pseudoephedrine from Clandestine Laboratories

Samples from five different clandestine laboratory cases were analysed. A range of samples including acidic, alkaline, solvents, alcohol rinses and two-layered liquids were screened using LOC and subsequently cross checked by gas chromatography–mass spectrometry (GC-MS). The LOC and the corresponding GC-MS results for each sample are summarised in Table 15.3. On each chip, a standard mixture of ephedrine, pseudoephedrine and methamphetamine was run to account for any inter-chip migration time variations. The results from LOC are in good agreement with those from GC-MS analysis (Table 15.3) with the exception of a single non-aqueous (two-layered) sample where LOC indicated both pseudoephedrine and methamphetamine, but only pseudoephedrine was confirmed by GC-MS (data summarised in Table 15.3). This indicates that LOC is suitable for rapid screening of samples prior to confirmatory analysis. In addition, this method distinguished between ephedrine and pseudoephedrine, which is valuable information, and which requires a far lengthier derivatisation procedure for determination by GC (Jonsson et al. 1996).

15.5.2 Interferents

Many by-products and intermediates generated during the methamphetamine manufacturing process (usually ketones or tertiary amines) do not contain an accessible functional group for derivatisation with FITC. This minimises the risk of false positives.

Table 15.3 A comparison of analyses of clandestine laboratory samples by LOC and GC–MS.

Sample type	Result	
	LOC	GC–MS
Acidic		
1	PSE, METH	METH, PSE
2	METH	METH
3	NEG	NEG
4	NEG	NEG
5	METH	METH
Basic		
1	PSE	PSE, chlorpheniramine
2	NEG	NEG
Solvent		
1	PSE, METH	METH, PSE/EPH, chlorpheniramine
2	NEG	NEG
3	NEG	NEG
4	NEG	NEG
5	NEG	NEG
Alcohol rinse		
1	EPH, PSE	PSE/EPH, chlorpheniramine
2	EPH, PSE	PSE/EPH, chlorpheniramine
3	NEG	NEG
4	NEG	NEG
5	NEG	NEG
6	NEG	NEG
7	METH	METH
8	NEG	NEG
Solid		
1	Pseudo	PSE
2	NEG	NEG
3	NEG	NEG
Two-layered liquid		
1	PSE	PSE
2	PSE, METH	PSE
3	EPH, PSE, METH	METH, PSE/EPH
4	PSE	PSE/EPH, dextromethorphan, chlorpheniramine, codeine

Table 15.3 (Continued)

Sample type	Result	
5	NEG	NEG
6	METH	METH, P2P, *N*-dimethylamphetamine and chloropheniramine
7^{a)}	PSE	TL = chloropheniramine
	PSE	BL = PSE and chloropeniramine
8	METH	TL = METH
9	PSE	METH, PSE/EPH

BL, bottom layer; EPH, ephedrine; METH, methamphetamine; NEG, no results; P2P, 1-phenyl-2-propanone; PSE, pseudoephedrine; PSE/EPH, either or both pseudoephedrine and ephedrine (i.e. not confirmed by further analysis); TL, top layer.
a) Two-layered liquid, both layers pH 6 therefore both layers analysed.
Source: Reproduced with permission from Lloyd et al. (2013).

Other impurities associated with this method of manufacture include amphetamine and methamphetamine dimer, which have different migration times to the target analytes. Their presence in relatively trace amounts in comparison with methamphetamine and pseudoephedrine minimises the risk of interference. Whilst the casework samples presented here are all from clandestine laboratories employing the hydriodic acid reduction of pseudoephedrine, the sampling procedure(s) and extraction methods are applicable to other manufacturing methods. Regardless of the starting products and the chosen route of synthesis, similar overall procedures (e.g. extraction, crystallisation) are involved in the synthesis as can been seen when methamphetamine is synthesised from 1-phenyl-2-propanone by reductive amination or Leuckart routes (Kunalan et al. 2009).

15.5.3 Simulated Surface Swabs

Swabs are commonly collected from various surfaces at clandestine laboratory sites to obtain evidence of manufacture. In this case study, the authors proposed a very simple and rapid procedure for extraction of methamphetamine with swabs from representative surfaces at clandestine laboratories, prior to analysis by LOC. The time taken from sampling to obtaining a result was 15 minutes per sample, which is around 4–10 times faster than that for the current standard method of GC-MS using base/solvent extraction (Lim Abdullah and Miskelly 2010). LOC was successfully able to analyse methamphetamines from surfaces including formica, ceramic tile, gloss painted wood and metal. The average recovery was 24.1% and the presence of the target analyte in these samples was confirmed by GC-MS.

15.6 Conclusions

CE has several advantages when compared with other separation techniques and should be more often applied to routine forensic investigations. Improvements in equipment and controlled capillary conditioning make CE today as reproducible as HPLC or GC. As can be seen in Table 15.1 CE-MS can be an attractive option for routine automated forensic analysis. ME has reached maturity in the past decade and the instrumentation is constantly being developed. Misconceptions regarding lack of resolution in ME are also disappearing with new complex sample separations being successfully performed.

Acknowledgements

The research in the case study was financially supported by the Commonwealth of Australia through the National Security Science and Technology Centre within the Defence Science and Technology Organisation, and the United States of America Department of Defence through the Combating Terrorism Technical Support Office.

References

Airado-Rodríguez, D., Cruces-Blanco, C., and García-Campaña, A.M. (2015). Ultrasensitive analysis of lysergic acid diethylamide and its C-8 isomer in hair by capillary zone electrophoresis in combination with a stacking technique and laser induced fluorescence detection. *Analytica Chimica Acta* 866: 90–98.

Alnajjar, A., Idris, A.M., Multzenberg, M., and McCord, B. (2007). Development of a capillary electrophoresis method for the screening of human urine for multiple drugs of abuse. *Journal of Chromatography B* 856 (1–2): 62–67.

Baker, D.R. and Foret, F. (1995). Capillary electrophoresis: techniques in analytical chemistry series. *Analytical Biochemistry* 231 (1): 274–274.

Baldacci, A., Theurillat, R., Caslavska, J. et al. (2003). Determination of γ-hydroxybutyric acid in human urine by capillary electrophoresis with indirect UV detection and confirmation with electrospray ionization ion-trap mass spectrometry. *Journal of Chromatography A* 990 (1–2): 99–110.

Ban, E. and Song, E.J. (2013). Recent developments and applications of capillary electrophoresis with laser-induced fluorescence detection in biological samples. *Journal of Chromatography B* 929: 180–186.

Belder, D., Tolba, K., and Nagl, S. (2011). Rapid quantitative determination of ephedra alkaloids in tablet formulations and human urine by microchip electrophoresis. *Electrophoresis* 32 (3–4): 440–447.

Brito-Neto, J.G.A., Fracassi da Silva, J.A., Blanes, L., and do Lago, C.L. (2005a). Understanding capacitively coupled contactless conductivity detection in capillary and microchip electrophoresis. Part 1. Fundamentals. *Electroanalysis* 17 (13): 1198–1206.

Brito-Neto, J.G.A., Fracassi da Silva, J.A., Blanes, L., and do Lago, C.L. (2005b). Understanding capacitively coupled contactless conductivity detection in capillary and microchip electrophoresis. Part 2. Peak shape, stray capacitance, noise, and actual electronics. *Electroanalysis* 17 (13): 1207–1214.

Casamento, S., Kwok, B., Roux, C. et al. (2003). Optimization of the separation of organic explosives by capillary electrophoresis with artificial neural networks. *Journal of Forensic Sciences* 48 (5): 2003010.

Catai, J.R., Sastre Toraño, J., Jongen, P.M.J.M. et al. (2007). Analysis of recombinant human growth hormone by capillary electrophoresis with bilayer-coated capillaries using UV and MS detection. *Journal of Chromatography B* 852 (1–2): 160–166.

Chiu, C.-W., Liang, H.-H., and Huang, H.-Y. (2007). Analyses of alkaloids in different products by NACE-MS. *Electrophoresis* 28 (22): 4220–4226.

Coltro, W.K.T., de Jesus, D.P., da Silva, J.A.F. et al. (2010). Toner and paper-based fabrication techniques for microfluidic applications. *Electrophoresis* 31 (15): 2487–2498.

Coltro, W.K.T., Lima, R.S., Segato, T.P. et al. (2012). Capacitively coupled contactless conductivity detection on microfluidic systems — ten years of development. *Analytical Methods* 4 (1): 25–33.

Dahlén, J. and Vriesman, T. (2002). Simultaneous analysis of γ-hydroxybutyric acid, γ-butyrolactone, and 1,4-butanediol by micellar electrokinetic chromatography. *Forensic Science International* 125 (2–3): 113–119.

Duffy, D.C., McDonald, J.C., Schueller, O.J.A., and Whitesides, G.M. (1998). Rapid prototyping of microfluidic systems in poly(dimethylsiloxane). *Analytical Chemistry* 70 (23): 4974–4984.

Epple, R., Blanes, L., Beavis, A. et al. (2010). Analysis of amphetamine-type substances by capillary zone electrophoresis using capacitively coupled contactless conductivity detection. *Electrophoresis* 31 (15): 2608–2613.

Esteve-Turrillas, F.A., Bicker, W., Lämmerhofer, M. et al. (2006). Determination of ethyl sulfate – a marker for recent ethanol consumption – in human urine by CE with indirect UV detection. *Electrophoresis* 27 (23): 4763–4771.

Fang, C., Liu, J.-T., and Lin, C.-H. (2003). On-line identification of lysergic acid diethylamide (LSD) in tablets using a combination of a sweeping technique and micellar electrokinetic chromatography/77 K fluorescence spectroscopy. *Electrophoresis* 24 (6): 1025–1030.

Felhofer, J.L., Blanes, L., and Garcia, C.D. (2010). Recent developments in instrumentation for capillary electrophoresis and microchip-capillary electrophoresis. *Electrophoresis* 31 (15): 2469–2486.

Fiorini, G.S. and Chiu, D.T. (2005). Disposable microfluidic devices: fabrication, function, and application. *BioTechniques* 38 (3): 429.

Fu, L.M., Yang, R.J., Lee, G.B., and Liu, H.H. (2002). Electrokinetic injection techniques in microfluidic chips. *Analytical Chemistry* 74 (19): 5084–5091.

Gillette, R., Doyle, J.M., Miller, M.L. et al. (2006). Capillary electrophoresis screening of poisonous anions extracted from biological samples. *Journal of Chromatography B* 831 (1–2): 190–195.

Gottardo, R., Mikšík, I., Aturki, Z. et al. (2012). Analysis of drugs of forensic interest with capillary zone electrophoresis/time-of-flight mass spectrometry based on the use of non-volatile buffers. *Electrophoresis* 33 (4): 599–606.

Guijt, R.M., Baltussen, E., van der Steen, G. et al. (2001). Capillary electrophoresis with on-chip four-electrode capacitively coupled conductivity detection for application in bioanalysis. *Electrophoresis* 22 (12): 2537–2541.

He, X., Chen, Q., Zhang, Y., and Lin, J.-M. (2014). Recent advances in microchip-mass spectrometry for biological analysis. *TrAC Trends in Analytical Chemistry* 53: 84–97.

Hezinová, V., Aturki, Z., Klepárník, K. et al. (2012). Simultaneous analysis of cocaine and its metabolites in urine by capillary electrophoresis–electrospray mass spectrometry using a pressurized liquid junction nanoflow interface. *Electrophoresis* 33 (4): 653–660.

Hooijschuur, E.W., Kientz, C.E., and Brinkman, U.A.T. (2001). Application of microcolumn liquid chromatography and capillary electrophoresis with flame photometric detection for the screening of degradation products of chemical warfare agents in water and soil. *Journal of Chromatography A* 928 (2): 187–199.

Huhn, C., Pütz, M., Martin, N. et al. (2005). Determination of tryptamine derivatives in illicit synthetic drugs by capillary electrophoresis and ultraviolet laser-induced fluorescence detection. *Electrophoresis* 26 (12): 2391–2401.

Hutchinson, J.P., Johns, C., Breadmore, M.C. et al. (2008). Identification of inorganic ions in post-blast explosive residues using portable CE instrumentation and capacitively coupled contactless conductivity detection. *Electrophoresis* 29 (22): 4593–4602.

Jonsson, J., Kronstrand, R., and Hatanpää, M. (1996). A convenient derivatization method for the determination of amphetamine and related drugs in urine. *Journal of Forensic Sciences* 41 (1): 13914J.

Karlinsey, J.M. (2012). Sample introduction techniques for microchip electrophoresis: a review. *Analytica Chimica Acta* 725: 1–13.

Kim, M.J., Lee, S.C., Pal, S. et al. (2011). High-content screening of drug-induced cardiotoxicity using quantitative single cell imaging cytometry on microfluidic device. *Lab on a Chip* 11 (1): 104–114.

Kitagishi, K. (1996). Introduction. In: *Handbook of Capillary Electrophoresis Application* (eds. H. Shitani and J. Polonský), 1–13. New York: Blackie Academic and Professional.

Klepárník, K. (2015). Recent advances in combination of capillary electrophoresis with mass spectrometry: methodology and theory: CE and CEC. *Electrophoresis* 36 (1): 159–178.

Kohler, I., Schappler, J., and Rudaz, S. (2013). Highly sensitive capillary electrophoresis-mass spectrometry for rapid screening and accurate quantitation of drugs of abuse in urine. *Analytica Chimica Acta* 780: 101–109.

Křivánková, L., Caslavska, J., Malášková, H. et al. (2005). Analysis of ethyl glucuronide in human serum by capillary electrophoresis with sample self-stacking and indirect detection. *Journal of Chromatography A* 1081 (1): 2–8.

Kubáň, P. and Hauser, P.C. (2011). Capacitively coupled contactless conductivity detection for microseparation techniques – recent developments. *Electrophoresis* 32 (1): 30–42.

Kubáň, P. and Hauser, P.C. (2013). Contactless conductivity detection for analytical techniques: developments from 2010 to 2012: CE and CEC. *Electrophoresis* 34 (1): 55–69.

Kula, A., Król, M., Wietecha-Posłuszny, R. et al. (2014). Application of CE-MS to examination of black inkjet printing inks for forensic purposes. *Talanta* 128: 92–101.

Kunalan, V., Nic Daéid, N., Kerr, W.J. et al. (2009). Characterization of route specific impurities found in methamphetamine synthesized by the Leuckart and reductive amination methods. *Analytical Chemistry* 81 (17): 7342–7348.

Landers, J.P. (ed.) (2007). Introduction to capillary electrophoresis. In: *Handbook of Capillary and Microchip Electrophoresis and Associated Microtechniques*, 3rd edn., 25–96. CRC Press.

Le Roux, D., Root, B.E., Reedy, C.R. et al. (2014). DNA analysis using an integrated microchip for multiplex PCR amplification and electrophoresis for reference samples. *Analytical Chemistry* 86 (16): 8192–8199.

Lim Abdullah, A.F. and Miskelly, G.M. (2010). Recoveries of trace pseudoephedrine and methamphetamine residues from impermeable household surfaces: implications for sampling methods used during remediation of clandestine methamphetamine laboratories. *Talanta* 81 (1–2): 455–461.

Liotta, E., Gottardo, R., Bonizzato, L. et al. (2009). Rapid and direct determination of creatinine in urine using capillary zone electrophoresis. *Clinica Chimica Acta* 409 (1–2): 52–55.

Liu, P., Li, X., Greenspoon, S.A. et al. (2011). Integrated DNA purification, PCR, sample cleanup, and capillary electrophoresis microchip for forensic human identification. *Lab on a Chip* 11 (6): 1041.

Lloyd, A., Russell, M., Blanes, L. et al. (2013). Lab-on-a-chip screening of methamphetamine and pseudoephedrine in samples from clandestine laboratories. *Forensic Science International* 228 (1–3): 8–14.

Lucio do Lago, C., da Silva, H.D.T., Neves, C.A. et al. (2003). A dry process for production of microfluidic devices based on the lamination of laser-printed polyester films. *Analytical Chemistry* 75 (15): 3853–3858.

Manz, A., Graber, N., and Widmer, H.M. (1990a). Miniaturized total chemical analysis systems: a novel concept for chemical sensing. *Sensors and Actuators B: Chemical* 1 (1–6): 244–248.

Manz, A., Miyahara, Y., Miura, J. et al. (1990b). Design of an open-tubular column liquid chromatograph using silicon chip technology. *Sensors and Actuators B: Chemical* 1 (1–6): 249–255.

Martinez, A.W., Phillips, S.T., Butte, M.J., and Whitesides, G.M. (2007). Patterned paper as a platform for inexpensive, low-volume, portable bioassays. *Angewandte Chemie International Edition* 46 (8): 1318–1320.

Martinez, A.W., Phillips, S.T., Carrilho, E. et al. (2008). Simple telemedicine for developing regions: camera phones and paper-based microfluidic devices for real-time, off-site diagnosis. *Analytical Chemistry* 80 (10): 3699–3707.

McDonald, J.C. and Whitesides, G.M. (2002). Poly(dimethylsiloxane) as a material for fabricating microfluidic devices. *Accounts of Chemical Research* 35 (7): 491–499.

McDonald, J.C., Duffy, D.C., Anderson, J.R. et al. (2000). Fabrication of microfluidic systems in poly(dimethylsiloxane). *Electrophoresis* 21 (1): 27–40.

McNaught, A.D. and Wilkinson, A. (1997). *IUPAC Compendium of Chemical Terminology*. Blackwell Science.

Morales, E.B. and Vázquez, A.L. (2004). Simultaneous determination of inorganic and organic gunshot residues by capillary electrophoresis. *Journal of Chromatography A* 1061 (2): 225–233.

Pascali, J.P., Bortolotti, F., Sorio, D. et al. (2011). Improved capillary electrophoresis determination of carbohydrate-deficient transferrin including on-line immunosubtraction. *Medicine, Science and the Law* 51 (1): 26–31.

Pejchinovski, M., Hrnjez, D., Ramirez-Torres, A. et al. (2015). Capillary zone electrophoresis on-line coupled to mass spectrometry: a perspective application for clinical proteomics. *Proteomics – Clinical Applications* 9 (5–6): 453–468.

Piccin, E., Dossi, N., Cagan, A. et al. (2009). Rapid and sensitive measurements of nitrate ester explosives using microchip electrophoresis with electrochemical detection. *The Analyst* 134 (3): 528–532.

Polettini, A., Gottardo, R., Pascali, J.P., and Tagliaro, F. (2008). Implementation and performance evaluation of a database of chemical formulas for the screening of pharmaco/toxicologically relevant compounds in biological samples using electrospray ionization-time-of-flight mass spectrometry. *Analytical Chemistry* 80 (8): 3050–3057.

Ramautar, R., Somsen, G.W., and de Jong, G.J. (2009). CE-MS in metabolomics. *Electrophoresis* 30 (1): 276–291.

Rittgen, J., Pütz, M., and Pyell, U. (2008). Identification of toxic oligopeptides in Amanita fungi employing capillary electrophoresis-electrospray ionization-mass spectrometry with positive and negative ion detection. *Electrophoresis* 29 (10): 2094–2100.

Saito, R.M., Neves, C.A., Lopes, F.S. et al. (2007). Monitoring the electroosmotic flow in capillary electrophoresis using contactless conductivity detection and thermal marks. *Analytical Chemistry* 79 (1): 215–223.

Saito, R.M., Coltro, W.K.T., and de Jesus, D.P. (2012). Instrumentation design for hydrodynamic sample injection in microchip electrophoresis: a review: microfluidics and miniaturization. *Electrophoresis* 33 (17): 2614–2623.

Shang, F., Guihen, E., and Glennon, J.D. (2012). Recent advances in miniaturisation – the role of microchip electrophoresis in clinical analysis. *Electrophoresis* 33 (1): 105–116.

da Silva, J.A.F., do Lago, C.L., de Jesus, D.P., and Coltro, W.K.T. (2013). Capacitively coupled contactless conductivity detection (C4D) applied to capillary electrophoresis (CE) and microchip electrophoresis (MCE). In: *Capillary Electrophoresis and Microchip Capillary Electrophoresis* (eds. C.D. García, K.Y. Chumbimuni-Torres, E. Carrilho, et al.), 145–160. Hoboken, NJ: Wiley.

Stefan, A.R., Dockery, C.R., Baguley, B.M. et al. (2009). Microextraction, capillary electrophoresis, and mass spectrometry for forensic analysis of azo and methine basic dyes from acrylic fibers. *Analytical and Bioanalytical Chemistry* 394 (8): 2087–2094.

Su, H.-L., Feng, L.-I., Jen, H.-P., and Hsieh, Y.-Z. (2008). Determination of cocaine and its metabolites using cation-selective exhaustive injection and sweeping-MEKC. *Electrophoresis* 29 (20): 4270–4276.

Szafarska, M., Wietecha-Posłuszny, R., Woźniakiewicz, M., and Kościelniak, P. (2011). Examination of colour inkjet printing inks by capillary electrophoresis. *Talanta* 84 (5): 1234–1243.

Wongwan, S. and Scriba, G.K.E. (2010). Impurity profiling of dexamphetamine sulfate by cyclodextrin-modified microemulsion electrokinetic chromatography. *Electrophoresis* 31 (17): 3006–3011.

Xu, Z., Doi, T., Timerbaev, A.R., and Hirokawa, T. (2008). Sensitive determination of anions in saliva using capillary electrophoresis after transient isotachophoretic preconcentration. *Talanta* 77 (1): 278–281.

Zhang, C.-X. and Manz, A. (2001). Narrow sample channel injectors for capillary electrophoresis on microchips. *Analytical Chemistry* 73 (11): 2656–2662.

Index

a

Accreditation   41, 255, 269–270, 283, 393
Accuracy   39, 40, 72
Acetonitrile   93, 376, 377, 379
Acid ionisation constant, K_a 80
Activity level II evidence   16, 205–206
Adhesive tape analysis   152, 155
Adjusted retention time (t'_R) 305
Adjusted retention volume (V'_R) 305
Admissibility of evidence   25, 26, 28–30
Adsorption chromatography   298
Affinity chromatography   301, 384
Alcoholic potassium hydroxide test   60
Analysis   4, 5, 11, 17, 31, 36
   accuracy of   39–40, 72
   of blank samples   37
   destructive and non-destructive   11
   cost considerations   13, 51
   error   18–19, 39–40
   method development   41, 72
      experimental design   42
   precision of   39–40, 72
   record keeping   40
   repeatability   40
   reproducibility   41
   stages of   36
   standards, use in   37, 41
      certified standards   41, 103–104, 247
      external standards   37, 353
      internal standards   38, 83, 353–354, 393
      standard addition   38
   strategy for   11, 73–74
   uncertainty   18
Analysis of Variance (ANOVA)   45

Analyte   71, 73
   confirmation   51, 68, 255–256, 382, 394–395
Analytical chemistry   30
Analytical techniques
   capillary and microchip electrophoresis   407
   fluorescence spectroscopy   115
   gas chromatography (GC)   327
   high performance liquid chromatography (HPLC)   365
   infrared (IR) spectroscopy   145
   isotope ratio mass spectrometry   267
   mass spectroscopy (MS)   219
   Raman spectroscopy   161
   scanning electron microscopy   185
   UV-Vis spectroscopy   115
Apparent mobility   408
ASTM guidelines and methods   40, 356
Attenuated total reflectance (ATR)   148
Atmospheric pressure chemical ionisation (APCI)   231–232, 388

b

Backscattered electrons (BSE)   187
Baseline   303, 310, 317
Bayes theorem   20
   likelihood ratio (LR)   20, 204–205, 210
   posterior odds   20
   prior odds   20
Beer–Lambert Law   113, 120, 121
   concentration range   121
Blank samples   37
Bonded phases   298

Analytical Techniques in Forensic Science, First Edition.
Edited by Rosalind Wolstenholme, Sue Jickells and Shari Forbes.
© 2021 John Wiley & Sons Ltd. Published 2021 by John Wiley & Sons Ltd.

Braggs' equation   191
Breath alcohol testing   61–62
Buffers   375–377

C

Calibration curve   37, 47
Canines, use in detection   67–68
Cannabis presumptive testing   52, 56
Capacitatively coupled contactless
   conductivity detector (C4D)
   407
Capillary columns   327–330, 338–341
Capillary electrochromatography (CEC)   407
Capillary electrophoresis (CE)   407
   case studies   412
      clandestine laboratory samples
         412–418
   data outputs
      electropherogram   409, 414
      migration time   409
   forensic applications   408, 412–418
   instrumentation
      capillary column   407–410
      detectors
         capacitively coupled contactless
            conductivity (C4D)   407
         electrochemical detectors   411
         laser induced fluorescence (LIF)   407
         mass spectrometry   232, 407, 411
         ultraviolet (UV) detectors   407
      sample injection
         electrokinetic injection   409, 410
         hydrodynamic injection   409
   separation process   408
      apparent mobility, of analytes   408
      electrical double layer   408
      electro-osmotic flow (EOF)   408, 410
      electrophoretic mobility   408
         buffer viscosity   408
         hydrated radius of ion   408
         ion charge   408
   types of CE
      capillary electrochromatography (CEC)
         407
      capillary gel electrophoresis (CGE)   407
      capillary zone electrophoresis (CZE)
         407
      isoelectric focusing (IEF)   407
      isotachophoresis (ITP)   407
      micellar electrokinetic capillary
         chromatography (MEKC)
         407, 412
      microchip electrophoresis (ME)   407,
         410
      microcapillary-electrophoresis
         (ME)/lab-on-a-chip (LOC)
         410, 412–416
         sample injection   410–411
         detection   411
Capillary electrophoresis-mass spectrometry
   (CE-MS)   232
Capillary gel electrophoresis (CGE)   407
Capillary zone electrophoresis (CZE)
   407
Carbon coating in SEM   194, 198, 208, 210,
   213
Carbon cycle   274–275
Carrier gas   298, 321–322, 327–329, 332, 348
Case studies
   gas chromatography (GC)   256–258
   high performance liquid chromatography
      (HPLC)   396
   isotope ratio mass spectrometry (IRMS)
      281
   lab-on-a-chip (LOC) screening of
      clandestine drug lab samples
      412, 414–418
   mass spectrometry (MS)   257–258
   Raman spectroscopy   180
   scanning electron microscopy-energy
      dispersive X-ray (SEM-EDX)
      207–213
   ultraviolet–visible (UV-Vis) spectroscopy
      138–140
Chain of custody   8–11, 255
Certified standards and reference materials
   41, 103–104, 268–270
Chemical ionization (CI)   227, 346
Chiral chromatography   385
Chromatogram *see under* Chromatography
Chromatography columns *see* Capillary
   chromatography;
   Chromatography; Gas
   chromatography; High
   performance liquid
   chromatography (HPLC)

Chromatography, definition of 298
  chromatogram 303
    in HPLC 386
    in MS 245, 247–249, 251–252
  chromatographic separation processes 300
    adsorption/desorption 301
    distribution 300
      distribution constant K_c 302
      definition of 302
      in separations 302–303
      molecular interactions in 303, 366
    partition 298
  classifications used 297
    elution chromatography 297
      column chromatography 298
      definition of 300
    planar chromatography 298
      definition of 299
      thin layer chromatography (TLC) 299
      normal phase separations 92–93
      reversed phase separations 92–93
  columns
    chiral 385
    ion exchange 92, 95
    packed 300
    silica gel 86, 93, 314, 371, 373, 379
    wall coated open tubular (WCOT) 300
    column length L 319
  mobile phase
    gas 298
    liquid 298
    supercritical fluid 298
    flow rate 320
  optimising separation
    column length 310
    mobile phase flow rate 320, 321
    molecular interactions 318, 319
    particle size 313, 318, 319, 322
    use of pressure equation in 319
  parameters derived from
    adjusted retention time t'_R 305
      adjusted retention volume V'_R 305
      baseline 303, 310, 317
      hold-up time t_M 304
      hold-up volume V_M 305
      peaks 303, 310, 317, 322
      resolution, R_s 317, 319
      retention time t_R 305
        importance in analysis 305
        use of standards to verify 305
      retention volume, V_R 305
    related parameters
      retention factor k 306–307, 317, 320
        calculating 306
        importance in analysis 306
        separation factor α 307
    peak asymmetry 310
  optimising separation 316–322
  stationary phases 82–95, 97, 99, 298
    form
      bonded phase 298
      gel 298
      liquid 298
      solid 298
    theory of 307
      efficiency of separation 297, 307, 308, 310, 315, 317, 322, 373
        band broadening 308, 309, 311, 313, 314, 316, 320
        extra column volume (ECV) and 316
      measuring efficiency 308–310
        plate number N 308, 310, 317, 319
        plate height, H 310
      height equivalent to a theoretical plate (HETP) 308, 310, 311, 313, 315, 320
      plate theory 308
      rate theory 311–316
        van Deemter equation 311, 314, 320, 322
          eddy diffusion/dispersion 311–313, 321
          longitudinal diffusion 311, 313–314, 320, 328, 385
          mass transfer coefficient 311, 314
          theoretical plate 308
          and UHPLC 315, 316
Class characteristics 10
Cold trapping 335–336
Collision induced dissociation (CID) 241–242, 247
Column bleed 340
Column chromatography 298, 300

Column length (L)   319, 322, 340
Colour tests   51, 56
Competent evidence   25–26
Compound specific isotope analyses (CSIA)   279
Confirmatory tests   51, 68, 382
Contemporaneous notes   9
Controls   51
  negative   51
  positive   51
  samples   9, 41, 355
Core shell technology   372
Crime scene investigator   6

d

Delta δ notation   269
Delta δ values   272–276
Diamond anvil cell   148, 151
Direct analysis in real time-mass spectrometry (DART-MS)   234
  sample preparation   75
Data analysis   36
Databases   18, 205, 280
Defence proposition   203
Defender's fallacy   20
Demonstrative evidence   26–27
Derivatisation   102, 126, 128, 348, 349, 354
Desorption ionization-mass spectrometry (DESI-MS)   75, 234
Detectors *see* Capillary electrophoresis; Gas Chromatography; High Performance Liquid Chromatography; Infrared spectroscopy; Isotope Ratio Mass Spectrometry; Raman spectroscopy; Scanning Electron Microscopy; Ultraviolet–visible (UV-Vis) spectroscopy
Diastereoisomers   385
Diffuse reflectance   149
Diphenylamine (DPA) test   60
Dipole moment   145, 164
Direct immersion-solid phase microextraction (DI-SPME)   98–99
Distribution   300
Distribution constant (K_c)   77, 302, 303, 310, 318, 336

Document analysis
  infrared spectroscopy   152, 156
Documentary evidence   26–27
Driving under the influence (DUI)   61, 335, 398–399
Drugs
  colour tests   56
  illicit manufacture   281–287, 412–418
  infrared spectroscopy   155, 156, 158
  lab-on-a-chip analysis   407, 412–418
  new psychoactive substances   55, 253
  presumptive tests for   51–57
  Raman spectroscopy   176
  sampling for analysis   51, 71

e

Eddy diffusion/dispersion   311–313, 321, 385
Electrochemical detector   411
Electrical double layer   408
Electrokinetic injection   409, 410
Electromagnetic radiation   111
Electromagnetic spectrum (EMS)   111, 112
  Beer–Lambert law   113
  electromagnetic radiation   111
  energy-dispersive X-ray spectrometry (EDX)   112
  energy (E)   111
  frequency (v)   111
  infrared (IR) spectroscopy   112
  mass spectrometry (MS)   113, 114
  Raman spectroscopy   112
  scanning electron microscopy (SEM)   113
  spectra   113
    peaks   113
    quantitation   113
  spectroscopic techniques   112–113
  ultraviolet–visible (UV-Vis) spectroscopy   112
  vibrational spectroscopy   112
  wavelength (λ)   111
  wavenumber   112
Electron beam   185, 189
Electron capture detector (ECD)   343–345, 351
Electronic detectors
  presumptive testing   65
    explosives and illicit drugs   66
    ignitable liquid residues   65

Electron ionization   225, 346
Electron multiplier   243–244
Electron scattering   187
Electro-osmotic flow (EOF)   408, 410
Electropherogram   409, 414
Electrophoretic mobility   408
Electrospray ionization (ESI)   230–231, 388
Elution chromatography   297
Elutropic strength $\varepsilon°$ 379
Enantiomers   385
Emission spectroscopy   120, 124
Energy (E)   111
Energy-dispersive X-ray spectrometry (EDX)   112
Environmental scanning electron microscopy (ESEM)   185
Error   18
  in analysis   18–19
    estimating   39–40
    systematic and random   18, 39–40
  in interpreting evidence
    rates   18–19
    types I and II   18
ESI *see* Electrospray ionization
Even-electron ions   229
Evidence   1, 3–11, 14, 23–25, 203
  admissibility   25, 26, 28–30
  biological   6
  chain of custody   8–11, 255
  chemical   6
  class characteristics   10
  competent   25–26
  contamination   8, 10, 11
  control sample   9, 18
  defence proposition   16, 203
  definition   3
  demonstrative   26–27
  documentary   26–27
  evidential value   15, 17, 19, 21
  examination of   11
  Frye standard   29
  'Golden hour'   10
  hearsay   26, 27
  inadmissible   25–26
  individual characteristics   10
  integrity of   8
  interpretation   13–14, 18
  defender's fallacy   20
  glass analysis by SEM-EDX   203–206
  hierarchy of propositions   16
    level I, source   16
    level II, activity   16, 205–206
    level III, offence   16
  prosecutor's fallacy   19–20
  verbal scale of support   19, 207
  packaging   8, 11
  pattern   6
  physical   6
  prejudicial   26
  prosecution proposition   16, 203
  real   26, 27
  recovery   10
  relevant   25–26
  rules of   25
  statistical interpretation   18–19
    Bayes' theorem   20
  testimonial   26, 27
  trace   6
  transfer   7, 206
Evidential value   15, 17, 19, 21
Excitation spectroscopy   120, 124
Expert witness   14–16, 19, 23, 27–29
Explosives   59
  infrared spectroscopy   152, 156
  ion chromatography   384
  isotope ratio mass spectrometry (IRMS)   267, 282, 288
  micellar electrokinetic capillary chromatography (MEKC)   407
  portable instruments, testing for   61, 66–67
  presumptive testing   60–61, 66–67
  sampling for   60
External standard   37, 353
Extra-column volume   316, 375
Extracted ion chromatogram   248–249, 251, 387
Extraction   75

f

False negative   51
False positive   51
Fibre analysis
  infrared spectroscopy   155, 157

Fibre analysis (*contd.*)
　microspectrophotometry　134–135, 138–139
　scanning electron microscopy　195
Film thickness　320, 339
Filtered light examination (FLE)　127–128
Firearms discharge residue (FDR) *see* Gunshot residue (GSR)
Flame ionization detector (FID)　65, 341, 343–345, 351
Flame photometric detector (FPD)　343, 344
Fluorescence spectroscopy　115, 118–119
　applications in forensic science
　　fibre examination　134
　　writing ink examination　132
　derivatization　119, 128
　efficiency　120
　excitation and emission　120, 124
　filtered light examination (FLE)　127
　fluorescence spectrometers　123
　fluorogram　125
　fluorometers　123
　fluorophores　119, 124
　high performance liquid chromatography (HPLC)　126
　hyperspectral imaging　126
　photobleaching　120
　quantification　120–121
　sample analysis in solution　129
　self-quenching　121
　sensitivity of　120
　spectra　119
　　interpreting　137
　　solvent, pH and temperature influence on　119
　Stokes shift　124
　transmission　123
Fluorogram　125
Fluorescence spectrometer　123
Fluorometer　123
Fluorophores　119, 124
Flow rate　320–321, 328–329, 332, 340, 379, 380
Footwear examination　203
Forensic　3
　investigation　3, 6

science　3, 17
scientist　5, 14, 16–17, 22–23, 28
specialists　6
Forensic Isotope Ratio Mass Spectrometry (FIRMS) network　282–283
Fourier transform-ion cyclotron resonance (FT-ICR)　239
Frequency (v)　111
Fourier transform-infrared spectrometry (FT-IR)　146
Frye standard　29

g

Gas chromatography (GC)　327
　analyte considerations
　　confirming identification
　　　mass spectrometry, use of　351
　　　retention time match　350
　　　standards, use of　350
　　derivatisation　348–350
　　polarity　348
　　thermal stability　348
　　volatility　348
　applications　327, 335, 347, 356–357, 360, 362
　case studies　356–358
　　toxicological findings in buried bodies　356–357
　　combustion compounds in fire victims　357–358
　chromatographic separation
　　columns　327–330, 339–341
　　　capillary　327–330, 338–341
　　　column bleed　340
　　　efficiency of　329, 339–340
　　　internal diameter　339–340
　　　length　319, 322, 340
　　　overloading　331, 334, 336
　　　packed column　300
　　　sample capacity　339, 340
　　　selection of　340
　　　stationary phase　300–301, 327, 340–341, 343
　　　film thickness　320, 339
　　　support-coated open tubular (SCOT)　339

wall coated open tubular (WCOT) 300, 322, 339
  efficiency of 329, 330
  mobile phase/carrier gas 327, 328, 348
    choice 321–322, 327–329, 334
    flow rate 321, 328, 329, 332, 340
  optimising separation 343
  temperature 306, 319, 322
    isothermal 341, 342
    optimising 319, 341–343
    temperature programming 319, 341–343
forensic and legal considerations 454
  MS as a detector 354
  mass spectral libraries 356
  QA and QC 354, 356
    blank samples 355
    control samples 355
    reference materials 355
    standards 355
  methods used 356
gases and volatile compounds, analysis of
  gas sampling valve 334–335
  gas tight syringe 334, 336
  headspace analysis 335
    dynamic/purge-and-trap 336–337
    headspace-solid phase microextraction (HS-SPME) 97–101, 336
  static 335–336
  trapping 335–338
instrumentation
  data system 328, 346, 353
  detectors 327, 328, 341, 343, 344
    electron capture detector (ECD) 343–345, 351
    flame ionization detector (FID) 65, 341, 343–345, 351
    flame photometric detector (FPD) 343, 344
    mass spectrometric detector (MSD) 341, 343, 344, 346
    nitrogen phosphorous detector (NPD) 343–345, 351
    photoionization detector (PID) 66
  injector 329
    autosampler 330
    on-column injector 329, 332
    cold on-column 333
    port 327, 331
    vapourising injector 329
    autosampler 330
    liner 329–330
    microsyringe 329
    programmed temperature vaporisation (PTV) injector 334
    volume injected 330
      split injection 331–332
      splitless injection 331–333
  oven 340, 343
  multidimensional GC 358–360
  portable instruments 66, 67, 361–362
  principles of analysis 327, 348
  pyrolysis-GC 338–339
  quantification 351–352, 355
    area normalisation 353
    calibration curve 353
    errors in and avoiding 352
    external standards 353
    internal standards 353
      advantages of 354
      isotopically labelled 354
      quantifying with 354
      requirements for 354
    samples and sampling 352
    standards, use of 350
Gas chromatography-mass spectrometry (GC-MS) 226, 230, 346–347
Gas sampling valve 334–335
Gas tight syringe 334, 336
GC oven 340, 343
Generalized linear models (GLM) 45, 48
Geographic origin 267, 276, 281–285
Glass examination 199, 204, 205, 209
  bulbs 212
'Golden hour', the 10
Gradient elution 372, 375, 378–379, 381
Gunshot residue (GSR)/firearms discharge residue (FDR)
  analysis of 13, 196–199, 208–209
  interpretation 198, 206–207
  morphology 12–13, 57, 196–199, 208–209
  presumptive testing 58–59
  sampling for 57–58

h

Headspace analysis 335–338
Headspace-solid phase microextraction, (HS-SPME) 98–99
Height equivalent to a theoretical plate (HETP) 308, 310, 311, 313, 315, 320
Hearsay evidence 26, 27
Hierarchy of propositions 16–17
High performance liquid chromatography (HPLC) 365
  analyte partitioning 379
  applications 367, 390–393
    acidic analytes 390
    basic analytes 390
    chiral compounds 392
    DNA 391
    driving under the influence of drugs (DUID) 399
    drugs and drug metabolites 393
    post-mortem death investigation 396–398
    proteins 391
  back pressure 319, 322, 379
  case studies 396
    post-mortem death investigations 396–398
    driving under the influence of drugs (DUID) 399–400
  chromatographic separation
    columns 370
      dimensions 373
      fittings 374
      guard column 373
      stationary phases 87–94, 371–373
        alumina-based 371
        chiral phases 385
        core shell technology 372
        hydrophilic interaction liquid chromatography (HILIC) 92, 372–373
        ion exchange 92, 95, 372
        mixed-mode 372
        modified ionic functional groups 372
        modified phases 371–373
        monolithic 300, 371
        residual silanols 86, 92, 366
          endcapping 366
        silica gel 86, 92–93, 105, 314, 371, 373, 379
      particle size 373
      pressure/backpressure 373, 379
  efficiency of 379
    band broadening 385
    core shell technology 372
    extra-column volume 375
    flow rate 379
    HPLC versus UHPLC 374
    particle size 373
  elution
    gradient 372, 375, 378–379, 381
    isocratic 375, 377
  interactions with mobile and stationary phases 366
  mobile phase 375–379
    additives to adjust ionization state 230, 376
    additives to adjust pH 375–376
    analyte partitioning 379
    buffers 375–377
    flow rate 379, 380
    phases A and B 376
    optimising 381
    organic solvents in 93, 318, 376–379
      acetonitrile 93, 376, 377, 379
      effect on retention factor k 379
      elutropic strength $\varepsilon°$ of 379
      methanol 93, 376, 377, 379
      other solvent 378
      solvent selectivity 379
      viscosity 378
    pH 376
    phase modifiers 376–377
    pKa 376, 380
    in RP separations 375
    water in 375
  reversed phase 366, 372, 375, 376, 379
  temperature, effect on 379–380
  data outputs 386
    chromatogram 386
    relative retention time (RRT) 386

retention index (RI)   386
  with tandem MS   387
    total ion chromatogram   387
    extracted ion chromatograms   387
forensic and legal considerations   394
  accreditation   393–394
  appropriate analytes   395
  case circumstances   396
  chromatographic resolution   395
  identification points   394–395
  interpreting toxicological findings   396
  ion ratios in MS   394
  method validation   393, 394
instrumentation   368
  detectors   368, 380
    mass spectrometric   380
    photo-diode array detector (PDA)   383
    ultraviolet (UV) detector   126, 382–383
  fittings   374–375
    extra-column volume   375
    UHPLC   375
  injector   369
    autosamplers   370
    6-port valve   370
    sample loop   370
  pump and mixer   368–369
internal standards   393
related techniques   384–385
  affinity chromatography   384
  chiral chromatography   384
  ion chromatography   384
sensitivity   367
UHPLC   373–374
ultra high performance liquid chromatography (UHPLC)
  efficiency   315
  efficiency   315, 373–374
  resolution   373–374
  sensitivity   373–374
Hold-up time t_M 304
Hold-up volume, V_M 305
Hydrodynamic injection   409
Hydrological cycle   275, 281

Hydrophilic interaction liquid chromatography (HILIC)   92, 372–373
Hyperspectral imaging (HSI)   126–127, 131, 132
Hyphenated techniques
  capillary electrophoresis-mass spectrometry (CE-MS)   232
  gas chromatography-mass spectrometry (GC-MS)   226, 230, 346–347
  liquid chromatography-mass spectrometry (LC-MS)   226, 230–231, 388–389
Hypothesis   35

i

Identification points   394–395
Inadmissible evidence   25–26
Inductively coupled plasma-atomic emission spectroscopy (ICP-AES)   103
Inductively coupled plasma-mass spectrometry (ICP-MS)   103, 104
Identification points   394
Ignitable liquid residues (ILR), analysis of
  controls   355
  gas chromatography-mass spectrometry   347, 350, 357–358
  headspace GC   335, 337
  isotope ratio mass spectrometry (IRMS)   287
  multidimensional GC   360
  portable GC   362
  presumptive testing   64–67
  reference materials   355–356
Individual characteristics   10
Infrared analysis   151, 152
  interpretation and law   155
  sampling   151
  spectral library databases   154
  spectrum   152
Infrared imaging   150
Infrared microscopy   150–151, 157
  infrared imaging   150
Infrared (IR) spectrometers
  Fourier transform infrared (FTIR) spectrometry   146

Infrared microscopy (contd.)
  interferogram   147
  attenuated total reflectance (ATR)   148
  reflection-absorption   148, 149
  diamond anvil cell   148, 149, 152
  diffuse reflectance   148, 149
  portable and handheld infrared spectrometers   151, 156, 158
Infrared (IR) spectroscopy   145
  case study, acrylic fibres   157
  forensic applications
    adhesive tape   152, 155
    documents   152, 156
    drugs   155, 156, 158
    explosives   152, 156
    fibres   155, 157
    paint   151, 155
    polymers   148, 152, 155, 157
  molecular vibrations   145
    dipole moment   145
    selection rule   145
  quantitative analysis   15
  reflectance spectroscopy   148
  transmission spectroscopy   148
Interference in analysis   75
Interferogram   146, 147
Internal standard   38, 83, 352–354, 393
Ion chromatography   384
Ion exchange   92, 95, 372
Ion mobility spectrometry (IMS)   67, 267
Ion ratios   394
Ion trap   237
Isocratic elution   375, 377
Isoelectric focusing (IEF)   407
Isoscapes   280–281
Isotachophoresis (ITP)   407
Isothermal analysis   341–342
Isotopic abundance   270, 272
Isotopic fractionation   281
Isotopically labelled standards   354
Isotope ratio mass spectrometry (IRMS)   267
  applications   283
    doping   284
    environmental contaminants   285
    explosives   267, 282, 288
    geographical origin   267, 276, 283, 285
      food   281, 284
      humans   272, 281, 284
      illicit drugs   281–286
    ignitable liquids   287
    illicit drug manufacture   283–287
      microorganisms   286
      MTBE   285
      oil   280, 285
      paint   281
      plastic films   287
      poisons   287
      tapes, films, and paper   271, 287
  case studies   281–283
  forensic and legal considerations   280–283
    case databases   280
    isoscapes   280–281
    isotopic fractionation   281
    material databases   280
  instrumentation   272, 276–281
    dual inlet   276
    continuous flow   276
    ion detection   276–277
      Faraday cup   276
  isotopes   268, 278
    carbon   262–265
      carbon cycle   274, 275
    delta (δ) notation   269
      delta values   272–276
    hydrogen   272
      hydrological cycle   275, 281
    isotopic abundance variability   270, 272
    isotopic fractionation   270, 281
      kinetic   270
      thermodynamic   271
    isotopologues   278
    isotopomer   276
    isoscapes   280
    neutrons   268
    protons   268
    oxygen   275, 278
    sulfur   278
  laboratory accreditation   269, 270, 283
  measurements
    bulk/whole sample   277
      high temperature combustion   278–279
      high temperature conversion   279

compound specific   279
  GC-IRMS   279
  interpretation   280–281
  reference materials and databases   268–270, 282–283
Isotopic fractionation   270–271, 280
Isotopologues   278
Isotopomer   276

k
Kastle–Meyer test   51

l
Laser induced fluorescence detector (LIF)   407
Legal system
  adversarial   24–25
  civil court   22–24
  civil law   3, 22–23
  criminal law   22–24
  expert witness testimony   28, 29
  federal law   22
  Frye vs. United States   28
  inquisitorial   24–25
  opinion testimony   28
  trier of fact   22
Likelihood ratio (LR)   20–21, 204–205, 210
Limit of detection (LOD)   38, 42, 305, 308
Limit of quantification (LOQ)   38
Liquid chromatography-mass spectrometry (LC-MS)   226, 230–231, 388–389
Liquid-liquid extraction (LLE)/liquid-liquid distribution (LLD)   75, 77, 104, 334
  pH and ionization   80–81
  repeat extraction   78
  solvent and sample volumes   79
Liquid phase microextraction (LPME)   79
  hollow fibre-LPME   80
  single drop-LPME   79–80
Locard's exchange principle   7
Longitudinal diffusion   311, 313–314, 320, 328, 385

m
Magnetic sector   236
Marijuana/cannabis
  presumptive testing   52–54, 56
Mass spectrometry (MS)   219
  analysis   224, 225
  ion detection   243
    electron multiplier   243–244
    Faraday cup   244
  ion formation/ionization   225
    ambient   232, 234
    atmospheric pressure   225, 230
    atmospheric pressure chemical ionization (APCI)   231
    chemical ionization (CI)   227
    desorption   232, 234
    desorption electrospray ionization (DESI)   234
    direct analysis in real time (DART)   234
    electron ionization (EI)   225
    electrospray ionization (ESI)   230
    gas phase   225, 227, 230
    liquid phase   230
    matrix assisted laser desorption ionization (MALDI)   232
    secondary ion mass spectrometry (SIMS)   234
    use of chemical modifiers in forming   230
  ions formed
    even-electron ions   229
    molecular ion   226, 228, 229, 232, 245
      quasi-molecular ion   232
    multiply charged species   231, 233, 244
    negative ions   228, 230, 232, 233, 235
    odd-electron ions   226
    positive ions   226–229, 232, 233, 235
    precursor ion   241, 242
      detection   252
      precursor-product ion transitions   256
    product ion   241–243
      mass spectrum   248
      precursor-product ion transitions   256

Mass spectrometry (MS) (contd.)
  mass analysis/ion separation   225, 235
    mass analysers
      Fourier transform-ion cyclotron resonance (FT-ICR)   239
      ion trap   237
      magnetic sector   236
      Orbitrap   243, 260
      quadrupole   236
      tandem mass analysers   242
    full scan   250
    mass range   235, 237–239, 245, 346
    mass-to-charge (m/z) ratio   219, 227, 229, 231, 235–238, 244, 246
    multiple reaction monitoring (MRM)/selected reaction monitoring (SRM)   251, 389
    sample introduction   224
    selected ion monitoring (SIM)   249–250
  analyte identification   255
    chromatographic criteria   256
    mass spectrometric criteria   256
    retention time   256
  applications in forensic science   66, 221–223, 252–254, 257–258, 260–261
  case studies
    serial killing by poisoning   257
    surreptitious insulin administration   257
  data outputs
    chromatogram
      extracted ion chromatogram (EIC)   248–249, 251
      multiple reaction monitoring (MRM)   251
        qualifier and quantifier ions   252
      selection ion chromatogram   249–250
        qualifier and quantifier ions   252
      total ion chromatogram (TIC)   245, 247
    mass spectrum   227, 236, 244, 245, 305
      base peak   245
      of certified reference standards   247
      collision induced dissociation (CID) spectra   247
      determining analyte structure, use in   247
      electron ionization (EI) spectra   247–248
      electrospray ionization (ESI) spectra   246
      fragment region/fragment ions   246–248
      full scan acquisition   247–248, 250–251
      from 'hard ionization' techniques   233, 246, 247
      ion abundance/ion intensity   244, 245, 256
      isotopic molecular ions   246
        use in determining carbon atoms   246
        use in determining elemental composition   239, 246, 260
      mass spectrometry (MS) libraries   227, 247
      molecular ion   246
        fragmentation of in EI   247
        quasi-molecular ion   246
      molecular ion region   246
      multiply charged ions/species   246
      nitrogen atoms, determining   246
      'rogue' ions in   247, 248
      from 'soft' ionization techniques   233, 241, 246
      tandem MS   247–248
        multiple reaction monitoring (MRM)/selected reaction monitoring (SRM)   251, 389
        MS/MS spectra   247
        product ion spectra   248
        QQQ precursor ion spectra and neutral loss scanning   252, 389
  forensic considerations
    acceptance criteria for MS data   255–256
    accreditation   255
    analyte identification   252, 255
    chain of custody   255
    standards and codes of conduct   255
  hyphenated-MS   221

Index | 437

capillary electrophoresis-mass
spectrometry (CE-MS)   232
gas chromatography-mass spectrometry
(GC-MS)   226, 230
liquid chromatography-mass
spectrometry (LC-MS)   226,
230, 231, 388–389
mass resolution   236–240, 246
exact mass/theoretical mass   236
mass accuracy   236–241, 243
high resolution/high mass
spectrometry (HRMS)   259,
390
mass resolving power   236, 238, 239
nominal mass   236, 246
mobile/portable mass spectrometers   66,
260
sensitivity   237, 241, 243, 252
tandem MS   241–24
collision induced dissociation (CID)
241–242
CID spectra   247
mass spectrometry/mass spectrometry
(MS/MS)   241, 389
multiple reaction monitoring (MRM)
251, 389
multistage MS (MS^n)   241
precursor ions   241–242, 252
product ions   241–243, 248
product ion spectra   248
use of hybrid instruments in   242
use of ion traps in   242
use of quadrupole-Orbitrap in   243
use of quadrupole-time of flight in
(Q-TOF)   260
use of triple-quadrupole (QQQ) in   242
Matrix assisted laser desorption
ionization-mass spectrometry
(MALDI-MS)   232, 253
MALDI-time of flight-mass spectrometry
(MALDI-TOF-MS)   253
Mass analyzer/mass analysis   237–243
Mass accuracy   236–241, 243, 259, 390
Mass resolution   236–240, 246
Mass spectral libraries   356
Mass spectrum *see* Mass spectrometry

Mass to charge (m/z) ratio   219, 227, 229,
231, 235–238, 246
Methanol   93, 376–377, 379
Method development   41
experimental design   42
backwards elimination strategy   45
categorical variables   43, 48
factorial designs   42–43, 47
general linear models (GLMs)   45, 47,
48
reduced designs   44
response surface designs   42, 46, 48
surface response surface designs   48
variables in
categorical variables   43, 48
critical variables   42
dependent variables   42
non-significant   46
significant   46
validation of   49, 72, 352, 393
Micellar electrokinetic capillary
chromatography (MEKC)
407, 412
Microchip electrophoresis (ME)   407, 410
Microcapillary-electrophoresis
(ME)/lab-on-a-chip (LOC)
410, 412–416
Microspectrophotometry (MSP)   126
acquiring spectra   131
applications in forensic science
fibre analysis   133, 138–139
writing ink examination   132
liquid crystal tunable filter MSP   134
photobleaching   131
reflectance mode   126
sample preparation and analysis   129–130
transmission mode   126, 130
Migration time   409
Mixed mode phases   372
Mobile phase *see under* Chromatography; Gas
Chromatography; and High
performance-liquid
chromatography
Modified Griess test   60
Molecular ion   233, 245–246, 346
Molecular interactions   303, 366

Molecularly imprinted polymers (MIPS)   75, 92, 95–96, 372
Molecular vibrations   145, 163
Monochromator   122, 168
Monolithic stationary phases   300, 371–372, 401
Moseley's law   188
Multidimensional gas-chromatography   358–360
Multiple reaction monitoring (MRM)   251–252, 256, 259, 367, 387, 389, 400
Multiply charged ions/species   231, 233, 244

n

National Institute of Standards and Technology (NIST)   269
Negative controls   51
Negative ions   228, 230, 232–233, 235
Neutral Loss (NL) scanning   252, 389
New psychoactive substances (NPS)   55, 253
Nitrogen phosphorous detector (NPD)   343–345
Noise signal   38, 40, 192
Nominal mass   236, 246
Normal phase (NP)   92–93

o

Odd electron ions   226
Offence Level III evidence   16
On-column injector   329, 332, 333
Optimising separations   316–322
Orbitrap   243, 260
Organic modifier   93, 318, 376–379
Overloading   331, 334, 336

p

Packaging   8
Packed columns   300
Paint analysis   12
  infrared spectroscopy.   151   155
  isotope ratio mass spectrometry (IRMS)   281
  Raman spectroscopy   175
  scanning electron microscopy (SEM)   203
Palynology   203
Particle size   313, 318, 319, 322

Partition chromatography   298
Partition ratio, K_D 77–79
Peak asymmetry   310
Peak, chromatographic   304, 352
  peak area   351
  peak capacity   358
  peak height   351
Phase collapse   94
Photobleaching   120
Photo diode array detector (PDA)/diode array detector (DAD)   122, 383
Photoionization detector (PID)   66
Photomultiplier   190
Planar chromatography   298–299
Plate theory   307
  plate height H 310
  plate number N 308
Polymer analysis   148, 152, 155, 157, 287
Portable GC   66, 361–362
Portable and handheld infrared spectrometers   151, 156, 158
Portable Raman systems   170, 174
6-Port valve   370
Posterior odds   20
Positive controls   51
Positive ions   226–229, 232–233, 235
Precision   39–40, 72
Precursor ion   241, 252, 256
Prejudicial evidence   26
Pressure equation   319
Presumptive testing   51
  amphetamines   53
  barbiturates   53
  benzodiazepines   53
  breath alcohol (BrAc)   61–62
  cocaine   53
  colour tests for drugs   56
  drugs   52, 53, 56, 66
  electronic detectors   64–66
  ethanol   63–64
  explosives   66
  gunshot residue (GSR)   58–59
  ignitable liquid residues   65
  LSD   53
  marijuana/cannabis   52
  microcrystal tests for drugs   56
  opiates   53

portable GC-MS   66
portable ion mobility spectrometry (IMS)
   67
thin layer chromatography (TLC)   56, 61
use of canines   67–68
Primary transfer   206
Prior odds   20
Product ion   241–243, 248, 256
Programmed temperature vaporisation (PTV)
   injector   334
Prosecution proposition   203
Prosecutor's fallacy   19–20
Purge-and-trap   336–337
Pyrogram   338
Pyrolysis-GC   338–339

q

Quadrupole   236
Quadrupole-Orbitrap   243
Quadrupole-time of flight (Q-TOF)-MS   260
Qualitative analysis   37, 350
Quality assurance (QA)   39–40, 281, 354, 356
Quality control (QC)   39–40, 270, 354
Quantification/quantitation   37–38, 350–354
   area normalisation method   353
   blank samples, use in   38
   calibration curve   37–38
      range   38
      response factor   38
   external standard method   37, 352
   in fluorescence spectroscopy   121
      gas chromatography (GC)   350–354
      Raman spectroscopy   170
      scanning electron microscopy (SEM)
         188, 189
      UV-Vis spectroscopy   118, 120–121
   internal standards, use of in   352, 354
   limit of detection (LOD)   38, 42, 351
   limit of quantification (LOQ)   38
   standard addition method   38
QuEChERS   75, 101
Questioned document examination   201

r

Raman spectroscopy   112, 161
   advantages and disadvantages   173
   case study   180

   paint analysis   180
   vehicle paint   180
in conjunction with infrared (IR)
      spectroscopy   161, 164, 165
forensic applications   175
   drugs   176
   paint   175
   pen inks   175
instrumentation   166
molecular vibrations   163
   dipole moment   164
   polarisability   164
   Raman active   164
portable Raman systems   170, 174
problems with fluorescence   166, 169, 171,
      173, 175
quantitation   170
Raman spectrometers   167
   dispersive   167, 168
   Fourier-Transform (FT)   167, 169
   monochromators   122, 168
   notch/edge filter   168
   lasers   166
      near infrared (NIR)   166
      visible excitation   166
Raman microscopy   169
   imaging and mapping   169
Raman scattering   161, 164, 165
Raman shift   165
Raman spectra   165
   acquiring   174
   interpreting   177
   spatial resolution   167, 168
   spectral resolution   167, 170
   wavenumber   162, 165, 168, 177
resonance Raman spectroscopy (RRS)   171
spatially offset Raman spectroscopy (SORS)
      172
surface enhanced Raman spectroscopy
      (SERS)   171
surface enhanced resonance Raman
      spectroscopy (SERRS)   171
Random error   18, 39–40
Rate theory   311
Real evidence   26, 27
Record keeping   40
Reflectance spectroscopy infrared   148

Relative retention time (RRT)   386
Relevant evidence   25–26
Repeatability   40
Reproducibility   41
Residual silanol groups   86, 92, 366
Resolution R_s in chromatography   317, 319
Resolution in scanning electron microscopy (SEM)   192
Response factor (F)   38, 353
Retention factor, (k)   306–307, 317, 320
Retention index (RI)   386
Retention time (t_R)   305, 328, 340
  adjusted retention time t'_R 305
  analyte identification in MS   256
  relative retention time (RRT)   386
Retention volume V_R 305
  adjusted retention volume, V'_R 305
Reversed phase (RP)   93
Rules of evidence   25

S

Sample capacity   339, 340
Sample loop   370
Sample preparation   71–73, 75–79, 84, 96, 97, 103, 277, 352, 354
  centrifugation   76
  derivatisation   102, 348–349, 354
  for DNA profiling   105–106
  extraction   75, 78
  filtration   76
  for inorganic elements   73, 75, 102–104
    sample handling   103
    solid-phase extraction for   104
    solid samples   104
    speciation   102–103, 105
  liquid-liquid extraction (LLE)   77, 78, 80, 81, 97
  liquid-phase microextraction (LPME)   79
  for liquids   72
  for Raman spectroscopy   75
  for scanning electron microscopy (SEM)   193–195, 197–198
  solid phase extraction (SPE)   82–85
  solid-phase microextration (SPME)   97–101
  solvent evaporation   101–102
  solvent extraction   73, 75, 76

Sampling   12, 36, 41, 51, 60, 71–73, 76–77
Scale of support   19, 21, 207
Scanning electron microscopy (SEM)   113, 185
  applications
    fibre and hair   195
    forgery   201
    glass   199, 209
    gunshot residue (GSR)/firearms discharge residue (FDR)   196, 207
  case studies   207
    glass   209
    gunshot residues (GSR)   207
  forensic and legal considerations   203–205
    defense proposition   203
    evaluating activity level evidence   205
    evaluating source level evidence   203
    probability density function   205
    prosecution proposition   203
  instrument components and operation
    electron detection   190
    electron gun   186
      electron beam   185, 186
        electron penetration depth   190
        scanning   190
        scan speed   192
        spot size   193
      electron scattering   187
        backscattered electrons (BSE)   187
          scintillation/photomultiplier detector (Robinson detector)   190
        secondary electrons (SE)   187
          Everthart–Thornley detector   190
        X-rays   188
          energy dispersive detector   191
          wave dispersion detector   191
    imaging   190
      magnification   192
      resolution   192
        signal/noise (S/N)   192
        and scan speed   192
        and aperture   193
      working distance   193
  instrument types

environmental scanning electron
    microscopy (ESEM)   185
  low vacuum   185
  high vacuum   185
  SEM coupled to energy dispersive X-ray
    detector (SEM-EDX)   185
particle characterisation
  composition   185
  morphology   185
  topography   185
sample preparation   193–195
  embedding procedure   200
  GSR/FDR   198
  use of stubs in   195, 198, 201, 208, 210
  coating by vacuum evaporation   194
Scanning electron microscopy–energy
    dispersive X-ray (SEM-EDX)
    185
Scientific method   35
Scientific working group for gunshot residues
    (SWGGSR)   198
Secondary electrons (SE)   187, 190
Secondary ion mass spectrometry (SIMS)
    234
Secondary transfer   206
Selected ion monitoring (SIM)   249–250
Selected reaction monitoring (SRM) *see*
    Multiple reaction monitoring
    (MRM)
Selection rule   145
Selectivity   41
Self-quenching   121
Semi-conductive detectors   191
Sensitivity   42
Separation factor (α)   307
Signal to noise ratio (S/N)   41, 192, 302
Silica, as stationary phase   86, 92–93
  end capping   93, 366
  hydrolysis and pH   86
  pore size   373
  silanol groups   86, 92, 366
  Types A, B and C   93
Solid phase extraction (SPE)   82–85
Solid-phase microextraction (SPME)
    97–101, 334, 336
  direct immersion-SPME   98–99
  headspace-SPME   98–99

Solvent extraction   73, 75–76
Solvent selectivity   379
Solvent viscosity   378
Source Level I evidence   16
Split injector   331–332
Splitless injector   331–333
Stages of analysis   36
Standard addition   38
Standard operating procedure (SOP)   40–41,
    72
Standard reference materials   103
Stationary phases   87–91
  alumina based   88, 371
  aminopropyl   88
  bonded   93, 298
  C_4 87
  carbowax   100
  cation exchange   90
  chelating resins   104, 105
  chiral   385
  cyanopropyl   88
  diol   88
  graphitized carbon   89
  for HILIC   92, 372–373
  immobilized   298
  ion exchange   92, 95, 372
  magnesium silicate   89
  mixed-mode   90, 372
  modified phases   371–373
  molecularly imprinted polymers (MIPS)
    75, 92, 95–96
  monolithic   300–371
  liquid   298
  octadecyl   87
  octyl   87
  phase collapse   94
  phenyl   87
  polyamide   90
  polydimethylsiloxane   100
  polymeric   90–92
  polystyrene-divinylbenzene   90–92
  silica *see under* Silica
Statistical interpretation of evidence   18–19
Stokes shift   124
Support coated open tubular (SCOT) columns
    339

Surface enhanced Raman spectroscopy (SERS)   171
Surface enhanced resonance Raman spectroscopy (SERRS)   171
Synthetic cannabinoids   55
Systematic error   18, 39–40

t

Tandem mass spectrometry   241, 251, 389
Temperature programming   319, 341–343
Testimonial evidence   26, 27
Thermal desorption   337
Thin layer chromatography (TLC)   56, 61, 299, 303
Total internal reflection (TIR)   190
Total ion chromatogram (TIC)   245, 247, 387
Trace evidence   6
Transfer   7, 206
Transmission spectroscopy infrared   120, 128, 131, 148
Triple quadrupole (QQQ)   242
Type I and II errors   18

u

Ultra high pressure/performance liquid chromatography (UHPLC)   315, 373–374
Ultraviolet–visible (UV-Vis) spectroscopy   112, 115
　absorbance (A)   120
　analysis in solution   128
　　concentration   128
　　cuvettes   128
　　solvents & UV cut-off   128–129
　　temperature   129
　applications in forensic science
　　fibre examination   134–135, 138–139
　　writing ink examination   132–133
　Beer–Lambert law   120, 121
　case studies   138–140
　chemometric methods   132, 134, 136
　chromophores   117, 135, 136
　derivatization   128
　electronic transitions   11
　molecular structure and   117
　organic molecules and   116–117
　hyphenation, HPLC-UV   126, 382
　instrumentation   122
　　microspectrophotometers   115
　　UV-Vis spectrometers   122–123
　percentage transmittance, (%T)   120
　quantification   118, 120–121, 131
　reflectance mode   131
　spectra
　　acquiring   130
　　interpreting   135–138
　theory   115–118
　transmission   120, 128, 131
　transmittance, T   120
　use of interpretation and law   134–135, 138
　UV range   116
　visible range   116
Uncertainty   18

v

van Deemter equation   311, 314, 320–322, 328, 385
van der Waals forces   366
Vaporisation chamber   329, 333
Vaporising injector   329
Vibrational spectroscopy   112
Volatile organic compounds, analysis of   334–337

w

Wall coated open tubular (WCOT) columns   300, 322, 339
Wave dispersion detectors   191
Wavelength (λ)   111
Wavenumber   112
Writing ink analysis
　Raman spectroscopy   175
　UV-visible spectroscopy   132

x

X-Ray detection   191
X-Rays   188